Contributions to Probability and Statistics

Dedicated to our friend and mentor

Ingram Olkin

On the occasion of his 65th birthday

Leon Jay Gleser Michael D. Perlman
S. James Press Allan R. Sampson
Editors

Contributions to Probability and Statistics

Essays in Honor of Ingram Olkin

With 50 Illustrations

Springer-Verlag
New York Berlin Heidelberg
London Paris Tokyo Hong Kong

Leon Jay Gleser
Department of Statistics
Purdue University
West Lafayette, IN 47907

S. James Press
Department of Statistics
University of California
Riverside, CA 92521

Michael D. Perlman
Department of Statistics
University of Washington
Seattle, WA 98195

Allan R. Sampson
Department of Mathematics and Statistics
University of Pittsburgh
Pittsburgh, PA 15260

Library of Congress Cataloging-in-Publication Data
Contributions to probability and statistics : essays in honor of
Ingram Olkin / Leon Jay Gleser, Michael D. Perlman, S. James Press, Allan R. Sampson, editors.
p. cm.
Includes index.
ISBN 0-387-97076-2 (alk. paper)
1. Probabilities. 2. Mathematical statistics. 3. Olkin, Ingram.
I. Gleser, Leon Jay. II. Perlman, Michael D. III. Press, S. James.
IV. Sampson, Allan R. V. Olkin, Ingram.
QA273.18.C683 1989 89-11470
519.2—dc20

Mathematical Subject Classification: 62–06, 60–06.

Printed on acid free paper.

Camera-ready copy supplied using LaTEX.
Printed and bound by R.R. Donnelley & Sons, Harrisonburg, Virginia.
Printed in the United States of America.

9 8 7 6 5 4 3 2 1

ISBN 0-387-97076-2 Springer-Verlag New York Berlin Heidelberg
ISBN 3-540-97076-2 Springer-Verlag Berlin Heidelberg New York

Preface

It is with great pleasure that this book of original articles is dedicated to Ingram Olkin on the occasion of his sixty-fifth birthday. All four co-editors of this volume have been closely associated with Ingram, first as his Ph.D. students and later as his collaborators. Our understanding of statistics and our careers have benefited greatly from his guidance and assistance. His *joie de vivre* and insightful judgement have strongly influenced our personal views of life. He has been a friend and mentor not only to us individually, but also to our families and to our own Ph.D. students. His enthusiasm for statistics sparked and fueled our interest in the field, and his positive outlook and energy have always served as a source of support. We know that Ingram takes great pleasure in visiting new places. It is our hope that the "trip" provided by this volume through topics that have interested him throughout his career will be as enjoyable and satisfying as any of his many adventures in life.

When we started planning this volume, it was clear that our most difficult task would be to limit the number of contributors without hurting the feelings of the many statisticians who through their attachment to Ingram would have wanted to contribute. We finally agreed to limit invitations to those researchers who had been his Ph.D. students or collaborators, or had been most closely associated with him as a colleague. For whatever oversights we may have made in this selection, we sincerely apologize. As it is, the enthusiastic acceptance of our invitations by nearly all of the individuals whom we contacted caused us to fear for a while that we would have too much material for one book.

Since Ingram has been an outspoken and vigorous advocate of high quality standards in publishing, we decided that every paper submitted would be carefully refereed, and that in cases where the referees did not recommend publication, we would abide by their recommendations. Although it was often painful to do so, we kept to this resolve. Accepted papers were required to be revised along the lines of comments from the referees, sometimes more than once. For those readers who do not personally know Ingram, particularly those who enter the field after this book is published, we decided to include an interview with Ingram in this volume, and to illustrate this volume with pictures of Ingram at various stages of his life.

We hope that these features will provide an informative and stimulating introduction to a person who has made, and continues to make, so many vital contributions to the field of statistics.

The editorial task we set for ourselves required the full participation of all four co-editors in serving as associate editors for the articles submitted, finding and dealing with the publishers, and preparing Ingram's biography, bibliography and interview. Even so, we could not have succeeded without the support of the following referees, who cheerfully and enthusiastically devoted their time to this project:

M. Aitkin (Educational Testing Service)
Y. Amemiya (Iowa State Univ.)
T.W. Anderson (Stanford Univ.)
B.C. Arnold (U. California, Riverside)
S.F. Arnold (Pennsylvania State Univ.)
R. Darrell Bock (U. Chicago)
G.C. Casella (Cornell Univ.)
A. Cohen (Rutgers Univ.)
D. Conway (U. Southern California)
S. DasGupta (Indian Statistical Instit.)
M. DeGroot (Carnegie-Mellon Univ.)
C. Derman (Columbia Univ.)
M.L. Eaton (U. Minnesota)
R.M. Elashoff (U. California, Los Angeles)
T.S. Ferguson (U. California, Los Angeles)
A.E. Gelfand (U. Connecticut)
C. Genest (U. Laval)
S. Ghosh (U. California, Riverside)
N.C. Giri (U. Montreal)
P.K. Goel (Ohio State Univ.)
D.V. Gokhale (U. California, Riverside)
S.S. Gupta (Purdue Univ.)
I. Guttman (U. Toronto)
L.R. Haff (U. California, San Diego)
A.S. Hedayat (U. Illinois, Chicago)
P.W. Holland (Educational Testing Service)
K. Joag-Dev (U. Illinois)
H. Joe (U. British Columbia)
J.H.B. Kemperman (Rutgers Univ.)
H.C. Kraemer (Stanford Univ.)
K-S. Lau (U. Pittsburgh)
S.Y. Lee (Chinese U. Hong Kong)
G.L. Lieberman (Stanford Univ.)
A. Madansky (U. Chicago)
C.N. Morris (U. Texas)
G.S. Mudholkar (U. Rochester)
J.G. Nicholls (Purdue Univ.)
J. Oosterhoff (Free Univ., Amsterdam)
S. Panchapakesan (S. Illinois Univ.)
S.K. Perng (Kansas State Univ.)
F. Proschan (Florida State Univ.)
T.R.C. Read (Hewlett-Packard Co.)
Y. Rinott (Hebrew Univ.)
J. Sacks (U. Illinois)
M.J. Schervish (Carnegie-Mellon Univ.)
J. Sethuraman (Florida State Univ.)
M. Shaked (U. Arizona)
K. Shigemasu (Tokyo Institute of Technology)
R.L. Smith (U. Surrey)
S.M. Stigler (U. Chicago)
D.S. Stoffer (U. Pittsburgh)
W.E. Strawderman (Rutgers Univ.)
T.W.F. Stroud (Queen's Univ.)
K.W. Tsui (U. Wisconsin)
D.E. Tyler (Rutgers Univ.)
V.R. Uppuluri (OakRidge Natl. Lab.)
J.S. Verducci (Ohio State Univ.)
L. Wolstenholme (U. Surrey)
G.Y. Wong (Sloane-Kettering Cancer Ctr.)

We would also like to acknowledge the assistance of Anita Olkin in obtaining most of the photographs that illustrate the volume, and Mary Epperson, Betty Gick, Diane Hall, Norma Lucas and Teena Seele for their assistance in preparing the manuscript.

The Editors

Contents

Part I

An Appreciation

A Brief Biography and Appreciation of Ingram Olkin

Ingram Olkin, known affectionately to his friends in his youth as "Red," was born July 23, 1924 in Waterbury, Connecticut. He was the only child of Julius and Karola (Bander) Olkin. His family moved from Waterbury to New York City in 1934. Ingram graduated from the Bronx's DeWitt Clinton High School in 1941, and began studying statistics in the Mathematics Department at the City College of New York. After serving as a meteorologist in the Air Force during World War II (1943–1946), achieving the rank of First Lieutenant, Ingram resumed his studies at City College. He received his B.S. in mathematics in 1947.

Ingram then began graduate study in statistics at Columbia University, finishing his M.A. in mathematical statistics in 1949. He completed his professional training at the University of North Carolina, Chapel Hill, by obtaining a Ph.D. in mathematical statistics in 1951.

During his tour of duty in the Air Force, Ingram met Anita Mankin. They were married on May 19, 1945. Their daughters Vivian, Rhoda and Julia were born, respectively, in 1950, 1953 and 1959. Ingram and Anita now are the proud grandparents of three grandchildren.

Ingram began his academic career in 1951 as an Assistant Professor in the Department of Mathematics at Michigan State University. He early on demonstrated his penchant for "visiting" by spending 1955–1956 at the University of Chicago and 1958–1959 at Stanford University. Ingram was promoted to Professor at Michigan State, but left in 1960 to become the Chairman of the Department of Statistics at the University of Minnesota. Shortly afterward in 1961 he moved to Stanford University to take a joint position, which he holds to this day, as Professor of Statistics and of Education. From 1973–1976, he was also Chairman of the Department of Statistics at Stanford.

Ingram's professional accomplishments span a broad spectrum, and have made and continue to make a significant impact upon the profession of statistics. He is an outstanding and prolific researcher and author, with nearly thirty Ph.D. students in both statistics and education. The professional societies in statistics and their journals have greatly benefited from his leadership and guidance. His contributions at the federal level include his work with the National Research Council, National Science Foundation, Center for Educational Statistics, and the National Bureau of Standards.

Over one hundred publications, five authored books, six edited books and two translated works are included in his bibliography. Although his prime research focus is multivariate statistics, his research contributions cover an unusually wide range from pure mathematics to educational statistics. Many of his papers and books are virtually classics in their fields — notably his work with Al Marshall on majorization and related distributional and inequality results. His statistical meta-analysis research and book with Larry Hedges are also extremely influential. His text books on probability and on ranking and selection have made novel pedagogical contributions, bringing statistics to a broader nontechnical audience. Also of substantial value to the profession has been his editing of the *Annals of Statistics Index* and the three volume set *Incomplete Data in Sample Surveys* which derived from the Panel on Incomplete Data, which he chaired (1977–1982) for the National Research Council.

Among Ingram's significant contributions to the statistical profession has been his fostering of the growth of quality journals of statistics. He was a strong proponent of splitting the *Annals of Mathematical Statistics* into the *Annals of Statistics* and the *Annals of Probability.* He oversaw this transition as the last editor (1971–1972) of the *Annals of Mathematical Statistics* and the first editor (1972–1974) of the *Annals of Statistics.* As President of the Institute of Mathematical Statistics (1984–1985), he was instrumental in initiating the journal *Statistical Science* and has served in the capacity of co-editor since its inception. He was also influential in introducing the IMS Lecture Notes — Monograph Series. Furthermore, he was heavily involved in the establishment of the *Journal of Educational Statistics*, for which he served as Associate Editor (1977–1985) and as Chair of the ASA/AERA Joint Managing Committee. In all these and numerous other editorial activities, he strongly supports and encourages the major statistics journals to publish applications of statistics to other fields and to build ties with other scientific societies' publications.

Ingram's activities also extend to his work on governmental committees. He was the first Chair of the Committee on Applied and Theoretical Statistics (1978–1981) of the National Research Council, and also was a member for six years of the Committee on National Statistics (1977–1983). He currently is involved with a major project to construct a national data base for educational statistics.

As Ingram will happily admit, he is a prolific traveler. He has given seminars at more than sixty American and Canadian universities, and at numerous universities in twenty five other countries. He also has attended statistical meetings throughout the world, and has been a visiting faculty member or research scientist at Churchill College (Cambridge University), Educational Testing Service (Princeton, NJ), Imperial College, The University of British Columbia, the University of Copenhagen (as a Fulbright Fellow), Eidgenüssische Technische Hochschule (Switzerland), the National Bureau of Standards, Hebrew University, and the Center for Educational Statistics. Anyone wishing to call Ingram has to be prepared to be forwarded from one phone number to another.

In his travels, Ingram has tirelessly promoted and advanced the discipline of statistics. On an outside review committee at a university, he will convince the dean to take steps to form a new department of statistics. On a governmental panel, he will persuade an agency to seek input from statisticians. He has been an effective advocate for increased interdisciplinary ties both in universities and in government, and has been equally successful in convincing deans and statistics department heads of the need to reward statistical consulting. At most statistics meetings, you will find Ingram in constant conversation — perhaps promoting a new journal, encouraging progress of a key committee, or giving advice about seeking grants or allocating funds. His public accomplishments are many and impressive, but equally important are his behind-the-scenes contributions.

Ingram flourishes when working with others. Many of his published papers are collaborations, and his collaborative relationships tend to be long lasting. Ingram is always bursting with new ideas and projects, and delighted when a common interest develops. His enthusiasm is contagious, and his energy and positive outlook (which are legendary in the field of statistics) are tremendously motivating to all around him.

In describing Ingram, one cannot simply list his personal accomplishments. He is above all a remarkably charming and unpretentious person, who gives much of himself to his family, friends and colleagues. For his former students and the many young statisticians he has mentored, he is a continual source of wisdom, guidance and inspiration. All of us whose lives have been touched by Ingram view him with deep personal affection and great professional admiration.

A Conversation with Ingram Olkin

Early in 1986, a new journal *Statistical Science* of the Institute of Mathematical Statistics appeared. This is a journal Ingram Olkin was intimately involved in founding. One of the most popular features of *Statistical Science* is its interviews with distinguished statisticians and probabilists. In the spirit of those interviews, the Editors of this volume wanted to include an interview with Ingram. However, one does not "interview" Ingram; one simply starts him talking, and sits back to listen and enjoy.

The following conversation took place at the home of S. James Press in Riverside, California in November of 1988.

Press: I am pleased to have this opportunity to interview you. How did you initially get interested in the subject of statistics?

Olkin: To tell the truth, I'm not quite sure. What I do know is that in my high school year book dated 1941 each student listed the profession that he wanted to follow; mine was listed as a statistician. I am quite sure that at that time I did not know what a statistician did, nor what kind of profession it was.

I was a mathematics major in DeWitt Clinton High School, which was an all male school, and then went to CCNY — The College of the City of New York, now called City University of New York. At City College I was a mathematics major and took a course in mathematical statistics. This was taught by Professor Selby Robinson, who became quite well known for having indoctrinated many of the statisticians who are currently at various universities, in government, or in industry.

It was through this course that I became interested in the subject. Selby was not a great teacher, but he was a lovely person who somehow managed to communicate an interest in the field. It may have been that I was challenged to find out more about the subject.

Press: I would like to hear more about Selby Robinson, and your courses with him.

Olkin: I believe that he got his degree at Iowa. He did publish a paper in 1937 on the chi-square distribution. The book we used in class was Kenney and Keeping, which was one of the few mathematically oriented texts. In the applications course we used Croxton and Cowden, which was a classic applied statistics text.

Anyone who was at CCNY and took a course in mathematical statistics probably studied with Selby; Kenneth Arrow, Herman Chernoff, Milton Sobel, Herbert Solomon, and many others were students in his class. I don't know how he managed to instill such an interest in statistics, but I'm grateful that he did.

Some years ago I learned that Selby had retired to California. Several of us invited Selby and his wife for a weekend to Stanford at a time that the Berkeley-Stanford Colloquium was scheduled. He and his wife had a marvelous time with us.

College Days

Press: Tell me more about City College, and how statistics was taught there.

Olkin: Statistics was not taught in a single department at City College. It was taught in part by the Mathematics Department. As a matter of fact, the name of one statistics course taught by the Economics Department was "Unattached, 15.1." The terminology "unattached" indicated its status at City College, that is, it was not basically part of a structured departmental discipline. It was the first in a sequence of three discrete courses, all of an applied nature. I left CCNY in 1943 in my junior year, during the war, and became a meteorologist in what was then the United States Army Air Force. (Shortly thereafter the Air Force became a separate branch of the military.) I returned from the service in 1946 and finished my bachelor's degree at City College. In 1947 I went to Columbia University to continue my studies, because by then I knew I was interested in statistics, and Columbia was a major center.

Press: Was there a Statistics Department at Columbia at that time?

Olkin: The Department of Mathematical Statistics was formed formally about 1946. The faculty at Columbia consisted of Ted Anderson, Howard Levene, Abraham Wald, and Jack Wolfowitz. I had most of my courses from Wald and Wolfowitz and a number of visitors; Anderson was on leave during my stay. That was a heyday for visitors. Henry Scheffé, Michel Loève, R.C. Bose, and E.J.G. Pitman were visitors about that time.

Press: How long were you at Columbia?

Olkin: I stayed at Columbia for my master's degree, and then went to Chapel Hill to continue my studies for the doctorate. Harold Hotelling started his career at Stanford University from 1924-1931, at which time he moved to Columbia. In 1946 he moved to Chapel Hill to form a new department. I left Columbia for Chapel Hill in 1948.

Press: Why did you go to Chapel Hill?

Olkin: It was partially for personal reasons. I was married to Anita in 1945 while I was in the service. When we returned to New York after my discharge from the army, the country was faced with a severe housing shortage. In fact, it was almost impossible to find an apartment at that time. Even telephones were rationed after the war. If you were a doctor you could get a telephone, but there was a very long waiting list for the general public.

My parents had a small apartment, but Anita's parents had an extra bedroom, so we lived with her parents in Manhattan for about two years. After living in California for our first year of marriage, we were not as enamored with New York as before. This prompted me to look for an alternative to Columbia, and I learned that Chapel Hill was another major center. I was offered a Rockefeller Fellowship at Chapel Hill which made such a move very attractive. But despite our desire to leave New York, I was not at all disenchanted with Columbia. Quite to the contrary. We had started a graduate student group that generated a sense of community among the students. There were virtually no books on statistics at this time, certainly not on advanced topics, and one of our accomplishments was the publication of class lecture notes. So I have fond memories of Columbia.

Press: Tell me about Chapel Hill.

Olkin: In 1948 there were very few places where you could get a Ph.D. in statistics. Berkeley didn't have a department, though you could get a doctorate in statistics. Iowa State had a department; Chicago had a program, but not a department. Princeton, though small, generated an amazing number of doctorates within the mathematics department. Chapel Hill had an Institute of Statistics with two departments, one at Chapel Hill and one at Raleigh. It had a galaxy of stars on the faculty. On the East Coast, Columbia and Chapel Hill were really the large centers and there was a lot of interaction between the two.

Press: So you ended up following Hotelling?

Olkin: In a certain sense, that's right. The faculty at Chapel Hill in 1948 when I arrived, consisted of Hotelling as chair, R.C. Bose, Wassily Hoeffding, P.L. Hsu, William Madow, George Nicholson, and Herbert Robbins. Gertrude Cox was Director of the Institute.

Hsu was on the faculty, but was on leave in China for a year. He never did return, and S.N. Roy joined the department the following year. The faculty together with visitors formed a phenomenally large group. At Raleigh, there was a Department of Experimental Statistics, with Bill Cochran and many others. The Chapel Hill-Raleigh group was really one of the great faculties.

Press: So you spent about three years there?

Olkin: Yes, from 1948 until 1951 when I graduated.

The Doctoral Dissertation at Chapel Hill

Press: What was the subject of your dissertation?

Olkin: Well, there is a story to my dissertation. I had planned to take a class in multivariate analysis from P.L. Hsu, but he was in China. That year Hoeffding gave a beautiful set of lectures in multivariate analysis, after which I wanted to continue working in this area. A fellow colleague, Walter Deemer, and I asked Hotelling about continuing our studies as a reading course. He suggested that we use student notes from previous courses given by Hsu. My memory is vague on this, but I recall that we had notes from Al Bowker and Ralph Bradley who had previously taken such a course. Walter and I formalized the material on Jacobians of matrix transformations, and extended many of the results. This was the basis of my joint paper with Walter Deemer on Jacobians of matrix transformations, and really set the stage for my later work. The next year when S.N. Roy arrived, I continued my work with him and with Hotelling on multivariate distribution theory. The object was to develop a methodology for deriving a variety of multivariate distributions. I was able to obtain new derivations for the distribution of the rectangular coordinates, for various beta-type distributions related to the Wishart distribution; for the joint distribution of singular values of a matrix and for the characteristic roots of a random symmetric matrix.

The singular value decomposition was not used much at that time, but this has now become a common decomposition used by numerical analysts. I believe that this was one of the earliest statistical uses of singular values.

Press: The dissertation was formally under Roy and Hotelling?

Olkin: They were both readers, but Roy served as principal advisor.

Press: What else can you tell me about Columbia and Chapel Hill?

Olkin: Both Columbia and Chapel Hill had great students. You have to remember that these were the first post-war classes. So there was a tremendous backlog of individuals who had been away during the war and were returning immediately thereafter. If you catalog the statisticians who received doctorates at both Columbia and Chapel Hill during those early years, you will find a large number who are leaders in the field today. It was a very exciting period at Chapel Hill, both in terms of faculty and in terms of what the students were doing.

Press: Who were some of your fellow students?

Olkin: The list of students at Columbia and Chapel Hill was very long, and my memory is not good enough to remember everyone. But I do recall many with whom I interacted.

At Columbia the list includes Raj Bahadur, Robert Bechhofer, Allan Birnbaum, Thelma Clark, Herbert T. David, Cyrus Derman, Charles Dunnett, Harry Eisenpress, Lillian Elveback, Peter Frank, Mina Haskind, Leon Herbach, Stanley Isaacson, Seymour Jablon, William Kruskal, Roy Kuebler, Gottfried Noether, Monroe Norden, Ed Paulson, G.R. Seth, Rosedith Sitgreaves, Milton Sobel, Henry Teicher, and Lionel Weiss.

At Chapel Hill-Raleigh there were Raj Bahadur, Isadore Blumen, Colin Blyth, Ralph Bradley, Uttam Chand, Willard Clatworthy, William Connor, Meyer Dwass, Sudhish Ghurye, Bernard Greenberg, Max Halpern, Jim Hannan, Gopinath Kallianpur, Marvin Kastenbaum, Paul Minton, Sutton Munro, D.N. Nanda, Joan (Raup) Rosenblatt, Shared Shrikhande, Morris Skibinsky, Paul Somerville, Robert Tate, Milton Terry, Geoffrey Watson, and Marvin Zelen.

Press: Did you do any statistics during the war, before you returned?

Olkin: No, I did not. I was trained at MIT and Chanute Air Force Base to be a meteorologist, and subsequently was a weather forecaster at several airports. At one point I thought of combining the two fields, since a variety of statistical procedures were being used to forecast weather. But somehow this merger did not materialize. Actually quite a number of statisticians and mathematicians were in the meteorology program — for example, those I remember are Kenneth Arrow, Jim Hannan, Gil Hunt, Selmer Johnson, Jack Kiefer, Sam Richmond, and Charles Stein, but I am sure there were many others.

Press: Did subjectivity enter weather forecasting at that time?

Olkin: Not in a formal way. Some of the good forecasters were old timers, who happened to remember similar weather patterns from previous years. They were able to retrieve information from old maps and use that as a basis for forecasting. As you may know, it is rather difficult to beat a forecast of continuity, that is, forecast for tomorrow what the weather is today. How to evaluate weather forecasts in terms of accuracy is also an interesting area.

Early Years

Press: Can we shift gears a bit and have you tell me about your childhood and your family?

Olkin: I was born in Waterbury, Connecticut. My father came to the United States from Vilna in Lithuania — probably to escape being inducted in the Tsarist Russian Army. This was a common sequence at that time. My mother was born and lived in Warsaw, and met my father there.

The move to Waterbury was primarily because some colleagues in my father's occupation — he was a jeweler — were in Waterbury and they had arranged a job for him. When the depression period in the early 1930's came, jewelry was one of the first professions to feel the financial pinch, because it was a luxury item. My family then moved to New York City. I suspect that the move to New York was also prompted by a concern about my future education. Connecticut did not have any tuition-free state universities. Of course, it had Yale University, but for immigrants Yale was totally out of the question, whereas City College was free. We moved to New York in 1934 and my formative years of high school and college were really there.

New York City was quite an exciting place. I went to DeWitt Clinton High School, which at that time had a mathematics team. There was also a football team, but I don't remember it. The math team was a good one. We used to have meets on Saturdays at one of the high schools, and two different high school mathematics teams would compete. It was very much like the Olympiad and Putnam competitions.

Press: What kind of high school was this?

Olkin: DeWitt Clinton was a very large school with an enrollment of about 4000. It was located in the Bronx bordering on a park area. My graduating class of 1941 boasts of James Baldwin, who wanted to be a writer and became a distinguished one, Julius Irving, who became Managing Director of the Vivian Beaumont Theater at Lincoln Center, and Charles Silberman, who wrote several books, including "Crisis in the Classroom." I am sure that many others have become success stories. With such a large enrollment there were opportunities to pursue many different avenues.

Papers That I Like

Press: I'd like to discuss your publications. You've published more than 100 papers that I know about. Which ones would you regard as your particular favorite ones?

Olkin: That's a hard question, Jim. Certainly the first one with Walter Deemer was a favorite. Walter and I spent a lot of time together, and it was an invigorating, productive, and enjoyable collaboration. It also was my first paper, and that often is special. In retrospect, the papers I tend to like most are the ones that brought me into a new area, ones that I had not worked on before. There is a tendency to continue working in the same research area, and it is not easy to move into different fields.

Chronologically, probably the next paper that I like was the one with John Pratt on Chebychev-type inequalities. That started me in a research area that I continued with Albert Marshall for approximately ten years. My association with Al came about by accident. He had completed his dissertation at the University of Washington. His thesis was also on Chebychev inequalities, and was related to my work with Pratt. In 1958 Al was a post-doctoral fellow at Stanford and I was on sabbatical leave from Michigan State University. We had corresponded before we met, and we were both immersed in the ideas related to Chebychev inequalities. We had adjacent offices, which made it easy to work together. We wrote several papers that year and generated ideas for later work. That started a long history of collaboration. The paper on this subject that I like most is the one in which we were able to obtain multivariate Chebychev inequalities in a rather general framework.

Earlier on I had given some lectures at Michigan State University on independence properties and characterizations of distributions. This led me to think about multivariate versions, and it started a collaborative effort with Sudhish Ghurye and with Herman Rubin. The key point here is that multivariate characterizations often introduce an ingredient that is quite different from the univariate case. With Ghurye the multivariate characterizations dealt with the normal distribution, and with Rubin the Wishart distribution. Each of these papers had novel aspects in their multivariate versions.

The paper with Al Marshall on the multivariate exponential distribution seemed to fill a niche in terms of being a non-normal distribution that had some very nice properties. That paper has probably been referenced more than any of my other papers.

Press: Why is that?

Olkin: It may be because the problem of constructing bivariate distributions with given marginals is rather tantalizing. We generated this particular bivariate exponential distribution from several disparate points of view, and they all converged to the same result. The bivariate exponential distribution has now been applied in different contexts — in reliability theory, in hydrology, and in medicine. Recently Neils Keiding in Denmark has used our bivariate distribution as a model in which cancer can occur individually or simultaneously at several sites. I think that this will become an important application.

A long-time interest of mine has been matrix theory. I think this started when I took a course with Alfred Brauer at Chapel Hill. He was the kind of teacher who was able to command an interest and excitement about the field. At that time he had obtained some nice new results on estimating the eigenvalues of a matrix. I studied matrix theory rather extensively, used it in my dissertation, and subsequently in my work in multivariate analysis.

I've enjoyed trying to mesh some probabilistic results with matrix theory results. For example, a quadratic form can be considered as the first moment of a distribution on the eigenvalues of the matrix of the quadratic form. Consequently, Chebychev inequalities can provide estimates for the location of eigenvalues of a symmetric matrix. There have been several papers of that type; one in particular, with Al Marshall, dealt with scaling of matrices.

An area that I've probably spent the most time on is majorization. I am not sure how Al and I started, but I believe that it was a natural follow up of the work on Chebychev inequalities. From probabilistic inequalities we moved into a variety of real variable inequalities, such as the Hölder, Minkowski and matrix eigenvalue inequalities. At one time we thought of trying to update the Hardy, Littlewood and Pólya book on inequalities. It didn't take long to realize such a plan was rather presumptuous. But we did discover that majorization was a fundamental notion with a rich theory that could be applied to a wide range of topics. On and off we spent approximately 15 years in writing our book on majorization. The reception of this work and its continued use in different areas is most gratifying. The reviews of our book were very laudatory.

Another general area that I enjoy is to try statistically or probabilistically to model a practical problem and to develop procedures for handling the statistical analysis. That has been an ongoing process throughout my career. The applications have been in the behavioral and social sciences, including education. Many of my papers have a genesis in an application.

Most recently I've been working on meta-analysis, which, again, deals with a different area. I became intrigued in this through my connections in education. One of my colleagues, Nate Gage, who is an expert in teacher education, pointed out that in education you rarely see profound or strong effects. What you see are small or modest effects that are consistent over repeated studies. The question he posed was whether there was a way in which one could strengthen the conclusion of the composite of studies, even though each particular study was non-significant. At that time the procedure that might be called "vote counting" (that is, counting the number of significant results) was in vogue. My first results in 1972 dealt with the development of an alternative method of analysis to vote counting. Later in 1976–1978 Gene Glass coined the term meta-analysis and proposed a quantification in combining results from independent studies. This served as a catalyst to work more seriously in this area. Larry Hedges was a doctoral student at Stanford and in 1980 wrote his dissertation with me on meta-analysis. Subsequently he continued to work in this area and contributed a lot to the field. Meta-analysis had begun to be somewhat of a fad, and the statistical procedures available or used were not always rigorous. So we decided to write a book that focused on the statistical methods for meta-analysis. This book, titled *Statistical Methods for Meta-Analysis*, was completed in 1985, and has had extensive use, in education, psychology and medicine.

People I've Known

Press: Let's talk about individuals and your relationships with them. Which ones were the closest? From what you've said so far, it's clear that you've spent a great many years working with Al Marshall. But what about your early years? For example, who was your mentor when you got started, and what was your relationship with people such as S.N. Roy and Harold Hotelling?

Olkin: Hotelling certainly had a great influence. He valued research, and did not emphasize personalities. He was a very strong advocate for the profession at large, and I think that this characteristic rubbed off on me. I don't think I ever heard Hotelling diminish anyone's work. He always built up individuals if they were productive. Also, he fought very strongly for the teaching of statistics by statisticians. We see this most clearly from his articles which have been reproduced in *Statistical Science*. But it was hard for a student to become close to him on a personal level. In part this was because he was such an esteemed figure, and his manner was somewhat on the distant side. I would say that one could feel respectful, fond, loyal, and appreciative.

Roy was much more approachable. He had just come to the United States, and I was rather close to him. He and I wrote a joint paper immediately after my thesis. But then he died at quite an early age. This was quite a shock to me.

There were others who influenced me, in particular, Wolfowitz at Columbia, Bose, Hoeffding, and Robbins at Chapel Hill. This was in terms of their scholarship and as role models.

Shortly thereafter I became a faculty member at Michigan State University. From the beginning I have written a lot of joint papers and my collaborators became close associates. At one time I counted over thirty collaborators. Milton Sobel has been a continual collaborator. He and I, together with Jean Gibbons, did write a book titled *Selecting and Ordering Populations.*

I've really enjoyed the collaboration and the closeness with almost all of my students. With some I've written a number of papers after they completed their doctoral degrees. One of my first students was Leon Gleser. I've written a book with him titled *Probability Models and Applications* and a number of papers. Today we keep up socially and professionally and are still involved on several papers. You were the second student, and we have collaborated on several papers. This was also true with Joe Eaton, Mike Perlman, Allan Sampson, Tom Stroud and others. So I've continued both the collaboration and friendship with students throughout my career. It is pleasing to me that I have had a total of 28 doctoral students in statistics and education, many of whom have had very successful careers.

Probably in terms of the individuals with whom I've collaborated most, I don't think there's any question but that Al and I have the longest history. We've worked together for 30 years, which is quite a long time. Joe Eaton, Leon Gleser, Mike Perlman, and Milton Sobel are the others that I have worked with the longest.

Press: I don't hear you mention any one person whom I would call a mentor, people who drove you on, or from whom you sought advice.

Olkin: They were mentors in a different sense, more as friends who were supportive. For example, when I went to Michigan State University, Leo Katz was the senior faculty member. Leo visited Chapel Hill where we first met, and he recruited me. Statistics at Michigan State was part of the mathematics department in 1951; a separate statistics department was formed about five years later. Leo and I were quite close, though we only wrote one or two papers together. There was no question that in many ways Leo played the role of a mentor or a senior advisor with whom I could talk about a number of problems.

There were a number of people of that type, but I didn't necessarily work with them. This was true to some degree when I visited Stanford in 1958 and when I joined the faculty in 1961. The faculty were young with strong colleagial relationships. There were lots of discussions and you could get advice from colleagues. I was quite close to Bob Bechhofer during my sabbatical and later established a special closeness with Jack Kiefer until his death and with Jerry Lieberman, which has continued to this day.

There are other individuals who influenced me in different ways. For example, I took more courses from Jack Wolfowitz at Columbia than from any other single person. Again, I can't say that he was a mentor in the sense that I would ask him for advice. Jack was not that approachable. But he was a mentor in the sense of being a role model in his emphasis on publishing, on being active, and on having students. While I am reminiscing about Wolfowitz, I can think back to an incident that is now humorous, but wasn't at the time. It was not very easy to meet with either Wald or Wolfowitz. First of all, they were well-known and busy, and the secretary considered it her duty to keep them sheltered. Wald was a very kind person, but somewhat formidable for a young student.

At Columbia you had to write a dissertation for the master's degree. Wolfowitz was my adviser, so I occasionally needed to see him. He had office hours from 12:45 to 1:00, and so one would queue up for a long time in order to see him. I remember once waiting to discuss my dissertation with him. He invited me into his office and then there was something like a quartet in a Verdi opera, except that no one sang. Jack asked me to state my problem at the blackboard, which I did. As I was speaking, the phone rang. Jack started a conversation with the other person on the telephone and would periodically tell me to continue speaking. While this was going on, he was reading his mail. So Jack and the other person were speaking, I was talking to the blackboard, and Jack was reading his mail. This kind of interaction had a salutary effect. It kept students from coming back to see him, and certainly was successful in my case.

Chapel Hill was quite different from Columbia. It was much more intimate because it was not a commuting community. There was a little snack bar where the faculty and students could buy ice cream and sit around. And so the faculty were very approachable. There's no question that there was a lot of interaction between the faculty and the students.

Both Robbins and Hoeffding were very active researchers. Robbins frequently would call students to his office or to a classroom to discuss his current research. Often, colloquia speakers were invited to faculty houses and occasionally a few students were also invited. I felt a part of a community, and the atmosphere at Chapel Hill fostered this feeling of community.

Press: When you had questions about your career, to whom did you speak at North Carolina?

Olkin: Mostly my own peer group. The students were a closely-knit group, but I don't recall talking to the faculty about non-technical matters. There were not that many people in the field. Statistics was a new field, so there was not much previous experience or previous track record concerning career opportunities.

For example, when I graduated in 1951 there were very few job openings. This was before the many statistics departments were formed. So most positions were in mathematics departments. This meant that there were at most a few statisticians in each mathematics department, and it was not unusual to be the only statistician in the department. I felt that it was important to get a position in a university where there were plans for building a nucleus of statisticians.

Press: Is that why you went to Michigan State?

Olkin: As I mentioned, Leo Katz was there, and he wanted to build a group. Chuck Kraft was just getting his doctorate from Berkeley, and was an instructor at the time. A year later Ken Arnold came from Wisconsin, and then Jim Hannan came from Chapel Hill. Leo managed to attract a large number of visitors. For example, Alfred Rènyi visited, as did R.A. Fisher. And so, within a period of three or four years we had a critical mass on the faculty.

Press: After Michigan State you were at the University of Minnesota.

Olkin: I spent a year and a half at the University of Minnesota. That was also a very nice period, though it wasn't for long. While at Michigan State I was invited to join the faculty at Minnesota, which I did. At that time Leo Hurwicz, Palmer Johnson, and Richard Savage were on the faculty. A statistics department was being formed and I was asked to serve as Chair. Within a short period thereafter

Meyer Dwass, Sudhish Ghurye, Gopinath Kallianpur, and Milton Sobel joined the faculty.

Press: Then there was Stanford?

Olkin: As you know, my years at Stanford comprised a major part of my life and I will soon mark my thirtieth anniversary at Stanford. I visited Stanford in 1958 while on sabbatical, and then joined permanently in 1961. The Bay Area with Berkeley and Stanford was phenomenally active. You have to remember that this was almost 30 years ago. Many of us were in our 30's or early 40's and a tremendous amount of energy and electricity flowed in the two places.

Press: How about the Berkeley Symposium?

Olkin: The Berkeley Symposium, held every five years, was very strong. In addition, we had frequent joint Berkeley-Stanford colloquia. These were very exciting years. Berkeley and Stanford each was trying to build its department, and there was a lot of research activity. The students were first rate. I think it's an interesting commentary that so many of our students have become the statistical leaders at this time.

This was also a period of a lot of visitors, both in summer and during the academic year. If you waited on the steps of Sequoia Hall or the department at Berkeley, you probably would meet almost every statistician at some point. Over the years Anita and I have entertained a very large number of visitors. It is not surprising for someone to tell Anita that he and his family had dinner in our house twenty years ago.

The faculty at Stanford in 1961 was great: Bowker, Chernoff, Chung, Johns, Karlin, Lieberman, Miller, Moses, Parzen, Solomon, Stein. During the first 15 years at Stanford my main energy was devoted to my own research in both departments and to helping build the statistics department and my program to train educational statisticians in the School of Education.

Current Interests

Press: What are your current research interests? You've been involved with many different research directions.

Olkin: Two topics seem to follow me. I often receive letters on inequalities and majorization, and I think I'm ready to say that I don't want to stare another inequality in the face. But I must also confess that I cannot keep away from a new inequality. I also receive letters concerning meta-analysis, in which I still have a strong interest, especially the meta-analyses being conducted in medicine.

Al Marshall and I are becoming more involved in an area that we had worked on earlier, namely the bivariate exponential distribution. That was a specialized result, and we are now concerned with more general questions. For some time now we've been intrigued by the question of how to build dependencies into bivariate distributions. We have developed several unifying themes, some of which have been published, and we are contemplating writing a book that brings the subject into better focus. Since the last book took 15 years to write, we each are cautious about beginning a venture that could take a long time to complete. However, I suspect that subconsciously we each have a book in mind, but are reluctant to state so too openly.

Let me add a bit about this project about dependencies. A natural question that needs to be resolved is how to simulate or generate distributions with these given dependencies. Computer simulation is an area that has recently interested me. In addition to my work with Al, I have written joint papers with Ted Anderson and George Marsaglia on generating correlation matrices and generating random orthogonal matrices. In order to apply these multivariate distributions, we need to develop methods for computer simulations. But it is not always clear how to generate observations with a particular type of dependency. This is quite different from, say, fitting data using the Pearson families. The families that we have in mind arise more from models than from data.

That's our main current work. Al and I are trying to get together more often, which should make it easier to keep working on a single project.

Press: How have you and Al managed to work together so much?

Olkin: I must say that in retrospect, I don't quite know how he and I have managed to collaborate as much as we have, inasmuch as we are not at the same university. In our early collaborations we would visit each other for periods ranging from three to seven days. This would give us a chance to get started on a project.

These visits occurred approximately once every six weeks, so we really had a continuing connection with one another. This was particularly the case when Al was at the Boeing Scientific Research Laboratories, at the University of Rochester and the University of British Columbia, and I was at Stanford. We managed to meet this way over the years, but we also had longer periods of time together.

We spent one year together at Cambridge University and one year at Imperial College. I visited the University of British Columbia and Al visited Stanford for longer periods. We also spent three months in Zurich. During these periods we had an opportunity to work intensively. That has been the modus operandi — namely, working for short stretches, and then meeting for a longer period of time when we could put things into perspective and write things up.

Press: Do you have any other major books or projects under way at this time?

Olkin: The project with Al has a high priority. But there are several other projects that I have in mind.

As you know, I have a fine collection of photographs of well-known statisticians and probabilists that adorn Sequoia Hall. Over the years visitors have suggested that I publish these so that others might enjoy them. I plan to do this as a joint venture with H.O. Lancaster, who is more of a historian than I.

The medical profession has taken to meta-analysis. There often are many studies dealing with the same illness, but with varying conditions and patients, and meta-analysis offers a method for combining results. In my book with Larry Hedges on meta-analysis we did not include a discussion of medical applications, and we now plan a sequel to do that. Dr. Thomas Chalmers of the Harvard School of Public Health and Mt. Sinai Hospital will join us in this project. He has engaged in a number of medical meta-analyses, and will bring a first-hand knowledge base of medicine to the project.

Another project that I have had in the back of my mind goes back to my Chapel Hill days, namely to write a book on matrix theory applications in statistics. But now I would like to add applications in operations research and numerical analysis. This project was to be a collaborative effort with Richard Cottle in the Department of Operations Research, and Gene Golub in the Department of Computer Science. We started to meet on Saturdays to discuss this project, but you have to recognize that these three participants include some of the world's heaviest travelers. So the absences became more and more frequent, and we did not make much headway. But I like to imagine that the future will bring some free time to all of us and that this project will come to fruition.

Press: I notice that you have not mentioned your work with the Department of Education in Washington. Tell me about that connection.

Olkin: Thanks for bringing that up. I have concentrated on the statistics part of my activities. But I would like to tell you about some of the education activities.

As you know, I have a joint appointment between Statistics and the School of Education at Stanford. This has been the case from the time I came to Stanford. My role in the School of Education has been primarily in the doctoral program, to train educational statisticians, in a comparable manner to the training of biostatisticians, psychometricians, econometricians, or geo-statisticians. I have also been involved in research arising from an educational context. The meta-analysis work is an example of research that started from my education connections. But there have been many other instances where a problem has arisen and led to a research study. Several of these have dealt with correlational models; others have involved statistical inference arising from achievement test models.

Recently the American Statistical Association started a wonderful Fellows program designed to bring academia and government closer. This program is supported by the National Science Foundation. The first Fellows program was with the Census Bureau, and later was extended to the National Center for Education Statistics (NCES) and to the Bureau of Labor Statistics. There is a similar program in the Department of Agriculture, and the potential for one at the National Bureau of Standards.

I was invited to be a Fellow at the Center, and before accepting this opportunity, it was important to me that I be engaged in a project that could have an impact in education. It would have been easy to become involved in particular studies at the Center, but over the years I've been on so many panels and studies that I did not see going to Washington for yet another study, even though it might be an important one. The NCES collects data in the form of studies or surveys. For example, the NCES sponsors a School and Staffing Survey, A National Educational Longitudinal Study, and Common Core of Data. There is also a major longitudinal study in mathematics and reading called the National Assessment of Educational Progress (NAEP). ETS is the contractor for NAEP, and periodically issues a report to the nation on the state of mathematics and reading learning. I am currently a member of the NAEP Technical Advisory Committee.

In addition to these national data bases, considerable data is collected by the states. Much data is required to fulfill federal legislation requirements. Thus, the states and the federal government collect data, but there is little integration among states or with the federal government. Ultimately, if not immediately, we need to have enough information to permit us to answer broad issues about education, and to make policy intelligently. This all pointed to the need for a national education data base. The idea of designing such a system was intriguing. It was an area in which I could serve as a link in bringing together the academic and governmental constituencies.

There are 50 states, approximately 16,000 school districts, 100,000 schools and 4 million teachers. It seemed reasonable as a beginning to focus on states and school districts. Each of these constituencies is of a manageable size. This was the thread of my thinking.

It seemed to me that this general area of a statistical education data base would be an interesting challenge that could have a great impact in terms of the statistics for the future. It combined my interests in both statistics and education. It was clear from the outset that such a project was a very ambitious kind of program, and that it would be addressed better by a small group of individuals. I proposed to two colleagues, Ed Haertel, who is a test and measurement specialist at Stanford University, and Larry Hedges, an educational statistician at the University of Chicago, to join me in this project. They both accepted the challenge. We were fortunate in being able to obtain help and advice from a number of colleagues more knowledgeable than we are about data bases. In particular, Nancy Flournoy, who had been instrumental in setting up a multi-disciplinary medical data base information system at the Fred Hutchinson Cancer Research Center, helped a lot. John McCarthy of Lawrence Laboratory, Berkeley, had developed a meta data base for the military, and his experiences were very informative. We Fellows have been holding a series of conferences to help us more fully understand the information and policy needs in a data base system and some of the caveats to worry about. The first conference was with the educational constituency — teachers, principals, administrators, educators, etc. The second conference was with data base specialists, and third was with state data base representatives. We are now in the process of amalgamating this information. But already we have had an effect in bringing some of these constituencies together.

Press: How did you get into the field of education 27 years ago, or earlier?

Olkin: The Department of Statistics at Stanford was modeled on some structural principles. You may have read about this in the interview with Al Bowker that appeared in *Statistical Science*. Some of the structure came from the Statistical Research Group at Columbia University during World War II, and some came from local needs. In effect, there was a strong outreach program. The word "outreach" is in vogue today, but in 1961 it wasn't. But the basic idea was that statistics should be intimately connected to substantive fields. During the 1960's approximately nine out of the sixteen faculty members in the Department of Statistics at Stanford had joint appointments with other fields: three with the medical school, one with operations research, one with electrical engineering, one with economics, one with education, one with geology, and one with mathematics. Later we had a connection with the linear accelerator center. These joint appointments were a guiding principle. Each joint appointee was supposed to develop a program in the other department.

An opportunity arose about 1958-1960 for a joint appointment with the School of Education. Since my work had dealt with multivariate analysis and with models in the social sciences, I was one of the candidates who could fit into both the Statistics Department and the School of Education. The key was to try to find individuals who would be acceptable to two departments as varying as the ones I've mentioned. That's been a very difficult task. We have been successful until now, and I hope that we continue in this vein. In a certain sense this guarantees the development of cross-disciplinary research.

Press: You have had many special satisfactions during your career. Which ones do you value or appreciate the most?

Olkin: There are several aspects of my professional career that I really have enjoyed. I think that what I've enjoyed most of all is the connection with the students I've had and the collaborations I've had. I've enjoyed keeping up with individuals, and I've enjoyed being able to work with them. It's been fruitful in many ways.

In general, I have enjoyed the opportunity to create some programs and to effect changes. Being President of the Institute of Mathematical Statistics (IMS) afforded me an opportunity to tackle a number of needs for the profession. For example, the creation of *Statistical Science* came out of that period, and I hope that this journal remains as a successful legacy. I think it is fair to say that *Statistical Science* came about from the concerted efforts of Morrie DeGroot, Bruce Trumbo, who was treasurer of the IMS at that time, and myself. Morrie DeGroot deserves a tremendous amount of credit for making the journal a success during its formative years.

Those are activities that have given me a lot of satisfaction over the years.

Press: How about your editorships?

Olkin: The editorship of *The Annals* is the kind of job where it is best if you're not asked to be Editor, but it is hard to say no if you are asked. It was a phenomenal amount of work. I was Editor at the time when the two journals — *The Annals of Statistics* and *The Annals of Probability* were still combined into the one, *The Annals of Mathematical Statistics.* I used to receive two new submissions every day. This meant that there were approximately 700 new papers a year to handle, not to mention the ones that were revised several times.

It was a monumental task, and it became quite clear that it was too much for a single Editor to deal with in a responsible way, and that a split was in order. I was pleased to be able to have an effect in starting the two offspring journals. At that time there was a lot of controversy as to whether a split was reasonable, and there were valid arguments on each side. But I doubt that many would want to combine the two *Annals* at this time. Each journal has become a leading publication and is a success. In fact, each journal is now sufficiently large that it is becoming a burden on each Editor. My impression is that professional societies have been too conservative in their publication policies. Because of the tremendous rise in both population and research we need more journal outlets.

Comments About Statistics

Press: I'd like to move on to a very broad question for the field of statistics as you see it. What is your assessment of the current state of the health of the field of statistics, and where do you see the field heading?

Olkin: I'm a bit worried about statistics as a field. As you know, I come from the mathematical community and I've always liked the mathematics of statistics. But I think that the connection with applications is an essential ingredient at this time. I say that because applications are crying out for statistical help. We currently produce approximately 300 Ph.D.'s in statistics and probability per year. This is a small number considering the number of fields of application that need statisticians. Fields such as geostatistics, psychometrics, education, social science statistics, newer fields such as chemometrics and legal statistics generate a tremendous need that we are not fulfilling. Inevitably this will mean that others will fulfill those needs. If that happens across fields of application, we will be left primarily with the mathematical part of statistics, and the applied parts will be carried out by others not well-versed in statistics. Indeed, I think that a large amount of statistics is now being carried out by non-statisticians who learn their statistics from computer packages and from short courses.

So I worry about this separation between theory and practice and the fact that we are not producing the number of doctorates to fulfill needs in all of these other areas.

Press: Has the number of doctorates been going up or down?

Olkin: The number is going up. There were approximately 150 Ph.D.'s in statistics and probability in 1970, 240 in 1975, and there are now about 300. Thus we have doubled in about 15 years. But this growth is not commensurate with the needs and growth of other fields.

Press: Do you mean that we are not producing enough doctorates to meet the demand?

Olkin: We are definitely not producing enough doctorates to meet the demand. An example of an area of high demand is the drug companies, which could use almost all the doctoral students produced each year. Education is another area in which the number of statisticians involved is relatively small, but for which there is a large potential. Statistical needs are growing at a rapid rate in the legal profession, and I don't think there are many law schools that have either statisticians on the faculty or connections with statistics departments. That's an area I would like to see statisticians become involved in early on, to avoid the inevitable turf battle as to who teaches statistics for law.

Press: Is the field of statistics heading now toward more applications?

Olkin: I think a number of individuals in the profession are heading more towards applications, but the field as a whole does not have enough faculty and students working in applied areas. Biostatistics is probably our only big success story in the sense that there are a lot of statisticians in medical schools, though perhaps not enough. This came about in part because of the prevalence of training grants, and in part because of federal regulations mandating clinical trials or other statistical procedures. But there are few, if any, statisticians in law schools, in social science departments, pharmacy, dentistry, education, business, and so on.

Industry used to be a big user of statistics; this diminished considerably about twenty years ago, and now has become a high demand area. Sample surveys are used a lot but this specialization is totally undernourished. The number of universities that teach sampling is small, and we have trained few experts. I am sure that we could expand the research effort and doctorate production in sampling theory.

There is still an excitement in the field, but my impression is that, except for a few places, the growth in statistics departments has reached a plateau. I believe that this is true because we do not have a natural mechanism for statistics departments to create strong links to other departments of the academic community. Academic institutions have not been designed for cross-disciplinary research, and indeed may actually be antagonistic to cross-disciplinary research.

Press: Whereas, by nature, statistics tends to be used and needed in other fields?

Olkin: Yes, indeed. It is particularly important because problems are now becoming much larger. For example, the study of a large scale problem such as pollution or acid rain with a small group of researchers is really not very realistic. We will need a lot of connections with other disciplines. Except for a few places, we are not fostering that connection. The National Science Foundation has recognized this need by creating centers that have a strong cross-disciplinary component, but these do not have strong statistical components. I think that the time has come for the profession to have a Statistical Sciences Institute that would focus on cross-disciplinary research.

It's interesting to note that when I was first at Chapel Hill, Gertrude Cox was a strong advocate for learning a substantive field. We were all encouraged, almost pushed, to become not only statisticians, but to gain a knowledge base in biology, sociology, political science, etc. — any area in which we could apply statistics. Except for the medical field, not many took this route. Remember that starting in the late 1940's the decision theory orientation was strong so that many of us studied mathematics, rather than a substantive field, and we became mathematical statisticians rather than socio-statisticians or geo-statisticians. That was fine up to a point, but the needs in 1990 center much more on our connections and usefulness in these other fields. We need to expand our vision.

Press: How about the direction of growth of the field with computers and data analysis? Are we moving in that direction, are we moving enough or too much?

Olkin: I am more comfortable with the previous general question than I am with the question about statistics and computers. I am not well versed about the field as a whole, but I do have some general impressions. I believe that statistical packages have had serious positive and negative components. The positive, of course, is that people now can carry out more sophisticated analyses than they would be able to if they had to learn programming on their own. There's no doubt that it's been a great service. On the other hand, there is a tendency for people not to learn the statistical underpinnings, but only to learn how to use a statistical computer package. Indeed, my experience in reading doctoral dissertations is that the availability of packages is what drives the choice of analysis. The availability of statistical packages also drives the curriculum and may emphasize how to generate numbers, rather than interpretation and understanding.

More recently there's been a strong development in statistical graphics and resampling schemes. Again I think that in principle these are positive developments. What worries me is that they will be overly used and subsequently abused, as is the case with almost any new area for which there is a lot of use. It doesn't take long before there's a certain amount of abuse, and it becomes a serious problem.

The result of the availability of computers and packages is that statistical analyses are being carried out by non-statisticians much more than ever before. This is fine when done well, but this is not always the case — probably not the case most of the time — so that the public may be faced with erroneous conclusions. The statistical community is not intimately involved in this growth of the use of computers. We provide some of the packages and some of the theory, but, in effect, its big use is elsewhere.

To illustrate the high use by non-statisticians, not long ago Stu Geman gave a talk on image processing, an area in which he and others have been working. Stu mentioned that he had published a paper in an I.E.E.E. (Institute of Electrical and Electronics Engineers) journal, for which he received well over a thousand reprint requests. It's hard to imagine any statistician of my acquaintance who publishes in any of the standard statistical journals receiving that many reprint requests. I usually get six reprint requests — especially if I ask my relatives to write for them. Of course, photocopying confounds these numbers, but the fact remains that little theory is translated to usable methods except through packages.

Press: That certainly makes the point.

Olkin: That's a good example of an area in which the statistical community had a large input, but it is being developed, extended, and used by other fields.

In Spain I Am a Bayesian

Press: Here's a difficult question. Are you Bayesian or not?

Olkin: I think I'm a part-time Bayesian. My inclination in dealing with a problem is to use classical procedures, but when I get more deeply involved and need to obtain information about the parameters I do not hesitate to incorporate some of the Bayesian ideas. I have found that some problems can be formulated in a manner that calls for a Bayesian approach. In other instances, this is not the case and a Bayesian approach would seem forced. I am not a Bayesian in the sense that I feel compelled to use a Bayesian approach, nor am I a classicist in not using a Bayesian procedure.

I suspect that I am begging the question a bit. I don't have too many papers in which the word "Bayesian" appears in the title, but since I want to be invited to the Bayesian conferences, especially when they are in Spain, I will say I'm 75 percent non-Bayesian and 25 percent Bayesian.

Press: You've certainly written papers with avowed Bayesians so you cannot be anti-Bayesian. Have you become increasingly sympathetic over the years?

Olkin: I am not sure how to answer that. Recently I have been working with Irwin Guttman on a model for interlaboratory differences. I had previously written a paper with Milton Sobel in which we used a ranking and selection procedure. At that time this seemed to be a reasonable formulation, but ranking and selection procedures have not been accepted very much in applied work. With Irwin we looked at an alternative formulation that led naturally to a Bayesian point of view, and I was interested to know how different the answers would be. I have no antipathy in using either approach, and try to understand what one gains from each method.

Visits to Other Universities

Press: How about some of your travels. What are some of the universities that you visited, and some of the people there?

Olkin: In 1955 I spent a year at the University of Chicago. At that time Allen Wallis had a program in which they invited two visitors every year, and Don Darling and I spent the year at Chicago. This was a very productive year for me. I taught a course in multivariate analysis and in sampling theory, as I recall, and this gave me an opportunity to renew old acquaintances and to begin new ones.

Chicago had a small but good student body; for example, Herb T. David, Morrie DeGroot, Al Madansky, and Jack Nadler were students at that time. The faculty consisted of Raj Bahadur, Pat Billingsley, Alec Brownlee, Leo Goodman, John Pratt and Dave Wallace. Bill Kruskal was on leave that year, and we lived in his house. Allen Wallis was there as Dean of the School of Business. Also, Meyer Dwass and Esther Seiden were at Northwestern, and we used to get together quite often.

Press: Tell me about your visits to other Universities. I know that you travel a lot.

Olkin: In 1967 I was on sabbatical leave from Stanford and was an Overseas Fellow at Churchill College, Cambridge. That was a great year. Al Marshall and Mike Perlman were also at Cambridge, and Alfred Rényi was a visitor for one quarter. Cambridge had a vigorous group led by David Kendall and Peter Whittle. We had a seminar on inequalities that got us much more deeply into the field. In fact, that year was a very active one in England with lots of visitors in London. We used to visit London quite often for seminars.

In 1971 I spent a year at the Educational Testing Service. Fred Lord, one of the leading researchers in tests and measurement, was head of a very active group. He invited visitors to participate in the program, and that year in addition to myself, Murray Aitkin, Leon Gleser, Karl Jöreskog, and Walter Kristof were in the group. There were some lively discussions. Also, since Princeton was nearby we were able to interact with some of their faculty. This was a period when Leon and I were able to work closely, and we wrote several papers and completed our book.

I again visited England in 1976-77, this time at Imperial College. For several years thereafter Anita and I tended to spend a month every year in London. These visits gave us an opportunity to maintain and to renew European contacts. I was a Fulbright Fellow during the fall of 1979 at the University of Copenhagen. It was a thriving place with Steen Andersson, Hans Brons, Anders Hald, Martin Jacobsen, Soren Johansen, Neils Keiding, Stefan Lauritzen and others. I gave some lectures there and was able to start some new projects.

In the spring of 1981 I was a visitor at the Eidgenüssische Technische Hochschule (ETH) in Zurich. Frank Hampel and his group, Elvezio Ronchetti, Peter Rousseeuw, Werner Stahel, were working on robust estimation. Others on the faculty were Hans Buhlmann and Hans Foellmer. Chris Field from Dalhousie, Bob Staudte from Melbourne, and Al Marshall were visitors. We each gave individual lectures, and I gave a series of lectures on multivariate analysis.

In the early 1950's the National Bureau of Standards was a center for applied mathematics and statistics. They had many postdoctoral and student visitors during those early days. The Bureau was trying to revitalize this program of visitors, and in the fall of 1983 I spent a quarter there. As a consequence of interactions with John Mandel, I again became involved in finding the expected value and covariances of the ordered characteristic roots of a random Wishart matrix. This moved me in the direction of some numerical work that was new to me. Recently, together with Vesna Luzar, who was a Fulbright visitor from Yugoslavia, we have compared several alternative modes of computation.

It was at the Bureau that I met Cliff Spiegelman, and he and I later started a collaboration on semi-parametric density estimation. This is work that we are both continuing.

I visited Hebrew University as a Lady Davis Fellow in the spring of 1984. Our family had visited Israel in 1967, and this gave me an opportunity to renew acquaintances. The statistics group was very lively with lots of activity. The faculty consisted of Louis Guttman, Yoel Haitovsky, Gad Nathan, Samuel Oman, Danny Pfeffermann, Moshe Pollak, Adi Raveh, Yosef Rinott, Ester Samuel-Cahn, Gideon Schwarz, and Josef Yahav. Larry Brown was a visitor for the year. I gave some lectures on inequalities, which generated a collaboration with Shlomo Yitzhaki, who is on the faculty of the Economics Department. This work had a basis in economics and the primary focus was on Lorenz curves and subsequently, on concentration curves. This also introduced me more intimately with some of the results of Gini, which we were able to use to greater advantage. My collaboration with Shlomo has continued, especially when he visits the U.S.

My Family

Press: You haven't had a chance to speak about your immediate family. Can you tell me more about them and bring me up to date?

Olkin: Anita and I were married in 1945 while I was in service. I was being transferred from LaGuardia Airport to San Francisco at the time, so we spent our honeymoon on the train from New York to San Francisco. We very much enjoyed our stay in California, and I am not sure why we returned to New York after my discharge. It never occurred to me to continue my studies at Berkeley. In any case, we did return and I attended CCNY, Columbia, and UNC. Anita and I now look back to our three years at Chapel Hill as a very happy period. We were one of few married couples, and Anita made our house available to many of the graduate students. Also, at Michigan State University we were very much involved in a University community, and our house was often a meeting place for visitors.

We have three daughters. The oldest, Vivian, was born in Chapel Hill in 1950. She and her husband, Sim, live in Austin, Texas and have two children. Vivian was a career counselor at the University of Texas, and is now getting her master's degree in Human Resources Development. Sim received a doctorate from Stanford and is now on the faculty of the Graduate School of Business, University of Texas. Our second daughter, Rhoda, was born in 1953 when I was at Michigan State University. She and her husband, Michael, live in Walnut Creek, California and have one child. Rhoda received a doctorate in counseling psychology from the University of California, Santa Barbara. She is now on the faculty of California State University, Sacramento. Michael is a bio-medical engineer with a company housed in Berkeley. Our youngest, Julia, was born in California in 1959. She and her husband, Juan, live in Castro Valley, California. Julia and Juan both received doctorates in mathematical sciences from Rice University, and are each working as numerical analysts — Julia at SRI and Juan at Sandia in Livermore. My family of females has taught me and trained me in the women's movement, and I have been sensitized to difficulties that women have in the workplace, and the prejudices that exist.

Other Activities

Press: Let me move away from statistics. What do you like to do when you're not doing statistics?

Olkin: I do enjoy traveling, which I think is well-known to a number of people. Whenever I do travel, I generally try to find museums, symphonies, operas, and theaters. I almost always do that, wherever I go. You also know that I'm generally a people person, which is one of the reasons why I've enjoyed students and collaborators. Over the years, the professional contacts have merged with the personal contacts. I enjoy hiking. We used to go to Yosemite regularly when the children were young. More recently I have visited state parks near meeting places.

Press: Do you hike alone?

Olkin: I have tried to entice colleagues to join me, and years ago when we went to Yosemite our youngest daughter Julia would always go with me. But more recently I have gone alone.

Press: How about sports?

Olkin: I enjoy tennis and swimming. The tennis is a social event, but I am a non-social swimmer.

Press: I have a vague recollection of having been told that you were the ping-pong champion on a ship. Tell me about that.

Olkin: That was so long ago that I had totally forgotten it. I mentioned that I was born in Waterbury, Connecticut. There's a resort near New Haven called Woodmont, and even though we were relatively poor, we used to go to Woodmont and rent a room during the summer time. There was a ping-pong room in one of the hotels, for which they used to charge an hourly fee. But if you would help clean up, you had access to the ping-pong tables when they were not being rented, so I used to play a lot when I was young. In 1967 we went to England on the Dutch ship, Rotterdam, which had table tennis contests in both tourist and first class. I played ping-pong in tournaments for both classes and won both. At Stanford there were ping-pong tables in the student union, but the building was altered and there are no tables there at this time, so I haven't played for years.

From 1989 On

Press: What does the future hold for you?

Olkin: The word retirement is a curious word in that it implies that you will stop working at a certain date. We need an alternative descriptor. As I understand the current California and federal laws, if you were born after August 31, 1923 you do not have to retire. The law may change in 1994, but as of now I will not have to retire. Of course, it may not make sense financially or intellectually not to retire. However, I don't see retirement as a problem. I have several projects that I'd like to work on, and I don't have enough time for these without giving up other activities. So retirement is one way of reallocating one's time to the activities that one likes. I am also deeply involved with the Center for Education Statistics project and even though the fellowship will be over within the near future, I would like to continue my involvement. If one has a high metabolic rate, it's difficult not to continue working. All my retired colleagues tell me that I will probably be doing more, rather than less.

Press: Thanks very much for the opportunity to review this part of your history. It was of interest to me and I am sure that it will be of great interest to many of our colleagues.

Bibliography of Ingram Olkin

BOOKS

1973

A Guide to Probability and Applications (with C. Derman and L.J. Gleser), Holt, Rinehart and Winston, Inc.: New York.

1977

Selecting and Ordering Populations: A New Statistical Methodology (with J.D. Gibbons and M. Sobel), John Wiley & Sons: New York.

1979

Theory of Majorization and Its Applications (with A.W. Marshall), Academic Press: New York. (Translated into Russian by G.P. Gavrilov, V.M. Kruzlov and V.G. Mirantsev, MIR: Moscow (1983).)

1980

Probability Models and Applications (with L.J. Gleser and C. Derman), Macmillan: New York.

1985

Statistical Methods for Meta-analysis (with L.V. Hedges), Academic Press: New York.

BOOKS EDITED

1960

Contributions to Probability and Statistics. Essays in Honor of Harold Hotelling (ed. by I. Olkin, S.G. Ghurye, W. Hoeffding, W.G. Madow, and H.B. Mann), Stanford University Press: Stanford CA.

1962

The Annals of Mathematical Statistics, Indexes to Volumes 1–31, 1930–1960 (ed. by J. Arthur Greenwood, Ingram Olkin, and I. Richard Savage), Institute of Mathematical Statistics: Hayward, CA.

1983

Incomplete Data in Sample Surveys, Vol. 1, Report and Case Studies (ed. by W.G. Madow, H. Nisselson and I. Olkin), Academic Press: New York.

Incomplete Data in Sample Surveys, Vol. 2, Theory and Bibliographies (ed. by W.G. Madow, I. Olkin and D.B. Rubin), Academic Press: New York.

Incomplete Data in Sample Surveys, Vol. 3, Proceedings of the Symposium (ed. by W.G. Madow and I. Olkin), Academic Press: New York.

1986

Inequalities in Statistics and Probability (ed. by Y.L. Tong with the cooperation of I. Olkin, M.D. Perlman, F. Proschan and C.R. Rao), IMS Lecture Notes Monograph Series, Volume 5. Institute of Mathematical Statistics: Hayward, CA.

RESEARCH PAPERS

1951

The Jacobians of certain matrix transformations useful in multivariate analysis (with Walter Deemer, Jr.), *Biometrika*, **38**, 345–367.

1953

Properties and factorizations of matrices defined by the operation of pseudo-transposition (with Leo Katz), *Duke Math. J.*, **20**, 331–337.

Note on "The Jacobians of certain matrix transformations useful in multivariate analysis", *Biometrika*, **40**, 43–46.

1954

On multivariate distribution theory (with S. N Roy), *Ann. Math. Statist.*, **25**, 329–339.

1958

Unbiased estimation of certain correlation coefficients (with John W. Pratt), *Ann. Math. Statist.*, **29**, 201–211.

On a multivariate Tchebycheff inequality (with John W. Pratt), *Ann. Math. Statist.*, **29**, 226–234.

Multivariate ratio estimation for finite populations, *Biometrika*, **45**, 154–165.

An inequality satisfied by the gamma function, *Skand. Akt.*, **41**, 37–39.

1959

On inequalities of Szegö and Bellman, *Proc. Nat'l. Acad. Sciences*, **45**, 230–231.

A class of integral identities with matrix argument, *Duke Math. J.*, **26**, 207–213.

Inequalities for the norms of compound matrices, *Arch. der Math.*, **10**, 241–242.

Extrema of quadratic forms with applications to statistics (with K.A. Bush), *Biometrika*, **46**, 483–486. *Corrigenda*, **48**, 474–475.

1960

A bivariate Chebyshev inequality for symmetric convex polygons (with Albert W. Marshall), *Contributions to Probability and Statistics. Essays in Honor of Harold Hotelling*, (ed. by I. Olkin, S.G. Ghurye, H. Hoeffding, W. G. Madow, H.B. Mann), Stanford University Press: Stanford, CA, 299–308.

A one-sided inequality of the Chebyshev type (with Albert W. Marshall), *Ann. Math. Statist.*, **31**, 488–491.

Multivariate Chebyshev inequalities (with Albert W. Marshall), *Ann. Math. Statist.*, **31**, 1001–1014.

1961

Extrema of functions of a real symmetric matrix in terms of eigenvalues (with K.A. Bush), *Duke Math. J.*, **28**, 143–152.

Multivariate correlation models with mixed discrete and continuous variables (with R.F. Tate), *Ann. Math. Statist.*, **32**, 448–465.

Game theoretic proof that Chebyshev inequalities are sharp (with Albert W. Marshall), *Pacific J. Math.*, **11**, 1421–1429.

1962

A characterization of the multivariate normal distribution (with S.G. Ghurye), *Ann. Math. Statist.*, **33**, 533–541.

A characterization of the Wishart distribution (with Herman Rubin), *Ann. Math. Statist.*, **33**, 1272–1280.

Reliability testing and estimation for single and multiple environments (with S.K. Einbinder), *Proceedings of the Seventh Conference on the Design of Experiments in Army Research Development and Testing*, Report No. 62-2, 261–291.

Evaluation of performance reliability (with S.K. Einbinder), *Proceedings of the Eighth Conference on the Design of Experiments in Army Research Development and Testing*, Report No. 63-2, 473–501.

1964

Multivariate beta distributions and independence properties of the Wishart distribution (with Herman Rubin), *Ann. Math. Statist.*, **35**, 261–269.

Inclusion theorems for eigenvalues from probability inequalities (with Albert W. Marshall), *Numer. Math.*, **6**, 98–102.

Reversal of the Lyapunov, Hölder, and Minkowski inequalities and other extensions of the Kantorovich inequality (with Albert W. Marshall), *J. Math. Anal. Appl.*, **8**, 503–514.

1965

Norms and inequalities for condition numbers (with Albert W. Marshall), *Pacific J. Math.*, **15**, 241–247.

On the bias of characteristic roots of a random matrix (with T. Cacoullos), *Biometrika*, **52**, 87–94.

Integral expressions for tail probabilities of the multinomial and negative multinomial distributions (with M. Sobel), *Biometrika*, **52**, 167–179.

1966

A K-sample regression model with covariance (with L.J. Gleser), *Multivariate Analysis*, (ed. by P.R. Krishnaiah), Academic Press: New York, 59–72.

Correlations Revisited, *Improving Experimental Design and Statistical Analysis* (ed. by J. Stanley), Rand McNally: Chicago, 102–156.

1967

Monotonicity of ratios of means and other applications of majorization (with A.W. Marshall and F. Proschan), *Proceedings of the Symposium on Inequalities*, (ed. by O. Shisha), Academic Press: New York, 177–190.

A multivariate exponential distribution (with A.W. Marshall), *J. Amer. Statist. Assoc.*, **62**, 30–44.

A generalized bivariate exponential distribution (with A.W. Marshall), *J. Applied Prob.*, **4**, 291–302.

1968

A general approach to some screening and classification problems, with discussion (with A.W. Marshall), *J. Roy. Statist. Soc., Ser. B*, 407–433.

Scaling of matrices to achieve specified row and column sums (with A.W. Marshall), *Numer. Math.*, **12**, 83-90.

1969

Testing for equality of means, equality of variances and equality of covariances under restrictions upon the parameter space (with L.J. Gleser), *Ann. Inst. Statist. Math.*, **21**, 33–48.

Testing and estimation for a circular stationary model (with S. J. Press), *Ann. Math. Statist.*, **40**, 1358–1373.

Approximate confidence regions for constraint parameters (with A. Madansky), *Multivariate Analysis II* (ed. by P.R. Krishnaiah), Academic Press: New York, 261–268.

Unbiased estimation of some multivariate probability densities and related functions (with S.G. Ghurye), *Ann. Math. Statist.*, **40**, 1261–1271.

Norms and inequalities for condition numbers, II, *Linear Algebra Appl.*, **2**, 167–172.

1970

An extension of Wilks' test for the equality of means (with S.S. Shrikhande), *Ann. Math. Statist.*, **41**, 683–687.

Linear models in multivariate analysis (with L.J. Gleser), *Essays in Probability and Statistics* (ed. by R.C. Bose, I. M. Chakravarti, P.C. Mahalanobis, C.R. Rao, and K.J.C. Smith), University of North Carolina Press: Chapel Hill, NC, 267–292.

Chebyshev bounds for risks and error probabilities in some classification problems (with A.W. Marshall), *Proceedings of the International Symposium on Nonparametric Analysis*, (ed. by M. Puri), Cambridge University Press, 465–477.

1971

A minimum-distance interpretation of limited information estimation (with A.S. Goldberger), *Econometrica*, **39**, 635–639.

1972

Estimation and testing for difference in magnitude or displacement in the mean vectors of two multivariate normal populations (with C.H. Kraft and C. Van Eeden), *Ann. Math. Statist.*, **43**, 455–467.

Jacobians of matrix transformations and induced functional equations (with A.R. Sampson), *Linear Algebra Appl.*, **5**, 257–276.

Applications of the Cauchy-Schwarz inequality to some extremal problems (with M.L. Eaton), *Proceedings of the Symposium on Inequalities III*, (ed. by O. Shisha), Academic Press: New York, 83–91.

Monotonicity properties of Dirichlet integrals with applications to the multinomial distribution and the analysis of variance, *Biometrika*, **59**, 303–307.

Estimation for a regression model with an unknown covariance matrix (with L.J. Gleser), *Proceedings of the Sixth Berkeley Symposium on Mathematical Statistics and Probability*, Vol. I, University of California Press: Berkeley, CA, 541–568.

Inequalities on the probability content of convex regions for elliptically contoured distributions (with S. Das Gupta, M.L. Eaton, M. Perlman, L.J. Savage, M. Sobel), *Proceedings of the Sixth Berkeley Symposium on Mathematical Statistics and Probability*, Vol. III, University of California Press: Berkeley, CA, 241–265.

1973

Norms and inequalities for condition numbers, III (with A.W. Marshall), *Linear Algebra Appl.*, **7**, 291–300.

Testing and estimation for structures which are circularly symmetric in blocks, *Proceedings of the Symposium on Multivariate Analysis*, Dalhousie, Nova Scotia, 183–195.

Multivariate statistical inference under marginal structure, I (with L.J. Gleser), *British J. Math. Statist. Psych.*, **26**, 98–123.

Identically distributed linear forms and the normal distribution (with S.G. Ghurye), *Adv. in Appl. Prob.*, **5**, 138–152.

1974

Inference for a normal population when the parameters exhibit some structure, *Reliability and Biometry: Statistical Analysis of Lifelength*, (ed. by F. Proschan and R.J. Serfling), SIAM, 759–773.

Majorization in multivariate distributions (with A.W. Marshall), *Ann. Statist.*, **2**, 1189–1200.

1975

A note on Box's general method of approximation for the null distributions of likelihood criteria (with L.J. Gleser), *Ann. Inst. Statist. Math.*, **27**, 319–326.

Multivariate statistical inference under marginal structure, II (with L.J. Gleser), *A Survey of Statistical Design and Linear Models*, (ed. by J.N. Srivastava), North-Holland Publishing Co.: Amsterdam, 165–179.

1976

Asymptotic distribution of functions of a correlation matrix (with M. Siotani), *Essays in Probability and Statistics*, (ed. by S. Ikeda), Shinko Tsusho Co., Ltd., Tokyo, Japan, 235–251.

1977

Estimating covariances in a multivariate normal distribution (with J.B. Selliah), *Statistical Decision Theory and Related Topics, II*, (ed. by S.S. Gupta), Academic Press: New York, 313–326.

Correlational analysis when some variances and covariances are known (with M. Sylvan), *Multivariate Analysis, IV*, (ed. by P.R. Krishnaiah), North-Holland Publishing Co.: Amsterdam, 175–191.

A study of X chromosome linkage with field dependence and spatial visualization (with D.R. Goodenough, E. Gandini, L. Pizzamiglio, D.Thayer, H.A. Witkin), *Behavior Genetics*, **7**, 373–387.

1978

An extremal problem for positive definite matrices (with T.W. Anderson), *Linear and Multilinear Algebra*, **6**, 257–262.

Baseball competitions—are enough games played? (with J.D. Gibbons and M. Sobel), *American Statistician*, **32**, 89–95.

1979

Admissible and minimax estimation for the multinomial distribution and for k independent binomial distributions (with M. Sobel), *Ann. Statist.*, **7**, 284–290.

A subset selection technique for scoring items on a multiple choice test (with J.D. Gibbons and M. Sobel), *Psychometrika*, **44**, 259–270.

Matrix extensions of Liouville-Dirichlet type integrals, *Linear Algebra Appl.*, **28**, 155–160.

An introduction to ranking and selection (with J.D. Gibbons and M. Sobel), *American Statistician*, **33**, 185–195.

1980

Vote-counting methods in research synthesis (with L.V. Hedges), *Psychological Bulletin*, **88**, 359–369.

1981

Unbiasedness of invariant tests for MANOVA and other multivariate problems (with M. Perlman), *Ann. Statist.*, **8**, 1326–1341.

A new class of multivariate tests based on the union-intersection principle (with J.L. Tomsky), *Ann. Statist.*, **8**, 792–802.

Entropy of the sum of independent Bernoulli random variables and of the multinomial distribution (with I.A. Shepp), *Contributions to Probability*, (ed. by J. Gani and V.K. Rohatgi), Academic Press: New York, 201–206.

A comparison of n-estimators for the binomial distribution (with J. Petkau and J. Zidek), *J. Amer. Statist. Assoc.*, **76**, 637–642.

The asymptotic distribution of commonality components (with L.V. Hedges), *Psychometrika*, **46**, 331–336.

Maximum likelihood estimation in a two-way analysis of variance with correlated errors in one classification (with M. Vaeth), *Biometrika*, **68**, 653–660.

Range restrictions for product-moment correlation matrices, *Psychometrika*, **46**, 469–472.

1982

A model for aerial surveillance of moving objects when errors of observations are multivariate normal (with S. Saunders) *Statistics and Probability: Essays in Honor of C.R. Rao*, (ed. by G.B. Kallianpur, P.R. Krishnaiah and J.K. Ghosh), North-Holland Publishing Co.: Amsterdam, 519–529.

The distance between two random vectors with given dispersion matrices (with F. Pukelsheim), *Linear Algebra Appl.*, **48**, 257–263.

Bounds for a k-fold integral for location and scale parameter models with applications to statistical ranking and selection problems (with M. Sobel and Y.L. Tong), *Statistical Decision Theory and Related Topics, III*, (ed. by S.S. Gupta), Academic Press: New York, 193–211.

A convexity proof of Hadamard's inequality (with A.W. Marshall), *Amer. Math. Monthly*, **89**, 687–688.

Analyses, reanalyses, & meta-analysis, (with Larry V. Hedges), *Contemporary Education Review*, **1**, 157–165.

1983

A sampling procedure and public policy (with G.J. Lieberman and F. Riddle), *Naval Research Logistics Quarterly*, **29**, 659–666.

Domains of attraction of multivariate extreme value distributions (with A.W. Marshall), *Ann. Prob.*, **11**, 168–177.

Regression models in research synthesis (with L.V. Hedges), *American Statistician*, **37**, 137–140.

Clustering estimates of effect magnitude from independent studies (with L.V. Hedges), *Psychological Bulletin*, **93**, 563–573.

Inequalities via majorization — an introduction (with A.W. Marshall), *General Inequalities 3*, (ed. by E.F. Beckenbach and W. Walter), Birkhäuser Verlag: Basel, 165–187.

Adjusting p-values to account for selection over dichotomies (with G. Shafer), *J. Amer. Statist. Assoc.*, **78**, 674–678.

An inequality for a sum of forms (with A.W. Marshall) *Linear Algebra Appl.*, **52/53**, 529–532.

Generating correlation matrices (with G. Marsaglia), SIAM, *J. Sci. Statist. Comput.*, **5**, 470–475.

1984

Joint distribution of some indices based on correlation coefficients (with L.V. Hedges), *Studies in Econometrics, Time Series, and Multivariate Statistics*, (ed. by S. Karlin, T. Amemiya, and L.A. Goodman), Academic Press: New York, 437–454.

Academic statistics: Growth, change and federal support (with D.S. Moore), *American Statistician*, **38**, 1–7.

Multidirectional analysis of extreme wind speed data (with E. Simiu, E.M. Hendrickson, W.A. Nolan and C.H. Spiegelman), *Engineering Mechanics in Civil Engineering*, Volume 2, (ed. by A.P. Boresi and K.P. Chong), American Society of Civil Engineers: New York, 1196–1199.

Nonparametric estimators of effect size in meta-analysis (with L. V. Hedges). *Psychological Bulletin*, **96**, 573–580.

Estimating a constant of proportionality for exchangeable random variables (with I. Guttman), *Design of Experiments, Ranking and Selection* (ed. by T.J. Santner and A.C. Tamhane), Marcel Dekker: New York, 279–285.

1985

A probabilistic proof of a theorem of Schur. *American Mathematical Monthly*, **92**, 50–51.

A family of bivariate distributions generated by the bivariate Bernoulli distribution (with A.W. Marshall), *J. Amer. Statist. Assoc.*, **80**, 332–338.

Estimating the Cholesky decomposition, *Linear Algebra Appl.*, **67**, 201–205.

Statistical inference for constants of proportionality (with D. Y. Kim and I. Guttman), *Multivariate Analysis-VI*, (ed. by P.R. Krishnaiah), Elsevier Science Publishers: New York, 257–280.

Multivariate exponential distributions, Marshall-Olkin (with A. W. Marshall), *Encyclopedia of Statistical Sciences, Volume 6*, (ed. by S. Kotz and N.L. Johnson), John Wiley Sons: New York, 59–62.

Maximum likelihood estimation of the parameters of a multivariate normal distribution (with T.W. Anderson). *Linear Algebra Appl.*, **70**, 147–171.

Inequalities for the trace function (with A.W. Marshall), *Aequationes Mathematicae*, **29**, 36–39.

1986

Maximum likelihood estimators and likelihood ratio criteria in multivariate components of variance (with B.M. Anderson and T.W. Anderson), *Ann. Statist.*, **14**, 405–417.

Meta Analysis: A review and a new view (with L.V. Hedges), *Educational Researcher*, 14–21.

1987

A semi-parametric approach to density estimation (with C. Spiegelman), *J. Amer. Statist. Assoc.*, **82**, 858–865.

Generation of random orthogonal matrices (with T.W. Anderson and L. Underhill), SIAM *Journal of Scientific and Statistical Computing*, **8**, 625–629.

A model for interlaboratory differences (with M. Sobel), *Advances in Multivariate Statistical Analysis*, (ed. by A.K. Gupta), D. Reidel: Boston, 303–314.

A conversation with Morris Hansen, *Statistical Science*, **2**, 162–179.

Best invariant estimators of a Cholesky decomposition (with M.L. Eaton), *Ann. Statist.* **15**, 1639–1650.

Statistical inference for the overlap hypothesis (with L.V. Hedges), *Foundations of Statistical Inference*, (ed. by I. B. McNeill and G.J. Umphrey), D. Reidel: Boston, 63–72.

1988

Families of multivariate distributions (with A.W. Marshall), *J. Amer. Statist. Assoc.*, **83**, 834–841.

TRANSLATIONS

Linnik, Ju. V. **Linear Forms and Statistical Criteria**, I. Ukrain Mat. Zurnal 5 (1933), 207–243. Translated jointly with M. Gourary, B. Hannan, in *Selected-Translations in Mathematical Statistics and Probability, Vol. 3*, American Mathematical Society, 1–40.

Azlarov, T.A. and Volodin, N.A., **Problems Associated with the Exponential Distribution**. (Translated by Margaret Stein and Edited by Ingram Olkin), Springer-Verlag: New York.

BOOK REVIEW PAPERS

1977

Review of **Characterization Problems in Mathematical Statistics** (translated from the Russian by B. Ranchandran), by A.M. Kagan, Yu. V. Linnik and C. Radhakrishna Rao; John Wiley & Sons, (with S.G. Ghurye and P. Diaconis), *Ann. Statist.*, **5**, 583–592.

1982

Analysis, Reanalyses, and Meta-Analysis (with L.V. Hedges). Review of *Meta-Analysis in Social Research* by G. V. Glass, B. McGraw, and M.L. Smith; *Contemporary Education Review*, Vol. 1, 157–165.

DISSERTATIONS WRITTEN UNDER THE PRIMARY DIRECTION OF PROFESSOR OLKIN

1. R.N.P. Bhargava, Multivariate tests of hypotheses with incomplete data, August 1962.
2. L.J. Gleser, The comparison of multivariate tests of hypotheses by means of Bahadur efficiency, July 1963.
3. J.B. Selliah, Estimation and testing problems in a Wishart distribution, January 1964.
4. S.J. Press, Some hypothesis testing problems involving multivariate normal distribution with unequal and intraclass structured covariance matrices, June 1964.
5. M.L. Eaton, Some optimal properties of ranking procedures with applications in multivariate analysis, March 1966.
6. M.D. Perlman, One-sided problems in multivariate analysis, July 1967.
7. H. Nanda, Factor and analytic techniques for inter-battery comparisons and their application to some psychometric problems, August 1967.
8. T.W.F. Stroud, Comparing conditional distributions under measurement errors of known variances, May 1968.
9. M. Sylvan, Estimation and hypothesis testing for Wishart matrices when part of the covariance matrix is known, June 1969.
10. S. Arnold, Products of problems and patterned covariance matrices that arise from interchangeable random variables, August 1970.
11. A. Sampson, An asymptotically efficient stepwise estimator for exponential families with applications to certain multivariate normal distributions, August 1970.
12. J.E. Lockley, A comparative study of some cluster analytic techniques with application to the mathematics achievement in junior high schools, 1970.
13. W. Zwirner, The Procrustes model in factor analysis: Evaluation of two alternative criteria, 1970.
14. D.Y. Kim, Statistical inference for constants of proportionality between covariance matrices, August 1971.

15. V. Thanyamanta, Tests for the equality of mean vectors when some information concerning the covariance matrix is available, August 1974.

16. J.L. Tomsky, A new class of multivariate tests based on the union-intersection principle, August 1974.

17. T.H. Szatrowski, Testing and estimation of multivariate normal mean and covariance structures which are compound symmetric in blocks, April 1976.

18. M. Viana, Combined estimates and tests of hypotheses for the correlation coefficient, August 1978.

19. T. Perl, Discriminating factors and sex differences in electing mathematics, January 1979.

20. L. Brower, The use of academic and non-academic variables in predicting grade point average profiles of black students, June 1979.

21. D. Conway, Multivariate distributions with specified marginals, August 1979.

22. L. Hedges, Combining the results of experiments using difference scales of measurement, February 1980.

23. J. Verducci, Discriminating between two populations on the basis of ranked preferences, August 1982.

24. J. Ekstrand, Methods of validating learning hierarchies with applications to mathematics learning, January 1983.

25. D.K. Ahn, Loglinear and Markov chain models of change in college academic major, August 1983.

26. M. Huh, Regression analysis of multicollinear data, April 1984.

27. W. Bricken, Analyzing errors in elementary mathematics, February 1986.

28. E. Holmgren, Utilizing p-p Plots in meta-analysis as general measures of treatment effects, March, 1989.

1970

1971

1985

At the 50th Anniversary of the IMS (1985).

Part II

Contributions to Probability and Statistics

Part A

Probability Inequalities and Characterizations

1

A Convolution Inequality

Gavin Brown[1]
Larry Shepp[2]

ABSTRACT We establish an elementary convolution inequality which appears to be novel although it extends and complements a famous old result of W.H. Young. In the course of the proof we are led to a simple interpolation result which has applications in measure theory.

1 The Convolution Inequality

Theorem 1.1

(i) Suppose that $\infty > p \geq s \geq q \geq 1$ *and let* $t^{-1} = p^{-1} + q^{-1} - s^{-1}$. *If* $f, g \in L^p(\Re) \cap L^q(\Re)$ *are nonnegative*

$$\left\| (f^p * g^p)^{1/p} \right\|_q \geq \|f\|_s \|g\|_t \geq \left\| (f^q * g^q)^{1/q} \right\|_p .$$

Unless f *or* g *is null, equality holds only when* $p = q = s = t$ *and then we have*

$$\left\| (f^p * g^p)^{1/p} \right\|_p = \|f\|_p \|g\|_p$$

for all $f, g \in L^p(\Re)$.

(ii) Suppose that $s^{-1} + t^{-1} = 1$, $s > 1, t > 1$. *If* f, g *are continuous with compact support then*

$$\int \sup_y |f(x-y)g(y)| \, dx \geq \|f\|_s \|g\|_t \geq \sup_x \int |f(x-y)g(y)| \, dy.$$

Equality holds if and only if f *or* g *is null.*

Remarks

1. The result will be seen to transfer to general LCA groups — with the proviso that non-null constant functions can give equality.

---

[1]Mathematics Department, University of New South Wales
[2]AT&T Bell Laboratories

2. If we allow $\{s,t\} = \{1,\infty\}$ in (ii), then the right hand inequality can be an equality for non-null functions (e.g. take $f = g = X_E$, the indicator of a set of positive measure).

3. If we delete the middle term in each inequality then we have a special case of a well known result about repeated means (Jessen (1931)).

4. The theorem of W.H. Young (1913), (see also B. Jessen (1931) p. 199), mentioned in the abstract, is the statement that for $u > 1, v > 1$, $u^{-1} + v^{-1} > 1$,

$$\sum z_n^{uv/(u+v-uv)} < \left(\sum x_i^u\right)^{v/(u+v-uv)} \left(\sum y_j^v\right)^{u/(u+v-uv)}$$

where

$$z_n = \sum_{i+j=n} x_i y_j, \qquad \text{and the } x_i,\ y_j \text{ are positive.}$$

This follows from the special case of the right hand inequality (i) of our theorem in which $q = 1, p = uv(u+v-uv)^{-1}$, $s = u$, $t = v$, $f(t) = \sum_i x_i \delta(t-i)$ and $g(t) = \sum_j y_j \delta(t-j)$. The integral analogue of Young's result is explicitly discussed on p. 201 of Hardy, Littlewood, and Pólya (1951).

5. In view of the homogeneity of the inequalities it is enough to consider the seemingly weaker result in which p, q and s, t are pairs of conjugates — simply raise f and g to the same suitable power.

Lemma 1.2 *With p, q, as in the Theorem,*

$$\left\| (f^p * g^p)^{1/p} \right\|_q \geq \|f\|_q \|g\|_p ,$$

with strict inequality unless f or g is null.

Proof. Let $h(x)(y) = h(x,y) = f(x-y)^q g(y)^q$

$$\int \left(\int f(x-y)^p g(y)^p \, dy \right)^{q/p} dx = \int \|h(x)\|_{p/q} \, dx.$$

By Minkowski's inequality, since $p/q \geq 1$, we have

$$\begin{aligned} \int \|h(x)\|_{p/q} \, dx &\geq \left\| \int h(x) \, dx \right\|_{p/q} \\ &= \left[\int \left(\int f(x-y)^q g(y)^q \, dx \right)^{p/q} dy \right]^{q/p} \end{aligned}$$

$$= \left[\left(\int f(x)^q \, dx \right)^{p/q} \int g(y)^p \, dy \right]^{q/p}$$

$$= \int f^q \, dx \left(\int g^p \, dy \right)^{q/p}.$$

Thus

$$\left\| (f^p * g^p)^{1/p} \right\|_q \geq \|f\|_q \, \|g\|_p$$

(and, of course, the corresponding inequality holds with f, g interchanged). Equality requires that $p = q$ or that $h(x, y)$ be of the form $k(x)\mathcal{L}(y)$.

The equation $f(x - y)g(y) = k(x)\mathcal{L}(y)$, when g is not null, forces

$$\frac{f(x)}{f(0)} \frac{f(y)}{f(0)} = \frac{f(x + y)}{f(0)} \tag{1.1}$$

or else f is null. If (1.1) holds then f is a multiple of a (non-negative) group character and hence constant. For the case of the real line, this shows of course that $f \notin L^p$ and we conclude that f is indeed null.

It remains to 'interpolate' for s, t. This is not quite trivial because $\log \|f\|_p$ need not be convex as a function of p and it is not clear how to describe the behavior of $\log \|f\|_q$ as a function of p. The next lemma shows that, although $\log \|f\|_p + \log \|f\|_q$ need not be U-shaped as a function of p, at least its graph has no caps.

Lemma 1.3 *Let $s_1 < s < s_2$ and let t_1, t, t_2 be conjugate to s_1, s, s_2 respectively. Then*

$$\|f\|_s \, \|g\|_t \leq \max(\|f\|_{s_1} \|g\|_{t_1}, \|f\|_{s_2} \|g\|_{t_2}).$$

Proof. Choose α_1, α_2 positive such that $\alpha_1 + \alpha_2 = 1$ and $s = \alpha_1 s_1 + \alpha_2 s_2$. By convexity of $\log \|f\|_s^s$ (i.e. by Hölder's inequality)

$$\|f\|_s^s \leq \|f\|_{s_1}^{\alpha_1 s_1} \, \|f\|_{s_2}^{\alpha_2 s_2}. \tag{1.2}$$

We choose

$$\beta_1 = \alpha_1 \frac{s_1}{s} \frac{t}{t_1}, \qquad \beta_2 = \alpha_2 \frac{s_2}{s} \frac{t}{t_2}$$

and observe that

$$\begin{aligned} \beta_1 + \beta_2 &= \left(\alpha_1 \frac{s_1}{t_1} + \alpha_2 \frac{s_2}{t_2} \right) \frac{t}{s} \\ &= \big(\alpha_1(s_1 - 1) + \alpha_2(s_2 - 1)\big) t/s \\ &= (s - 1)\frac{t}{s} = 1. \end{aligned}$$

Again by convexity of $\log \|g\|_t^t$,

$$\|g\|_t^t \leq \|g\|_{t_1}^{\beta_1 t_1} \, \|g\|_{t_2}^{\beta_2 t_2}. \tag{1.3}$$

But

$$\frac{\beta_1 t_1}{t} = \frac{\alpha_1 s_1}{s}, \qquad \frac{\beta_2 t_2}{t} = \frac{\alpha_2 s_2}{s}. \tag{1.4}$$

Combining (1.2), (1.3), (1.4) we find

$$\|f\|_s \|g\|_t \le (\|f\|_{s_1} \|g\|_{t_1})^{\alpha_1 s_1/s} (\|f\|_{s_2} \|g\|_{t_2})^{\alpha_2 s_2/s}.$$

Since $\alpha_1 s_1/s + \alpha_2 s_2/s = 1$, the result follows.

The problem of establishing the lower bound for $\|f\|_s \|g\|_t$ is a little easier (cf. Hardy, Littlewood and Pólya (1951), p. 199).

Lemma 1.4 *Let p, q, s, t, f, g be as in Theorem 1.1 (with p, q conjugate) and choose s', t' such that*

$$\frac{s'}{q} + \frac{s}{p} = \frac{t'}{q} + \frac{t}{p} = 1.$$

Then

$$(f^q * g^q)^{1/q} \le \|f\|_s^{s'/q} \|g\|_t^{t'/q} (f^s * g^t)^{1/p}.$$

Proof. Since

$$(1 - qt^{-1}) + (1 - qs^{-1}) + (q - 1) = 1,$$

Hölder's inequality gives

$$\|h_1 h_2 h_3\|_1 \le \|h_1\|_{(1-qt^{-1})^{-1}} \|h_2\|_{(1-qs^{-1})^{-1}} \|h_3\|_{(q-1)^{-1}},$$

where

$$h_1(y) = f(x-y)^{s'}, \quad h_2(y) = g(y)^{t'}, \quad h_3(y) = (f(x-y)^s g(y)^t)^{q-1}.$$

Because

$$s'(1 - qt^{-1})^{-1} = s, \quad t'(1 - qs^{-1})^{-1} = t,$$

this is the statement that

$$f^q * g^q(x) \le \|f\|_s^{s'} \|g\|_t^{t'} (f^s * g^t)^{q-1}(x),$$

and the required result follows at once.

It is now clear how to piece together the proof of the theorem. Part (ii) can be obtained by a limiting argument — but it also yields to a very quick direct proof that bypasses Lemma 1.3.

2 Application

Lemma 1.3 makes no use of the special properties of Lebesgue measure, so it remains valid for $\|f\|_{s,\mu} = (\int f^s \, d\mu)^{1/s}$, $\|g\|_{s,v} = (\int g^s dv)^{1/s}$ where μ, v are arbitrary probability measures. Moreover the choice of t, t_1, t_2 as conjugates of s, s_1, s_2 was unnecessarily restrictive. What we used in the proof is the fact that there exists a ϕ function $\phi(s) = s/(s-1)$ with $t = \phi(s)$, $t_1 = \phi(s_1)$, $t_2 = \phi(s_2)$ and, the crucial condition, that $s/\phi(s)$ is affine.

It turns out that these are just the properties we need for the application in mind. In fact Oberlin (preprint) recently proved that

$$\int \sup_y f(x-y)g(y)\, d\lambda(x) \geq \int f\, d\mu \Big(\int g^q \, dx \Big)^{1/q},$$

where f, g are nonnegative continuous functions on the circle, λ is Haar measure, μ is Cantor measure and $q = \log 3/(\log 3 - \log 2)$. We extend his result as follows:

Theorem 2.1 *Suppose that f, g are nonnegative continuous functions on the circle and that $s, t \geq 1$ and satisfy*

$$1 = s^{-1} + \left(\frac{\log 2}{\log 3} \right) t^{-1}.$$

Then

$$\int \sup_y f(x-y)g(y)\, d\lambda(x) \geq \Big(\int f^s \, d\lambda \Big)^{1/s} \Big(\int g^t \, d\mu \Big)^{1/t}$$

Corollary 2.2 *For s, t as above and arbitrary Borel sets E, F it is true that*

$$\lambda(E+F) \geq \mu(F)^{1/t} \lambda(E)^{1/s}$$

Proof of Theorem 2.1. Oberlin proved the limit case with $t = 1$. Let's consider (the much easier) limit case with $t = \infty$. We must prove

$$\int \sup_y g(x-y)f(y)\, d\lambda(x) \geq \|f\|_\infty \int g(y)\, d\lambda(y).$$

Without loss of generality we assume that $\|f\|_\infty = 1$ and that f vanishes outside a small neighborhood N of y_0. As N shrinks,

$$\int \sup_y g(x-y)f(y)\, d\lambda(x) \to \int g(x-y_0)\, d\lambda(x) = \|g\|_1 \, ,$$

as required.

In view of the remarks at the beginning of this section we can now use Lemma 1.4 to interpolate to obtain the full force of the theorem.

The proof of the corollary is straightforward.

Now that the utility of the sharpened form of Lemma 1.4 has been demonstrated it seems worthwhile to state it in the form of a postscript to the discussion of convexity of r-th order means in §16 of Beckenbach and Bellman (1971).

Recall that the *mean of order* t is

$$M_t(x,\alpha) = \left(\sum_{i=1}^{n} \alpha_i x_1^t\right)^{1/t},$$

while the *sum of order* t is

$$S_t(x) = \left(\sum_{i=1}^{n} x_i^t\right)^{1/t}.$$

It is, of course, implicit in these statements that $x = (x_i)_{i=1}^n$ and $\alpha = (\alpha_i)_{i=1}^n$ have positive entries and that $\sum \alpha_i = 1$. We will consider also $y = (y_i)_{i=1}^n$, $\beta = (\beta_i)_{i=1}^m$ where m may differ from n.

Lemma 2.3 *Suppose that, for* $i = 0, 1, 2,$ $s_i \geq 1, t_i \geq 1$ *and* $as_i^{-1} + bt_i^{-1} = 1$, *for constants* a, b. *If* $s_1 \leq s_0 \leq s_2$ *then*

(i) $$M_{s_0}(x,\alpha)M_{t_0}(y,\beta) \leq \max_{i=1,2} M_{s_i}(x,\alpha)M_{t_i}(y,\beta),$$

and, if further $a : b = \log n : \log m$,

(ii) $$S_{s_0}(x)S_{t_0}(y) \leq \max_{i=1,2} S_{s_i}(x)S_{t_i}(y).$$

Proof The remarks at the beginning of this section establish (i). For (ii), we normalize by dividing both sides by $n^{1/s_0}m^{1/t_0}$. The condition on $a : b$ enables us to apply (i). □

Further applications of Lemma 2.3 are given in Brown (preprint).

References

Beckenbach, E.F. and Bellman, R. (1971). *Inequalities*, Springer-Verlag, Berlin, Heidelberg, New York.

Brown, G. Some inequalities that arise in measure theory. preprint.

Hardy, G.H., Littlewood, J.E., and Pólya, G. (1951). *Inequalities.* Cambridge University Press, London.

Jessen, B. (1931). Om Uligheder imellem Potensmiddelvaerdier. *Mat Tidsskrift*, B No. 1.

D.H. Oberlin, The size of sum sets, II. preprint.

Young, W.H. (1913). On the determination of the summability of a function by means of its Fourier constants. *Proc. London Math. Soc.* **2** (12), 71–88.

2

Peakedness of Weighted Averages of Jointly Distributed Random Variables

Wai Chan[1]
Dong Ho Park[2]
Frank Proschan[3]

ABSTRACT This note extends the Proschan (1965) result on peakedness comparison for convex combinations of i.i.d. random variables from a PF_2 density. Now the underlying random variables are jointly distributed from a Schur-concave density. The result permits a more refined description of convergence in the Law of Large Numbers.

1 Introduction

Proschan (1965) shows that:

Theorem 1.1 *Let f be PF_2, $f(t) = f(-t)$ for all t, $X_1, \ldots, X_n$ independently distributed with density f, $\vec{a} \overset{m}{\geq} \vec{b}$; $\vec{a}, \vec{b}$ not identical, $\sum_{i=1}^n a_i = \sum_{i=1}^n b_i = 1$. Then $\sum_{i=1}^n b_i X_i$ is strictly more peaked than $\sum_{i=1}^n a_i X_i$.*

Definitions of majorization ($\vec{a} \overset{m}{\geq} \vec{b}$), PF_2 density, and peakedness are presented in Section 2. The Law of Large Numbers asserts that the average of a random sample converges to the population mean under certain conditions. Roughly speaking, Theorem 1.1 states that a *weighted* average of i.i.d. random variables converges more rapidly in the case in which weights are close together as compared with the case in which the weights are diverse.

In the present note, we extend the basic univariate result to the multivariate result in which the underlying random variables have a joint Schur-concave density. Theorem 2.4 presents the precise statement of the multivariate extension.

---

[1]Ohio State University, Department of Statistics, Columbus, Ohio 43210
[2]University of Nebraska, Department of Statistics, Lincoln, Nebraska 68588
[3]Florida State University, Department of Statistics, Tallahassee, Florida 32306

2 Peakedness Comparisons

The theory of majorization is exploited in this section to obtain more general versions of the result of Proschan (1965). We begin with some definitions. The definition of peakedness was given by Birnbaum (1948).

Definition 2.1 *Let X and Y be real valued random variables and a and b real constants. We say that X* is more peaked about a than Y about b *if $P(|X-a| \geq t) \leq P(|Y-b| \geq t)$ for all $t \geq 0$. In the case $a=b=0$, we simply say that X is more peaked than Y.*

Next we define the ordering of majorization among vectors. The standard reference on the theory of majorization is the book by Marshall and Olkin(1979).

Definition 2.2 *Let $a_1 \geq \cdots \geq a_n$ and $b_1 \geq \cdots \geq b_n$ be decreasing rearrangements of the components of the vectors $\vec{a}$ and $\vec{b}$. We say that $\vec{a}$ majorizes $\vec{b}$ (written $\vec{a} \overset{m}{\geq} \vec{b}$) if $\sum_{i=1}^n a_i = \sum_{i=1}^n b_i$ and $\sum_{i=1}^k a_i \geq \sum_{i=1}^k b_i$ for $k=1,\ldots,n-1$.*

Definition 2.3 *A real valued function f defined on $\mathcal{R}^n$ is said to be a* Schur-concave function *if $f(\vec{a}) \leq f(\vec{b})$ whenever $\vec{a} \overset{m}{\geq} \vec{b}$.*

A nonnegative function f on $(-\infty,\infty)$ is called a *Pólya frequency function of order 2* (PF_2) if $\log f$ is concave. If f is a PF_2 function then $\phi(\vec{x}) = \prod_{i=1}^n f(x_i)$ is Schur-concave. Thus the random vector $\vec{X} = (X_1,\ldots,X_n)$ has a Schur-concave density under the conditions of Theorem 1.1. A function f defined on $\mathcal{R}^n$ is said to be sign-invariant if $f(x_1,\ldots,x_n) = f(|x_1|,\ldots,|x_n|)$. In the following theorem, we give a peakedness comparison for random variables with a sign-invariant Schur-concave density.

Theorem 2.4 *Suppose the random vector $\vec{X} = (X_1,\ldots,X_n)$ has a sign-invariant Schur-concave density. Then for all $t \geq 0$,*

$$\phi(a_1,\ldots,a_n) = P\left(\sum_{i=1}^n a_i X_i \leq t\right)$$

is a Schur-concave function of $\vec{a} = (a_1,\ldots,a_n), a_i \geq 0$ for all i. Equivalently, $\sum_{i=1}^n b_i X_i$ is more peaked than $\sum_{i=1}^n a_i X_i$ whenever $\vec{a} \overset{m}{\geq} \vec{b}$.

Proof Without loss of generality, we may further assume that $\sum a_i = 1$. We first consider the case n = 2.

Let $\vec{a} = (a_1,a_2), \vec{b} = (b_1,b_2), \vec{a} \overset{m}{\geq} \vec{b}$. Since X_1, X_2 are exchangeable, we may assume that $a_1 > b_1 \geq 1/2 \geq b_2 > a_2$. To show that

$$P(a_1X_1 + a_2X_2 \leq t) \leq P(b_1X_1 + b_2X_2 \leq t)$$

FIGURE 1.

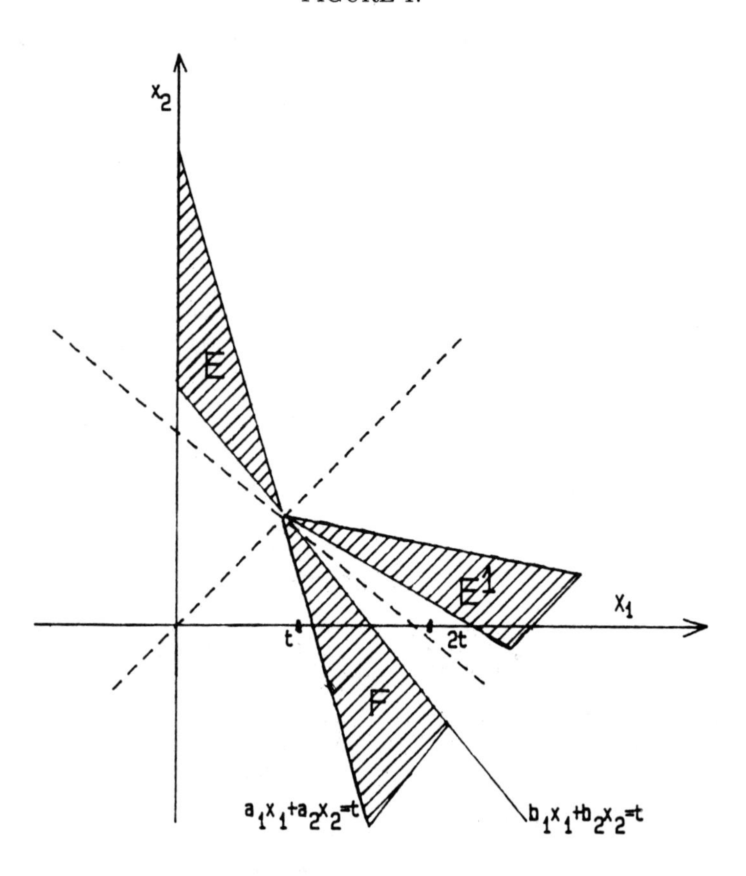

for $t \geq 0$, consider the lines $a_1x_1 + a_2x_2 = t$ and $b_1x_1 + b_2x_2 = t$ in Figure 1.

Since $a_1 > b_1 \geq 1/2$, both lines intersect the x_1-axis in the interval $[t, 2t]$ and they intersect the 45 degree line at the point (t, t) $(a_1+a_2 = b_1+b_2 = 1)$. We must show that $P(E) \leq P(F)$. Now reflect E across the 45 degree line to form the wedge E'. Then $P(E) = P(E')$ because the joint density f is invariant under permutation. For $k \geq 0$, the line $x_1 - x_2 = k$ intersects E' at the line segment joining $(t + b_1k, t - b_2k)$ and $(t + a_1k, t - a_2k)$, and it intersects F at the line segment joining $(t+a_2k, t-a_1k)$ and $(t+b_2k, t-b_1k)$. Note that both segments are of equal length. But f sign-invariant and Schur-concave implies that

$$\begin{aligned} f(t + b_1k, t - b_2k) &= f(t + b_1k, b_2k - t) \\ &\leq f(t + b_2k, b_1k - t) = f(t + b_2k, t - b_1k). \end{aligned}$$

This last fact then clearly implies that $P(E') \leq P(F)$ by conditioning on $X_1 - X_2$.

The result for $n \geq 3$ now follows since

$$P\left(\sum_{i=1}^{n} a_i X_i \leq t\right) = E\left[P\left(a_1 X_1 + a_2 X_2 \leq t - \sum_{i=3}^{n} a_i X_i\right) \,\Big|\, X_3, \ldots, X_n\right]$$

and the conditional density $f(x_1, x_2 | x_3, \ldots, x_n)$ is also Schur-concave and sign-invariant. □

For an example of a Schur-concave density that is also sign-invariant, consider the multivariate Cauchy density:

$$f(x_1, \ldots, x_n) = \Gamma\left(\frac{n+1}{2}\right)\left(\pi + \pi \sum_{i=1}^{n} x_i^2\right)^{-(n+1)/2}.$$

The following result is an immediate consequence of Theorem 2.4.

Corollary 2.5 *Let $X_1, \ldots, X_n$ be random variables with a Schur-concave and sign-invariant joint density f. Then $\sum_{i=1}^{k} X_i/k$ is increasing in peakedness as k increases from 1 to n.*

Proof Let $\vec{a_1} = (1, 0, \ldots, 0), \vec{a_2} = (\frac{1}{2}, \frac{1}{2}, 0, \ldots, 0), \ldots$, and $\vec{a_n} = (\frac{1}{n}, \ldots, \frac{1}{n})$, where each vector contains n components. Then $\vec{a_1} \stackrel{m}{\geq} \cdots \stackrel{m}{\geq} \vec{a_n}$. The result follows from Theorem 2.4. □

Suppose $\vec{X} = (X_1, \ldots, X_n)$ and $\vec{Y} = (Y_1, \ldots, Y_n)$ are independently distributed with Schur-concave and sign-invariant densities f and g. Then Theorem 2.4 implies that $\sum_{i=1}^{n} b_i(X_i + Y_i)$ is more peaked than $\sum_{i=1}^{n} a_i(X_i + Y_i)$ whenever $\vec{a} \stackrel{m}{\geq} \vec{b}$. This is true because the convolution of Schur-concave functions is Schur-concave. Now suppose that $Y_1, \ldots, Y_n$ are i.i.d. Cauchy, then the joint density g given by

$$g(y_1, \ldots, y_n) = \left(\frac{a}{\pi}\right)^n \Big/ \prod_{i=1}^{n}(1 + a^2 y_i^2), \qquad a > 0, \tag{2.1}$$

is not Schur-concave. In Theorem 2.6 below, we show that $\sum_{i=1}^{n} b_i(X_i + Y_i)$ is more peaked than $\sum_{i=1}^{n} a_i(X_i + Y_i)$ whenever $\vec{a} \stackrel{m}{\geq} \vec{b}$. This result identifies a different class of densities for which the conclusion of Theorem 2.4 holds.

Theorem 2.6 *Suppose that the random vector $\vec{X} = (X_1, \ldots, X_n)$ has a sign-invariant Schur-concave density f. Let $Y_1, \ldots, Y_n$ be i.i.d. Cauchy with joint density g as given in (2.1). Let $\vec{X}$ and $\vec{Y} = (Y_1, \ldots, Y_n)$ be independent, and $\vec{a} \stackrel{m}{\geq} \vec{b}$ where $a_i \geq 0, b_i \geq 0$ for all i and $\sum_{i=1}^{n} a_i = \sum_{i=1}^{n} b_i = 1$. Then $\sum_{i=1}^{n} b_i(X_i + Y_i)$ is more peaked than $\sum_{i=1}^{n} a_i(X_i + Y_i)$.*

Proof Since f is sign-invariant, both $\sum_{i=1}^{n} a_i X_i$ and $\sum_{i=1}^{n} b_i Y_i$ are symmetric random variables. We use the fact that $\sum_{i=1}^{n} a_i Y_i$ and $\sum_{i=1}^{n} b_i Y_i$ have the same distribution as does Y_1. The result now follows from Theorem 2.4 and the Lemma of Birnbaum (1948) by noting that Y_1 has a symmetric and unimodal density. □

Acknowledgments: Research sponsored by the Air Force Office of Scientific Research under Grant AFOSR 88-0040. The U.S. Government is authorized to reproduce and distribute reprints for Governmental purposes notwithstanding any copyright notation thereon.

References

Birnbaum, Z.W. (1948). On random variables with comparable peakedness. *Annals of Mathematical Statistics.* **19**, *76-81.*

Marshall, A.W. and Olkin, I. (1979). *Inequalities: Theory of Majorization and Its Applications.* Academic Press, New York.

Proschan, F. (1965). Peakedness of distribution of convex combinations. *Annals of Mathematical Statistics.* **36**, *1703-1706.*

3

Multivariate Majorization

Somesh Das Gupta[1]
Subir Kumar Bhandari[1]

ABSTRACT The concept of univariate majorization plays a central role in the study of Lorenz dominance for income distribution comparisions in economics. The first part of this paper reviews different conditions which are equivalent to Lorenz dominance. The second part of the present paper poses the question whether such equivalences extend to the multivariate case. Some concepts of multivariate majorization are presented along with a few new results. For economic applications, the notion of a concave utility function on vector observations appears to play a crucial role in multivariate majorization. It is shown that such concavity follows from some easily understandable axioms.

1 Introduction

The concept of univariate majorization plays a central role in the study of Lorenz dominance for comparing two income distributions. Some alternative relations equivalent to univariate majorization or Lorenz dominance in the stochastic set-up are scattered in the literature. These results are missing in the excellent treatise on majorization by Marshall and Olkin (1979). The first part of this paper reviews the different conditions equivalent to Lorenz dominance, citing the appropriate references. This aspect has also been briefly mentioned in a recent monograph by Arnold (1987).

The second part of the paper poses the problem whether the equivalent conditions in relation to univariate majorization could be extended to the multivariate case. The problem of comparing two communities in which individuals are characterized by a set of socio-economic attributes has not received much attention among the economists. The difficulty is not merely due to the non-uniqueness of the concept of "rich-poor," but also due to the fact that the effects of different socio-economic attributes or measurements on total utility or total social welfare are not easily comprehensible. One may, of course, reduce the data to a single measurement by using some appropriate weights, although the validity of such a simplification may

[1]Indian Statistical Institute, Calcutta

be questionable. On the other hand, the use of multivariate majorization in a purely mathematical framework is extremely limited in economics. Within these limitations some concepts of multivariate majorization have been presented along with a few new results. It appears that the notion of a concave utility function on $\Re^m$ plays a crucial role in multivariate majorization. It is shown in the appendix that the concavity of the utility function results from some easily understandable axioms.

2 Univariate Majorization

Given a vector $x = (x_1, \ldots, x_n)$ in $\Re^n$, let $x_{(1)} \leq \cdots \leq x_{(n)}$ be the ordered values of the x_i 's. For two vectors $x = (x_1, \ldots, x_n)$ and $y = (y_1, \ldots, y_n)$ in $\Re^n$, consider the following conditions:

(a) *Rearrangement Condition.* x is majorized by y, written as $x \prec y$, i.e.

$$\sum_1^k x_{(i)} \geq \sum_{i=1}^k y_{(i)} \text{ for } k = 1, \ldots, n, \text{and} \sum_1^n x_i = \sum_1^n y_i.$$

(b) *Structural Condition (Schur, 1923).* There exists a doubly stochastic matrix P such that $x = yP$.

(c) *Convexity Condition.* For all continuous convex functions φ

$$\sum_1^n \varphi(x_i) \leq \sum_1^n \varphi(y_i).$$

(d) *Residual Condition.* For all real a

$$\sum_1^n (x_i - a)^+ \leq \sum_1^n (y_i - a)^+, \qquad \sum_1^n x_i = \sum_1^n y_i,$$

where $(a)^+ = \max(a, 0)$.

It is well known that the above four conditions are mutually equivalent; see Marshall and Olkin(1979).

Now suppose that the components of x and y are non-negative with $(\sum_1^n x_i)(\sum_1^n y_i) > 0$. Then the Lorenz-curve (Lorenz, 1905) corresponding to x lies above the Lorenz-curve corresponding to y if, and only if, $x/(\sum_1^n x_i) \prec y/(\sum_1^n y_i)$; such a relation is often stated as $x \prec_L y$ (Arnold, 1987; Nygard and Sandstrom, 1981). It is clear that if $\sum_1^n x_i = \sum_1^n y_i$ then $x \prec_L y$ is equivalent to $x \prec y$.

The Lorenz-order $\prec_L$ has been extended to the stochastic set-up as follows (Arnold, 1987; Nygard and Sandstrom, 1981). Let X and Y be two non-negative random variables with distribution functions F and G, respectively, having finite non-zero means. For $0 \leq u \leq 1$, define

$$L_X(u) = \int_0^u F^{-1}(p)\,dp \Big/ \int_0^1 F^{-1}(p)\,dp.$$

Then $X \prec_L Y$, by definition, if $L_X(u) \geq L_Y(u)$ for every $u \in [0,1]$; $\prec_L$ is denoted by $\leq_L$ by Arnold (1987).

The above concept leads to the extension of the rearrangement condition (a) to the stochastic set-up as follows:

Condition (A) $X \prec_L Y$, $E(X) = E(Y)$.

It is now natural to ask whether conditions (B), (C) and (D) could be extended to the stochastic set-up, and whether these conditions, when adequately defined, would be equivalent to condition (A). The convexity condition (C) and the residual condition (D) have straightforward extensions as follows:

Condition (C) $E\varphi(X) \leq E\varphi(Y)$, for all continuous convex functions φ for which the above expectations exist.

Condition (D) $E(X-a)^+ \leq E(Y-a)^+$ for all real a, and $E(X) = E(Y)$.

The structural condition (B) does not have a straightforward extension. Ryff (1965) has introduced a doubly stochastic operator to get an extension of (B); a different development is given by Rothschild and Stiglitz (1970). However, the most satisfactory version of the structural condition in the stochastic set-up may be formulated following the work of Strassen (1965) as follows:

Condition (B) There exists a probability space and associated random variables U and V such that the distribution of U is the same as that of X, the distribution of V is the same as that of Y, and $E(V|U = u) = u$ almost surely.

The above development of stochastic majorization is included in Bhandari (1987), and it was pointed out in Bhandari (1987) that Conditions (A), (B), (C) and (D) are equivalent. This result is also included in Arnold (1987); however, adequate references and a detailed proof are not provided there. Atkinson (1970) proved that (A) $\leftrightarrow$ (D), Strassen's work (1965) contains the results (B) $\leftrightarrow$ (C), and the result (C) $\leftrightarrow$ (D) has been proved by Karamata (1932) and Ross (1983).

The above development can also be extended to weak majorization. For example, if we start out with the definition of weak sub-majorization $x \overset{w}{\prec} y$ (Marshall and Olkin, 1979), then corresponding conditions in the stochastic set-up may be stated as follows:

Condition (A′)

$$\int_u^1 F_X^{-1}(p)\,dp \leq \int_u^1 F_Y^{-1}(p)\,dp, \qquad \text{for all } 0 \leq u \leq 1.$$

Condition (B′) There exists a probability space and random variables U and V associated with it, such that the distribution of U is the same as that of X, the distribution of V is the same as that of Y, and $E(V|U=u) \geq u$ almost surely.

Condition (C′) $E\varphi(X) \leq E\varphi(Y)$ for all non-decreasing convex functions φ for which the above expectations exists.

Condition (D′) $E(X-a)^+ \leq E(Y-a)^+$ for all real a.

The equivalence of the above four conditions can be easily derived from the work in Atkinson (1970), Karamata (1932), Stoyan (1983), and Strassen (1965). It may be noted in this connection that the condition (D′) has been used by Stoyan (1983) to define a convex ordering $\overset{c}{\prec}$. Conditions similar to the above, and the corresponding result on their equivalence, can also be developed for weak super-majorization (Marshall and Olkin, 1979).

3 Multivariate Majorization

The ordering of univariate populations does not have a straightforward generalization to the case when the ordering is based on observations on multiple characteristics of the experimental units. The basic difficulty in extending the Lorenz curve, in particular, is due to the fact that there is no unique way to define an ordinal scale (poor to rich) to describe the units to start with, although attempts for such an extension have been made by Taguchi (Arnold, 1983; Taguchi, 1972a; and Taguchi, 1972b). It seems that any such concept of ordering should depend on the objectives and possible uses of such a study; besides, the physical nature of the problem as manifested in concrete situations may call for some specific types of ordering on the basis of relevant auxiliary information. Any abstract formulation of the concept of ordering would primarily be a mathematical exercise, although such a formulation often may give insight into various underlying relations.

In order to compare different communities with respect to social welfare, Sen (1976) has suggested the criterion $C = \sum_{i=1}^n ie_{[i]}$, where $e_{[1]} \geq \cdots \geq e_{[n]}$ are the ordered components of the income-vector of n individuals in a community, based on a given market-price of the commodities and the consumption-matrix for these commodities. Sen has shown (1976) that C is approximately proportional to $\bar{e}(1-G)$ for large n, where $\bar{e}$ is the average income and G is the Gini-index of income. Following Sen (1976), a community 1 is said to be socially better than another community 2 of equal size if the C-value of community 1 is more than that of community 2 with respect to the price-vector prevailing in community 1. Note that this ordering is not even a preordering, and use of C-values for comparing different communities may lead to inconsistency. Social Welfare surely depends on the consumption of some basic commodities; however, it also depends on a variety of individual and social attainments and the prevailing socio-political norms.

From a mathematical viewpoint, multivariate majorization has been briefly discussed in Marshall and Olkin (1979), as well as in Arnold (1987). Some new results on concepts of multivariate majorization are given below.

When comparisons have to be made based on measurements of m characteristics on each individual in a population, partial ordering may be introduced with respect to each of the characteristics separately. Following Marshall and Olkin (1979), $X : n \times m$ is said to be column-majorized by $Y : n \times m$, written as $X \overset{\text{col}}{\prec} Y$, when $X_i^c \prec Y_i^c$ $(i = 1, \ldots, m)$, where X_i^c and Y_i^c denote the vectors of measurements of the ith characteristic — that is, the ith column-vectors of X and Y, respectively. This can be viewed as the structural condition.

We now define a convexity condition for column-majorization and show that it is equivalent to the structural condition. For a row vector $Z = (Z_1, \ldots, Z_m)$ define

$$g(Z) = g_1(Z_1) + \cdots + g_m(Z_m),$$

where the g_i's are convex functions. For a matrix $X : n \times m$ with rows $X_1^R, \ldots, X_n^R$, define

$$\varphi(X) = \sum_{i=1}^{n} g(X_i^R).$$

Let Φ be the set of all such functions φ. The following theorem can be proved easily using the developments in Marshall and Olkin (1979).

Theorem 3.1 *For any two $n \times m$ real matrices X and Y, $X \overset{\text{col}}{\prec} Y$ if, and only if, $\varphi(X) \leq \varphi(Y)$ for all φ in Φ.*

In relation to the partial ordering defined by column-majorization, one may define appropriate inequality measures for X which would preserve such an ordering. For example, let

$$I(X) = h\big(\varphi_1(X_1^c), \ldots, \varphi_m(X_m^c)\big),$$

where h is increasing, the φ_i's are Schur-concave, and X_i^c are columns of X. Then $X <^{\text{col}} Y$ implies $I(X) \geq I(Y)$. In particular, for $X \geq 0$ (elementwise) one may consider

$$I_0(X) = \frac{1}{m} \sum_{j=1}^{m} \frac{1}{n-1} \frac{\sum\sum_{i \neq i'} |X_{ij} - X_{i'j}|}{\sum_i X_{ij}}.$$

Marshall and Olkin (1979) have considered the following extension of majorization to the multivariate case. A matrix $X : n \times m$ is said to be majorized by another matrix $Y : n \times m$, written as $X \prec Y$, if there exists a doubly stochastic matrix D such that $X = DY$. It is clear that $X \prec Y$ implies $X \overset{\text{col}}{\prec} Y$. The following is a useful necessary condition for matrix majorization:

Theorem 3.2 *For two $n \times m$ matrices X and Y, $X \prec Y$ implies $Y'Y - X'X$ is non-negative definite.*

The above theorem follows from the fact that $I - D'D$ is non-negative definite for any stochastic matrix D.

Matrix majorization calls for the same type of averaging for every commodity; on the other hand, column-majorization calls for averaging separately for each commodity. It may be noted in this connection that redistribution of the average amount of farm equipment without any change in the distribution of farm land may not result in an increase in social welfare. On the other hand, matrix majorization seems to be quite restrictive, and apparently presupposes a definite relationship among the commodities.

The equivalence between the structural condition and the convexity conditions has been proved by Karlin and Rinott (1983), using the general result on dilations; their result is stated below.

Theorem 3.3 *For any two $n \times m$ real matrices X and Y, the following conditions are equivalent:*

(i) $X = DY$ *for some doubly stochastic matrix* D,

(ii) $\sum_{i=1}^{n} f(X_i^R) \leq \sum_{i=1}^{n} f(Y_i^R)$

for every continuous convex function $f : \Re^m \to \Re$, where X_i^R and Y_i^R denote the ith rows of X and Y, respectively.

The above theorem seems to be useful from the viewpoint of economics. Suppose X and Y denote two consumption matrices of n individuals on m commodities. Suppose that the total welfare for X is given by $\omega(X) = \sum_1^n U(X_i^R)$, where U is a concave function. Then $X \prec Y$ is equivalent to $\omega(X) \geq \omega(Y)$, for all concave functions U. However, it is conceptually difficult to understand a concave utility function U on $\Re^m$. In the appendix we shall pose some easily understandable axioms for U, which in turn would imply its concavity.

The following inequality measures preserve the partial order of matrix majorization for $X \geq 0$:

$$\eta(X) = \sum_{i=1}^{n} \bar{X}_i^R A(\bar{X}_i^R)', \tag{3.1}$$

where

$$\bar{X}_i^R = X_i^R \cdot \operatorname{diag}(r_1, \ldots, r_m)$$
$$r_j = \frac{1}{\sum_{i=1}^n X_{ij'}}$$

and A is a positive semidefinite matrix.

$$\eta(X) = h\big(\varphi_1(\bar{X}\alpha_1), \ldots, \varphi_k(\bar{X}\alpha_k)\big), \tag{3.2}$$

where $\alpha_i \in \Re^m, \alpha_i \geq 0$ (componentwise) for all i, h is an increasing function, the φ_i's are univariate inequality measures which preserve the partial order of majorization, and $\bar{X}$ is the matrix with X_i^R defined in (3.1).

We have pointed out that $X \prec Y$ is equivalent to $\sum_{i=1}^n U(X_i^R) \geq \sum_{i=1}^n U(Y_i^R)$ for all concave functions $U : \Re^m \to \Re$. A comparison between X and Y can also be made by comparing $\big(U(X_1^R), \ldots, U(X_n^R)\big)$ with $\big(U(Y_1^R), \ldots, U(Y_n^R)\big)$, as given in the following theorem:

Theorem 3.4 *For two $n \times m$ matrices X and Y, the following conditions are equivalent:*

(i) *For all increasing concave functions U on $\Re^m$*

$$\big(U(X_1^R), \ldots, U(X_n^R)\big) \overset{\omega}{\prec} \big(U(Y_1^R), \ldots, U(Y_n^R)\big).$$

(ii) *There exists a doubly stochastic matrix D such that $X \geq DY$, componentwise.*

First note that from the viewpoint of economics the condition (i) in the above theorem appears to be quite reasonable for the comparison of two populations. Secondly, the above theorem is a generalization of Theorem 5.A.2, part (iii) in Marshall and Olkin (1979). The proof of the above theorem depends on the following result which can be easily obtained from the development in Karlin and Rinott (1983).

Theorem 3.5 *For two $n \times m$ matrices X and Y, the following conditions are equivalent:*

(i) *$X \geq DY$ for some doubly stochastic matrix D.*

(ii) $\sum_1^n U(X_i^R) \geq \sum_1^n U(Y_i^R)$

for all increasing concave functions U on $\Re^m$.

Comparison of two populations of different sizes has been considered by Fischer and Holbrook (1980), and later generalized to the multivariate case by Karlin and Rinott (1983).

Comparisons of multivariate populations can also be done by using the concept of univariate majorization and some suitable transformations. Following Marshall and Olkin (1979), a matrix $X : n \times m$ is said to be linearly majorized by a matrix $Y : n \times m$ if $Xa \prec Ya$ for all vectors a in $\Re^m$. Bhandari (1988) has studied the relation between matrix majorization and linear majorization.

4 Multivariate Majorization: Stochastic Case

Following the definition of multivariate majorization in the non-stochastic case, one may say that a random vector $X : 1 \times m$ is majorized by another random vector $Y : 1 \times m$, written as $X \prec Y$, if

$$Eh(X) \leq Eh(Y)$$

for all convex functions h for which the above expectations exist. Such a relation is called Lorenz dominance and denoted by $\leq_L$ by Arnold (1987). It follows from Strassen's work (1965) that the convexity condition defined above is equivalent to the structural condition (B′) with random variables replaced by random vectors.

One may also define majorization in the stochastic case by requiring $\varphi(X) \prec_L \varphi(Y)$ for a class Φ of functions $\varphi : \Re^m \to \Re$.

Marshall and Olkin (1979) have given a class of definitions for stochastic majorization; however, such definitions are quite restrictive, especially from the point of view of economics, since they are expressed in terms of symmetric or Schur functions. Moreover, such definitions do not reduce to Lorenz order in the univariate case.

In order to ensure $X = (X_1, \ldots, X_m) \prec Y = (Y_1, \ldots, Y_m)$, one needs a much stronger condition than $X_i \prec_L Y_i$ for $i = 1, \ldots, m$. However, when the X_i's are independent and so are the Y_j's, them $X_i \prec_L Y_i$ for $i = 1, \ldots, m$ is equivalent to $X \prec Y$. This result follows from the theorem below.

Theorem 4.1 *Let $X_1, \ldots, Y_m$ be a set of m independent random variables, and $Y_1, \ldots, Y_m$ be another set of m independent random variables. Suppose $X_i \prec_L Y_i$ for $i = 1, \ldots, m$. Then*

$$Eh(X_1, \ldots, X_m) \leq Eh(Y_1, \ldots, Y_m)$$

for every real-valued function h, separately convex in each argument, whenever the above expectations exist.

The proof of this theorem can easily be obtained from a closely related result of Ross (1983). It is interesting to note that under the assumptions of the above theorem

$$c_1 X_1 + \cdots + c_m X_m \prec c_1 Y_1 + \cdots + c_m Y_m,$$

for any c_i's, and

$$X_1 X_2 \dots X_m \prec_L Y_1 Y_2 \dots Y_m.$$

However, the above theorem does not yield $\max(X_i) \prec_L \max(Y_i)$. Bhandari (1987) has shown the following weaker result for comparison between $\max(X_i)$ and $\max(Y_i)$.

Theorem 4.2 *Let $(X_1, \dots, X_m)$ and $(Y_1, \dots, Y_m)$ be two sets of m non-negative independent random variables such that $X_i \prec_L Y_i$ for $i = 1, \dots, m$. Then there exists a non-negative random variable Z with $E(\max(Y_i)) = E(Z)$ such that*

$$\max(X_1) \leq^{St} Z, \qquad Z \prec_L \max(Y_i),$$

where $\leq^{St}$ denotes stochastic order (Marshall and Olkin, 1979).

A1 Concavity of Utility Function

First we shall consider the case $m = 2$. We postulate the following axioms.

Axiom A1.1 *U is strictly increasing.*

Axiom A1.2 *U is concave in the positive direction, i.e., for $x \geq y$ and $0 \leq \lambda \leq 1$*

$$U(\lambda x + (1 - \lambda)y) \geq \lambda U(x) + (1 - \lambda)U(y).$$

Axiom A1.3 *U is continuously twice differentiable.*

Axiom A1.4 *Given $x_1^* > x_1^{**}$, $x_2, \Delta x_2 > 0$, define Δx_1^* and Δx_1^{**} by*

$$U(x_1^*, x_2) = U(x_1^* + \Delta x_1^*, x_2 - \Delta x_2)$$

$$U(x_1^{**}, x_2) = U(x_1^{**} + \Delta x_1^{**}, x_2 - \Delta x_2).$$

Then $\Delta x_1^* > \Delta x_1^{**}$.

Theorem A1.5 *Under the Axioms A1.1–A1.4, U is a concave function on $\Re^2$.*

Proof Let

$$U_{ij} = \frac{\partial^2 U(x_1, x_2)}{\partial x_i \partial x_j}; \qquad i,\ j = 1,\ 2. \tag{A1.1}$$

It is sufficient to show that the matrix U_{ij} is negative semidefinite for all x_1, x_2.

Axiom A1.2 implies that for fixed $x \geq y$,

$$H(\lambda) \equiv U(\lambda x + (1 - \lambda)y) \tag{A1.2}$$

is a concave function of λ in [0,1]. This, in turn, implies that $H''(\lambda) \leq 0$. It can be easily seen that

$$H''(\lambda) = \sum_i \sum_j (x_i - y_i)(x_j - y_j) \left. \frac{\partial^2 U(\omega)}{\partial \omega_i \partial \omega_j} \right|_{\omega = \lambda x + (1-\lambda) y} . \tag{A1.3}$$

Given $\omega > 0$ and $a > 0$ there exist x, y with $x \geq y$ and $0 < \lambda < 1$ such that

$$\omega = \lambda x + (1 - \lambda) y, \tag{A1.4}$$

and

$$\sum_i \sum_j a_i a_j \frac{\partial^2 U(\omega)}{\partial \omega_i \partial \omega_j} \leq 0. \tag{A1.5}$$

To see this, note that there exists $\varepsilon > 0$ such that $x = \omega + \varepsilon a > 0, y = \omega - \varepsilon a > 0$ and use the fact that $H''(\lambda) \leq 0$. It follows from the above development that (A1.5) also holds for $a < 0$.

We want to show that (A1.5) holds for all a. Suppose $U_{12} > 0$ for $x = \omega$. Since U is concave in each argument (by Axiom A1.2), $U_{11} < 0, U_{22} < 0$. Hence (A1.5) holds when $a_1 a_2 < 0$.

Now suppose that $U_{12} < 0$ for $x = \omega$. It follows from Axiom A1.4 that

$$\frac{\partial}{\partial x_1} \left(-\frac{U_2}{U_1} \right) < 0$$

where

$$U_i = \frac{\partial U(x_1, x_2)}{\partial x_i}$$

Thus

$$U_2 U_{11} - U_1 U_{12} < 0.$$

Reversing the role of x_1 and x_2 we get

$$U_1 U_{22} - U_2 U_{12} < 0.$$

From the above two relations we get

$$U_{11} U_{22} > U_{12}^2,$$

since $U_{12} < 0, U_{11} < 0, U_{22} < 0, U_1 > 0$, and $U_2 > 0$. Thus the proof of the theorem is complete. □

Next we consider the case $m > 2$. We define a new characteristic (or a commodity) by a mixture of the m given characteristics (or commodities) in fixed proportions. We modify Axioms A1.1–A1.4 so that they hold for *any* such two new characteristics. Under these modified axioms the utility function U is concave on $\Re^m$.

To verify the above claim, take any two fixed points x and y in $\Re^m$, and consider the plane $\underline{P}$ passing through x and y and the origin O. It is now sufficient to prove that U is concave on the plane $\underline{P}$. Consider the convex cone which is the intersection of the plane $\underline{P}$ and the positive orthant, and let Q_1 and Q_2 be the unit vectors corresponding to the two extreme rays of this cone. All points on $\underline{P}$ can be considered as linear combinations of Q_1 and Q_2, i.e., for $p \in \underline{P}$

$$p = p_1 Q_1 + p_2 Q_2.$$

Thus any such point p in $\underline{P}$ can be represented by (p_1, p_2). It is now sufficient to show that the modified Axioms A1.1–A1.4 in terms of $(p1, p_2)$. This fact trivially follows for Axioms A1.1, A1.2, and A1.4. To see axiom A1.2, take any two points u and v on $\underline{P}$. Let Q_1 and Q_2 have coordinates $(i_1, \ldots, i_m)$, and $(j_1, \ldots, j_m)$, respectively, and let

$$u = u_1 Q_1 + u_2 Q_2,$$

$$v = v_1 Q_1 + v_2 Q_2.$$

Suppose now $(u_1, u_2) \leq (v_1, v_2)$. Then $u \leq v$. Thus, U is concave on the line joining u and v.

Acknowledgments: The author is thankful to the referees and a member of the Editorial Board for useful suggestions.

REFERENCES

Arnold, B.C. (1987). *Majorization and the Lorenz Order: A Brief Introduction.* Springer-Verlag, New York.

Arnold, B.C. (1983). *Pareto Distributions.* International Cooperative Publishing House, Fairland, Maryland, U.S.A.

Atkinson, S.B. (1970). On the measurement of inequality. *J. Econ. Theory,* **5**, 244–263.

Bhandari, S.K. (1987). *Some Aspects of Majorization and their Statistical Applications.* Ph. D. Dissertation (Unpublished), Indian Statistical Institute.

Bhandari, S.K. (1988). Multivariate and directional majorization: Positive results. To appear in *Sankhyā*, June, 1988.

Fischer, P. and Holbrook, J.A. (1980). Balayage defined by non-negative convex functions. *Proc. Amer. Math. Soc.* **79**, 445–448.

Karamata, J. (1932). Sur une inegalite aux functions convexes. *Publ. Math. Univ. Belgrade* **1**, 145–148.

Karlin, S. and Rinott, Y. (1983). Comparison of measures, multivariate majorization, and applications to statistics. *Studies in Econometrics,*

Time Series and Multivariate Statistics. (Ed. S. Karlin, T. Amemiya and Leo A. Goodman.) Academic Press, New York.

Lorenz, M.O. (1905). Methods of measuring concentration of wealth. *J. Amer. Statist. Assoc.* **9**, 209–219.

Marshall, A.W. and Olkin, I. (1979). *Inequalities: Theory of Majorization and its Applications.* Academic Press, New York.

Nygard, F. and Sandstrom, A. (1981). *Measuring Income Inequalities.* Almquist and Wiksell International, Stockholm.

Ross, S.M. (1983). *Stochastic Processes.* Wiley, New York.

Rothschild, M. and Stiglitz, J.E. (1970). Increasing risk: A definition. *J. Econ. Theory*, 6.

Ryff, J.V. (1965). Orbits of L'functions under doubly stochastic transformations. *Trans. Amer. Math. Soc.* **117**, 92–100.

Schur, I. (1923). Uber eine Klasse von Mittelbildungen mit Anwendungen die Determinanten. *Theorie Sitzungsber. Berlin Math. Gesellschaft* **22**, 9–20.

Sen, A. (1976). Real national income. *Rev. Econ. Theory* **63**, 19–39.

Stoyan, D. (1983). *Comparison Methods for Queues and other Stochastic Models.* Wiley, New York.

Strassen, V. (1965). The existence of probability measures with given marginals. *Ann. Math. Statist.* **36**, 423–439.

Taguchi, T. (1972a). On the two-dimensional concentration surface and extensions of concentration coefficient and Pareto distribution to the two-dimensional case — I. *Ann. Inst. Statist. Math.* **24**, 355–382.

Taguchi, T. (1972b). On the two-dimensional concentration surface and extensions of concentration coefficient and Pareto distribution to the two-dimensional case — II. *Ann. Inst. Statist. Math.* **24**, 599–619.

4

Some Results on Convolutions and a Statistical Application

M.L. Eaton[1]
L.J. Gleser[2]

ABSTRACT Classes of distributions, of both discrete and continuous type, are introduced for which the right tail of the distribution is nonincreasing. It is shown that these classes are closed under convolution, thus providing sufficient conditions for nonincreasing right tails to be preserved under convolution. A start is made on verifying a conjecture concerning the extension to the left of nondecreasing right tails under successive convolution. The results give properties of the distributions of random walks on the integers. A statistical application is the verification of a conjecture of Sobel and Huyett (1957) concerning the minimal probability of correct selection for the usual indifference zone procedure for selecting the Bernoulli population with the largest success probability.

1 Introduction

A well known result of Wintner (1938, pp. 30, 32) asserts that the class of symmetric (about 0) unimodal densities on the real line R is closed under convolution. The corresponding result for symmetric unimodal probability mass functions on the integers is proved by Gupta and Sobel (1960). Consequently, for symmetric distributions the property of having a nonincreasing right (or left) tail is preserved under convolution.

In the present paper, a larger class of distributions is introduced in which convolution preserves nonincreasing right tails. In Section 2, the following two theorems are proved.

Theorem 1.1 *For any integer m, let $\mathcal{P}(m)$ be the class of probability mass functions $p(\cdot)$ defined on the integers which satisfy*

$$p(m-j) \geq p(m+j), \qquad j = 0, 1, 2, \ldots, \tag{1.1a}$$

$$p(j) \geq p(j+1), \qquad \textit{for } j = m, m+1, \ldots . \tag{1.1b}$$

[1]University of Minnesota
[2]Purdue University

Then if $p_i(\cdot)$ belongs to $\mathcal{P}(m_i)$, $i = 1, 2$, the convolution $p_1 * p_2(\cdot)$ of $p_1(\cdot)$ and $p_2(\cdot)$ belongs to $\mathcal{P}(m_1 + m_2)$.

Theorem 1.2 *For any real number m, let $\mathcal{F}(m)$ be the class of density functions $f(\cdot)$ defined on the real line R which satisfy*

$$f(m-t) \geq f(m+t) \qquad \text{for } t \geq 0, \tag{1.2a}$$

$$f(x) \geq f(y) \qquad \text{when } m \leq x \leq y. \tag{1.2b}$$

Then, if $f_i(\cdot)$ belongs to $\mathcal{F}(m_i)$, $i = 1, 2$, the convolution $f_1 * f_2(\cdot)$ of $f_1(\cdot)$ and $f_2(\cdot)$ belongs to $\mathcal{F}(m_1 + m_2)$.

Note that symmetric (about 0) unimodal probability mass functions belong to $\mathcal{P}(0)$, and that symmetric unimodal densities belong to $\mathcal{F}(0)$; for these distributions the inequalities in (1.1a) and (1.2a) are actually equalities.

Suppose that a probability mass function $p(\cdot)$ belongs to $\mathcal{P}(0)$, and that $p(\cdot)$ is not symmetric about 0. Theorem 1.1 says that for every $n \geq 1$ the n-fold convolution $p_{(n)}(\cdot)$ of $p(\cdot)$ with itself satisfies

$$p_{(n)}(j) \geq p_{(n)}(j+1), \qquad \text{all } j = 0,\ 1,\ 2,\ \ldots\ .$$

Thus, $p_{(n)}(\cdot)$ has a nonincreasing right tail beginning with $j = 0$ for all $n \geq 1$. However, under these circumstances the mean (if it exists) of the distribution defined by $p(\cdot)$ is negative. Hence, the weak law of large numbers implies that the probability mass of $p_{(n)}(\cdot)$ moves to minus infinity as $n \to \infty$. Also, if the variance of $p(\cdot)$ exists, the Central Limit Theorem suggests that $p_{(n)}(j)$, $j = 0, \pm 1, \pm 2, \ldots$, becomes part of the right (decreasing) tail of the standard normal distribution as $n \to \infty$. These observations lead to the following conjecture.

Conjecture 1.3 *There exists a nondecreasing sequence $\{n_i\colon i = 1, 2, \ldots\}$ of positive integers such that $p_{(n)}(-i) \geq p_{(n)}(-i+1)$, all $n \geq n_i$, for $i = 1$, 2, ...*

In Section 3, a special case of this conjecture is verified. Suppose that $p(\cdot)$ has support on the integers $-1, 0, 1$. That is

$$p(-1) > p(1), \qquad p(0) \geq p(1),$$

and $p(j) = 0$ for $j \neq -1, 0, 1$. In this case, it is shown that

$$p_{(n)}(-1) \geq p_{(n)}(0), \qquad n \geq \max\left\{3, \frac{1}{\alpha} - 1\right\}, \tag{1.3}$$

where $\alpha = p(-1) - p(1)$. The proof depends upon the following result, which is of independent interest.

Theorem 1.4 *Let $q(\cdot)$ be the probability mass function of the uniform distribution on $\{-1, 0, 1\}$, and let $q_{(n)}(\cdot)$ be the n-fold convolution of $q(\cdot)$ with itself. Then $q_{(n)}(0) - q_{(n)}(1)$ is nonincreasing in n for all $n \geq 2$.*

Both (1.3), in the case where $p(\cdot)$ has support $\{-1, 0, 1\}$, and also Theorem 1.4 give properties of the distribution of the n-th stage of a random walk on the integers.

Finally, in Section 4, the above probability results are applied to the problem of choosing the Bernoulli population with the largest probability of success when independent random samples of size n are chosen from each of two Bernoulli populations. If the two probabilities of success differ by at least an amount Δ, $0 < \Delta < 1$, it is shown that the probability of correct choice for the standard procedure (Sobel and Huyett, 1957) is, for all $n \geq \max\{4, \Delta^{-1}\}$, minimized when the smaller probability of success is $\frac{1}{2}(1 - \Delta)$ and the larger probability of success is $\frac{1}{2}(1 + \Delta)$. This verifies a conjecture of Sobel and Huyett (1957).

2 Proofs of Theorems 1.1 and 1.2

The following is a sketch of the main steps in the proof of Theorem 1.1:

Step 1. If X_i has mass function $p_i(\cdot)$ in $\mathcal{P}(m_i)$, $i = 1, 2$, it is easily seen that $Y_i = X_i - m_i$ has mass function in $\mathcal{P}(0)$. Further, $X_1 + X_2$ has mass function in $\mathcal{P}(m_1 + m_2)$ if and only if $Y_1 + Y_2$ has mass function in $\mathcal{P}(0)$. Hence, to prove Theorem 1.1 it is sufficient to show that $\mathcal{P}(0)$ is closed under convolution.

Step 2. $\mathcal{P}(0)$ is closed under convex linear combinations. That is, if $p_i(\cdot)$ belongs to $\mathcal{P}(0)$, $i = 1, 2, \ldots, k$, and a_i, $1 \leq i \leq k$, are nonnegative constants satisfying $\sum_{i=1}^{k} a_i = 1$, then $\sum_{i=1}^{k} a_i p_i(\cdot)$ belongs to $\mathcal{P}(0)$. This assertion is straightforwardly verified from the definition of $\mathcal{P}(0)$.

Step 3. Let $\mathcal{S}(0)$ be the collection of all symmetric (about 0) unimodal mass functions on the integers, and let $\mathcal{N}(0)$ be the collection of all mass functions having support on the negative integers $-1, -2, \ldots$. It has been previously noted that $\mathcal{S}(0)$ is a subcollection of $\mathcal{P}(0)$, and it is easily seen that $\mathcal{N}(0)$ is also a subcollection of $\mathcal{P}(0)$. Any mass function $p(\cdot)$ in $\mathcal{P}(0)$ can be written as a convex linear combination

$$p(\cdot) = \alpha n(\cdot) + (1 - \alpha)s(\cdot) \tag{2.1}$$

of a mass function $n(\cdot)$ in $\mathcal{N}(0)$ and a mass function $s(\cdot)$ in $\mathcal{S}(0)$. To see this, let $\alpha = \sum_{i=1}^{\infty} \big(p(-i) - p(i)\big)$,

$$s(i) = \begin{cases} (1-\alpha)^{-1} p(|i|), & \text{if } i = -1, -2, \ldots, \\ (1-\alpha)^{-1} p(i), & \text{if } i = 0, 1, 2, \ldots, \end{cases}$$

and

$$n(i) = \begin{cases} \alpha^{-1}\big(p(i) - p(-i)\big), & i = -1, -2, \ldots, \\ 0, & i = 0, 1, 2, \ldots. \end{cases}$$

Step 4. Let $p_1(\cdot)$ and $p_2(\cdot)$ belong to $\mathcal{P}(0)$. Then by Step 3,

$$p_i(\cdot) = \alpha_i n_i(\cdot) + (1-\alpha_i) s_i(\cdot), \qquad i = 1, 2,$$

where $n_1(\cdot)$ and $n_2(\cdot)$ belong to $\mathcal{N}(0)$, $s_1(\cdot)$ and $s_2(\cdot)$ belong to $\mathcal{S}(0)$, and $0 \le \alpha_1,\ \alpha_2 \le 1$. Note that

$$\begin{aligned} p_1 * p_2(\cdot) = {} & \alpha_1\alpha_2\big(n_1 * n_2(\cdot)\big) + \alpha_1(1-\alpha_2)\big(n_1 * s_2(\cdot)\big) \\ & + (1-\alpha_1)\alpha_2\big(s_1 * n_2(\cdot)\big) + (1-\alpha_1)(1-\alpha_2)\big(s_1 * s_2(\cdot)\big). \end{aligned} \tag{2.2}$$

It is shown by Gupta and Sobel (1960) that $s_1 * s_2(\cdot) \in \mathcal{S}(0) \subset \mathcal{P}(0)$. Further, it is clear that $n_1 * n_2(\cdot) \in \mathcal{N}(0) \subset \mathcal{P}(0)$. If it can be shown that $n_1 * s_2(\cdot)$ and $s_1 * n_2(\cdot)$ belong to $\mathcal{P}(0)$, then it will follow from (2.2) and Step 2 that $p_1 * p_2(\cdot) \in \mathcal{P}(0)$.

The proof of Theorem 1.1 is thus completed by the following lemma.

Lemma 2.1 *Both $n_1 * s_2(\cdot)$ and $s_1 * n_2(\cdot) = n_2 * s_1(\cdot)$ belong to $\mathcal{P}(0)$.*

Proof We will show that $n_1 * s_2(\cdot) \in \mathcal{P}(0)$. The proof that $n_2 * s_1(\cdot) \in \mathcal{P}(0)$ is similar. For $j \ge 0$,

$$\begin{aligned} n_1 * s_2(j) &= \sum_{i=-\infty}^{\infty} n_1(i) s_2(j-i) = \sum_{i=1}^{\infty} n_1(-i) s_2(j+i) \\ &\ge \sum_{i=1}^{\infty} n_1(-i) s_2(j+1+i) = \sum_{i=-\infty}^{\infty} n_1(i) s_2(j+1-i) \\ &= n_1 * s_2(j+1). \end{aligned}$$

Also, for $j \ge 1$,

$$n_1 * s_2(j) = \sum_{i=j+1}^{\infty} n_1(-i) s_2(j+i) + \sum_{i=1}^{j} n_1(-i) s_2(j+i)$$

$$\leq \sum_{i=j+1}^{\infty} n_1(-i)s_2(-j+i) + \sum_{i=1}^{j} n_1(-i)s_2(j-i)$$
$$= \sum_{i=j+1}^{\infty} n_1(-i)s_2(-j+i) + \sum_{i=1}^{j} n_1(-i)s_2(-j+i)$$
$$= \sum_{i=-\infty}^{\infty} n_1(i)s_2(-j-i) = n_1 * s_2(-j).$$

Thus, $n_1 * s_2(\cdot)$ obeys properties (1.1b) and (1.1a), respectively, defining $\mathcal{P}(0)$. □

The proof of Theorem 1.2 follows the same steps as the proof of Theorem 1.1, substituting densities for mass functions and integrals for sums. Verification that $s_1 * s_2(\cdot) \in \mathcal{S}(0)$ follows from Wintner (1937).

3 Proof of (1.3)

Let $p(\cdot)$ be a probability mass function on the integers, with

$$p(-1) > p(1), \qquad p(0) \geq p(1), \tag{3.1}$$

and $p(j) = 0$ for $j \neq -1, 0, 1$. Thus, $p(\cdot)$ belongs to $\mathcal{P}(0)$, and by Theorem 1.1 the n-fold convolution,

$$p_{(n)}(\cdot) = p * p * \cdots * p(\cdot),$$

of $p(\cdot)$ with itself also belongs to $\mathcal{P}(0)$. The goal of the present section is to verify the conjecture (1.3) in this special case.

Let $\alpha = p(-1) - p(1)$,

$$s(j) = \begin{cases} (1-\alpha)^{-1}p(1), & \text{if } j = -1, 1, \\ (1-\alpha)^{-1}p(0), & \text{if } j = 0, \\ 0, & \text{otherwise,} \end{cases}$$

and

$$n(j) = \begin{cases} 1, & \text{if } j = -1, \\ 0, & \text{otherwise.} \end{cases}$$

Then

$$p(j) = (1-\alpha)s(j) + \alpha n(j),$$

and for any $m \geq 1$,

$$p_{(m)}(j) = \sum_{i=0}^{m} \binom{m}{i} (1-\alpha)^i \alpha^{m-i} s_{(i)} * n_{(m-i)}(j),$$

where $s_{(k)}(\cdot)$ and $n_{(k)}(\cdot)$ are, respectively, the k-fold convolutions of $s(\cdot)$, $n(\cdot)$ with themselves. (Define $s_{(0)}(\cdot)$ and $n_{(0)}(\cdot)$ to be mass functions placing probability 1 on $j = 0$.) It is easily seen that for $0 \le i \le m$,

$$n_{(m-i)}(j) = \begin{cases} 1, & \text{if } j = -(m-i), \\ 0, & \text{otherwise,} \end{cases}$$

so that

$$s_{(i)} * n_{(m-i)}(j) = s_{(i)}(j + m - i).$$

Consequently for $m \ge 1$,

$$p_{(m)}(j) = \sum_{i=0}^{m} \binom{m}{i} (1-\alpha)^i \alpha^{m-i} s_{(i)}(m + j - i), \tag{3.2}$$

for $j = 0, \pm 1, \pm 2, \ldots$. Note that from (3.1), and from the definition of $s(\cdot)$, it follows that $s(\cdot) \in \mathcal{S}(0)$. Thus $s_{(i)}(\cdot) \in \mathcal{S}(0)$, $i = 2, 3, \ldots$.

Lemma 3.1 *For all* $m \ge 2$,

$$p_{(m)}(-1) - p_{(m)}(0) \ge (1-\alpha)^{m-1}\{m\alpha\big(s_{(m-1)}(0) - s_{(m-1)}(1)\big) - (1-\alpha)\big(s_{(m)}(0) - s_{(m)}(1)\big)\}.$$

Proof It follows from (3.2) and the symmetry about 0 and unimodality of each $s_{(i)}(\cdot)$ that

$$\begin{aligned} &p_{(m)}(-1) - p_{(m)}(0) \\ &= (1-\alpha)^m \big(s_{(m)}(-1) - s_{(m)}(0)\big) + m\alpha(1-\alpha)^{m-1}\big(s_{(m-1)}(0) - s_{(m-1)}(1)\big) \\ &\quad + \sum_{i=0}^{m-2} \binom{m}{i} (1-\alpha)^i \alpha^{m-i} \big(s_{(i)}(m-1-i) - s_{(i)}(m-i)\big) \\ &\ge -(1-\alpha)^m \big(s_{(m)}(0) - s_{(m)}(1)\big) + m\alpha(1-\alpha)^{m-1}\big(s_{(m-1)}(0) - s_{(m-1)}(1)\big), \end{aligned}$$

from which the stated inequality directly follows. □

Let

$$\beta = \frac{p(0) - p(1)}{1-\alpha}, \quad r(j) = \begin{cases} 1, & \text{if } j = 0, \\ 0, & \text{otherwise,} \end{cases}$$

and

$$q(j) = \begin{cases} \frac{1}{3}, & j = -1, 0, 1, \\ 0, & \text{otherwise.} \end{cases}$$

It is easily seen that

$$s(\cdot) = \beta r(\cdot) + (1-\beta) q(\cdot),$$

and thus that

$$s_{(k)}(\cdot) = \sum_{i=0}^{k} \binom{k}{i} \beta^{k-i} (1-\beta)^i q_{(i)}(\cdot), \tag{3.3}$$

where $q_{(0)}(\cdot) \equiv r(\cdot)$. Define

$$\Delta(i) = \begin{cases} q_{(3)}(0) - q_{(3)}(1), & i = 0, 1, 2, \\ q_{(i)}(0) - q_{(i)}(1), & i \geq 3, \end{cases}$$

and for $i \geq 1$ let J_i denote a random variable having a binomial distribution with sample size i and probability of success $1 - \beta$.

If Theorem 1.4 is correct, $\Delta(i)$ is a nonincreasing function of i, all $i = 0$, 1, 2, ... Since

$$q_{(0)}(0) - q_{(0)}(1) = 1, \qquad q_{(1)}(0) - q_{(1)}(1) = 0,$$
$$q_{(2)}(0) - q_{(2)}(1) = \frac{1}{9}, \qquad q_{(3)}(0) - q_{(3)}(1) = \frac{1}{27},$$

it follows from (3.3) that for $m \geq 2$,

$$\begin{aligned}\left(s_{(m-1)}(0) - s_{(m-1)}(1)\right) &- \left(s_{(m)}(0) - s_{(m)}(1)\right) \\ &= E\{\Delta(J_{m-1}) - \Delta(J_m)\} + R(m, \beta),\end{aligned} \tag{3.4}$$

where

$$\begin{aligned}R(m,\beta) = \frac{26}{27}(P\{J_{m-1} = 0\} - P\{J_m = 0\}) &- \frac{1}{27}\Big(P\{J_{m-1} = 1\} \\ &- P\{J_m = 1\}\Big) \\ + \frac{2}{27}(P\{J_{m-1} = 2\} &- P\{J_m = 2\}).\end{aligned}$$

For fixed β, it is known that J_k is stochastically increasing in k. Thus, $\Delta(J_k)$ is stochastically nonincreasing in k, and

$$E\{\Delta(J_{m-1}) - \Delta(J_m)\} \geq 0.$$

Some algebra shows that

$$\begin{aligned}R(m,\beta) &= \frac{\beta^{m-3}(1-\beta)}{27}\left\{(26 + m^2)\left(\beta - \frac{(m-1)(2m-1)}{2(m^2+26)}\right)^2 \right. \\ &\qquad \left. + \frac{9(m-1)}{4(m^2+26)}(11m - 23)\right\} \\ &\geq 0\end{aligned}$$

for $m \geq 3$. Thus, it follows from (3.4), assuming that Theorem 1.4 is true, that

$$s_{(m-1)}(0) - s_{(m-1)}(1) \geq s_{(m)}(0) - s_{(m)}(1) \tag{3.5}$$

for all $m \geq 3$.

Hence, if Theorem 1.4 is true, it follows from Lemma 3.1 and (3.5) that

$$p_{(m)}(-1) - p_{(m)}(0) \geq 0, \qquad \text{all } m \geq \max\left\{3, \ \frac{1}{\alpha} - 1\right\},$$

which, since $\alpha = p(-1) - p(1)$, verifies (1.3). Note that, as one would intuitively expect, the nonincreasing right tail of $p_{(m)}(\cdot)$ moves one step to the left (from $j = 0$ to $j = -1$) at a rate depending inversely on the difference α in probability mass between the left tail and right tail of $p(\cdot)$.

It remains to prove Theorem 1.4.

Proof of Theorem 1.4 The characteristic function of $q(\cdot)$ is

$$\phi(t) = \frac{1}{3}(1 + 2\cos(t)),$$

so

$$\phi^m(t) = \left(\frac{1}{3}\right)^m (1 + 2\cos(t))^m$$

is the characteristic function of $q_{(m)}(\cdot)$. Using the Fourier inversion formula (see Feller, 1966, p. 484) and the fact that $\phi(t)$ is real-valued, we have

$$\begin{aligned} q_{(m)}(j) &= \frac{1}{2\pi}\int_{-\pi}^{\pi} \phi^m(t) e^{-ijt}\, dt \\ &= \frac{1}{2\pi}\int_{-\pi}^{\pi} \left(\frac{1}{3}(1 + 2\cos(t))\right)^m \cos(jt)\, dt \end{aligned}$$

for $j = 0, \pm 1, \pm 2, \ldots$. Therefore,

$$\begin{aligned} \omega_m &= \left(q_{(m-1)}(0) - q_{(m-1)}(1)\right) - \left(q_{(m)}(0) - q_{(m)}(1)\right) \\ &= \frac{1}{2\pi}\int_{-\pi}^{\pi} \left(\left(\frac{1 + 2\cos(t)}{3}\right)^{m-1} - \left(\frac{1 + 2\cos(t)}{3}\right)^m\right)(1 - \cos(t))\, dt \\ &= \frac{4}{2\pi 3^m}\left(\int_0^{\pi} (1 + 2\cos(t))^{m-1}(1 - \cos(t))^2\, dt\right) \end{aligned} \tag{3.6}$$

which is obviously nonnegative when m is an odd integer ($m = 1, 3, 5, \ldots$). For $m = 4$, direct computation of the probabilities $q_{(3)}(0), q_{(3)}(1), q_{(4)}(0), q_{(4)}(1)$, or use of (3.6), yields

$$\omega_4 = 0.$$

Note that ω_m is nonnegative if and only if

$$\tau_m = \frac{3^m 2\pi}{4}\omega_m \tag{3.7}$$

is nonnegative. Also, $\tau_4 = 0$. We now show that τ_{2k} is nondecreasing in k, $k \geq 2$, and this will complete the proof that $\omega_m \geq 0$ for all even $m \geq 4$.

From (3.6), for $k \geq 2$,

$$\tau_{2k+2} - \tau_{2k} = \int_0^{\pi} \big(1+2\cos(t)\big)^{2k-1}\big(1-\cos(t)\big)^2\Big(\big(1+2\cos(t)\big)^2 - 1\Big)\,dt$$
$$= 4\int_0^{\pi} \big(1+2\cos(t)\big)^{2k-1}\sin^2(t)\cos(t)\big(1-\cos(t)\big)\,dt$$
$$\geq 4\int_{\frac{1}{3}\pi}^{\frac{2}{3}\pi} \big(1+2\cos(t)\big)^{2k-1}\sin^2(t)\cos(t)\big(1-\cos(t)\big)\,dt, \quad (3.8)$$

since $1+2\cos(t)$ and $\cos(t)\big(1-\cos(t)\big)$ have the same sign for t in $[0, \frac{1}{3}\pi] \cup [\frac{2}{3}\pi, \pi]$.

Let

$$H_k(t) = \big(1+2\cos(t)\big)^{2k-1}\sin^2(t)\cos(t)\big(1-\cos(t)\big). \quad (3.9)$$

From (3.8),

$$\tau_{2k+2} - \tau_{2k} \geq 4\int_{\frac{1}{3}\pi}^{\frac{2}{3}\pi} H_k(t)\,dt = 4\int_0^{\frac{1}{6}\pi} \big(H_k(\frac{1}{2}\pi - u) + H_k(\frac{1}{2}\pi + u)\big)\,du.$$
$$= 4\int_0^{\frac{1}{6}\pi} \sin^2(\frac{1}{2}\pi + u)\sin(u)\big\{\big(1+2\sin(u)\big)^{2k-1}\big(1-\sin(u)\big) - \big(1-2\sin(u)\big)^{2k-1}\big(1+\sin(u)\big)\big\}\,du \quad (3.10)$$

since $\sin^2(\frac{1}{2}\pi - u) = \sin^2(\frac{1}{2}\pi + u)$, and

$$\cos\left(\frac{1}{2}\pi - u\right) = \sin(u) = -\cos\left(\frac{1}{2}\pi + u\right).$$

Noting that $0 \leq \sin(u) \leq \frac{1}{2}$ for $u \in [0, \frac{1}{6}\pi]$, and that for $x \in [0, \frac{1}{2}]$, $k \geq 2$,

$$(1+2x)^{2k-1}(1-x) \geq (1+2x)(1-x) \geq (1-2x)(1+x) \geq (1-2x)^{2k-1}(1+x),$$

it follows that the right-hand side of (3.10) is nonnegative, all $k \geq 2$. This completes the proof of Theorem 1.4, and verifies the result (1.3). □

A proof of Conjecture 1.3 made in Section 1, even in the special case of $p(\cdot)$ considered in this section, appears to be extremely difficult. It is possible that the methods used to prove (1.3) can be extended, but such an approach appears cumbersome. A more promising attack on the problem may be through the characteristic function argument used to prove Theorem 1.4.

4 A Statistical Application

In the indifference zone formulation for the problem of ranking Bernoulli parameters (Sobel and Huyett, 1957), independent random samples of size

n are obtained from each of k Bernoulli populations. The goal is to choose the population with the largest probability of success, but there is concern about a correct choice only when the largest probability of success exceeds the second largest probability of success by at least Δ, $0 < \Delta < 1$, where Δ is a prespecified constant.

When $k = 2$ Bernoulli populations are being compared, the procedure usually recommended is to compare the observed numbers X_1, X_2 of successes in the two samples, and conclude that population 1 has the largest probability of success if $X_1 > X_2$ and population 2 has the largest probability of success if $X_2 > X_1$. If $X_1 = X_2$, a population is either randomly selected (without loss of generality by a mechanism that does not depend upon the common observed value of X_1 and X_2), or else the population believed *a priori* to have the largest probability of success is chosen. Attention then concentrates on determining the smallest sample size n such that the probability of correctly choosing the population with the highest probability of success is no less than a prespecified constant γ, $0 < \gamma < 1$.

Let Y denote the number of successes in the sample obtained from the population with the largest probability of success, and let X denote the number of successes in the remaining sample. Under the given assumptions,

$$\begin{aligned} &X \text{ and } Y \text{ are statistically independent,} \\ &X \sim \text{ binomial } (n, p), \\ &Y \sim \text{ binomial } (n, p + d), \end{aligned} \tag{4.1}$$

where

$$0 \le p \le 1 - d, \qquad \Delta \le d \le 1,$$

and p, d are unknown.

Let θ be the (conditional) probability of selecting the population of Y when $Y = X$ $(0 \le \theta \le 1)$. Note that $\theta = 1$ corresponds to always selecting Y when $X = Y$, while $\theta = 0$ corresponds to always selecting X in such a situation. Since selecting the population of Y is the correct choice, the probability of correct selection is

$$\begin{aligned} PCS(p, d, n) &= P\{Y > X\} + \theta P\{Y = X\} \\ &= \theta P\{Y - X \ge 0\} + (1 - \theta) P\{Y - X \ge 1\}. \end{aligned} \tag{4.2}$$

In order that PCS is never less than γ, n must be chosen so that

$$\inf_{\Delta \le d \le 1} \inf_{0 \le p \le 1 - d} PCS(p, d, n) \ge \gamma.$$

However, Sobel and Huyett (1957) show that $PCS(p, d, n)$ is (strictly) decreasing in d for fixed p, n, θ. Thus, it can be assumed that $d = \Delta$, and n is determined to satisfy

$$\inf_{0 \le p \le 1 - \Delta} PCS(p, \Delta, n) \ge \gamma. \tag{4.3}$$

Since $PCS(p,\Delta,n)$ is for fixed Δ, n, θ, a continuous function of p, and p takes values in a closed interval $[0, 1-\Delta]$, the infimum in (4.3) is achieved. The value

$$p^* = p^*(\Delta, n, \theta)$$

which achieves the infimum (minimum) is said to be *least favorable*. In general, p^* depends upon n and θ, as well as on Δ. However, Sobel and Huyett (1957) use the large sample normal approximation to the distribution of $Y - X$ to show that for fixed θ, Δ,

$$\lim_{n\to\infty} p^*(\Delta, n, \theta) = \frac{1-\Delta}{2}. \tag{4.4}$$

Using both normal approximations and exact calculations, they give a table of the smallest values of n needed to assure that

$$PCS\left(\frac{1-\Delta}{2}, \Delta, n\right) \geq \gamma$$

when $\theta = \frac{1}{2}$. They remark that some exact calculations suggest that the limit in (4.4) is approached rapidly, so that their table gives a good approximation to an exact solution for determining the sample size n for the randomized selection rule with $\theta = \frac{1}{2}$. They also indicate how to adjust their table to find n when $\theta = 0$ or 1.

In this section, it is shown that

$$p^*(\Delta, n, 1) = \frac{1}{2}(1-\Delta), \qquad \text{all } n \geq 1, \tag{4.5}$$

and that for $0 \leq \theta < 1$,

$$p^*(\Delta, n, \theta) = \frac{1}{2}(1-\Delta), \qquad \text{all } n \geq \max\{4, \Delta^{-1}\}. \tag{4.6}$$

These results permit exact determination of the sample size n for both randomized $(0 < \theta < 1)$ and nonrandomized selection rules.

Define

$$G(j;\ p, \Delta, n) = P\{X - Y \geq j\},$$

for $j = 0, \pm 1, \pm 2, \ldots, \pm n$. Note from (4.1) and the above discussion that

$$\begin{aligned}
&\inf_{\Delta\leq d\leq 1}\ \inf_{0\leq p\leq 1-d} PCS(p, d, n) \\
&\quad = \min_{0\leq p\leq 1-\Delta} PCS(p, \Delta, n) \\
&\quad = \min_{0\leq p\leq 1-\Delta} \big(1 - (1-\theta)G(1;\ p, \Delta, n) - \theta G(0;\ p, \Delta, n)\big).
\end{aligned} \tag{4.7}$$

Theorem 4.1 *Fix $\Delta, 0 < \Delta < 1$. For all $n \geq 1$, $j \geq 1$, $G(j;\ p, \Delta, n)$ is unimodal in p. Further, $G(0;\ p, \Delta, n)$ is unimodal in p for $n \geq \max\{4, \Delta^{-1}\}$. The mode in both cases is $p^* = \frac{1}{2}(1-\Delta)$.*

Proof Let $v = p - \frac{1}{2}(1-\Delta)$. Then from (4.1),

$$X \sim \text{ binomial } \big(n, v + \frac{1}{2}(1-\Delta)\big),$$
$$Y \sim \text{ binomial } \big(n, v + \frac{1}{2}(1+\Delta)\big),$$

and X, Y are independent. Further since $0 \le p \le 1-\Delta$,

$$-\frac{1}{2}(1-\Delta) \le v \le \frac{1}{2}(1-\Delta).$$

Also

$$G(j;\ p, \Delta, n) = G(j;\ v + \frac{1}{2}(1-\Delta), \Delta, n)$$

so that $G(j;\ p, \Delta, n)$ is unimodal in p, $0 \le p \le 1-\Delta$, if and only if $G(j;\ v + \frac{1}{2}(1-\Delta), \Delta, n)$ is unimodal in v, $|v| \le \frac{1}{2}(1-\Delta)$.

Since X and Y are independent binomials,

$$X \sim \sum_{i=1}^{n} X_i, \qquad Y \sim \sum_{i=1}^{n} Y_i, \qquad X - Y = \sum_{i=1}^{n} Z_i$$

where $X_1, \ldots, X_n$, $Y_1, \ldots Y_n$ are independent Bernoulli variables with

$$X_i \sim \text{ Bernoulli } \big(v + \frac{1}{2}(1-\Delta)\big), \qquad Y_i \sim \text{ Bernoulli } \big(v + \frac{1}{2}(1+\Delta)\big).$$

Thus,

$$Z_i = X_i - Y_i, \qquad i = 1, \ldots, n,$$

are i.i.d. random variables with common mass function

$$p(z) = \begin{cases} \frac{1}{4}(1+\Delta)^2 - v^2, & \text{if } z = -1, \\ \frac{1}{2}(1-\Delta^2) + 2v^2, & \text{if } z = 0, \\ \frac{1}{4}(1-\Delta)^2 - v^2, & \text{if } z = 1, \\ 0, & \text{otherwise.} \end{cases} \tag{4.8}$$

It follows that

$$G(j;\ v + \frac{1}{2}(1-\Delta), \Delta, n) = P\Big\{\sum_{i=1}^{n} Z_i \ge j\Big\} \tag{4.9}$$

depends upon v only through v^2, and is thus an even function of v. Consequently, (4.9) is unimodal in v if and only if it is nonincreasing as a function of v^2, in which case the mode occurs at $v = 0$. [Note that $v = 0$ corresponds to $p = \frac{1}{2}(1-\Delta)$.]

It is now convenient to change notation. Let $t = v^2$ and

$$H(j;\ t, n) = P\Big\{\sum_{i=1}^{n} Z_i \ge j\Big\}, \tag{4.10}$$

where our notation suppresses the dependence of this probability on Δ. (Recall that Δ is held fixed.) Let

$$p(\ell;t) = P\{Z_i = \ell\}, \qquad \ell = -1, 0, 1,$$

and note from (4.8) that

$$\begin{aligned} p(1;t) &= \frac{1}{4}(1-\Delta)^2 - t, \\ p(0;t) &= \frac{1}{2}(1-\Delta^2) + 2t, \\ p(-1;t) &= \frac{1}{4}(1+\Delta)^2 - t. \end{aligned} \tag{4.11}$$

Finally, if $p_{(n)}(\cdot;t)$ is the n-fold convolution of $p(\cdot;t)$ with itself, then

$$p_{(n)}(i;\ t) = P\left\{\sum_{i=1}^{n} Z_i = i\right\},$$

and

$$H(j;\ t,n) = \sum_{i=j}^{n} p_{(n)}(i;\ t).$$

In an appendix, it is shown that for all $i = 0, \pm 1, \pm 2, \ldots,$

$$\frac{d}{dt}p_{(n)}(i;t) = n\big(2p_{(n-1)}(i;t) - p_{(n-1)}(i-1;t) - p_{(n-1)}(i+1;t)\big).$$

Consequently,

$$\begin{aligned} \frac{d}{dt}H(j;t,n) &= \sum_{i=j}^{n} \frac{d}{dt}p_{(n)}(i;t) \\ &= n\big(2H(j;t,n-1) - H(j-1;t,n-1) - H(j+1;t,n-1)\big) \\ &= n\big(p_{(n-1)}(j;t) - p_{(n-1)}(j-1;t)\big), \end{aligned}$$

and $(d/dt)H(j;t,n)$ will be less than or equal to 0 for $0 \le t \le \frac{1}{4}(1-\Delta)^2$ if and only if

$$p_{(n-1)}(j-1;t) \ge p_{(n-1)}(j;t), \text{ all } 0 \le t \le \frac{1}{4}(1-\Delta)^2. \tag{4.12}$$

Note from (4.8), or (4.11), that

$$p(-1;t) - p(1;t) = \frac{1}{4}(1+\Delta)^2 - \frac{1}{4}(1-\Delta)^2 = \Delta > 0,$$

$$\begin{aligned} p(0;t) - p(1;t) &= \frac{1}{2}(1-\Delta^2) - \frac{1}{4}(1-\Delta)^2 + 3t \\ &= \frac{1}{4}(1-\Delta)(1+3\Delta) + 3t \ge 0. \end{aligned}$$

for $0 \le t \le \frac{1}{4}(1-\Delta)^2$. Theorem 1.1 now applies to show that (4.12) holds for $j \ge 1$, all $n \ge 1$. Hence, $H(j;\ t,n)$ is nonincreasing in t for all $j \ge 1$, $n \ge 1$; and consequently, for all $n \ge 1$, $G(j;\ p,\Delta,n)$ is unimodal in p with mode at $p = \frac{1}{2}(1-\Delta)$.

For $j = 0$, the result (1.3) can be applied to show that (4.12) holds for

$$n - 1 \ge \max\{3,\ \Delta^{-1} - 1\}.$$

Thus, when $n \ge \max\{4, \Delta^{-1}\}$, $G(0;\ p,\Delta,n)$ is unimodal in p with mode at $p = \frac{1}{2}(1-\Delta)$. □

The asserted results (4.5) and (4.6) now follow immediately from (4.7) and Theorem 4.1.

A1 Derivatives of Convolutions

For any functions $p(\cdot), q(\cdot)$ mapping the integers $0,\ \pm 1,\ \pm 2, \ldots$ into the real line, define the convolution $p * q(\cdot)$ by

$$p * q(j) = \sum_{i=-\infty}^{\infty} p(i)q(j-i),$$

provided the infinite sum exists. It is easily seen that

$$\begin{aligned} p * q(\cdot) &= q * p(\cdot), \\ p * (q * r)(\cdot) &= (p * q) * r(\cdot), \\ (ap + bq) * r(\cdot) &= a\big(p * r(\cdot)\big) + b\big(q * r(\cdot)\big), \end{aligned} \tag{A1.1}$$

for real constants $a,\ b$.

For each t in an interval (t_L, t_U), let $p(\cdot;\ t)$ and $q(\cdot;\ t)$ map the integers into the real line, and assume that for every integer j the derivatives

$$\frac{d}{dt}p(j;t), \quad \frac{d}{dt}q(j;t)$$

exist for all t in (t_L, t_U). If

$$p * q(j;t) = \sum_{i=-\infty}^{\infty} p(i;t)q(j-i;t) \tag{A1.2}$$

exists for all $j = 0,\ \pm 1,\ \pm 2,\ \ldots$, all t in (t_L, t_U), then under the usual conditions for interchange of summation and differentiation, we have

$$\left(\frac{d}{dt}\ (p * q)\right)(\cdot;t) = \left(\left(\frac{d}{dt}\ p\right) * q\right)(\cdot;t) + \left(p * \left(\frac{d}{dt}\ q\right)\right)(\cdot;t). \tag{A1.3}$$

Lemma A1.1 *Let $p_{(n)}(\cdot;t)$ be the n-fold convolution of $p(\cdot;t)$, where $p(j;t)$ has a derivative with respect to t for all integers j, all t in (t_L, t_U). Then, assuming we can interchange summation and derivative, for all $n \geq 1$, all integers j,*

$$\frac{d}{dt}\, p_{(n)}(j;t) = n\left(p_{(n-1)} * \left(\frac{d}{dt}\, p\right)\right)(j;t).$$

Proof Using (A.1) and (A.3),

$$\begin{aligned}\frac{d}{dt}\, p_{(2)}(\cdot;t) &= \left(\left(\frac{d}{dt}\, p\right) * p\right)(\cdot;t) + \left(p * \left(\frac{d}{dt}\, p\right)\right)(\cdot;t)\\ &= 2\left(p * \left(\frac{d}{dt}\, p\right)\right)(\cdot;t).\end{aligned}$$

The stated result now follows by use of (A1.1), (A1.3) and induction on n. □

An important application of Lemma A1.1 is to the case where $p(\cdot;t)$ is linear in t. If

$$p(i;t) = a(i) + b(i)t$$

$i = 0, \pm 1, \pm 2, \ldots$, then $(d/dt)\, p(i;t) = b(i)$ and

$$\frac{d}{dt}\, p_{(n)}(j;t) = n \sum_{i=-\infty}^{\infty} p_{(n-1)}(i;t)b(j-i).$$

In particular, if $p(j;t)$ is given by (4.11), then

$$b(i) = \begin{cases} 2, & i = 0, \\ -1, & i = -1, 1, \\ 0, & \text{otherwise,} \end{cases}$$

and

$$\frac{d}{dt}\, p_{(n)}(i;t) = n\big(2p_{(n-1)}(i;t) - p_{(n-1)}(i-1;t) - p_{(n-1)}(i+1;t)\big).$$

Acknowledgments: We gratefully acknowledge the comments of Mary Ellen Bock and Keith Crank which led to the proof of Theorem 1.4 given in this paper. M.L. Eaton's research was supported in part by NSF Grant DMS 8319924. L.J. Gleser's research was supported in part by NSF Grant DMS 8501966.

References

Feller, W. (1966). *An Introduction to Probability Theory and its Applications, Volume II.* Wiley, New York.

Gupta, S.S. and Sobel, M. (1960). Selecting a subset containing the best of several populations. In *Contributions to Probability and Statistics* (I. Olkin, ed.), Stanford University Press, 224–248.

Sobel, M. and Huyett, M.J. (1957). Selecting the best one of several binomial populations. *Bell System Tech. J.* **36**, 537–576.

Wintner, A. (1938). *Asymptotic Distributions and Infinite Convolutions.* Edwards Brothers, Ann Arbor, MI.

5

The $X + Y$, X/Y Characterization of the Gamma Distribution

George Marsaglia[1]

ABSTRACT We prove, by elementary methods, that if X and Y are independent random variables, not constant, such that $X+Y$ is independent of X/Y then either X, Y or $-X, -Y$ have gamma distributions with common scale parameter. This extends the result of Lukacs, who proved it for positive random variables, using differential equations for the characteristic functions. The aim here is to use more elementary methods for the X, Y positive case as well as elementary methods for proving that the restriction to positive X, Y may be removed.

1 Introduction

We say that X is a gamma-a variate if X has the standard gamma density $x^{a-1}e^{-x}/\Gamma(a), x > 0$. If X and Y are independent gamma-a and gamma-b variates then $X + Y$ is independent of X/Y. This article is concerned with the converse: if $X + Y$ is independent of X/Y for independent X, Y, what can be said about X and Y? The general result is this: ruling out the case where X and Y are constants, in which case any function of X and Y is independent of any other, if $X + Y$ is independent of X/Y then there is a constant c and positive constants a and b such that cX is gamma-a and cY is gamma-b. The constant c may be negative, but neither X nor Y can take both positive and negative values; either X, Y or $-X, -Y$ are pairs of positive gamma variates with a common scale parameter and possibly different gamma parameters.

In 1955, Lukacs (1955) proved the basic result under the assumption that X and Y were positive. His method was to show that the characteristic functions of X and Y satisfied differential equations whose only solutions were characteristic functions of gamma variates with a common scale parameter. A few years after that, in the late 1950's, I needed the

[1]Supercomputer Computations Research Institute and Department of Statistics, The Florida State University

$X+Y, X/Y$ characterization in developing computer methods for generating random points on surfaces by means of projections of points with independent coordinates. But the coordinates could be negative as well as positive, so I set out to extend Lukacs' result by removing the restriction that X and Y be positive. I was able to do this, using elementary methods, but I still needed Lukacs' result for the positive case. I put the matter aside until I could find an elementary argument that established that case as well, perhaps motivated by a sentiment attributed to Herman Rubin: If you have to use characteristic functions you don't really understand what is going on.

I was not able to find an elementary proof of the X, Y positive case, and the matter sat for years, until I was sent a manuscript, by Findeisen, which contained a clever device that might be used to establish Lukacs' result without resorting to characteristic functions. In the form that Findeisen's result was published, Findeisen (1978), there is a disclaimer suggested by the referees, to the effect that characteristic function results are implicit in parts of Findeisen's arguments.

And there the matter rests today, the point of departure for this article. In it, I will use a variation of Findeisen's device, together with my earlier proof that the X, Y positive restriction can be removed, to provide a complete treatment of the $X+Y, X/Y$ characterization of the gamma distribution by elementary methods. Opinions differ on what is elementary, of course. In the development below, the most advanced result that I need is the fact that a distribution on [0,1] is determined by its moments. This was a deep result when first proved by Hausdorff, but it may now, thanks to Feller, be considered elementary, as the elegant proof in Feller (1971) shows, using basic probability and limit arguments.

2 The Unrestricted Theorem

Theorem 2.1 *If X and Y are independent, non-degenerate (i.e. not constant) random variables such that $X+Y$ is independent of X/Y, then there are constants a, b and c such that cX has the gamma density $x^{a-1}e^{-x}/\Gamma(a)$ and cY has the gamma density $y^{b-1}e^{-y}/\Gamma(b)$.*

Note that use of the expression X/Y requires the implicit assumption that $\Pr(Y=0)=0$, but also note that X/Y independent of $X+Y$ for independent X, Y also requires that $\Pr(X=0)=0$, since, if $0<\Pr(X/Y=0)<1$,

$$\Pr(X/Y=0)\Pr(X+Y<s)=\Pr(X/Y=0, X+Y<s)=\Pr(Y<s).$$

As s grows, the left side approaches $\Pr(X/Y=0)$, the right goes to 1. Thus we need not be concerned with possibilities $X=0$ or $Y=0$ in the theorem or subsequent discussion.

Theorem 2.1 is the most general form of the $X+Y, X/Y$ characterization of the gamma distribution. It does not require that the variates be positive. Our proof depends on four propositions, each of which will be proved by elementary methods below. Two of the propositions depend on what we call the *exponential moments* of a non-negative random variable Z, defined as the sequence of values

$$E(Z^n e^{-Z})/E(e^{-Z}) \quad \text{for } n = 1, 2, 3, \ldots .$$

Evidently the exponential moments all exist, since $z^n e^{-z}$ is bounded for $z \geq 0$.

3 Four Propositions

Proposition 3.1 *If X and Y are independent random variables, not constant, such that $X+Y$ is independent of X/Y then either* $\Pr(X > 0, Y > 0) = 1$ *or* $\Pr(X < 0, Y < 0) = 1$.

Proposition 3.2 *If X and Y are independent positive random variables, not constant, such that $X+Y$ is independent of X/Y, then there are positive constants a, b and k such that the exponential moments of X and Y are those of gamma distributions with common scale parameter: for $n = 1, 2, 3, \ldots,$*

$$\frac{E(X^n e^{-X})}{E(e^{-X})} = \frac{k^n \Gamma(a+n)}{\Gamma(a)} \quad \textit{and} \quad \frac{E(Y^n e^{-Y})}{E(e^{-Y})} = \frac{k^n \Gamma(b+n)}{\Gamma(b)}.$$

Proposition 3.3 *Every distribution on $[0, \infty)$ is determined by its exponential moments.*

Proposition 3.4 *If X and Y are independent, positive random variables such that X/Y is independent of $X - Y$, then X and Y are both constants.*

These four propositions will be proved below, but first we show how they are combined to prove the main theorem.

Proof of the main theorem

We now have independent X and Y, not constant, with X/Y independent of $X+Y$. Assume the four Propositions. Then Proposition 3.1 ensures that either X, Y or $-X, -Y$ are pairs of positive variates. Proposition 3.2 then provides the exponential moments of X and Y, or $-X$ and $-Y$, and Proposition 3.3 ensures that, with the resulting gamma exponential moments, cX and cY are gamma-a and gamma-b for some constant c, possibly negative. Proposition 3.4 is not used directly, but is required for the proof of Proposition 3.1.

Proof of Proposition 3.1.

We have independent X and Y, not constant but otherwise unrestricted, such that $X+Y$ is independent of X/Y. We must prove that either $Pr(X > 0, Y > 0) = 1$ or $Pr(X < 0, Y < 0) = 1$. Let $p_x = \Pr(X > 0)$ and $p_y = \Pr(Y > 0)$. If $p_x p_y > 0$, let (X_+, Y_+) be the point (X, Y) conditioned by $X > 0$ and $Y > 0$:

$$\Pr(X_+ < x, Y_+ < y) = \frac{\Pr(0 < X < x, 0 < Y < y)}{p_x p_y}.$$

Evidently X_+ is independent Y_+ (a product measure is still a product measure when restricted to a product set), and, in fact, $X_+ + Y_+$ is independent of X_+/Y_+ because

$$\begin{aligned}\Pr(X_+ + Y_+ < r, X_+/Y_+ < s) &= \frac{\Pr(0 < X + Y < r, 0 < X/Y < s)}{p_x p_y} \\ &= \frac{\Pr(0 < X + Y < r)\Pr(0 < X/Y < s)}{p_x p_y}.\end{aligned}$$

Thus $X_+ + Y_+$ is independent of X_+/Y_+, since their joint distribution is a product. Propositions 3.2 and 3.3 apply: there is a positive constant c such that cX_+ is gamma-a and cY_+ is gamma-b.

This takes care of the positive quadrant, with measure $p_x p_y$. If $(1 - p_x)(1 - p_y) > 0$ then (X_-, Y_-) is well-defined and an argument similar to that for (X_+, Y_+) shows that cX_- and cY_- must be standard gamma variates for some negative constant c. Thus $0 < p_x < 1$ and $0 < p_y < 1$ and $X + Y$ independent of X/Y lead to four possibilities:

- X and Y each have densities that are (proper) mixtures of scaled "negative" and "positive" gamma densities.
- X_+, Y_+ are gamma and X_-, Y_- are constants.
- $-X_-, -Y_-$ are gamma and X_+, Y_+ are constants.
- Both X_+, Y_+ and X_-, Y_- are constants.

It is elementary to verify that for none of these four cases is $X + Y$ independent of X/Y.

Next, we eliminate the possibility that only one of p_x, p_y is between 0 and 1. Suppose, for example, that X is positive and Y can take both positive and negative values. Then $\Pr(X + Y < 0) > 0$ and $\Pr(Y/X > -1) > 0$. This leads to

$$0 < \Pr(X + Y < 0)\Pr(Y/X > -1) = \Pr(X + Y < 0, X + Y > 0) = 0,$$

with similar contradictions for $p_x = 0, 0 < p_y < 1$, etc.

Thus p_x is either 0 or 1, and p_y is either 0 or 1. That conclusion leads to four more possibilities:

$$\begin{array}{llll} (\mathbf{a}) & p_x = 1,\ p_y = 1 & (\mathbf{b}) & p_x = 0,\ p_y = 1 \\ (\mathbf{c}) & p_x = 0,\ p_y = 0 & (\mathbf{d}) & p_x = 1,\ p_y = 0. \end{array}$$

If conditions **(a)** or **(c)** hold, then X, Y or $-X, -Y$ are independent pairs of positive variates and Propositions 3.2 and 3.3 apply. If **(b)** holds, then $-X$ and Y satisfy the conditions of Proposition 3.4, so they must be constant; if **(d)** holds, then Proposition 3.4 shows that X and $-Y$ must be constant.

This completes the proof of Proposition 3.1: the support of (X, Y) must be either the first or the third quadrant. The proof used Propositions 3.2, 3.3 and 3.4, which we now proceed to prove.

PROOF OF PROPOSITION 3.2.

We have independent, positive non-constant X and Y with X/Y independent of $X+Y$. Consider the exponential moments of X, Y and $X+Y$:

$$R_n = E[X^n e^{-X}]/E[e^{-X}]$$
$$S_n = E[Y^n e^{-Y}]/E[e^{-Y}]$$
$$T_n = E[(X+Y)^n e^{-X-Y}]/E[e^{-X-Y}] = \sum_{i=0}^{n} \binom{n}{i} R_i S_{n-i}.$$

If X and Y were gamma variates with common scale parameter, then R_n, S_n, T_n would have the form, for some positive constants a, b and k:

$$R_n = k^n \frac{\Gamma(a+n)}{\Gamma(a)}, \quad S_n = k^n \frac{\Gamma(n+b)}{\Gamma(b)}, \quad T_n = k^n \frac{\Gamma(a+b+n)}{\Gamma(a+b)}. \tag{3.1}$$

The independence of $X/(X+Y)$ and $X+Y$ will be used to provide a pair of recurrence equations for R_{n+1} and S_{n+1} that will have a unique solution: the exponential moments of expression (3.1). Then, because by Proposition 3.3 the exponential moments determine the distribution, we will be led to gamma distributions.

The recursions may be derived by dividing, side for side, the relation

$$E[X^n (X+Y) e^{-X-Y}] = E\left[\left(\frac{X}{X+Y}\right)^n\right] E[(X+Y)^{n+1} e^{-X-Y}]$$

by the sides of

$$E[X^n e^{-X-Y}] = E\left[\left(\frac{X}{X+Y}\right)^n\right] E[(X+Y)^n e^{-X-Y}].$$

These relation follow easily from the independence of X, Y and of $X/Y, X+Y$. Upon division, side for side, we get the relation

$$\frac{R_{n+1}}{R_n} + S_1 = \frac{T_{n+1}}{T_n}. \tag{3.2}$$

Reversing the roles of X and Y then provides

$$\frac{S_{n+1}}{S_n} + R_1 = \frac{T_{n+1}}{T_n}. \tag{3.3}$$

Note that the gamma exponential moments in (3.1) satisfy (3.2) and (3.3). We must show that no others do. When $n = 1$, (3.2) and (3.3) lead to

$$\frac{R_2 - R_1^2}{R_1} = \frac{S_2 - S_1^2}{S_1}. \tag{3.4}$$

Now $R_2 - R_1^2$ is the variance of a non-degenerate random variable, (the W defined by

$$\Pr(W \le w) = \int_0^w e^{-x}\, dF(x) \Big/ \int_0^\infty e^{-x}\, dF(x),$$

with F the distribution of X). Thus there is a positive value k such that

$$R_2 = kR_1(1+R_1) \qquad \text{and} \qquad S_2 = kS_1(1+S_1). \tag{3.5}$$

A little algebra will verify that (3.2) and (3.3) give R_{n+1} and S_{n+1} uniquely in terms of $R_1, S_1, R_2, S_2, \ldots, R_n, S_n$. (Each pair (R_{n+1}, S_{n+1}) arises from a linear system with matrix having determinant $T_n(T_n - R_n - S_n)$, easily shown to be non-zero by induction.)

Since (5) provides R_2 and S_2 in terms of R_1 and S_1 and the common parameter k, the two sets of exponential moments for X and Y are determined by R_1, S_1 and the constant k. Specifically, given R_1, S_1 and the common value k required by (3.4), define a and b by the conditions $R_1 = ak$, $S_1 = bk$. Then $R_2 = ka(a+1)$, $S_2 = kb(b+1)$, $R_3 = ka(a+1)(a+2)$, $S_3 = kb(b+1)(b+2)$ and, in general,

$$R_n = k^n\Gamma(a+n)/\Gamma(a) \qquad \text{and} \qquad S_n = k^n\Gamma(b+n)/\Gamma(b)$$

provides the unique solution to conditions (3.2),(3.3) and (3.4) derived from the assumption of independent pairs X, Y and $X/Y, X+Y$.

Proof of Proposition 3.3

Let X be non-negative with distribution F and exponential moments

$$R_n = \frac{E(X^n e^{-X})}{E(e^{-X})} = \int_0^\infty x^n e^{-x}\, dF(x) \Big/ \int_0^\infty e^{-x}\, dF(x), \qquad n = 1, 2, 3, \ldots.$$

We must show that the R's determine F. To do this, let W be the random variable with distribution G defined by

$$G(w) = \Pr(W \le w) = \int_0^w e^{-x}\, dF(x) \Big/ \int_0^\infty e^{-x}\, dF(x).$$

Then the exponential moments of X are the regular moments of W:

$$E(W^n) = \int_0^\infty w^n \, dG(w) = \int_0^\infty x^n e^{-x} \, dF(x) \Big/ \int_0^\infty e^{-x} \, dF(x).$$

The distribution of X determines that of W, and *vice versa*; indeed, $F(x) = \int_0^x e^w \, dG(w) / \int_0^\infty e^w \, dG(w)$.

It turns out that the moments of W determine its distribution, but that result requires analytic function theory, violating our proposed goal that proofs be elementary. We overcome this problem by converting W to a random variable Z on the unit interval. For such, an elementary proof that the moments determine the distribution is available—see Feller (1971), pages 225–227 for a beautiful elementary proof that for points of continuity z,

$$\Pr(Z \le z) = \lim_{n \to \infty} \sum_{j \le nt} \binom{n}{j} (-1)^{n-j} E[(Z^j(1-Z)^{n-j}].$$

So, let $Z = e^{-W}$. Then the distribution of Z is determined by its moments. To see that the moments of W, (the exponential moments of X), determine the moments of Z, write

$$E(Z^k) = E(e^{-kW}) = \int_0^\infty e^{-kx} e^{-x} \, dF(x) \Big/ \int_0^\infty e^{-x} \, dF(x).$$

$$= \int_0^\infty e^{-x} \left(1 - kx + \frac{(kx)^2}{2!} - \frac{(kx)^3}{3!} + \cdots \right) dF(x) \Big/ \int_0^\infty e^{-x} \, dF(x).$$

We may exchange the integral and summation operations to get

$$E(Z^k) = \sum_{j=0}^\infty \frac{(-1)^j}{j!} \int_0^\infty (kx)^j e^{-x} \, dF(x) \Big/ \int_0^\infty e^{-x} \, dF(x) = \sum_{j=0}^\infty \frac{(-k)^j}{j!} R_k.$$

Thus the exponential moments of X determine the moments of Z, which determine the distribution of Z, which determines the distribution of $W = -\ln(Z)$, which determines the distribution of X, and that sequence of implications provides proof of Proposition 3.3:

exp. moments of $X \Rightarrow$ moments of $Z \Rightarrow$
dist. of $Z \Rightarrow$ dist. of $W \Rightarrow$ dist. of X.

PROOF OF PROPOSITION 3.4.

We have X and Y independent, positive and X/Y independent of $X - Y$. Evidently this cannot hold if only one of X or Y is constant, so assume that neither is constant. We will develop a contradiction.

Let $p = \Pr(Y - X > 0) = \Pr(Y/X > 1)$. Then

$$p = \Pr(Y - X > 0, Y/X > 1) = \Pr(Y - X > 0) \Pr(Y/X > 1) = p^2.$$

Thus p is idempotent, $p^2 = p$, and p must be 0 or 1. Interchanging the roles of X and Y if necessary, we may assume that $p = \Pr(Y > X) = 1$. Then X must be bounded, and X, not constant, will have two points of increase $x_1 < x_2$ such that $\Pr(X > x_2) = 0$.

Since Y is not constant, it (or its distribution) has two points of increase $y_1 < y_2$. Now define two sets $\mathcal{A}$ and $\mathcal{B}$:

$$\mathcal{A} = \{(x,y) : y/x < y_2/x_2\}, \qquad \mathcal{B} = \{(x,y) : y - x > y_2 - x_2\}.$$

Then $\Pr(\mathcal{A}) > 0$, since $\mathcal{A}$ contains the point (x_2, y_1). Similarly, $\Pr(\mathcal{B}) > 0$ because $\mathcal{B}$ contains the point (x_1, y_2).

From the assumed independence of Y/X and $Y - X$,

$$\Pr(\mathcal{A} \cap \mathcal{B}) = \Pr(\mathcal{A}) \Pr(\mathcal{B}) > 0,$$

contradicting the fact that every point (x,y) in $\mathcal{A} \cap \mathcal{B}$ has $x > x_2$, so that

$$\Pr(\mathcal{A} \cap \mathcal{B}) \leq \Pr(X > x_2) = 0.$$

This proves Proposition 3.4: $Y - X$ independent of Y/X for positive independent X and Y requires that both be constant.

Proof of the four Propositions, and hence an elementary proof of the unrestricted $X + Y, X/Y$ characterization of the gamma distribution, is now complete.

References

Feller, William. (1971). *An Introduction to Probability Theory and its Applications.* Volume 2, Second Edition. Wiley, New York.

Findeisen, Peter. (1978). A simple proof of a classical theorem which characterizes the gamma distribution. *The Annals of Statistics.* **6** No.**5**, 1165-1167.

Lukacs, Eugene. (1955). A characterization of the gamma distribution. *Annals of Mathematical Statistics.* **26** 319-324.

6

A Bivariate Uniform Distribution

Albert W. Marshall[1]

ABSTRACT The univariate distribution uniform on the unit interval $[0,1]$ is important primarily because of the following characterization: Let X be a random variable taking values in $[0,1]$. Then the distribution of $X+U \pmod 1$ is the same as the distribution of X for all nonnegative random variables U independent of X if and only if X has a distribution uniform on $[0,1]$.

A natural bivariate version of this is the following: Let (X,Y) be a random vector taking values in the unit square. Then $(*)$ $(X+U \pmod 1, Y+V \pmod 1)$ has the same distribution as (X,Y) for every pair (U,V) of nonnegative random variables independent of (X,Y) if and only if X and Y are independent and uniformly distributed on $[0,1]$. But if $(*)$ is required to hold only when $U=V$ with probability one, then (X,Y) can have any one of a large class of bivariate uniform distributions which are given an explicit representation and studied in this paper.

1 Introduction

The literature abounds with examples of bivariate distributions having marginals uniform on $[0,1]$; indeed, any bivariate distribution with continuous marginals can be transformed to provide such an example. Bivariate distributions with uniform marginals can be regarded as "canonical forms" representing all bivariate distributions with marginals that are both continuous and have continuous inverses. In this context, they are sometimes called "copulas" or "dependence functions".

The purpose of this note is to take a different viewpoint, and to address this question: Which bivariate distributions with uniform marginals are important in their own right? This problem is naturally approached by starting with the property primarily responsible for the importance of the univariate uniform $[0,1]$ distribution. The property is a characterization most easily stated with the following notation:

(i) $x \oplus y := x + y - [x+y] = x + y \pmod 1, \quad x, y \geq 0,$

---

[1]University of British Columbia.

(ii) $U \overset{dist}{=} V$ means U and V are random variables with the same distribution,

where $[t]$ denotes the integer part of t.

Univariate characterization

Suppose that $P\{X \in [0,1]\} = 1$. Then

$$X \oplus U \overset{dist}{=} X \text{ for every random variable } U \geq 0 \text{ independent of } X \quad (1.1)$$

if and only if X has the uniform $[0,1]$ distribution.

The property (1.1) can be easily interpreted by considering $2\pi X$ to be a random direction. Reformulated as a functional equation in terms of the distribution F of X, (1.1) becomes

$$\int_0^1 F(x-y)\,dG(y) + \int_0^1 F(1+x-y)\,dG(y) - \int_0^1 F(1-y)\,dG(y) = F(x), \quad (1.2)$$

for $0 \leq x \leq 1$, for all distributions G with support in $[0,1]$. This functional equation is easily solved by standard methods; the only solution is $F(x) = x$, $0 \leq x \leq 1$.

2 Bivariate Versions

Perhaps the most straightforward two-dimensional version of (1.1) is the following.

First bivariate characterization

Suppose that $P\{(X,Y) \in [0,1]^2\} = 1$. Then

$$(X \oplus U,\ Y \oplus V) \overset{dist}{=} (X,Y) \quad \text{for every pair } U \geq 0,\ V \geq 0 \text{ of random variables independent of } (X,Y) \quad (2.1)$$

if and only if X and Y are independent and uniformly distributed on $[0,1]$.

It is not difficult to show that (2.1) is equivalent to

$$(X \oplus u,\ Y \oplus v) \overset{dist}{=} (X,Y) \text{ for all } u, v \in [0,1]. \quad (2.2)$$

Condition (2.2) can be reformulated as a functional equation for the distribution F of (X,Y) which is easily solved to yield $F(x,y) = xy$, $0 \leq x$, $y \leq 1$. Alternatively the characterization (2.1) can be obtained directly from the uniqueness of Haar measure.

The above result is reminiscent of a result of Marshall and Olkin (1967) concerning bivariate exponential distributions. They started with the univariate functional equation

$$P(X > x + y) = P(X > x)\, P(X > y), \quad x, y \geq 0,$$

extended it to two dimensions by taking X, x, and y to be vectors, and then solved to find that the components of X must be independent. More interesting solutions were admitted by weakening the functional equation, and the same procedure can be followed here.

Second bivariate characterization

Suppose that $P\{(X,Y) \in [0,1]^2\} = 1$. Then

$$\begin{array}{l}(X \oplus U,\ Y \oplus U) \stackrel{dist}{=} (X,Y) \text{ for every random variable } U \geq 0 \\ \text{independent of } (X,Y)\end{array} \tag{2.3}$$

if and only if (X,Y) have a joint distribution H of the form

$$H(x,y) = \int_0^1 F_\theta(x,y)\, dG(\theta), \tag{2.4}$$

where G is a distribution with support contained in $[0,1]$, and for fixed $\theta \in [0,1]$,

$$F_\theta(x,y) = \begin{cases} x & \text{if } 0 \leq \theta \leq y - x \\ y - \theta & \text{if } y - x \leq \theta \leq \min(1-x, y) \\ 0 & \text{if } y \leq \theta \leq 1 - x \\ x + y - 1 & \text{if } 1 - x \leq \theta \leq y \\ x + \theta - 1 & \text{if } \max(1-x, y) \leq \theta \leq 1 - x + y \\ y & \text{if } 1 - x + y \leq \theta \leq 1. \end{cases} \tag{2.5}$$

To verify this result, first note that (2.3) is equivalent to

$$(X \oplus u,\ Y \oplus u) \stackrel{dist}{=} (X,Y) \quad \text{for all} \quad u \in [0,1]. \tag{2.6}$$

Condition (2.6), when rewritten as a functional equation for the distribution H of (X,Y), becomes

$$\begin{aligned} H(x,y) = H(x-u,\ y-u) &+ [H(1+x-u,\ y-u) - H(1-u,\ y-u)] \\ &+ [H(x-u,\ 1+y-u) - H(x-u,\ 1-u)] \\ &+ [H(1+x-u,\ 1+y-u) - H(1-u,\ 1+y-u) \\ &\quad - H(1+x-u,\ 1-u) + H(1-u,\ 1-u)] \end{aligned} \tag{2.7}$$

for all $u \in [0,1]$.

FIGURE 1. The set S_θ and the value of the distribution function F_θ in various regions.

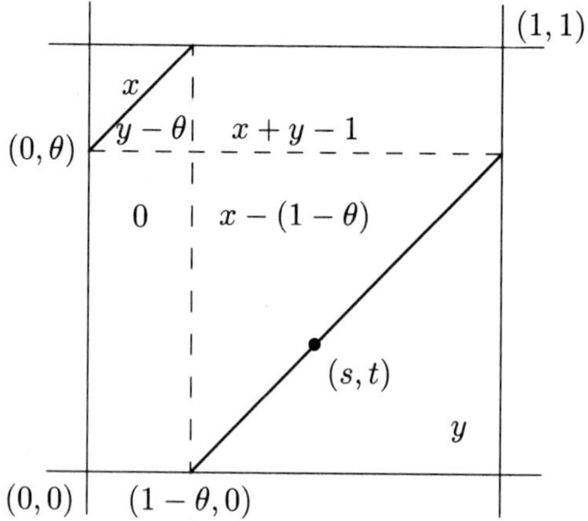

The functional equation (2.7) appears to be difficult to solve using standard methods. However, it is immediate from (2.7) that the set of all solutions is convex. A crucial step in solving (2.7) might be to identify the extreme point solutions.

Suppose that $H = F$ is a solution of (2.7) and that the point (s,t) belongs to the support of F. Let $\theta = t \oplus (1-s)$. It follows from (2.6) that all points in the set

$$\begin{aligned} S_\theta &= \{(x,y) : x = s \oplus z,\ y = t \oplus z,\ 0 \leq z \leq 1\} \\ &= \{(x,y) : x = z,\ y = z \oplus \theta,\ 0 \leq z \leq 1\} \end{aligned}$$

also belong to the support of F. Moreover, the conditional distribution F_θ of (X,Y) given $(X,Y) \in S_\theta$ must be uniform on S_θ. A somewhat tedious but straightforward calculation shows that F_θ is given by (2.5).

Now suppose that there are at least two points (s_1, t_1) and (s_2, t_2) in the support of F for which $\theta_1 = t_1 \oplus (1 - s_1) \neq \theta_2 = t_2 \oplus (1 - s_2)$. The above arguments lead to the conclusion that the conditional distribution of (X,Y) given $(X,Y) \in S_{\theta_1} \cup S_{\theta_2}$ must be a mixture of the distributions uniform on S_{θ_1} and S_{θ_2} so that $H = F$ cannot be an extreme point solution of (2.7). It follows from the Krein-Milman theorem that solutions of (2.7) must be of the form (2.4).

3 Properties of the Distribution H

Bivariate distributions of the form (2.4) have some quite nice properties in spite of the fact that the definition of F_θ is a bit awkward.

Uniqueness

The representation (2.4) is essentially unique, with non-uniqueness occurring only because $F_0 = F_1$. If G is not allowed to have point mass at 1, then G can be recovered from H through the equation

$$G(\theta) = P\{(X,Y) \in A_\theta\} \tag{3.1}$$

where (X,Y) has distribution H and

$$A_\theta = \{(x,y) : y = x \oplus a \text{ for some } a \in [0,\theta]\} = \cup_{a \leq \theta} S_a.$$

Random variable representation

If

$$(X,Y) = (X,\ X \oplus \theta), \tag{3.2}$$

where X has a uniform $[0,1]$ distribution, then (X,Y) has the distribution F_θ.

Covariances

From (3.2) it is easily verified that if (X,Y) has the distribution F_θ, then

$$EXY = \frac{1}{3} - \frac{1}{2}\theta(1-\theta), \quad \operatorname{cov}(X,Y) = \frac{1}{12} - \frac{1}{2}\theta(1-\theta),$$

and

$$\operatorname{corr}(X,Y) = 1 - 6\theta(1-\theta). \tag{3.3}$$

It follows from (3.3) that $-\frac{1}{2} \leq \operatorname{corr}(X,Y) \leq 1$, where the minimum correlation is achieved with $\theta = \frac{1}{2}$ and the maximum correlation is achieved with $\theta = 0$ or 1.

It also follows from (3.3) that if (X,Y) has the distribution H of (2.4), then

$$\begin{aligned}\operatorname{corr}(X,Y) &= \int [1 - 6\theta(1-\theta)]\, dG(\theta) \\ &= 1 - 6E\Theta + 6E\Theta^2,\end{aligned} \tag{3.4}$$

where Θ has the distribution G.

Regression

If (X, Y) has the distribution H of (2.4), it follows from (3.2) that

$$E(Y|X) = G(1-X) - (1-X) + E\Theta. \tag{3.5}$$

Convolutions (mod 1)

Suppose that (U_i, V_i) is a random vector with distribution K_i, $i = 1, 2$. Denote the distribution of $(U_1 \oplus U_2,\ V_1 \oplus V_2)$ by $K_1 \circledast K_2$. It can be seen with the aid of (3.2) that

$$F_{\theta_1} \circledast F_{\theta_2} = F_{\theta_1 \oplus \theta_2}, \quad \theta_1, \theta_2 \in [0,1]. \tag{3.6}$$

It follows that if $H_i(\cdot) = \int F_\theta(\cdot)\, dG_i(\theta)$, then

$$H_1 \circledast H_2(\cdot) = \int F_\theta(\cdot)\, d(G_1 \circledast G_2)(\theta). \tag{3.7}$$

Thus, the class of distributions of the form (2.4) is closed under convolution (mod 1), as well as under mixtures and weak limits.

The fact that (3.7) follows from (3.6) has an analog for ordinary convolutions which is best known under the condition that the mixing distribution G is infinitely divisible (Feller, 1971, p. 538). See also Keilson and Steutel (1974, Proposition 1.4) and Marshall and Olkin (1989, Lemma 2.6).

4 Some Examples

Independent marginals

It is easy to see that (2.4) yields $H(x, y) = xy$, $0 \le x,\ y \le 1$ if and only if $G(\theta) = \theta$, $0 \le \theta \le 1$, is itself uniform on $[0, 1]$.

Conditional waiting time in a Poisson process

If $\{N(t),\ t \ge 0\}$ is a Poisson process, it is well known that the conditional distribution of $T = \inf\{t : N(t) = 1\}$ given $N(1) = 1$ is uniform $[0, 1]$. This natural occurrence of the uniform distribution has a bivariate analog.

Let $\{(N_1(t),\ N_2(t)),\ t \ge 0\}$ be a bivariate Poisson process, i.e., $N_1(t) = M_1(t) + M_{12}(t)$ and $N_2(t) = M_2(t) + M_{12}(t)$, where M_1, M_2 and M_{12} are independent Poisson processes having respective parameters λ_1, λ_2 and λ_{12}, say. Let $T_i = \inf\{t : N_i(t) = 1\}$, $i = 1, 2$. Then the conditional distribution H of (T_1, T_2) given $N_1(1) = N_2(1) = 1$ must have uniform marginals. Elementary calculations show that

$$H_\alpha(x, y) = (1-\alpha)xy + \alpha \min(x, y), \quad 0 \le x,\ y \le 1, \tag{4.1}$$

where $\alpha = \lambda_{12}/(\lambda_1\lambda_2 + \lambda_{12})$. This distribution is of the form (2.4) where G puts mass α at 0 and mass $(1-\alpha)$ uniformly on $[0,1]$. Of course H_α is just a convex combination of the case of independence and the upper Fréchet bound.

The class of distributions of the form (4.1) is closed under convolutions (mod 1), and

$$H_{\alpha_1} \circledast H_{\alpha_2} = H_{\alpha_1\alpha_2}. \tag{4.2}$$

This fact can be obtained directly or from (3.7).

If (X,Y) has the distribution H_α of (4.1) then from (3.4) it follows that $\operatorname{corr}(X,Y) = \alpha$. From (3.5), it follows that

$$E(Y|X) = \alpha x + (1-\alpha)/2.$$

Moreover, distributions H of the form (2.4) which have linear regressions must be of the form H_α for some $\alpha \in [0,1]$. To see this, set $E(Y|X) = aX+b$ in (3.5) and use $G(1) = 1$ to conclude that $G(x) = (1-x)G(0) + x$, $0 \leq x \leq 1$; thus G is a convex combination of the distribution giving unit mass to the origin and the distribution uniform on $[0,1]$.

Additional examples

If G is uniform on $[0, \frac{1}{2}]$ and (X,Y) has the distribution H of (2.4), then X and Y are not independent. However, X and Y are uncorrelated (this follows from (3.4)) and $X+Y$ has the same distribution as it does in the case of independence (this is easily seen from a sketch of the density of H).

If $G(\theta) = \theta^\alpha$, $0 \leq \theta \leq 1$ where $\alpha > 0$, then (2.4) yields

$$H(x,y) = \begin{cases} x + \frac{1}{\alpha+1}\{y^{\alpha+1} + (1-x)^{\alpha+1} - (y-x)^{\alpha+1} - 1\} & 0 \leq x \leq y \leq 1 \\ y + \frac{1}{\alpha+1}\{x^{\alpha+1} + (1-y)^{\alpha+1} - (x-y)^{\alpha+1} - 1\} & 1 \geq x \geq y \geq 0. \end{cases}$$

5 Extension to Higher Dimensions

Conceptually, extension of the preceding results to higher dimensions involves no difficulties, but distributions uniform on sets of the form

$$S = \{(x_1, \ldots, x_n) : x_i = s_i \oplus z,\ i = 1, \ldots, n,\ 0 \leq z \leq 1\}$$

are not so easily written down. They can be parameterized by points in the set $[0,1]^{n-1}$, so that in the n-dimensional version of (2.4), G is a distribution on $[0,1]^{n-1}$. The n-dimensional versions of H in (2.4) have nice properties; e.g., their $(n-1)$ dimensional marginals are $(n-1)$ dimensional versions of H.

Acknowledgments: This work was supported in part by the National Science Foundation Grant DMS-87-08083 and the National Sciences and Engineering Research Council of Canada.

References

Feller, W. (1971). *An Introduction to Probability Theory and Its Applications.* Vol. II (second edition). Wiley: New York.

Keilson, J. and Steutel, F.W. (1974). Mixtures of distributions, moment inequalities and measures of exponentiality and normality. *Ann. Probab.* 2, 112–130.

Marshall, A.W. and Olkin, I. (1967). A multivariate exponential distribution. *J. Amer. Statist. Assoc.* 62, 30–44.

Marshall, A.W. and Olkin, I. (1989). Multivariate distributions generated from mixtures of convolution and product families. To appear in *Topics in Statistical Dependence*, ed. by H.W. Block, A.R. Sampson, and T.H. Savits. IMS Lecture Notes/Monograph Series.

7

Multinomial Problems in Geometric Probability with Dirichlet Analysis

Milton Sobel[1]

ABSTRACT A variety of new combinatorial results are obtained using the recently developed technique of Dirichlet Analysis, which utilizes the study of Dirichlet integrals. These results are stated as seventeen problems which are geometrical or combinatorial in nature.

1 Introduction

The following 17 problems (and solutions) are all geometrical and/or combinatorial in nature and illustrate the wide application of a recently developed method of analysis, called Dirichlet Analysis because of the usage of Dirichlet integrals. These integrals are introduced, studied and tabled in Volumes 4 and 9 of *Selected Tables of Mathematical Statistics* and we refer to these books below simply as Vol. 4 and Vol. 9. The main emphasis in these 17 problems is (1) to show the wide application of these integrals and (2) to solve the so-called faces problem, the edges problem and the vertices problem for the regular and for certain quasi- regular polyhedra with central symmetry. Assuming that a polyhedral die falls with a face upward (otherwise the bottom face is used), we see the top face (resp., all the edges associated with the top face, resp., all the vertices associated with the top face) and the problem is to find the expected number of tosses $E(T)$ needed to see all the faces (resp., all the edges; resp., all the vertices) of the given polyhedra. In many cases the Dirichlet analysis gives higher moments in addition to the first moment, so that variances can also be obtained. A summary table for the various polyhedra investigated is given as Table 1. In some cases exact fractional values of Dirichlet C-integrals are more useful than the decimal values in Vol. 9 and for this purpose a small exact C-table is included as an appendix to this paper; these values are used extensively throughout this paper.

It is assumed that readers are familiar with the notations $I_a^{(b)}(r;m)$ and

[1]University of California, Santa Barbara

TABLE 1. Expectation results for selected polyhedra

Polyhedron	Faces Problem	Edges Problem	Vertices Problem
Tetrahedron	8.33333	4.33333	2.33333
Cube	14.70000	8.10000	4.20000
Octahedron	21.51786	11.01905	4.14286
Dodecahedron	37.23853	21.60216	12.34699
Icosahedron	72.42848	37.50874*	14**
Cuboctahedron (with $p_2 = 0$, $p_1 = p_3 = \frac{1}{14}$)	45.52187	26.29376	8.46870
Cuboctahedron (with $p_2 = 0$, $6p_1 = 8p_3 = \frac{1}{2}$)	42.42012	n.c.	n.c.
Zonohedron ($n = 6$) (triacontahedron)	119.84961	60**	31**

n.c. means not computed

* the last 3 decimals may not be correct.

** estimated values.

$C_a^{(b)}(r;m)$ for the type-1 and type-2 Dirichlet integrals. Definitions, recurrence relations and examples of the use of such integrals in combinatorial problems are given in Vol. 4 and Vol. 9, respectively. The related type-2 Dirichlet integral $D_a^{(b)}(r;m)$, and generalizations of these integrals, such as $I_{(a_1,a_2)}^{(b_1,b_2)}(r_1,r_2;m)$ and $C_{(a_1,a_2)}^{(b_1,b_2)}(r_1,r_2;m)$, are also discussed in Vol. 9.

Let T_f (resp., T_e; resp., T_v) denote the number of tosses required to see (in the sense mentioned above) all the faces (resp., all the edges; resp., all the vertices) each at least once. Let $f_e = 2$ denote the number of faces associated with each edge and let f_v (variable) denote the number of faces associated with each vertex; e.g. $f_v = 3$ for the tetrahedron, the cube and the dodecahedron but $f_v = 4$ for the octahedron and $f_v = 5$ for the icosahedron. As a result of the exact calculations in the problems below (in terms of C-functions), the following double inequality (and approximation) arises as a conjecture since no proof has been found.

Conjecture 1.1 *For any given regular or quasi-regular polyhedra (with central symmetry)*

$$f_v E(T_v) \leq E(T_f) \leq f_e E(T_e). \tag{1.1}$$

Since we have a definitive formula only for ET_f, namely

$$ET_f = s\left(1 + \frac{1}{2} + \cdots + \frac{1}{s}\right) \tag{1.2}$$

where s is the number of faces of the given polyhedron, the inequalities in (1.1) provide an upper bound for $E(T_v)$ and a lower bound for $E(T_e)$. If the exact answer is not known, then (1.1) also provides the best approximation known to the author.

In rhombic and/or zonal polyhedra there can be two or more values of f_v and the question then arises how to use (1.1) as an approximation. Thus in a certain 30-sided zonohedron (with $n = 6$ in the notation of Coxeter (1973)), there are 32 vertices, 12 of which have 5 associated faces and the remaining 20 of which have 3 associated faces. Then the average number of faces associated with a vertex is $[12(5) + 20(3)]/32 = 3.75$. Hence for this zonohedron (called the triacontahedron) we give the answer to the faces problem as

$$E(T_f) = 30 \sum_{j=1}^{30} \frac{1}{j} = 30(3.994987) = 119.84961 \tag{1.3}$$

and approximate the other two answers by

$$E(T_e) \approx \frac{119.8497}{2} = 59.92, \tag{1.4}$$

$$E(T_v) \approx \frac{119.8497}{3.75} = 31.96. \tag{1.5}$$

The use of equality in (1.3) deserves some comment. This zonohedron has central symmetry and each of the 30 (rhombic) faces has two vertices associated with 5 faces and the other two associated with 3 faces. By symmetry we assume that the faces are equiprobable and (1.3) then holds.

The vertices problem for the icosahedron has not been done, so that the use of (1.1) is in order; this gives

$$E(T_v) \approx \frac{1}{5}(72.42848) = 14.5, \tag{1.6}$$

and we estimate the exact answer to be less than this value. In an expanded version of this paper (Sobel (1987), Problem 22), we get the exact answer for the vertices problem on the cuboctahedron, namely 8.46870, and the approximation gives $(45.52187)/4 = 11.4$. Although the approximation is not close, the inequality (1.1) still holds.

Problem 1.2 *"Seeing Double"* Suppose we have a fair die with $\binom{c}{2}$ sides and that each side has on it a different combination of two of the numbers $(1, 2, \ldots, c)$. We want the expected number of tosses $E(T)$ required to see each of the c numbers at least r times. Consider first the case $r = 1$ and do $r > 1$ later. How does ET vary as a function of c? For the special case $c = 4$, find the variance of T.

Remark 1.3 *This problem has applications to finding the number of boundary constraints of a feasible region and separating them from the redundant constraints . A random point in the feasible region and a random direction*

TABLE 2. #Faces, #Edges, #Vertices and f_v for the Polyhedra in Table 1

Polyhedron	#Faces	#Edges	#Vertices	f_v
Tetrahedron	4	6	4	3
Cube	6	12	8	3
Octahedron	8	12	6	4
Dodecahedron	12	30	20	3
Icosahedron	20	30	12	5
Cuboctahedron	14	24	12	4
Zonohedron ($n = 6$)				3 for 20 vertices
(triacontahedron)	30	60	32	5 for 12 vertices
				Average = 3.75

give rise to 2 constraints. If c were known and each pair was equally likely, then ET would be the expected number of these line-cut operations needed to see each of the boundary constraints at least once. If each pair had its own probability the same methods apply (c.f. next problem). If the number of constraints c is unknown then any reasonable stopping rule provides information about c.

Solution:

In this problem we can make more progress with Markovian methods than with Dirichlet integrals, but we consider both (along with other) techniques. Let x_j denote $E(T|j)$, the expected number of additional tosses needed after seeing j constraints (or numbers) each at least once, so that $x_0 = E(T|0)$ is the required ET. It is easily seen that the equations to be solved are: $x_0 = 1 + x_2$ and

$$
\begin{aligned}
x_2 &= 1 + \left\{\binom{2}{2}x_2 + 2(c-2)x_3 + \binom{c-2}{2}x_4\right\}\Big/\binom{c}{2} \\
x_3 &= 1 + \left\{\binom{3}{2}x_3 + 3(c-3)x_4 + \binom{c-3}{2}x_5\right\}\Big/\binom{c}{2} \\
&\ \ \vdots \\
x_{c-3} &= 1 + \left\{\binom{c-3}{2}x_{c-3} + (c-3)3x_{c-2} + \binom{3}{2}x_{c-1}\right\}\Big/\binom{c}{2} \\
x_{c-2} &= 1 + \left\{\binom{c-2}{2}x_{c-2} + (c-2)x_{c-1}\right\}\Big/\binom{c}{2} \\
x_{c-1} &= 1 + \binom{c-1}{2}x_{c-1}\Big/\binom{c}{2}
\end{aligned}
\tag{1.7}
$$

For small values of c from 3(1) 20 the solutions are given in Table 4 to 7 decimal places. If we disregard all the ones after the equal sign in (1.7) and write the result as $\underset{\sim}{x} = Q\underset{\sim}{x}$, then Q is the probability transition matrix for the $c-2$ transient states $(2, 3, \ldots, c-1)$. Hence from standard Markov chain theory about the fundamental matrix $(I-Q)^{-1}$, we have the result

$$E(T) = 1 + \left[(I-Q)^{-1}\underset{\sim}{1}\right]_1 \tag{1.8}$$

where $[\underset{\sim}{a}]_1$ is the first coordinate of the vector $\underset{\sim}{a}$, $\underset{\sim}{1}$ is a vector of ones and Q (as well as $I-Q$ and $(I-Q)^{-1}$) are upper triangular with at most 3 non-zero diagonals. To illustrate the solution (1.8), we consider $c=4$ where x_4 is already zero and, from (1.7)

$$Q = \frac{1}{6}\begin{pmatrix} 1 & 4 \\ 0 & 3 \end{pmatrix}, \quad (I-Q)^{-1} = 6\begin{pmatrix} 5 & -4 \\ 0 & 3 \end{pmatrix}^{-1} = \frac{2}{5}\begin{pmatrix} 3 & 4 \\ 0 & 5 \end{pmatrix}. \tag{1.9}$$

Hence for $c=4$ the desired result for $E(T)$ is $1 + \frac{14}{5} = 3.8$ as in Table 4 below. The same answer is easily obtained from (1.7), since $x_{c-1} = c/2$ from the last equation, which is 2 for $c=4$. From the next equation

$$5x_2 = 6 + 4x_3 = 6 + 8 = 14. \tag{1.10}$$

It follows that

$$ET = x_0 = 1 + 14/5 = 19/5 = 3.8. \tag{1.11}$$

A Dirichlet analysis of this problem for $c=4$ is based on the concept of a minimal covering set. Let the sides of the die numbered $(1,2)$, $(1,3)$, $(1,4)$, $(2,3)$, $(2,4)$ and $(3,4)$ be denoted by 1, 2, 3, 4, 5 and 6 respectively. Then there are seven minimal covering sets, namely $(1,6)$, $(2,5)$, $(3,4)$, $(1,2,3)$, $(2,4,6)$, $(1,4,5)$ and $(3,5,6)$. We need unions of $t \geq 1$ of these minimal covering sets in order to calculate the probability of at least 1 covering using inclusion-exclusion. For each t we want to break up the $\binom{7}{t}$ possible unions of t minimal covering sets according to the size of the union, i.e., the number of different pairs in the union. See Table 3.

We need the probability of terminating in any one of the seven minimal covering sets, i.e., in their union. We use inclusion-exclusion with $+1$ for odd values of t and -1 for even values of t. All sets of the same size are treated alike and hence we only need to consider a typical case. Thus $(1,6)$ is one of 3 cases all of size 2 and each occurs once, i.e., with multiplicity one. Since there are 7 minimal covering sets we have a row check of $\binom{7}{t}$ for the tth row. The last 2 rows have to add to 1 and 0, respectively, and are useful checks. From the last row (after factoring out the common 6) we obtain for the γth ascending factorial moment (for all $\gamma \geq 0$)

$$\begin{aligned} E(T^{[\gamma]}) = \frac{6\Gamma(\gamma+1)}{p^\gamma} &\left[C_1^{(1)}(1,\gamma+1) + 2C_1^{(2)}(1,\gamma+1) - 10C_1^{(3)}(1,\gamma+1)\right. \\ &\left. + 10C_1^{(4)}(1,\gamma+1) - 3C_1^{(5)}(1,\gamma+1)\right] \end{aligned} \tag{1.12}$$

TABLE 3. The Union of t of the Seven Minimal Covering Sets for $c = 4$

	Size 2	Size 3	Size 4	Size 5	Size 6	Absolute Value Check
$t = 1$	+3	+4				$\binom{7}{1} = 7$
$t = 2$			−15	−6		$\binom{7}{2} = 21$
$t = 3$				+24	+11	$\binom{7}{3} = 35$
$t = 4$				−6	−29	$\binom{7}{4} = 35$
$t = 5$					+21	$\binom{7}{5} = 21$
$t = 6$					−7	$\binom{7}{6} = 7$
$t = 7$					+1	$\binom{7}{7} = 1$
						Algebraic Value Check
Coef.	+3	+ 4	−15	+12	− 3	+1
Coef. × Size	+6	+12	−60	+60	−18	0

where $p = 1/6$. For $\gamma = 1$ and $\gamma = 2$ this yields (exactly)

$$E(T) = 36[3/4 + 2(11/18) - 10(25/48) + 10(137/300) - 3(49/120)] = 3.8, \tag{1.13}$$

$$\begin{aligned} E\{T(T+1)\} = 432[&.875 + 2(.78703704) - 10(.72048611) \\ &+ 10(.66772222) - 3(.62449074)] = 20.7200. \end{aligned} \tag{1.14}$$

It follows that for $c = 4$

$$\sigma^2(T) = 20.7200 - (3.8)(4.8) = 2.48, \tag{1.15}$$

which is quite small for problems of this type.

The problem with this analysis is that the number of minimal covering sets grows with c (it is 35 for $c = 5$) and the total number of subsets that have to be distributed in the table grows too rapidly with c.

Although we have now considered three different solutions of the problem, none of these are easy to generalize so that explicit formulas or tables are easily obtained. We therefore consider one more method which does lead to an explicit formula for any c-value and from which a table can easily be constructed.

The probability $P_c^{(n)}$ of covering all c digits for the first time on the nth toss is equal to the sum of two terms. One term is the probability of staying within and covering any specific set of exactly $c - 1$ digits in the first $n - 1$ tosses; this is multiplied by c and by the probability of getting the missing digit on the nth toss. The other term is the probability of staying within and covering any specific set of exactly $c-2$ digits in the first $n-1$ tosses; this is multiplied by $\binom{c}{2}$ and by the probability of getting the two missing digits on the nth toss. Hence, letting $[x]$ denote the integer part of x, we obtain

$$P_c^{(n)} = c\left(\frac{2}{c}\right)\left(\frac{\binom{c-1}{2}}{\binom{c}{2}}\right)^{n-1}\sum_{\alpha=0}^{c-3}(-1)^\alpha\binom{c-1}{\alpha}\left\{\frac{\binom{c-1-\alpha}{2}}{\binom{c-1}{2}}\right\}^{n-1} + \binom{c}{2}\left(\frac{1}{\binom{c}{2}}\right)\left(\frac{\binom{c-2}{2}}{\binom{c}{2}}\right)^{n-1}\sum_{\alpha=0}^{c-4}(-1)^\alpha\binom{c-2}{\alpha}\left\{\frac{\binom{c-2-\alpha}{2}}{\binom{c-2}{2}}\right\}^{n-1} \tag{1.16}$$

and hence the desired result for $E(T)$ is

$$E(T) = \sum_{n=[(c+2)/2]}^{\infty} 2n\left(\frac{\binom{c-1}{2}}{\binom{c}{2}}\right)^{n-1}\sum_{\alpha=0}^{c-3}(-1)^\alpha\binom{c-1}{\alpha}\left(\frac{\binom{c-1-\alpha}{2}}{\binom{c-1}{2}}\right)^{n-1} + \sum_{n=[c/2]}^{\infty} n\left(\frac{\binom{c-2}{2}}{\binom{c}{2}}\right)^{n-1}\sum_{\alpha=0}^{c-4}(-1)^\alpha\binom{c-2}{\alpha}\left(\frac{\binom{c-2-\alpha}{2}}{\binom{c-2}{2}}\right)^{n-1}. \tag{1.17}$$

Interchanging the order of summations, we obtain after some algebra

$$E(T) = 2\sum_{\alpha=0}^{c-3}(-1)^\alpha\left\{\binom{c-1}{\alpha}\left(\frac{\binom{c}{2}}{\binom{c}{2}-\binom{c-1-\alpha}{2}}\right)^2 + (-1)^c\binom{c-1}{\alpha+2}f_r(\alpha,c)\right\} + \sum_{\alpha=0}^{c-4}(-1)^\alpha\left\{\binom{c-2}{\alpha}\left(\frac{\binom{c}{2}}{\binom{c}{2}-\binom{c-1-\alpha}{2}}\right)^2 + (-1)^{c-1}\binom{c-1}{\alpha+2}f_{r-1}(\alpha,c)\right\}, \tag{1.18}$$

where $r=[c/2]$ and $f_r(\alpha,c)$ is given by

$$f_r(\alpha,c) = \left[1-(r+1)\left(\frac{\binom{\alpha+2}{2}}{\binom{c}{2}}\right)^r + r\left(\frac{\binom{\alpha+2}{2}}{\binom{c}{2}}\right)^{r+1}\right]\Big/\left(1-\left\{\frac{\binom{\alpha+2}{2}}{\binom{c}{2}}\right\}\right)^2. \tag{1.19}$$

From (1.18) and (1.19) we obtain Table 4 of ET-values. The fact that symmetrical dice with $\binom{c}{2}$ sides exist only for $c=4$ is of no serious concern, since the random experiment can be carried out otherwise, say with cards.

Regarding an asymptotic evaluation, it can be shown from the linear equations in (1.7) that the leading asymptotic term as $c\to\infty$ is $\frac{1}{2}c\log c$; we omit the details. It follows that any linear function of c obtained by solving a subset of the equations (1.7) and adding 1 for each additional equation will eventually be exceeded. A reasonable approximation for $E(T)$ in view of the above is

$$E(T) \approx \frac{c}{2}(\log c + .57), \tag{1.20}$$

TABLE 4. Values of $E(T)$ for Problem 1.4

c	$E(T)$	c	$E(T)$
3	2.5000000	12	18.0592579
4	3.8000000	13	20.0936347
5	5.3253968	14	22.1675957
6	6.9285714	15	24.2782575
7	8.6257576	16	26.4231256
8	10.3954727	17	28.6000215
9	12.2306693	18	30.8070265
10	14.1233619	19	33.0424380
11	16.0678397	20	35.3047349

which gives 35.65 for $c = 20$ and 2.50 for $c = 3$. The possibility that (1.20) is an upper bound for all c is consistent with our table above since the error is increasing slightly with c. On the other hand if we replace .57 in (1.20) by .50 the result appears to be a lower bound for all c. It is conjectured that the error in (1.20) will remain less than 1 at least up to $c = 50$.

Problem 1.4 Sliced Edges and Chopped Vertices Starting with an ordinary die with 6 sides, 12 edges and 8 vertices we modify it as follows. Cut a slice off each of the 12 edges so that the die can also stand (with equal probability) on any one of the 12 new edge-faces. At each of the 8 original vertices we chop off a piece so that the die can also stand (with equal probability) on any one of the 8 new vertex- faces. Let p_1, p_2 and p_3 denote respectively the new probabilities of each of the original faces, each of the 12 edge-faces, and each of the 8 vertex-faces. Under this symmetry the only algebraic restraint we have for the $p_i \geq 0$ $(i = 1, 2, 3)$ is

$$6p_1 + 12p_2 + 8p_3 = 1. \tag{1.21}$$

If the die lands on an original face we see only one number $j (1 \leq j \leq 6)$ as usual. If it lands on an edge-face, we see a pair of adjacent (or neighboring) sides of the original die. If it lands on a vertex-face, we see a neighboring triple, i.e., three sides of the original die that had a common vertex. Hence, we see i numbers with probability p_i $(i = 1, 2, 3)$, but only neighboring pairs and neighboring triples on the original die are possible. With the usual marking on a die (i.e., opposite sides adding to seven), the neighboring pairs exclude exactly those pairs that add to seven and the 8 neighboring triples exclude exactly those that contain a pair adding to seven. We make use of this geometric property to find the expected number of tosses $E(T)$ needed to see all six numbers, each at least once. The variance $\sigma^2(T)$ is also found as a special case of higher moments. It would be desirable to use a method that can also be applied to other regular or centrally symmetric polyhedra with faces having different shapes.

Solution:

The solution is based on a combination of conditioning and inclusion- exclusion which uses the geometry of the die. By virtue of the way dice are marked, the geometry consists of counting how many pairs of elements in a subset add to seven. Thus, the subsets $(6,4,3,1)$ and $(6,5,2,1)$ are equivalent but $(6,5,4,3)$ and $(6,4,3,1)$ are not.

We first condition on the event that exactly one number is not seen until the nth toss and all the others are covered at least once in the first $n-1$ tosses; this is multiplied by 6. In the second summation in (1.23) below we condition on the event that for one of the 12 neighboring pairs neither member is seen until the nth toss and all the others are covered at least once in the first $n-1$ tosses; this is multiplied by 12. In the third (and last) summation we do the analogous thing for each of the 8 neighboring triples and multiply the result by 8. The covering part is accomplished by inclusion-exclusion but in place of using $\binom{5}{1}$ for the number of sides that can be missed (when side #6 say is reserved for the nth toss), the pair $(6,1)$ has to be treated differently than the other four pairs. Similarly instead of $\binom{5}{2}$ we consider $4+6$ cases, etc. One definite advantage of our method is that we get in one calculation the γth ascending factorial moment for all $\gamma \geq 0$. The second factor outside each of the three summations is easily seen to be the probability of the required single, double and triple (respectively) needed on the nth set to terminate the tossing. It is also clear that the first two summations start with $n=3$ but the last one starts with $n=2$. However, corrections (after summing the infinite series) are needed only for $n=1$ due to the fact that we are omitting terms of the form 0^{n-1}, which equals one if and only if $n=1$ (and 0 otherwise). We thus obtain for any $\gamma \geq 0$

$$
\begin{aligned}
E(T^{[\gamma]}) = {} & 6(p_1+4p_2+4p_3)\sum_{n=3}^{\infty} n^{[\gamma]}(5p_1+8p_2+4p_3)^{n-1} \\
& \left\{1-4\left(\frac{4p_1+5p_2+2p_3}{5p_1+8p_2+4p_3}\right)^{n-1}\right. \\
& -\left(\frac{4p_1+4p_2}{5p_1+8p_2+4p_3}\right)^{n-1}+4\left(\frac{3p_1+3p_2+p_3}{5p_1+8p_2+4p_3}\right)^{n-1} \\
& +6\left(\frac{3p_1+2p_2}{5p_1+8p_2+4p_3}\right)^{n-1}-8\left(\frac{2p_1+p_2}{5p_1+8p_2+4p_3}\right)^{n-1} \\
& \left.-2\left(\frac{2p_1}{5p_1+ip_2+4p_3}\right)^{n-1}+5\left(\frac{p_1}{5p_1+8p_2+4p_3}\right)^{n-1}\right\} \\
& +12(p_2+2p_3)\sum_{n=3}^{\infty} n^{[\gamma]}(4p_1+5p_2+2p_3)^{n-1}
\end{aligned}
$$

$$
\begin{aligned}
&\cdot\left\{1-2\left(\frac{3p_1+3p_2+p_3}{4p_1+5p_2+2p_3}\right)^{n-1}\right.\\
&-2\left(\frac{3p_1+2p_2}{4p_1+5p_2+2p_3}\right)^{n-1}+5\left(\frac{2p_1+p_2}{4p_1+5p_2+2p_3}\right)^{n-1}\\
&\left.+\left(\frac{2p_1}{4p_1+5p_2+2p_3}\right)^{n-1}-4\left(\frac{p_1}{4p_1+5p_2+2p_3}\right)^{n-1}\right\}\\
&+8p_3\sum_{n=2}^{\infty} n^{[\gamma]}(3p_1+3p_2+p_3)^{n-1}\\
&\cdot\left\{1-\binom{3}{1}\left(\frac{2p_1+p_2}{3p_1+3p_2+p_3}\right)^{n-1}+\binom{3}{2}\left(\frac{p_1}{3p_1+3p_2+p_3}\right)^{n-1}\right\}.
\end{aligned}
\tag{1.22}
$$

For $\gamma = 1$, after summing each of the above series, combining like terms and correcting for $n = 1$ (since $n = 2$ needs no correction) we obtain

$$
\begin{aligned}
E(T) =\ &\frac{6(p_1+4p_2+4p_3)}{(1-5p_1-8p_2-4p_3)^2}-\frac{12(2p_1+7p_2+6p_3)}{(1-4p_1-5p_2-2p_3)^2}\\
&-\frac{6(p_1+4p_2+4p_3)}{(1-4p_1-4p_2)^2}+\frac{8(3p_1+9p_2+7p_3)}{(1-3p_1-3p_2-p_3)^2}\\
&+\frac{12(3p_1+10p_2+8p_3)}{(1-3p_1-2p_2)^2}-\frac{12(4p_1+11p_2+8p_3)}{(1-2p_1-p_2)^2}\\
&-\frac{12(p_1+3p_2+2p_3)}{(1-2p_1)^2}+\frac{6(5p_1+12p_2+8p_3)}{(1-p_1)^2}\\
&-2(3p_1+6p_2+4p_3),
\end{aligned}
\tag{1.23}
$$

where the last term is the correction for $n = 1$. For the $(\gamma+1)^{st}$ ascending factorial moment we simply replace in (1.23) all the squares (i.e., the exponents 2) by $\gamma + 1$ and multiply every term (including the correction) by γ! Hence, we need not rewrite the result (1.23) for general γ, although we use it below. This prescription also holds for $\gamma = 0$, and the answer must then be equal to 1, of course.

Consider the following eight different cases (or models) for the above.

$$
\begin{array}{lll}
\text{Case 1:} & p_1=p_2=p_3=1/26, & \\
\text{Case 2:} & 6p_1=12p_2=8p_3=1/3 & (\text{or } p_1+\frac{4}{72},\ p_1=\frac{2}{72},\ p_1=\frac{3}{72}),\\
\text{Case 3:} & p_1=0,\ 12p_2=8p_3=1/2 & (\text{or } p_1=0,\ p_2=\frac{2}{48},\ p_3=\frac{3}{48}),\\
\text{Case 4:} & p_2=0,\ 6p_1=8p_3=1/2 & (\text{or } p_1=\frac{4}{48},\ p_2=0,\ p_3=\frac{3}{48}),\\
\text{Case 5:} & p_3=0,\ 6p_1=12p_2=1/2 & (\text{or } p_1=\frac{2}{24},\ p_2=\frac{1}{24},\ p_3=0),\\
\text{Case 6:} & p_1=p_2=0,\ 8p_3=1, & \\
\text{Case 7:} & p_1=p_3=0,\ 12p_2=1, & \\
\text{Case 8:} & p_2=p_3=0,\ 6p_1=1; &
\end{array}
\tag{1.24}
$$

we calculate ET for each of these eight cases. [The reader may wish to use his intuition to rank the eight results before looking at the answers below.]

From (1.23) the expectation answers for these eight cases are

$$\begin{array}{llll} \text{Case 1:} & 6.42929 & \text{Case 5:} & 9.40574 \\ \text{Case 2:} & 6.66515 & \text{Case 6:} & 4.14286 \\ \text{Case 3:} & 5.21532 & \text{Case 7:} & 6.90433 \\ \text{Case 4:} & 6.52797 & \text{Case 8:} & 14.70000. \end{array} \tag{1.25}$$

The reader will of course recognize the last case as our usual die problem (waiting to see each side at least once) without any slicing or chopping; it serves as a standard to compare with the other cases.

For $\gamma = 2$ we again use (1.23) with the modification indicated above. For Case 1 we obtain for $\gamma = 2$ from (1.23) (with a factor of 2 and all exponents equal to 3)

$$E\{T(T+1)\} = 55.54164, \tag{1.26}$$

and hence for Case 1 we obtain the variance

$$\sigma^2(T) = 55.54164 - (6.42929)(7.42929) = 7.77658. \tag{1.27}$$

For Case 8 the same method yields 38.99 for the variance and this is a check since this is a known result, i.e., it has been obtained by other methods. The reader may wish to see if Case 6 with the smallest mean also has the smallest variance.

Problem 1.5 Do the faces, edges and vertices problems for the (regular) tetrahedron, i.e., find the expected number of tosses ET needed to see all the faces (resp., all the edges, resp., all the vertices) of the tetrahedron if on each toss we 'see' all elements associated with the face touching the ground.

Solution:

The faces problem has the usual analysis and answer in two different forms

$$E(T_f) = 16C_1^{(3)}(1,2) = 4\left(1 + \frac{1}{2} + \frac{1}{3} + \frac{1}{4}\right) = 8.33333, \tag{1.28}$$

and we omit the detailed derivation. For the edge problem we can only get one new edge on the nth (or last) toss. Hence we obtain (with $p = 1/4$)

$$P\{T_e = n\} = 6(2p)(1-2p)^{n-1} I^{(2)}_{\frac{p}{1-2p}}(1; n-1). \tag{1.29}$$

As a check we obtain by summing (1.29) on n

$$P\{T_e < \infty\} = 6C_{1/2}^{(2)}(1;1) = 6\Big/\binom{4}{2} = 1, \tag{1.30}$$

where the C-value can be obtained by using Vol. 9 (p. 106) or by a simple probability calculation using the probability interpretation of C in Vol. 9 (p. 13). If we multiply by n and then sum we obtain

$$E(T_e) = \frac{6}{2p} C_{1/2}^{(2)}(1;2) = 12\left(\frac{13}{36}\right) = 4.33333, \tag{1.31}$$

where the C-value is obtained from Vol. 9 (p. 106) or from Table 8 below.

For the vertex problem we can only get one new vertex on the nth toss and hence (with $p = 1/4$)

$$P\{T_v = n\} = 4(3p)(1-3p)^{n-1} I_{\frac{p}{1-3p}}^{(1)}(1; n-1). \tag{1.32}$$

The check, by summing on n, gives

$$P(T_v < \infty) = 4C_{1/2}^{(1)}(1;1) = 4\left(\frac{1}{4}\right) = 1. \tag{1.33}$$

If we multiply by n and then sum we obtain

$$E(T_v) = \frac{4}{3p} C_{1/3}^{(1)}(1;2) = \frac{16}{3}\left(\frac{7}{16}\right) = 2.33333, \tag{1.34}$$

where the C-value is obtained from Vol. 9 (p. 105) or from the exact C-table in the Appendix below.

Problem 1.6 Do the vertices problem for the cube.

Solution:

Another method was rejected in favor of the following analysis which emphasizes the common aspects of all the polyhedral problems. Consider the (at most) 4 disjoint events A_j where j is the number of new vertices seen on the nth (or last) toss. Let P_j (resp., T_j) denote the contribution for each j to $P\{T = n\}$ (resp., $E(T)$), so that $P\{T = n\}$ (resp., $E(T)$) is the sum of the four contributions. For $j = 1$ we have 8 possible vertices for the last one seen each associated with 3 faces and hence, letting $p = 1/6$,

$$P_1 = 8(3p)(1-3p)^{n-1} I_{\frac{p}{1-3p}}^{(3)}(1; n-1); \quad T_1 = \frac{8}{3p} C_{1/3}^{(3)}(1;2) = \frac{57}{25}, \tag{1.35}$$

using Vol. 9 (p. 105) or the exact C-table in the Appendix below. For $j = 2$ we have 12 possible edges each associated with 2 faces and hence

$$P_2 = 12(2p)(1-4p)^{n-1} I_{\frac{p}{1-4p}}^{(2)}(1; n-1); \quad T_2 = \frac{3}{2p} C_{1/4}^{(2)}(1;2) = \frac{37}{25}, \tag{1.36}$$

using Vol. 9 (p. 104) or the exact C-table in the Appendix. It should be noted that 2 opposite vertices (and also 3 new vertices) cannot occur for the first time on the nth toss, i.e., $P_3 = T_3 = 0$. Thus if the vertices on the last face seen are marked 1, 2, 3, 4 cyclically and if vertices 1, 2 and 3 were not seen on the first $n-1$ tosses then vertex 4 which has to be seen either with 1 or with 3 was also not seen on the first $n-1$ tosses. For $j=4$ we have 6 possible faces and hence

$$P_4 = 6(p)(1-5p)^{n-1} I^{(1)}_{\frac{p}{1-5p}}(1; n-1); \quad T_4 = \frac{6}{25p} C^{(1)}_{1/5}(1;2) = \frac{11}{25}. \quad (1.37)$$

It follows that

$$P\{T_v = n\} = P_1 + P_2 + P_4; \quad E(T_v) = \frac{57+37+11}{25} = \frac{21}{5} = 4.20000. \quad (1.38)$$

The check in this case gives

$$P\{T_v < \infty\} = 8C^{(3)}_{1/3}(1;1) + 6C^{(2)}_{1/4}(1;1) + \frac{6}{5}C^{(1)}_{1/5}(1;1) = \frac{8}{20} + \frac{6}{15} + \frac{6}{5}\left(\frac{1}{6}\right) = 1. \quad (1.39)$$

Problem 1.7 Do the edges problem for the octahedron.

Solution:

Consider the three disjoint (and exhaustive) events A_j ($j = 1, 2, 3$) where j is the number of new edges seen on the nth (or last) toss. For $j = 1$ two triangles are not seen on the first $n-1$ tosses but the four triangles surrounding these two are seen. In addition at least one of the 2 remaining triangles has to be seen in order to include the edge joining these two. Hence for $p = 1/8$

$$P_1 = 12(2p)(1-2p)^{n-1}\left[2I^{(5)}_{\frac{p}{1-2p}}(1; n-1) - I^{(6)}_{\frac{p}{1-2p}}(1; n-1)\right] \quad (1.40)$$

and multiplying by n and summing for $n \geq 6$

$$T_1 = \frac{12}{2p}[2C^{(5)}_{1/2}(1;2) - C^{(6)}_{1/2}(1;2)] = 48(.18069728) = 8.67346944. \quad (1.41)$$

For $j = 2$ we obtain by a straightforward similar analysis

$$P_2 = 24p(1-3)^{n-1} I^{(4)}_{\frac{p}{1-3p}}(1; n-1); \quad T_2 = \frac{8}{3p} C^{(4)}_{1/3}(1;2) = 1.99836736. \quad (1.42)$$

For $j = 3$ a similar analysis yields

$$P_3 = 8p(1-4p)^{n-1} I^{(3)}_{\frac{p}{1-4p}}(1; n-1); \quad T_3 = \frac{1}{2p} C^{(3)}_{1/4}(1;2) = .34721088. \quad (1.43)$$

The sum $P_1 + P_2 + P_3$ equals $P\{T_e = n\}$ and

$$E(T_e) = T_1 + T_2 + T_3 = 11.01904768. \quad (1.44)$$

The check for this problem is

$$P\{T_e < \infty\} = 12[2C_{1/2}^{(5)}(1;1) - C_{1/2}^{(6)}(1;1)] + 8C_{1/3}^{(4)}(1;1) + 2C_{1/4}^{(3)}(1;1)$$
$$= 12\left(\frac{2}{21} - \frac{1}{28}\right) + \frac{8}{35} + \frac{2}{35} = \frac{12}{7}\left(\frac{5}{12}\right) + \frac{2}{7} = 1. \quad (1.45)$$

Although we carried out the computation in (1.41) through (1.44) with decimals from Vol. 9, an alternate method would be to use the entries in the exact C-table and this furnishes a rational answer, which will agree with the above.

Problem 1.8 Do the edges problem for the dodecahedron which, according to Table 1, has 30 edges.

Solution:

Consider the disjoint (and exhaustive) events A_j where j is an index for the number and relative position of the new edges seen on the nth (or last) toss. Here we need at most 7 cases since for $j = 2$ and $j = 3$ we may want to consider whether or not the edges are connected; surprisingly it turns out that we only need 2 cases. For $j = 1$, since there are 30 edges and $p = 1/12$,

$$P_1 = 30(2p)\left[(1-2p)^{n-1} I_{\frac{p}{1-2p}}^{(8)}(1;n-1) + 2(1-3p)^{n-1} I_{\frac{p}{1-3p}}^{(9)}(1;n-1)\right]$$
$$= 30(2p)(1-2p)^{n-1}$$
$$\times \left[I_{\frac{p}{1-2p}}^{(8)}(1;n-1) + 2I_{\frac{p}{1-2p}}^{(9)}(1;n-1) - 2I_{\frac{p}{1-2p}}^{(10)}(1;n-1)\right]; \quad (1.46)$$

here we gave two equivalent forms, the latter of which will give us C-functions with the same subscript. The derivation of (1.46) follows from the fact that 2 pentagons are not seen on the first $n-1$ tosses and the 6 pentagons surrounding these two are seen. The remaining 4 pentagons are connected and the reader can then follow the argument with a simple sketch.

For $j = 2$ suppose 2 edges both seen for the first time on the nth (or last) toss are connected. This cannot occur since the third edge at the common vertex of these two cannot be seen (on the first $n-1$ tosses) without one of these two. Similarly the only other case is Case 2B where 2 edges on the same pentagon have one space between them. There are 60 such pairs and each implies that 3 connected pentagons are not seen on the first $n-1$ tosses and that 7 pentagons surrounding these are seen on the first $n-1$ tosses. The remaining 2 pentagons have 1 common edge and we see it by getting either one of these 2 pentagons or both. Hence for Case 2B

$$P_{2B} = 60(p)\left[(1-3p)^{n-1} I_{\frac{p}{1-3p}}^{(9)}(1;n-1) + 2(1-4p)^{n-1} I_{\frac{p}{1-4p}}^{(8)}(1;n-1)\right]$$
$$= 60(p)(1-3p)^{n-1}\left[2I_{\frac{p}{1-3p}}^{(8)}(1;n-1) - I_{\frac{p}{1-3p}}^{(9)}(1;n-1)\right]. \quad (1.47)$$

From the second expression in (1.46) we obtain

$$T_1 = \frac{15}{p}[C_{1/2}^{(8)}(1;2) + 2C_{1/2}^{(9)}(1;2) - 2C_{1/2}^{(10)}(1;2)] \tag{1.48}$$

and from the second expression in (1.47) we obtain

$$T_{2B} = \frac{20}{3p}[2C_{1/3}^{(8)}(1;2) - C_{1/3}^{(9)}(1;2)]. \tag{1.49}$$

As mentioned above the cases for $j = 3, 4$ and 5 cannot occur; the argument is the same as above for Case 2A. Hence

$$\begin{aligned} P\{T = n\} &= P_1 + P_{2B} \\ P\{\text{Case } 1\} &= 30[C_{1/2}^{(8)}(1;1) + 2C_{1/2}^{(9)}(1;1) - 2C_{1/2}^{(10)}(1;1)] \\ &= \frac{28}{33}; \qquad (1.50) \\ P\{\text{Case } 2\} &= 20[2C_{1/3}^{(8)}(1;1) - C_{1/3}^{(9)}(1;1)] = \frac{5}{33}, \qquad (1.51) \end{aligned}$$

and the check lies in the fact that the latter two add to one. If we multiply by n and then sum we obtain (with $p = 1/12$)

$$\begin{aligned} E(T_e) &= \frac{15}{p}[C_{1/2}^{(8)}(8)(1;2) + 2C_{1/2}^{(9)} - 2C_{1/2}^{(10)}] + \frac{20}{3p}[2C_{1/3}^{(8)}(1;2) - C_{1/2}^{(9)}] \\ &= 180\left[\left(\frac{4861}{56700}\right) + 2\left(\frac{55991}{762300}\right) - 2\left(\frac{58301}{914760}\right)\right] \\ &\quad + 80\left[2\left(\frac{42131}{1524600}\right) - \left(\frac{44441}{2032800}\right)\right] \\ &= \frac{1646733}{33(2310)} = \frac{49901}{2310} = 21.602165. \qquad (1.52) \end{aligned}$$

The use of the first expressions in (1.46) and (1.47) gives exactly the same result. In distinction to the method used for problem 1.7 where we used decimal expressions from Vol. 9, we have used here only C-values from the exact C-table in the Appendix below to illustrate the two different methodologies; these methods are of course interchangeable.

Consider the generalized die of Problem 1.4 with $p_2 = 0$ so that

$$6p_1 + 8p_3 = 1, \tag{1.53}$$

i.e., we have 6 squares (resp., 8 triangles) with common probability p_1 (resp., p_3). Among geometers the figure in question is known as a cuboctahedron and is classified as quasi-regular (cf. Coxeter, 1973, p. 18). [We prefer to work with p_1 and p_3 subject to (1.53) without specifying them since they may depend on factors other than the geometry of the polyhedron (such as the substance it is made of or the manner of tossing or etc.).]

Cases 4 and 6 of Problem 2 both pertain to this polyhedron, which has 24 edges if $p_3 > 0$ and is a cube with 12 edges if $p_3 = 0$. Since we cannot make a continuous change from the cuboctahedron to the cube by letting $p_3 \to 0$ without major changes in the geometry, we should not expect the answer to an edge problem for the former to necessarily yield the correct answer for the latter by simply setting $p_3 = 0$. We now investigate this point numerically in detail.

Problems 1.9 (A and B) Do the edges problem a) for the cuboctahedron with 24 edges and b) also for the cube with 12 edges. Check whether the answer to a) with $p_3 = 0$ agrees with the answer in b) or if it agrees with the 'magic' answer 14.7 for the faces of a die. If the latter, give an explanation.

Solution:

We consider 7 disjoint (and exhaustive) ways of termination which we call TT_j ($j = 1, 2, \ldots, 7$):

TT_1: with one new edge,

TT_2: with a pair of parallel new edges on the same square,

TT_3: with a connected pair of new edges on the same square,

TT_4: with 3 new edges on the same square,

TT_5: with 4 new edges on the same square,

TT_6: with 2 new edges on the same triangle,

TT_7: with 3 new edges on the same triangle.

For TT_1 if any one of the 24 edges is not seen until the nth (or last) toss then the associated square and triangle were not seen on the first $n-1$ tosses. This implies that 2 other square and 3 other triangles surrounding this pair were seen in the first $n-1$ tosses; this accounts for a total of 3 squares and 4 triangles. The remaining 3 squares and 4 triangles are sketched in Fig. 1, where arrows point to identifiable sides and S denotes edges included in the previous discussion.

There are eight edges not yet accounted for; to see all of these consider 8 disjoint subcases:

Subcase 1:	S_0, S_1, S_2	Subcase 5:	$S_0, \bar{S}_1, \bar{S}_2, T_0, T_1, T_3$
Subcase 2:	$\bar{S}_0, S_1, S_2, T_0, T_1, T_2, T_3$	Subcase 6:	$\bar{S}_0, S_1, \bar{S}_2, T_0, T_1, T_2, T_3$
Subcase 3:	$S_0, S_1, \bar{S}_2, T_0, T_3$	Subcase 7:	$\bar{S}_0, \bar{S}_1, S_2, T_0, T_1, T_2, T_3$
Subcase 4:	$S_0, \bar{S}_1, S_2, T_0, T_1$	Subcase 8:	$\bar{S}_0, \bar{S}_1, \bar{S}_2, T_0, T_1, T_2, T_3$

Here $\bar{S}$ denotes the absence of S in the first $n-1$ tosses. For TT_2 two triangles and one square are not seen until the nth (or last) toss and hence

FIGURE 1. Backside of the Cuboctahedron

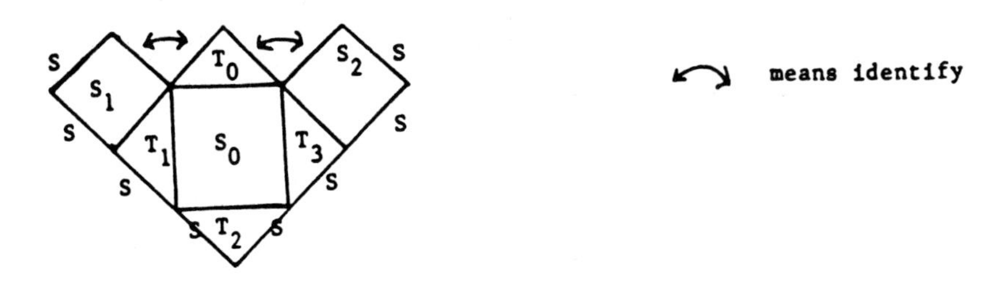

the 4 square and 2 triangles surrounding these are seen in the first $n-1$ tosses. We consider two disjoint (and exhaustive) subcases:

Subcase 1. The remaining square S_6 is seen among the first $n-1$ tosses.

Subcase 2. S_6 is not seen but the 4 remaining triangles are all seen in the first $n-1$ tosses.

For TT_3 two triangles and a square are not seen until the nth (or last) toss; it follows that 2 other triangles and 3 other squares are seen on the first $n-1$ tosses. Then we consider four disjoint (and exhaustive) subcases (cf. Fig. 2) where S indicates edges included in the previous discussion.

Subcase 1:	S_0, S_1	Subcase 3:	S_0, $\bar{S}_1$, T_1, T_2
Subcase 2:	$\bar{S}_0$, S_1, T_1, T_2, T_3, T_4	Subcase 4:	$\bar{S}_0$, $\bar{S}_1$, T_1, T_2, T_3, T_4.

For TT_4 we need two subcases as in TT_2 above.

For TT_5 we again need two subcases as in TT_2 above.

For TT_6 we use four subcases and Fig. 3.

FIGURE 2. Case TT_3 for the edges problem of the cuboctahedron

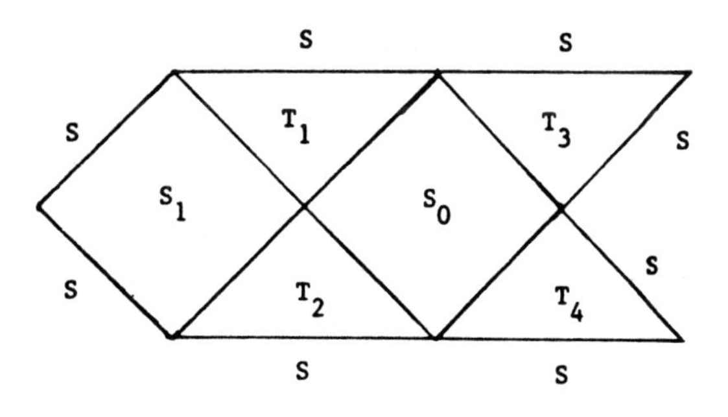

FIGURE 3. Case TT_6 for the edges problem of the cuboctahedron

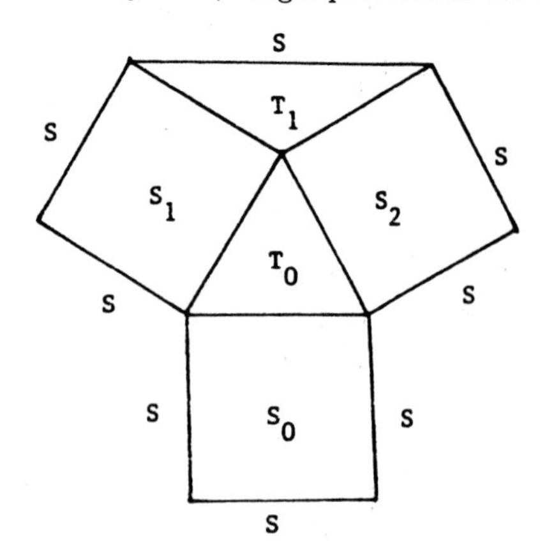

Subcase 1:	T_0, T_1	Subcase 3:	T_0, $\bar{T}_1$, S_1, S_2
Subcase 2:	$\bar{T}_0$, T_1, S_0, S_1, S_2	Subcase 4:	$\bar{T}_0$, $\bar{T}_1$, S_0, S_1, S_2

Finally for TT_7 we have two disjoint (and exhaustive) subcases:

Subcase 1. The one remaining triangle is seen in the first $n-1$ tosses.

Subcase 2. The one remaining triangle is not seen but all 3 remaining squares are seen in the first $n-1$ tosses.

Let P_i (resp., T_i) ($i = 1, 2, \ldots, 7$) denote the contribution of TT_i to $P\{T = n\}$ (resp., to $E_e(T)$). Then

$$\begin{aligned}
P_1 = 24(p_1+p_3)\Big[&(1-p_1-p_3)^{n-1} I^{(5,3)}_{\left(\frac{p_1}{1-p_1-p_3}, \frac{p_3}{1-p_1-p_3}\right)}(1,1;n-1) \\
&+ (1-2p_1-p_3)^{n-1} I^{(4,7)}_{\left(\frac{p_1}{1-2p_1-p_3}, \frac{p_3}{1-2p_1-p_3}\right)} \\
&+ 2(1-2p_1-p_3)^{n-1} I^{(4,5)}_{\left(\frac{p_1}{1-2p_1-p_3}, \frac{p_3}{1-2p_1-p_3}\right)} \\
&+ (1-3p_1-p_3)^{n-1} I^{(3,6)}_{\left(\frac{p_1}{1-3p_1-p_3}, \frac{p_3}{1-3p_1-p_3}\right)} \\
&+ 2(1-3p_1-p_3)^{n-1} I^{(3,7)}_{\left(\frac{p_1}{1-3p_1-p_3}, \frac{p_3}{1-3p_1-p_3}\right)} \\
&+ (1-4p_1-p_3)^{n-1} I^{(2,7)}_{\left(\frac{p_1}{1-4p_1-p_3}, \frac{p_3}{1-4p_1-p_3}\right)}\Big],
\end{aligned} \tag{1.54}$$

$$\begin{aligned}
P_2 = 12p_1\Big[&(1-p_1-2p_3)^{n-1} I^{(5,2)}_{\left(\frac{p_1}{1-p_1-2p_3}, \frac{p_3}{1-p_1-2p_3}\right)}(1,1;n-1) \\
&+ (1-2p_1-2p_3)^{n-1} I^{(4,6)}_{\left(\frac{p_1}{1-2p_1-2p_3}, \frac{p_3}{1-2p_1-2p_3}\right)}\Big],
\end{aligned} \tag{1.55}$$

$$P_3 = 24p_1\Big[(1-p_1-2p_3)^{n-1}I^{(5,2)}_{\left(\frac{p_1}{1-p_1-2p_3},\frac{p_3}{1-p_1-2p_3}\right)}(1,1;n-1)$$
$$+(1-2p_1-2p_3)^{n-1}I^{(4,6)}_{\left(\frac{p_1}{1-2p_1-2p_3},\frac{p_3}{1-2p_1-2p_3}\right)}$$
$$+(1-2p_1-2p_3)^{n-1}I^{(4,4)}_{\left(\frac{p_1}{1-2p_1-2p_3},\frac{p_3}{1-2p_1-2p_3}\right)}$$
$$+(1-3p_1-2p_3)^{n-1}I^{(3,6)}_{\left(\frac{p_1}{1-3p_1-2p_3},\frac{p_3}{1-3p_1-2p_3}\right)}\Big], \quad (1.56)$$

$$P_4 = 24p_1\Big[(1-p_1-3p_3)^{n-1}I^{(5,1)}_{\left(\frac{p_1}{1-p_1-3p_3},\frac{p_3}{1-p_1-3p_3}\right)}(1,1;n-1)$$
$$+(1-2p_1-3p_3)^{n-1}I^{(4,5)}_{\left(\frac{p_1}{1-2p_1-3p_3},\frac{p_3}{1-2p_1-3p_3}\right)}\Big], \quad (1.57)$$

$$P_5 = 6p_1\Big[(1-p_1-4p_3)^{n-1}I^{(5)}_{\frac{p_1}{1-p_1-4p_3}}(1;n-1)$$
$$+(1-2p_1-4p_3)^{n-1}I^{(4,4)}_{\left(\frac{p_1}{1-2p_1-4p_3},\frac{p_3}{1-2p_1-4p_3}\right)}(1,1;n-1)\Big], \quad (1.58)$$

$$P_6 = 24p_3\Big[(1-2p_1-p_3)^{n-1}I^{(1,7)}_{\left(\frac{p_1}{1-2p_1-p_3},\frac{p_3}{1-2p_1-p_3}\right)}(1,1;n-1)$$
$$+(1-2p_1-2p_3)^{n-1}I^{(4,6)}_{\left(\frac{p_1}{1-2p_1-2p_3},\frac{p_3}{1-2p_1-2p_3}\right)}$$
$$+(1-2p_1-2p_3)^{n-1}I^{(3,6)}_{\left(\frac{p_1}{1-2p_1-2p_3},\frac{p_3}{1-2p_1-2p_3}\right)}$$
$$+(1-2p_1-3p_3)^{n-1}I^{(4,5)}_{\left(\frac{p_1}{1-2p_1-3p_3},\frac{p_3}{1-2p_2-3p_3}\right)}(1,1;n-1)\Big], \quad (1.59)$$

$$P_7 = 8p_3\Big[(1-3p_1-p_3)^{n-1}I^{(7)}_{\frac{p_3}{1-3p_1-p_3}}(1;n-1)$$
$$+(1-3p_1-2p_3)^{n-1}I^{(3,6)}_{\left(\frac{p_1}{1-3p_1-2p_3},\frac{p_3}{1-3p_1-2p_3}\right)}(1,1;n-1)\Big]. \quad (1.60)$$

Letting $D_{ij} = ip_1 + jp_3$, the corresponding T-values are

$$T_1 = 24(p_1+p_3)\Big[\frac{1}{D_{11}^2}C^{(5,3)}_{\left(\frac{p_1}{D_{11}},\frac{p_3}{D_{11}}\right)}(1,1;2) + \frac{1}{D_{21}^2}C^{(4,7)}_{\left(\frac{p_1}{D_{21}},\frac{p_3}{D_{21}}\right)} + \frac{2}{D_{21}^2}C^{(4,5)}_{\frac{p_1}{D_{21}},\frac{p_3}{D_{21}}}$$
$$+\frac{1}{D_{31}^2}C^{(3,6)}_{\left(\frac{p_1}{D_{31}},\frac{p_3}{D_{31}}\right)} + \frac{2}{D_{31}^2}C^{(3,7)}_{\left(\frac{p_1}{D_{31}},\frac{p_3}{D_{31}}\right)} + \frac{1}{D_{41}^2}C^{(2,7)}_{\left(\frac{p_1}{D_{41}},\frac{p_3}{D_{41}}\right)}\Big], \quad (1.61)$$

$$T_2 = 12p_1\left[\frac{1}{D_{12}^2}C^{(5,2)}_{\left(\frac{p_1}{D_{12}},\frac{p_3}{D_{12}}\right)}(1,1;2) + \frac{1}{D_{22}^2}C^{(4,6)}_{\left(\frac{p_1}{D_{22}},\frac{p_3}{D_{22}}\right)}\right], \quad (1.62)$$

$$T_3 = 24p_1\Bigg[\frac{1}{D_{12}^2}C^{(5,2)}_{\left(\frac{p_1}{D_{12}},\frac{p_3}{D_{12}}\right)}(1,1;2) + \frac{1}{D_{22}^2}C^{(4,6)}_{\left(\frac{p_1}{D_{22}},\frac{p_3}{D_{22}}\right)}$$
$$+\frac{1}{D_{22}^2}C^{(4,4)}_{\left(\frac{p_1}{D_{22}},\frac{p_3}{D_{22}}\right)} + \frac{1}{D_{32}^2}C^{(3,6)}_{\left(\frac{p_1}{D_{32}},\frac{p_3}{D_{32}}\right)}\Bigg], \quad (1.63)$$

$$T_4 = 24p_1\left[\frac{1}{D_{13}^2}C^{(5,1)}_{\left(\frac{p_1}{D_{13}},\frac{p_3}{D_{13}}\right)}(1,1;2) + \frac{1}{D_{23}^2}C^{(4,5)}_{\left(\frac{p_1}{D_{23}},\frac{p_3}{D_{23}}\right)}\right], \tag{1.64}$$

$$T_5 = 6p_1\left[\frac{1}{D_{14}^2}C^{(5)}_{\frac{p_1}{D_{14}}}(1;2) + \frac{1}{D_{24}^2}C^{(4,4)}_{\left(\frac{p_1}{D_{24}},\frac{p_3}{D_{24}}\right)}(1,1;2)\right], \tag{1.65}$$

$$T_6 = 24p_3\left[\frac{1}{D_{21}^2}C^{(1,7)}_{\left(\frac{p_1}{D_{21}},\frac{p_3}{D_{21}}\right)}(1,1;2) + \frac{1}{D_{22}^2}C^{(4,6)}_{\left(\frac{p_1}{D_{22}},\frac{p_3}{D_{22}}\right)}\right.$$
$$\left. + \frac{1}{D_{22}^2}C^{(3,6)}_{\left(\frac{p_1}{D_{22}},\frac{p_3}{D_{22}}\right)} + \frac{1}{D_{22}^2}C^{(4,5)}_{\left(\frac{p_1}{D_{23}},\frac{p_3}{D_{23}}\right)}\right], \tag{1.66}$$

$$T_7 = 8p_3\left[\frac{1}{D_{31}^2}C^{(7)}_{\frac{p_3}{D_{31}}}(1;2) + \frac{1}{D_{32}^2}C^{(3,6)}_{\left(\frac{p_1}{D_{32}},\frac{p_3}{D_{32}}\right)}(1,1;2)\right]. \tag{1.67}$$

The sum of $T_1, T_2, \ldots, T_7$ is the desired expectation $E_e(T)$. For $p_1 = p_3 = 1/14$ the result is

$$\begin{aligned}E_e(T) = 168\Big[&C^{(8)}_{1/2}(1;2) + \frac{4}{9}C^{(11)}_{1/3} + \frac{8}{9}C^{(9)}_{1/3} + \frac{1}{4}C^{(9)}_{1/4} + \frac{1}{2}C^{(10)}_{1/4}\\
&+ \frac{4}{25}C^{(9)}_{1/5} + \frac{1}{9}C^{(7)}_{1/3} + \frac{1}{16}C^{(10)}_{1/4} + \frac{2}{9}C^{(7)}_{1/3} + \frac{1}{8}C^{(10)}_{1/4} + \frac{1}{8}C^{(8)}_{1/4}\\
&+ \frac{2}{25}C^{(9)}_{1/5} + \frac{1}{8}C^{(6)}_{1/4} + \frac{2}{25}C^{(9)}_{1/5} + \frac{1}{50}C^{(5)}_{1/5} + \frac{1}{72}C^{(8)}_{1/6} + \frac{2}{9}C^{(8)}_{1/3}\\
&+ \frac{1}{8}C^{(10)}_{1/4} + \frac{1}{8}C^{(9)}_{1/4} + \frac{2}{25}C^{(9)}_{1/5} + \frac{1}{24}C^{(7)}_{1/4} + \frac{2}{75}C^{(9)}_{1/5}\Big]\\
= 168&(.15651047) = 26.29376,\end{aligned} \tag{1.68}$$

where the C-values are all obtained from Vol. 9 except for one entry that appears in the exact C-table in the Appendix.

As a partial check we calculate the probabilities for each of the seven cases and see if they add to one, but this is done only for the special case $p_1 = p_3 = 1/14$. From (1.54) we obtain

$$\begin{aligned}P(\text{Case 1}) &= 24C^{(8)}_{1/2}(1;1) + 16C^{(11)}_{1/3} + 32C^{(9)}_{1/3} + 12C^{(9)}_{1/4} + 24C^{(10)}_{1/4} + \frac{48}{5}C^{(9)}_{1/5}\\
&= \frac{23072}{30030} = .76830,\end{aligned} \tag{1.69}$$

$$P(\text{Case 2}) = 4C^{(7)}_{1/3}(1;1) + 3C^{(10)}_{1/4} = \frac{4}{\binom{10}{3}} + \frac{3}{\binom{14}{4}} = \frac{1091}{30030} = .03633, \tag{1.70}$$

$$P(\text{Case 3}) = 8C^{(7)}_{1/3}(1;1) + 6C^{(8)}_{1/4} + 6C^{(10)}_{1/4} + \frac{24}{5}C^{(9)}_{1/5} = \frac{2618}{30030} = .08718, \tag{1.71}$$

$$P(\text{Case 4}) = 6C^{(6)}_{1/4}(1;1) + \frac{24}{5}C^{(9)}_{1/5} = \frac{6}{\binom{10}{4}} + \frac{24}{5\binom{14}{5}} = \frac{930}{30030} = .03097, \tag{1.72}$$

$$P(\text{Case } 5) = \frac{6}{5}C_{\frac{1}{5}}^{(5)}(1;1) + C_{1/6}^{(8)} = \frac{6}{5\binom{10}{5}} + \frac{1}{\binom{14}{6}} = \frac{153}{30030} = .00509, \quad (1.73)$$

$$P(\text{Case } 6) = 8C_{1/3}^{(8)}(1;1) + 6C_{1/4}^{(9)} + 6C_{1/4}^{(10)} + \frac{24}{5}C_{1/5}^{(9)} = \frac{1960}{30030} = .06527, \quad (1.74)$$

$$P(\text{Case } 7) = 2C_{1/4}^{(7)}(1;1) + \frac{8}{5}C_{1/5}^{(9)} = \frac{2}{\binom{11}{4}} + \frac{8}{5\binom{14}{5}} = \frac{206}{30030} = .00686, \quad (1.75)$$

and the sum of these is exactly one (which is our check).

If we set $p_3 = 0$ (and hence $p_1 = 1/16$ by (1.53)), then only one non-zero term appears and the result is

$$E\{T_e | p_3 = 0\} = 36C_1^{(5)}(1;2) = 14.70000. \quad (1.76)$$

This is not the answer for the edges problem on the cube. However suppose that the edges of the cube are directed (as in the graph theory), marked differently for each direction and that they are assigned one to each of the two faces associated with a given edge of the cube. Then the problem of seeing all the directed edges is equivalent to the problem of seeing all the faces of the cube and this has expectation 14.7. Since the six squares of the cuboctahedron have no common edges, it is reasonable that this should occur as $p_3 \to 0$.

For part b) dealing with the cube (which we will call Problem 1.9B) consider the disjoint (and exhaustive) events A_j where j is the number of new edges seen on the nth (or last) toss. For $j = 2$ only pairs of parallel edges are possible and $j = 3$ and 4 are also not possible. Hence we have only 2 cases to consider which we denote by 1 and 2B. Then (with $p = 1/6$)

$$P_1 = 12(2p)(1-2p)^{n-1} I_{\frac{p}{1-2p}}^{(4)}(1; n-1); \quad P_{2B} = 12p(1-3p)^{n-1} I_{\frac{p}{1-3p}}^{(3)}(1; n-1) \quad (1.77)$$

and $P(T = n)$ is the sum of these two. Hence

$$E(T_e) = \frac{6}{p}C_{1/2}^{(4)}(1;2) + \frac{4}{3p}C_{1/3}^{(3)}(1;2) = \frac{81}{10} = 8.10000. \quad (1.78)$$

In particular the result 14.7 is not the answer to the edges problem for the cube. The check for the latter computation gives

$$P(T < \infty) = 12C_{1/2}^{(4)}(1;1) + 4C_{1/3}^{(3)}(1;1) = \frac{12}{15} + \frac{4}{20} = 1 \quad (1.79)$$

and the probabilities of Cases 1 and 2B are $\frac{4}{5}$ and $\frac{1}{5}$, respectively.

Problem 1.10 Do the faces problem for the cuboctahedron, i.e., using the polyhedron in problem 1.4 with $p_2 = 0$, common p_1 for the 6 squares and common p_3 for the 8 triangles find the expected number of tosses $E(T_f)$ needed to see all the 14 faces.

Solution:

Consider 2 disjoint (and exhaustive) cases according to whether the last face seen is a square or a triangle. Then we obtain

$$P\{T=n\} = 6p_1(1-p_1)^{n-1} I^{(5,8)}_{\left(\frac{p_1}{1-p_1},\frac{p_3}{1-p_1}\right)}(1,1;n-1) + 8p_3(1-p_3)^{n-1} I^{(6,7)}_{\left(\frac{p_1}{1-p_3},\frac{p_3}{1-p_3}\right)}(1,1;n-1). \tag{1.80}$$

The check yields an identity for any p_1, p_3 with $6p_1 + 8p_3 = 1$, namely

$$6C^{(5,8)}_{1,p_3/p_1}(1,1;1) + 8C^{(6,7)}_{\left(\frac{p_1}{p_3},1\right)}(1,1;1) = 1, \tag{1.81}$$

whose proof follows from the probability interpretation of C. Multiplying (1.80) by n and summing for $n \geq 3$, the general result is

$$E(T_f) = \frac{6}{p_1} C^{(5,8)}_{1,p_3/p_1}(1,1;2) + \frac{8}{p_3} C^{(6,7)}_{\left(\frac{p_1}{p_3},1\right)}(1,1;2). \tag{1.82}$$

For $p_1 = p_3 = 1/14$ we obtain

$$E(T_f) = 196C^{(13)}_1(1;2) = 196(.23225445) = 45.52187, \tag{1.83}$$

which is exactly equal to $14\sum_{j=1}^{14} 1/j$. For $6p_1 = 8p_3 = 1/2$

$$\begin{aligned} E(T_f) &= 72C^{(5,8)}_{1,3/2}(1,1;2) + 128C^{(6,7)}_{(2/3,1)}(1,1;2) \\ &= 72\sum_{\beta=0}^{5}\sum_{\alpha=0}^{8}(-1)^{\alpha+\beta}\frac{\binom{8}{\alpha}\binom{5}{\beta}}{(1+\frac{3\alpha}{2}+\beta)^2} \\ &\quad + 128\sum_{\beta=0}^{7}\sum_{\alpha=0}^{6}(-1)^{\alpha+\beta}\frac{\binom{6}{\alpha}\binom{7}{\beta}}{(1+\frac{2}{3}\alpha+\beta)^2} \\ &= 72(.32153771) + 128(.15054223) \\ &= 42.42012, \end{aligned} \tag{1.84}$$

where the double sums were done on a computer by Jer-Yan Lin of University of California at Santa Barbara. It is interesting that the numerical answer in (1.84) is smaller than in (1.83) and this raises the question of finding a minimum, which we do not consider here.

Problem 1.11 Do the vertices problem for the dodecahedron, i.e., assuming that on each toss we see the five vertices on the top face (and only those),find the expected number of tosses $E(T)$ to see all twenty vertices of the dodecahedron. Whatever disjoint sets of cases are considered, find the probability of each set and check to see if they add to unity.

Solution:

In case j we see exactly j new vertices on the nth (or last) toss ($j = 1$, 2, 3, 4, 5). For $j > 1$ if these vertices are not neighbors (i.e., connected by an edge) then the case is not possible. Hence the case $j = 4$ does not occur and there is only one (type of) case for $j = 2$ and 3, i.e., we have a total of four (types of) cases, which we still call 1, 2, 3 and 5.

For $j = 1$ there are three pentagons that we do not see on the first $n - 1$ tosses and (cf. Fig. 4) hence there are three other pentagons that we must see on the first $n - 1$ tosses. This accounts for sixteen vertices. The back piece in Fig. 4 shows the four remaining vertices that have to be included in the first $n - 1$ tosses. The latter is accomplished by including (at least two Z's) or (exactly one Z and the opposite W).

FIGURE 4. Front piece and back piece of the dodecahedron for the case $j = 1$

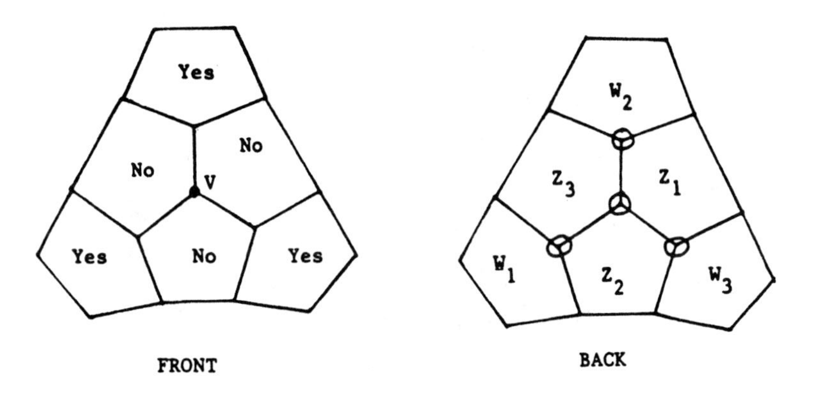

As before P_j (resp., E_j) is the contribution to $P(N = n)$ (resp., to ET) from Case j and P_j^* is the probability of Case j ($j = 1$, 2, 3, 5). Thus we have for Case 1, letting p denote 1/12,

$$P_1 = 20(3p)(1-3p)^{n-1}\left[3I^{(5)}_{\frac{p}{1-3p}}(1;n-1) - 2I^{(6)}_{\frac{p}{1-3p}}(1;n-1)\right] \\ + 20(3p)(1-5p)^{n-1}3I^{(5)}_{\frac{p}{1-5p}}(1;n-1), \tag{1.85}$$

$$P_1^* = 60C^{(5)}_{1/3}(1,1) - 40C^{(6)}_{1/3}(1,1) + 36C^{(5)}_{1/5}(1,1), \tag{1.86}$$

$$E_1 = \frac{20}{p}C^{(5)}_{1/3}(1;2) - \frac{40}{3p}C^{(6)}_{1/3}(1;2) + \frac{36}{5p}C^{(5)}_{1/5}(1;2). \tag{1.87}$$

For Case 2 the two vertices must be neighbors and there are four pentagons (marked No in Fig. 5) that we don't see on the first $n - 1$ tosses and another four pentagons (marked Yes) that we do see. This accounts for eighteen vertices and the remaining two are circled in Fig. 5 below. The latter two are included by including at least one Z pentagon or both W pentagons.

Hence we have for the Case $j = 2$

$$P_2 = 30(2p)\Big[(1-4p)^{n-1} I^{(6)}_{\frac{p}{1-4p}}(1; n-1) + 2(1-5p)^{n-1} I^{(5)}_{\frac{p}{1-5p}}(1; n-1) + (1-6p)^{n-1} I^{(6)}_{\frac{p}{1-6p}}(1; n-1)\Big], \tag{1.88}$$

$$P_2^* = 15C^{(6)}_{1/4}(1;1) + 24C^{(5)}_{1/5}(1,1) + 10C^{(6)}_{1/6}(1,1), \tag{1.89}$$

$$E_2 = \frac{15}{4p} C^{(6)}_{1/4}(1;2) + \frac{24}{5p} C^{(5)}_{1/5}(1;2) + \frac{5}{3p} C^{(6)}_{1/6}(1;2). \tag{1.90}$$

FIGURE 5. The case $j = 2$ for the dodecahedron vertices problem (connecting the front and back sides)

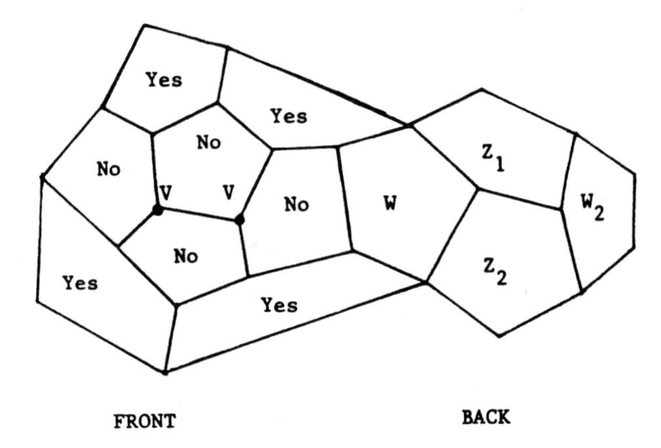

The reader should have little difficulty sketching the case $j = 3$ where we have three neighboring vertices that are not seen until the nth (or last) toss. Then five pentagons are not seen on the first $n-1$ tosses and four others are. This leaves 3 more pentagons which surround only one vertex not yet accounted for. We accomplish this by including at least one Z. Hence for $j = 3$ we have

$$P_3 = 60(p)(1-5p)^{n-1} \left[3I^{(5)}_{\frac{p}{1-5p}}(1; n-1) - 3I^{(6)}_{\frac{p}{1-5p}}(1; n-1) + I^{(7)}_{\frac{p}{1-5p}}(1; n-1)\right], \tag{1.91}$$

$$P_3^* = 12\left[3C^{(5)}_{1/5} - 3C^{(6)}_{1/5}(1;1) + C^{(7)}_{1/5}(1;1)\right], \tag{1.92}$$

$$E_3 = \frac{12}{5p}\left[3C^{(5)}_{1/5}(1;2) - 3C^{(6)}_{1/5}(1;2) + C^{(7)}_{1/5}(1;2)\right]. \tag{1.93}$$

For $j = 4$ we already noted that $P_4 = P_4^* = E_4 = 0$. For $j = 5$ we obtain similarly

$$P_5 = 12p(1-6p)^{n-1} I^{(5)}_{\frac{p}{1-6p}}(1; n-1), \tag{1.94}$$

$$P_5^* = 2C^{(5)}_{1/6}(1;1), \tag{1.95}$$

$$E_5 = \frac{1}{3p} C^{(5)}_{1/6}(1;2). \tag{1.96}$$

The sum of the P_j is $P\{N = n\}$. The sum of the P_j^* is

$$\begin{aligned}
&\left[\frac{60}{\binom{8}{3}} - \frac{40}{\binom{9}{3}} + \frac{36}{\binom{10}{5}}\right] + \left[\frac{15}{\binom{10}{4}} + \frac{24}{\binom{10}{5}} + \frac{10}{\binom{12}{6}}\right] \\
&\quad + \left[\frac{36}{\binom{10}{5}} - \frac{36}{\binom{11}{5}} + \frac{12}{\binom{12}{5}}\right] + [0] + \left[\frac{2}{\binom{11}{5}}\right] \\
&\quad = \left[\frac{31}{42}\right] + \left[\frac{41}{231}\right] + \left[\frac{37}{462}\right] + [0] + \left[\frac{1}{231}\right] = 1,
\end{aligned} \tag{1.97}$$

where the jth bracket is the probability of case j ($j = 1, 2, 3, 4, 5$). Finally the sum of E_j gives the desired expected number of tosses

$$\begin{aligned}
E(T) &= 240\left(\frac{1023}{15680}\right) - 160\left(\frac{3349}{70560}\right) + \frac{1152}{5}\left(\frac{2131}{127008}\right) + 45\left(\frac{2761}{132300}\right) \\
&\quad + 20\left(\frac{27767}{213444}\right) - \frac{432}{5}\left(\frac{25961}{2561328}\right) + \frac{144}{5}\left(\frac{28271}{4390848}\right) + 4\left(\frac{20417}{2134440}\right) \\
&= \frac{13179474}{1067220} = 12.346989.
\end{aligned} \tag{1.98}$$

Problem 1.12 Do the vertices problem for the octahedron (which has eight faces, twelve edges and six vertices). Let p denote 1/8.

Solution:

Consider the three disjoint cases for $j = 1, 2, 3$ where j is the number of new vertices seen on the nth (or last) toss. For $j = 1$ there are four triangles not seen on the first $n - 1$ tosses and using the notation in Fig. 6 we get all five remaining vertices by the events Y_1Y_4 or Y_2, Y_3 or both.

Hence for Case 1 we have

$$P_1 = 6(4p)(1-4p)^{n-1}\left[2I^{(2)}_{\frac{p}{1-4p}}(1; n-1) - I^{(4)}_{\frac{p}{1-4p}}(1; n-1)\right], \tag{1.99}$$

$$P_1^* = 12C^{(2)}_{1/4}(1;1) - 6C^{(4)}_{1/4}(1;1) = \frac{12}{\binom{6}{2}} - \frac{6}{\binom{8}{4}} = \frac{5}{7}, \tag{1.100}$$

$$E_1 = \frac{3}{p} C^{(2)}_{1/4}(1;2) - \frac{3}{2p} C^{(4)}_{1/4}(1;2). \tag{1.101}$$

FIGURE 6. The case $j = 1$ for the vertices problem of the octahedron

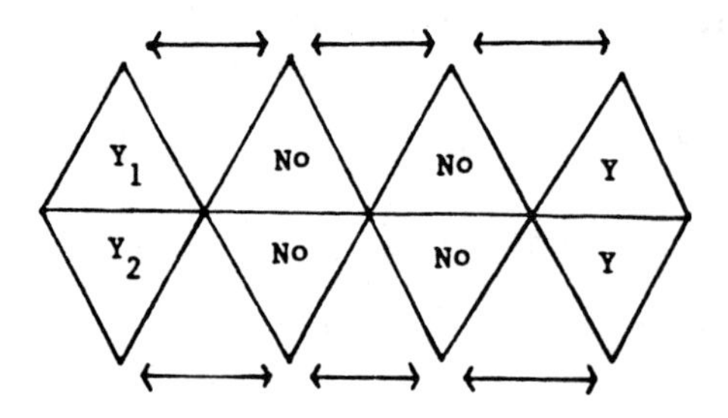

For $j = 2$ we mark another V at the box in Fig. 6 and change Y_3 and Y_4 to No. Then we need both Y_1 and Y_2 for the four remaining vertices. Hence we have

$$P_2 = 12(2p)(1-6p)^{n-1} I^{(2)}_{\frac{p}{1-6p}}(1; n-1), \tag{1.102}$$

$$P_2^* = 4C^{(2)}_{1/6}(1;1) = \frac{4}{\binom{8}{2}} = \frac{1}{7}, \tag{1.103}$$

$$E_2 = \frac{2}{3p} C^{(2)}_{1/6}(1;2). \tag{1.104}$$

For $j = 3$ we mark the top vertex as a V and make the additional change, Y_1 to No. This gives

$$P_3 = 8(p)(1-7p)^{n-1} I^{(1)}_{\frac{p}{1-7p}}(1; n-1), \tag{1.105}$$

$$P_3^* = \frac{8}{7} C^{(1)}_{1/7}(1;1) = \frac{8}{7}\frac{1}{\binom{8}{1}} = \frac{1}{7}, \tag{1.106}$$

$$E_3 = \frac{8}{49p} C^{(1)}_{1/7}(1;2). \tag{1.107}$$

The sum of the P_j is $P\{N = n\}$; the sum of P_j^* is unity. The sum of E_j is

$$\begin{aligned} E(T) &= \frac{8}{p}\left[\frac{3}{8}C^{(2)}_{1/4}(1;2) - \frac{3}{16}C^{(4)}_{1/4}(1;2) + \frac{1}{12}C^{(2)}_{1/6}(1;2) + \frac{1}{49}C^{(1)}_{1/7}(1;2)\right] \\ &= 64\left[\frac{3}{8}\left(\frac{37}{225}\right) - \frac{3}{16}\left(\frac{743}{14700}\right) + \frac{1}{12}\left(\frac{73}{784}\right) + \frac{1}{49}\left(\frac{15}{64}\right)\right] = \frac{29}{7} \\ &= 4.142857. \end{aligned} \tag{1.108}$$

Problem 1.13 Do the edges problem for the icosahedron, i.e., if on each toss you see all three edges associated with the top face, find the expected number of tosses $E(T)$ needed to see all thirty edges.

Solution:

Consider the three disjointed sets of cases according to whether j new edges are seen on the nth (or last) toss for $j(= 1, 2, 3)$. Actually we have sixty disjoint cases for $j = 1$ (since there are 2 possible faces for each of the 30 edges), 60 disjoint cases for $j = 2$ (since there are $\binom{3}{2}$ pairs of edges for each of twenty faces), and twenty disjoint cases for $j = 3$ (one for each face); below we also call these three sets Case I, Case II and Case III. Our sketches below make use of the fact that the icosohedron is a zonohedron with $n = 5$ (so that #faces $= n(n-1) = 20$), cf. Coxeter (1973). In each case we label one side the 'front' side (with the final face in the center of the sketch) and the other side the 'back' side (where most of our analysis applies). For Case I there is one final edge E, two N–faces not seen (resp., four Y–faces seen) on the first $n-1$ tosses.

FIGURE 7. Case I — Icosahedron (edge problem)

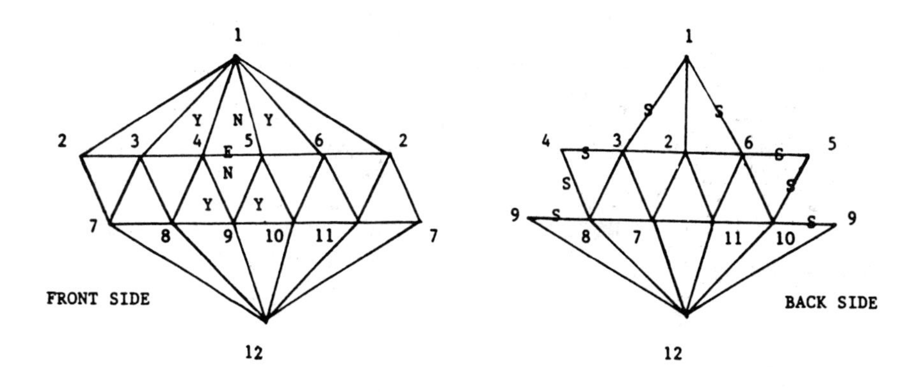

The four edges marked S are edges of the Y faces and are in a sense accounted for in our main analysis. For convenience we call the two top faces X_1, X_2, the five bottom faces Z_i ($i = 1, 2, 3, 4, 5$) and the seven middle zone faces W_j ($j = 1, 2, 3, 4, 5, 6, 7$), all going from left to right on the Back Side sketch in Fig. 7.

Let $\bar{X}$, $\bar{Z}$ and $\bar{W}$ denote that the corresponding faces are not seen on the first $n-1$ tosses. To concentrate on the middle zone and avoid worrying about the 5 horizontal (unmarked) edges, we consider the two sets

	Set 1	Set 2
X_1X_2	$Z_1Z_2Z_3Z_4Z_5$	$Z_1Z_3Z_4\bar{Z}_2\bar{Z}_5W_2$
$X_1\bar{X}_2W_5$	$Z_2Z_3Z_4Z_5\bar{Z}_1$	$Z_2Z_4Z_5\bar{Z}_1\bar{Z}_3W_4$
$\bar{X}_1X_2W_3$	$Z_1Z_2Z_3Z_4\bar{Z}_5$	$Z_2Z_3Z_5\bar{Z}_1\bar{Z}_4W_6$
	$Z_1Z_3Z_4Z_5\bar{Z}_2W_2$	
	$Z_1Z_2Z_4Z_5\bar{Z}_3W_4$	$Z_1Z_2Z_4\bar{Z}_3\bar{Z}_5W_4$
	$Z_1Z_2Z_3Z_5\bar{Z}_4W_6$	$Z_1Z_3Z_5\bar{Z}_2\bar{Z}_4W_2W_6$

whose cartesian product contains $3(11) = 33$ disjoint events, each containing all the 11 edges in the top and bottom zones and their boundaries. For the 6 remaining (unmarked) edges in the middle zone we use Table 5.

TABLE 5. W-Table for Case I

For Terms Containing	Multiply by (i.e., intersect with)
W_2 only (or W_6 only)	$W_4W_6 + W_4W_6W_3W_5 + W_6\bar{W}_4W_3W_5 + \bar{W}_4\bar{W}_6W_3W_5$
W_4 only	$W_2W_6 + W_2\bar{W}_6W_5W_7 + W_6\bar{W}_2W_1W_3 + \bar{W}_2\bar{W}_6W_1W_3W_5W_7$
W_3 only (or W_5 only)	$(W_1 + W_2 - W_1W_2)(W_4W_6 + W_4\bar{W}_6W_5W_7 + W_6\bar{W}_4W_5 + \bar{W}_4\bar{W}_6W_5W_7$
W_2W_6	$W_4 + \bar{W}_4W_3W_5$
W_2W_3 (or W_5W_6)	$W_4W_6 + W_4\bar{W}_6W_5W_7 + W_6\bar{W}_4W_5 + \bar{W}_4\bar{W}_6W_5W_7$
W_3W_4 (or W_4W_5)	$(W_1 + W_2 - W_1W_2)(W_6 + \bar{W}_6W_5W_7)$
W_3W_6 (or W_2W_5)	$(W_1 + W_2 - W_1W_2)(W_4 + W_5 - W_4W_5)$
$W_2W_5W_6$ (or $W_2W_3W_6$)	$W_3 + W_4 - W_3W_4$
no W's at all	$\begin{cases} W_2W_4W_6 + W_2W_4\bar{W}_6W_5W_7 + W_2W_6\bar{W}_4W_3W_5 + W_4W_6\bar{W}_2W_1W_3 \\ W_2\bar{W}_4\bar{W}_6W_3W_5W_7 + W_6\bar{W}_2\bar{W}_4W_1W_3W_5 + W_4\bar{W}_2\bar{W}_6W_1W_3W_5W_7 \\ +\bar{W}_2\bar{W}_4\bar{W}_6W_1W_3W_5W_7. \end{cases}$

For parenthetical cases (e.g. or W_6 only) the right side is obtained by a simple interchange of W_2 with W_6 and also of W_3 with W_5. After considerable algebra and simplification we obtain for Case I with $p = 1/20$

$$\begin{aligned} P_1 = 30(2p)\Big[&(1-2p)^{n-1}I^{(14)}_{\frac{p}{1-2p}}(1;n-1) + 5(1-3p)^{n-1}I^{(13)} \\ &+ 4(1-3p)^{n-1}I^{(14)} + (1-3p)^{n-1}I^{(15)} + 5(1-4p)^{n-1}I^{(12)} \\ &+ 22(1-4p)^{n-1}I^{(13)} + 2(1-4p)^{n-1}I^{(14)} + 6(1-4p)^{n-1}I^{(15)} \\ &- (1-4p)^{n-1}I^{(16)} + 22(1-5p)^{n-1}I^{(12)} + 7(1-5p)^{n-1}I^{(13)} \\ &+ 20(1-5p)^{n-1}I^{(14)} - 4(1-5p)^{n-1}I^{(15)} + 4(1-6p)^{n-1}I^{(12)} \\ &+ 20(1-6p)^{n-1}I^{(13)} - 4(1-6p)^{n-1}I^{(14)} - 4(1-7p)^{n-1}I^{(11)} \\ &+ 8(1-7p)^{n-1}I^{(12)} - 2(1-7p)^{n-1}I^{(13)}\Big] \end{aligned} \tag{1.109}$$

where the I-arguments $(1; n-1)$ and the appropriate subscripts $p/(1-jp)$

are omitted. Hence the contribution E_1 to $E(T_e)$ for Case 1 is given by

$$E_1 = \frac{60}{p}\Big[\frac{1}{4}C_{1/2}^{(14)}(1;2) + \frac{5}{9}C_{1/3}^{(13)} + \frac{4}{9}C_{1/3}^{(14)} + \frac{1}{9}C_{1/3}^{(15)} + \frac{5}{16}C_{1/4}^{(12)} + \frac{11}{8}C_{1/4}^{(13)} + \frac{1}{8}C_{1/4}^{(14)} + \frac{3}{8}C_{1/4}^{(15)} - \frac{1}{16}C_{1/4}^{(16)} + \frac{22}{25}C_{1/5}^{(12)} + \frac{7}{25}C_{1/5}^{(13)} + \frac{4}{5}C_{1/5}^{(14)} - \frac{4}{25}C_{1/5}^{(15)} + \frac{1}{9}C_{1/6}^{(12)} + \frac{5}{9}C_{1/6}^{(13)} - \frac{1}{9}C_{1/6}^{(14)} - \frac{4}{49}C_{1/7}^{(11)} + \frac{8}{49}C_{1/7}^{(12)} - \frac{2}{49}C_{1/7}^{(13)}\Big], \qquad (1.110)$$

where the common arguments $(1;2)$ are omitted after the first term.

FIGURE 8. Case II — Icosahedron (edge problem)

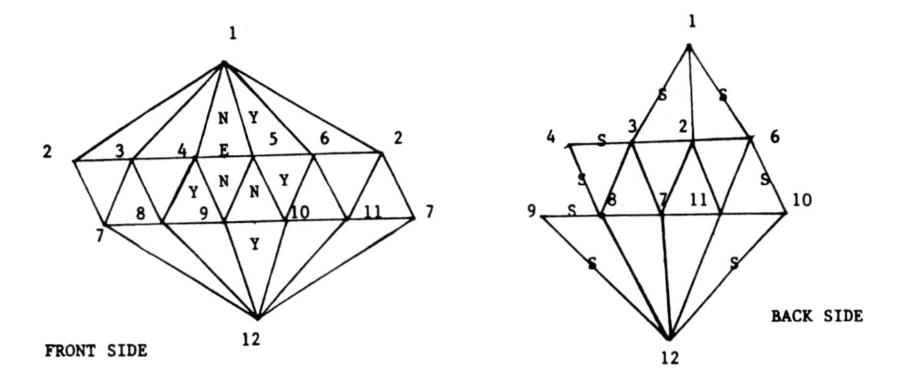

In a similar manner using Fig. 8 and a new analysis containing $3(8) = 24$ disjoint events for the top and bottom zones and a new W-table we obtain for Case II. In this case there are $3N$-faces, $5Y$-faces and 8 marked edges (on the FRONT SIDE) and 14 unmarked edges to account for. The result obtained for Case II is

$$P_2 = 20\binom{3}{2}(p)\Big[(1-3p)^{n-1}I^{(14)} + 4(1-4p)^{n-1}I^{(13)} + 3(1-4p)^{n-1}I^{(14)} + 2(1-4p)^{n-1}I^{(15)} + 6(1-5p)^{n-1}I^{(12)} + 14(1-5p)^{n-1}I^{(13)} + 6(1-5p)^{n-1}I^{(14)} + 4(1-5p)^{n-1}I^{(15)} + (1-6p)^{n-1}I^{(11)} + 13(1-6p)^{n-1}I^{(12)} + 11(1-6p)^{n-1}I^{(13)} + 11(1-6p)^{n-1}I^{(14)} + 7(1-7p)^{n-1}I^{(12)} + 10(1-7p)^{n-1}I^{(13)} + 3(1-8p)^{n-1}I^{(12)}\Big]. \qquad (1.111)$$

Hence the contribution E_2 of Case II to $E(T)$ is

$$E_2 = \frac{60}{p}\Big[\frac{1}{9}C_{1/3}^{(14)}(1;2) + \frac{1}{4}C_{1/4}^{(13)} + \frac{3}{16}C_{1/4}^{(14)} + \frac{1}{8}C_{1/4}^{(15)} + \frac{6}{25}C_{1/5}^{(12)} + \frac{14}{25}C_{1/5}^{(13)}$$
$$+ \frac{6}{25}C_{1/5}^{(14)} + \frac{4}{25}C_{1/5}^{(15)} + \frac{1}{36}C_{1/6}^{(11)}$$
$$+ \frac{13}{36}C_{1/6}^{(12)} + \frac{11}{36}C_{1/6}^{(13)} + \frac{11}{36}C_{1/6}^{(14)}$$
$$+ \frac{1}{7}C_{1/7}^{(12)} + \frac{10}{49}C_{1/7}^{(13)} + \frac{3}{64}C_{1/8}^{(12)}\Big]. \tag{1.112}$$

For Case III we use Fig. 9 with a new simpler analysis containing $3(5) = 15$ disjoint events for the top and bottom zones and also a new W-Table. In this case we have $4N$-faces, $6Y$-faces, 6 marked edges and 12 unmarked

FIGURE 9. Case III — Icosahedron (edge problem)

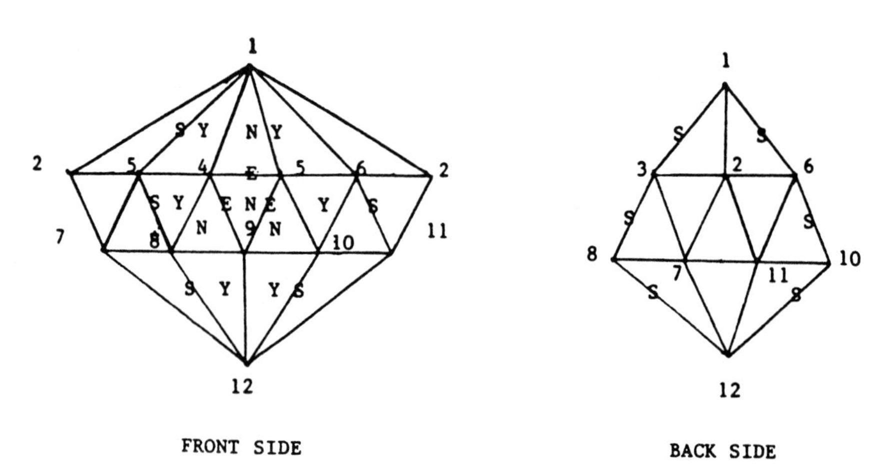

edges (on the BACK SIDE) to account for in the analysis. The result obtained for Case III is

$$P_3 = 20(p)[(1-4p)^{n-1}I^{(14)} + 5(1-5p)^{n-1}I^{(13)} + 2(1-5p)^{n-1}I^{(14)}$$
$$+ (1-5p)^{n-1}I^{(15)} + 9(1-6p)^{n-1}I^{(12)} + 5(1-6p)^{n-1}I^{(13)}$$
$$+ 6(1-6p)^{n-1}I^{(14)} + 6(1-7p)^{n-1}I^{(12)} + 7(1-7p)^{n-1}I^{(13)}$$
$$+ 2(1-8p)^{n-1}I^{(12)}]. \tag{1.113}$$

Hence the contribution E_3 to $E(T_e)$ for Case III is

$$E_3 = \frac{20}{p}\Big[\frac{1}{16}C_{1/4}^{(14)}(1;2) + \frac{1}{5}C_{1/5}^{(13)} + \frac{2}{25}C_{1/5}^{(14)} + \frac{1}{25}C_{1/5}^{(15)} + \frac{1}{4}C_{1/6}^{(12)}$$
$$+ \frac{5}{36}C_{1/6}^{(13)} + \frac{1}{6}C_{1/6}^{(14)} + \frac{6}{49}C_{1/7}^{(12)} + \frac{1}{7}C_{1/7}^{(13)} + \frac{1}{32}C_{1/8}^{(12)}\Big]. \tag{1.114}$$

By adding E_1 in (1.110), E_2 in (1.112) and E_3 in (1.114), the desired result is

$$\begin{aligned} E(T_e) = \frac{20}{p}\Big[&\frac{3}{4}C_{1/2}^{(14)}(1;2) + \frac{5}{3}C_{1/3}^{(13)} + \frac{5}{3}C_{1/3}^{(14)} + \frac{1}{3}C_{1/3}^{(15)} + \frac{15}{16}C_{1/4}^{(12)} \\ &+ \frac{39}{8}C_{1/4}^{(13)} + C_{1/4}^{(14)} + \frac{3}{2}C_{1/4}^{(15)} - \frac{3}{16}C_{1/4}^{(16)} + \frac{84}{25}C_{1/5}^{(12)} \\ &+ \frac{68}{25}C_{1/5}^{(13)} + \frac{16}{5}C_{1/5}^{(14)} + \frac{1}{25}C_{1/5}^{(15)} + \frac{1}{12}C_{1/6}^{(11)} + \frac{5}{3}C_{1/6}^{(12)} \\ &+ \frac{49}{18}C_{1/6}^{(13)} + \frac{3}{4}C_{1/6}^{(14)} - \frac{12}{49}C_{1/7}^{(11)} + \frac{3}{7}C_{1/7}^{(12)} + \frac{31}{49}C_{1/7}^{(13)} + \frac{11}{64}C_{1/8}^{(12)}\Big] \\ = 400\Big[&\frac{3}{4}(.03967882) + \frac{5}{3}(.01007533) + \frac{5}{3}(.00855685) + \frac{1}{3}(.00733496) \\ &+ \frac{15}{16}(.00340087) + \frac{39}{8}(.00269953) + .00217225 \\ &+ \frac{3}{2}(.00176925) - \frac{3}{16}(.001456682) + \frac{84}{25}(.00109585) \\ &+ \frac{68}{25}(.00082386) + \frac{16}{5}(.00062969) + \frac{1}{25}(.00048339) \\ &+ \frac{1}{12}(.00560545) + \frac{5}{3}(.000391653) + \frac{49}{18}(.000279612) \\ &+ \frac{3}{4}(.000155554) - \frac{12}{49}(.000031422) + \frac{3}{7}(.00015250) \\ &+ \frac{31}{49}(.000014805) + \frac{11}{64}(.000063817)\Big] = 37.50874, \end{aligned} \tag{1.115}$$

where C-values were taken from Vol. 9 or calculated exactly. Since the check sum added to slightly more than 1 (namely, 1.001) we cannot trust the last 2 or 3 decimals in the result. The amount of arithmetic needed made the solution difficult to pin down. The probabilities associated with Cases I, II and III, respectively are approximately .859, .135 and .007.

The structure of our answers for ET as a linear combination of C-functions is not unique and it is often desirable to get the answer with a common subscript on all the C-functions. We illustrate this by redoing the vertices problem for the cube (Problem 1.6) in this manner.

Problem 1.14 Do the vertices problem for the cube so that the subscripts on the C-functions are all equal (preferably to one).

Solution:

We are waiting to see any pair of opposite faces (i.e. adding to seven) for the first time. The terminating side (say face #1) can be any fixed one of the 6 faces. Then among the first $n-1$ tosses one specific side (face #6) has to appear at least once, and two combinations of 3 sides (namely 6, 2, 5 and 6, 3, 4) have to be ruled out, i.e., must not occur in the first $n-1$

tosses. Using inclusion-exclusion the full combination of all 5 remaining sides was ruled out twice and hence is added back once. Thus we have, letting $p = 1/6$ and N denote the random number of tosses needed,

$$P\{N = n\} = 6p(1-p)^{n-1}\left[I^{(1)}_{\frac{p}{1-p}}(1; n-1) - 2I^{(3)}_{\frac{p}{1-p}}(1; n-1) + I^{(5)}_{\frac{p}{1-p}}(1; n-1).\right] \tag{1.116}$$

Multiplying by n and summing yields

$$E(T) = 36\left[C^{(1)}_1(1;2) - 2C^{(3)}_1(1;2) + C^{(5)}_1(1;2)\right] = 36\left[\frac{3}{4} - \frac{25}{24} + \frac{49}{120}\right] = 4.2, \tag{1.117}$$

where all the C-values can be obtained from our exact C-table. As a check we sum (1.116) without multiplying by n and obtain the identity

$$P\{N < \infty\} = 6\left[C^{(1)}_1(1;1) - 2C^{(3)}_1(1;1) + C^{(5)}_1(1;1)\right] = 6\left(\frac{1}{2} - \frac{2}{4} + \frac{1}{6}\right) = 1. \tag{1.118}$$

Problem 1.15 Do the edge problem for the cube as in Problem 1.14.

Solution:

Assume that face #1 is the terminal face (which accounts for a factor of 6 below) and let $p = 1/6$. Going around the cube in cylindrical motion, we can see that the first $n-1$ tosses must include either sides $(3, 4, 6)$ or sides $(2, 5, 6)$ but not their union (which must therefore be subtracted twice). Thus

$$P\{N = n\} = 6p(1-p)^{n-1}\left[2I^{(3)}_{\frac{p}{1-p}}(1; n-1) - 2I^{(5)}_{\frac{p}{1-p}}(1; n-1)\right], \tag{1.119}$$

$$ET = \frac{6}{p}\left[2C^{(3)}_1(1;2) - 2C^{(5)}_1(1;2)\right] = 72\left(\frac{25}{48} - \frac{49}{120}\right) = 8.1, \tag{1.120}$$

and the check is simply that $6\left(\frac{2}{4} - \frac{2}{6}\right) = 1$.

Problem 1.16 Do the vertices problem for the octahedron so that the C-functions all have the same subscript (preferably one).

Solution:

To see all 6 vertices we have to wait either for a pair of opposite faces or for a set of 3 faces such that two are opposite on a pyramid (i.e., half of the octahedron) and the third one is anywhere on the other pyramid. Assume without loss of generality that face #1 in Figure 10 is the terminal face. Among the first $n-1$ tosses we need either face #8 or one of the combinations $(4, 6)$, $(4, 7)$, $(6, 7)$ but it can not contain any of the following $(2, 7)$, $(3, 6)$, $(4, 5)$, $(2, 3, 5)$, $(2, 3, 8)$, $(2, 5, 8)$, $(3, 5, 8)$ or $(4, 6, 7)$. After a

FIGURE 10. Pyramid structure of the octahedron (from above)

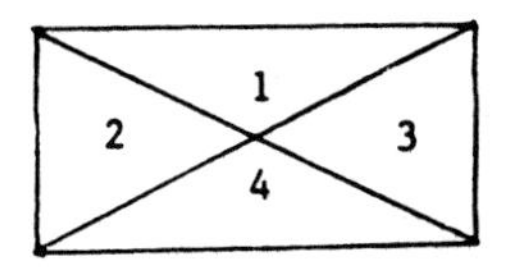

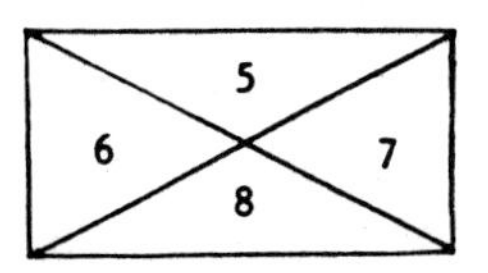

LOWER PYRAMID

careful application of inclusion-exclusion to all subsets of $(2,3,4,5,6,7,8)$ we obtain, letting $p = 1/8$,

$$
\begin{aligned}
P\{N = n\} = 8p(1-p)^{n-1}\Big[&I^{(1)}_{\frac{p}{1-p}}(1,n-1) + 3I^{(2)}_{\frac{p}{1-p}}(1;n-1)\\
&-18I^{(3)}_{\frac{p}{1-p}}(1;n-1) + 25I^{(4)}_{\frac{p}{1-p}}(1;n-1)\\
&-12I^{(5)}_{\frac{p}{1-p}}(1;n-1) + I^{(7)}_{\frac{p}{1-p}}(1;n-1)\Big]. \qquad (1.121)
\end{aligned}
$$

The sum of the coefficients is zero since the complete sum has to be removed. The first two coefficients are clear; the negative 18 is to remove $(2,4,7)$, $(2,6,7)$, $(3,4,6)$, $(3,6,7)$, $(4,5,6)$, $(4,5,7)$, $(2,3,8)$, $(2,5,8)$, $(3,5,8)$, $(4,6,8)$, $(4,7,8)$, $(6,7,8)$, $(2,7,8)$, $(3,6,8)$, $(4,5,8)$ all once and $(4,6,7)$ three times. We omit the other explanations. Multiplying by n and summing gives the desired result

$$
\begin{aligned}
E(T) &= \frac{8}{p}\Big[C^{(1)}_1(1;2) + 3C^{(2)}_1(1;2) - 18C^{(3)}_1(1;2) + 25C^{(4)}_1(1;2)\\
&\qquad -12C^{(5)}_1(1;2) + C^{(7)}_1(1;2)\Big]\\
&= 64\left[\frac{3}{4} + \frac{11}{6} - \frac{75}{8} + \frac{137}{12} - \frac{49}{10} + \frac{761}{2240}\right] = \frac{29}{7}\\
&= 4.142857, \qquad (1.122)
\end{aligned}
$$

where the exact table of C-values was again used. The same method can also be used to obtain higher moments. The check gives

$$
\begin{aligned}
P\{N < \infty\} &= 8\Big[C^{(1)}_1(1;1) + 3C^{(2)}_1(1;1) - 18C^{(3)}_1(1;1) + 25C^{(4)}_1(1;1)\\
&\qquad - 12C^{(5)}_1(1;1) + C^{(7)}_1(1;1)\Big]\\
&= 8\left[\frac{1}{2} + 1 - \frac{9}{2} + 5 - 2 + \frac{1}{8}\right] = 1. \qquad (1.123)
\end{aligned}
$$

Problem 1.17 Show that for the tetrahedron, letting $p = \frac{1}{4}$,

$$P\{N_f = n\} = 4p(1-p)^{n-1} I^{(3)}_{\frac{p}{1-p}}(1; n-1), \tag{1.124}$$

$$P\{N_v = n\} = 4p(1-p)^{n-1}\left[3I^{(1)}_{\frac{p}{1-p}}(1; n-1) - 6I^{(2)}_{\frac{p}{1-p}}(1; n-1) + 3I^{(3)}_{\frac{p}{1-p}}(1; n-1)\right], \tag{1.125}$$

$$P\{N_e = n\} = 4p(1-p)^{n-1}\left[3I^{(2)}_{\frac{p}{1-p}}(1; n-1) - 3I^{(3)}_{\frac{p}{1-p}}(1; n-1)\right] \tag{1.126}$$

and as a consequence show that

$$E(T_f) = \frac{4}{p} C_1^{(3)}(1; 2) = \frac{25}{3}, \tag{1.127}$$

$$E(T_v) = \frac{4}{p}\left[3C_1^{(1)}(1; 2) - 6C_1^{(2)}(1; 2) + 3C_1^{(3)}(1; 2)\right] = \frac{7}{3}, \tag{1.128}$$

$$E(T_e) = \frac{4}{p}\left[3C_1^{(2)}(1; 2) - 3C_1^{(3)}(1; 2)\right] = \frac{13}{3}. \tag{1.129}$$

Problem 1.18 Do the vertices problem for the cuboctahedron i.e., for the polyhedron in Problem 1.4 with $p_2 = 0$. Find the expected number of tosses to see all the vertices. Find the numerical answer for the special case $p_1 = p_3 = 1/14$.

Solution:

The cuboctahedron has 6 squares and 8 triangles as faces; it has 24 edges each separating a square from a triangle, and it has 12 vertices each associated with 2 square and 2 triangular faces. As in previous discussions we take p_1 (resp., p_3) for the probability of a square (resp., triangular) face so that $6p_1 + 8p_3 = 1$. The values of p_1 and p_3 may depend on the material from which the cuboctahedron is made, on the kind of flooring used to roll it and possibly on the manner of tossing. If the reader feels that these can be calculated (or estimated) then they can be inserted into our final formula to get numerical results.

There are six disjoint and exhaustive possibilities for the last toss (which we call cases 1, 2, ..., 6); these are

1. One new vertex,
2. Two new vertices on the same edge,
3. Two new opposite vertices on the same square,
4. Three new vertices on the same triangle,
5. Three new vertices on the same square,

6. Four new vertices on the same square.

Each of these 6 cases adds a contribution P_i (resp., E_i) to $P^* = P\{N < \infty\} = \sum_{i=1}^{6} P_i$ (resp., to $E(T) = \sum_{i=1}^{6} E_i$); $P^* = 1$ is a useful check and $E(T)$ is the desired result. The following (planar) sketch of the cuboctahedron is for Case 1 but with obvious modifications is useful for all 6 cases; vertices with the same label (and the corresponding edges connecting them) have to be identified.

FIGURE 11. Planar sketch of the cubohedron, case I (one new-vertex on the last toss)

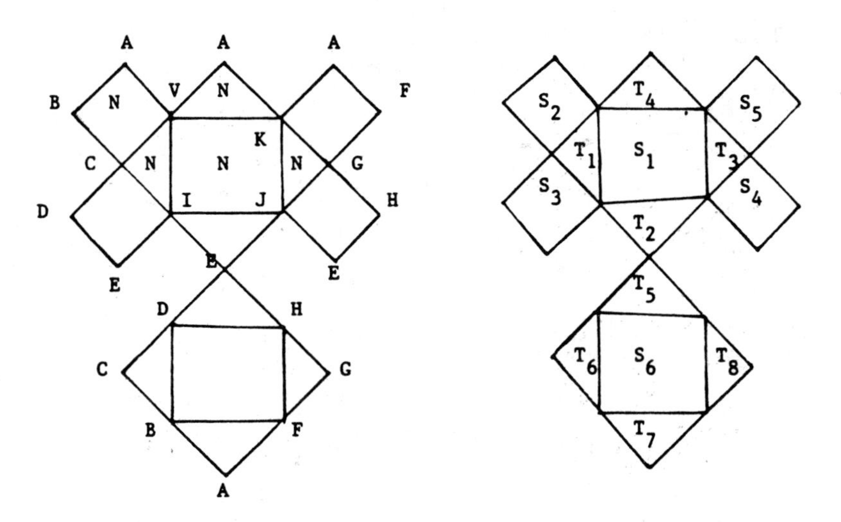

Using Fig. 11 we sketch the analysis for Case 1; similar analyses were used for each of the 6 cases. For the marked vertex (the new one on the last toss), the four faces marked N are not seen on the first $n-1$ tosses. This leaves $6-2=4$ squares, $8-2=6$ triangles and $12-1=11$ vertices to account for. Each of the 11 vertices gives us a condition but they reduce logically to the following conditions:

$$\begin{gathered} T_7 \text{ or } S_5,\ T_3 \text{ or } S_5,\ T_6 \text{ or } S_3, \\ T_2 \text{ or } T_3 \text{ or } S_4,\ T_6 \text{ or } T_7 \text{ or } S_6,\ T_5 \text{ or } T_8 \text{ or } S_4 \text{ or } S_6, \end{gathered} \tag{1.130}$$

where T (resp., S) stands for triangle (resp., square) and these are labeled in the second sketch of Fig. 11. The 4 squares to be accounted for in Case 1 are S_3, S_4, S_5 and S_6 and we list the $2^4 = 16$ possibilities: no S, S_3, S_4, S_5, S_6, S_3 and $S_4, \ldots, S_3$ and S_4 and S_5 and S_6. For each of these we list the possibilities with respect to the 6 triangles: T_2, T_3, T_5, T_6, T_7 and T_8 that are not yet accounted for. This gives us a total of 322 subcases satisfying (1.130) to each of which a pair (s,t) is associated indicating that s squares and t triangles are seen in the first $n-1$ tosses. The distribution of these 322 cases with respect to pairs (s,t) is shown in the column labeled $(\alpha = 1)$ in Table 6 below; the other cases $\alpha = 2$, 3, 4, 5 and 6 are also there. The constants in the body of Table 6 are called $K_{s,t}^{(\alpha)}$ in our formulas below; the item Q_α at the bottom of Table 6 is the probability associated with the last toss together with a factor for the total number of possibilities in each of the 6 cases.

TABLE 6. *Cuboctahedron (Vertex Problem)* $K_{s,t}^{(\alpha)}$*-Values.* Analysis of all possibilities satisfying (1.130) for Case α ($\alpha = 1$, 2, 3, 4, 5, 6)

(s,t)	$\alpha=1$	$\alpha=2$	$\alpha=3$	$\alpha=4$	$\alpha=5$	$\alpha=6$
$(0,3)$	0	0	0	1	0	0
$(0,4)$	0	1	0	1	0	1
$(0,5)$	2	1	0	0	0	0
$(0,6)$	1	0	0	0	0	0
$(1,2)$	0	0	0	0	1	0
$(1,3)$	4	2	0	3	2	0
$(1,4)$	12	5	0	3	1	1
$(1,5)$	12	3	0	0	0	0
$(1,6)$	4	0	0	0	0	0
$(2,1)$	0	1	0	0	0	0
$(2,2)$	4	4	4	3	1	0
$(2,3)$	24	8	4	6	2	0
$(2,4)$	37	8	1	3	1	0
$(2,5)$	24	3	0	0	0	0
$(2,6)$	6	0	0	0	0	0
$(3,0)$	0	0	1	1	0	0
$(3,1)$	4	1	4	4	0	0
$(3,2)$	20	4	6	6	0	0
$(3,3)$	40	6	4	4	0	0
$(3,4)$	40	4	1	1	0	0
$(3,5)$	20	1	0	0	0	0
$(3,6)$	4	0	0	0	0	0
$(4,0)$	1	0	0	0	0	0
$(4,1)$	6	0	0	0	0	0
$(4,2)$	15	0	0	0	0	0
$(4,3)$	20	0	0	0	0	0
$(4,4)$	15	0	0	0	0	0
$(4,5)$	6	0	0	0	0	0
$(4,6)$	1	0	0	0	0	0
Totals	322	52	25	36	8	2
Q_α	$12(2p_1+2p_3)$	$24(p_1+p_3)$	$6(2)p_1$	$8(p_3)$	$6(4)p_1$	$6(p_1)$

For the special case $p_1 = p_3$ we only need a summary of Table 6 arranged with respect to $F = s + t$, where F denotes the # of faces; this is given in Table 7 where p denotes the common value of $p_1 = p_3$. Using the weights

TABLE 7. Summary of Table 6

F	$\alpha = 1$	$\alpha = 2$	$\alpha = 3$	$\alpha = 4$	$\alpha = 5$	$\alpha = 6$	Weighted Sums
3	0	1	1	2	1	0	100p
4	13	8	8	11	3	1	1270p
5	64	18	10	15	3	1	4254p
6	105	17	5	7	1	0	5996p
7	88	7	1	1	0	0	4580p
8	41	1	0	0	0	0	2016p
9	10	0	0	0	0	0	480p
10	1	0	0	0	0	0	48p
Totals	322	52	25	36	8	2	445
Weights	48p	48p	12p	8p	24p	6p	

in the bottom row, the weighted sum of the six columns is given in the last column of Table 7; these eight constants are used for both $P^* = P\{N < \infty\}$ and $E(T)$ below for the special case $p_1 = p_3$.

The values $K_{st}^{(\alpha)}$ in Table 6 can be put into a rectangular array with $0 \leq s \leq 4$ and $0 \leq t \leq 6$. For convenience let $s' = 6 - s$ and $t' = 8 - t$ and let $F' = s' + t' = 14 - F$. Then our first result is

$$P\{T = n\} = \sum_{\alpha=1}^{6} \sum_{t=0}^{6} \sum_{s=0}^{4} K_{s,t}^{(\alpha)} Q_\alpha (1 - s'p_1 - t'p_3)^{n-1} \times I^{(s,t)}_{\frac{p_1}{1-s'p_1-t'p_3}, \frac{p_3}{1-s'p_1-t'p_3}}(1; 1; n-1). \qquad (1.131)$$

If we sum on n we obtain

$$P^* = P\{T < \infty\} = \sum_{\alpha=1}^{6} \sum_{t=0}^{6} \sum_{s=0}^{4} \frac{K_{s,t}^{(\alpha)} Q_\alpha}{s'p_1 + t'p_3} C^{(s,t)}_{\frac{p_1}{s'p_1+t'p_3}, \frac{p_3}{s'p_1+t'p_3}}(1, 1; 1). \qquad (1.132)$$

If we multiply by n and then sum (on n) we obtain

$$E(T) = \sum_{\alpha=1}^{6} \sum_{t=0}^{6} \sum_{s=0}^{4} \frac{K_{s,t}^{(\alpha)} Q_\alpha}{(s'p_1 + t'p_3)^2} C^{(s,t)}_{\frac{p_1}{s'p_1+t'p_3}, \frac{p_3}{s'p_1+t'p_3}}(1, 1; 2). \qquad (1.133)$$

For the special case $p_1 = p_3$ we use the constants in Table 7 and write out the eight terms in each result. The first result is

$$P\{T = n\} = 100p(1 - 11p)^{n-1} I^{(3)}_{\frac{p}{1-11p}}(1; n-1)$$

$$
\begin{aligned}
&+ 1270p(1-10p)^{n-1} I^{(4)}_{\frac{p}{1-10p}}(1;n-1) \\
&+ 4254p(1-9p)^{n-1} I^{(5)}_{\frac{p}{1-9p}}(1;n-1) \\
&+ 5996p(1-8p)^{n-1} I^{(6)}_{\frac{p}{1-8p}}(1;n-1) \\
&+ 4580p(1-7p)^{n-1} I^{(7)}_{\frac{p}{1-7p}}(1;n-1) \\
&+ 2016p(1-6p)^{n-1} I^{(8)}_{\frac{p}{1-6p}}(1;n-1) \\
&+ 480p(1-5p)^{n-1} I^{(9)}_{\frac{p}{1-5p}}(1;n-1) \\
&+ 48p(1-4p)^{n-1} I^{(10)}_{\frac{p}{1-4p}}(1;n-1).
\end{aligned} \tag{1.134}
$$

The second "result" is a check which "worked" to 12 decimals, namely

$$
\begin{aligned}
P\{T < \infty\} = {} & \frac{100}{11} C^{(3)}_{1/11}(1;1) + \frac{1270}{10} C^{(4)}_{1/10} + \frac{4254}{9} C^{(5)}_{1/9} + \frac{5996}{8} C^{(6)}_{1/8} \\
& + \frac{4580}{7} C^{(7)}_{1/7} + \frac{2016}{6} C^{(8)}_{1/6} + \frac{480}{5} C^{(9)}_{1/5} + \frac{48}{4} C^{(10)}_{1/4} = 1
\end{aligned} \tag{1.135}
$$

(independent of p). The last result is the desired numerical result for ET when $p_1 = p_3 = p$ (say) namely

$$
\begin{aligned}
E(T) = \frac{1}{p} \Big\{ & \frac{100}{121} C^{(3)}_{1/11}(1;2) + \frac{1270}{100} C^{(4)}_{1/10} + \frac{4554}{81} C^{(5)}_{1/9} + \frac{5996}{64} C^{(6)}_{1/8} \\
& + \frac{4580}{49} C^{(7)}_{1/7} + \frac{2016}{36} C^{(8)}_{1/6} + \frac{480}{25} C^{(9)}_{1/5} + \frac{48}{16} C^{(10)}_{1/4} \Big\}.
\end{aligned} \tag{1.136}
$$

Here $p = 1/14$ and using the formula at the bottom of our exact C-table

$$
C^{(13)}_{1/11}(1;2) = \frac{3875}{397488}, \qquad C^{(4)}_{1/10}(1;2) = \frac{25381}{6012006}, \tag{1.137}
$$

$$
C^{(5)}_{1/9}(1;2) = \frac{96163}{40080040}, \qquad C^{(6)}_{1/8}(1;2) = \frac{237371}{135270135};
$$

the remaining four C-values are given in our exact C table. Hence from (1.136)

$$
\begin{aligned}
ET &= 14 \left\{ \frac{100}{121} \left(\frac{3875}{397488} \right) + \cdots + 3 \left(\frac{511073}{90180090} \right) \right\} \\
&= 14(.6049068) = 8.46870.
\end{aligned} \tag{1.138}
$$

Acknowledgments: Thanks are due to Ingram Olkin for suggesting the problem dealing with cylindrical die and for suggesting that I look for the explicit expressions for certain partial sums. Thanks are also due to Herbert Solomon for asking me to prepare a paper on geometric probability for the 1986 Chicago meeting of the ASA, which led to this paper.

TABLE 8. Appendix: Exact (Fractional) Values of $C_a^{(b)}(1;2)$ for $b = 1(1)10$ and $a^{-1} = 1(1)$

$a \rightarrow$ / $b \downarrow$	1	1/2	1/3	1/4	1/5	1/6	1/7
1	$\frac{3}{4}$	$\frac{5}{9}$	$\frac{7}{16}$	$\frac{9}{25}$	$\frac{11}{36}$	$\frac{13}{49}$	$\frac{15}{64}$
2	$\frac{11}{18}$	$\frac{13}{36}$	$\frac{47}{200}$	$\frac{37}{225}$	$\frac{107}{882}$	$\frac{73}{784}$	$\frac{191}{2592}$
3	$\frac{25}{48}$	$\frac{77}{300}$	$\frac{57}{400}$	$\frac{319}{3675}$	$\frac{533}{9408}$	$\frac{56}{7056}$	$\frac{1207}{43200}$
4	$\frac{137}{300}$	$\frac{29}{150}$	$\frac{459}{4900}$	$\frac{743}{14700}$	$\frac{1879}{63504}$	$\frac{1627}{75600}$	$\frac{15797}{1306800}$
5	$\frac{49}{120}$	$\frac{223}{1470}$	$\frac{1023}{15680}$	$\frac{2509}{79380}$	$\frac{2131}{127008}$	$\frac{20417}{2134440}$	$\frac{18107}{3136320}$
6	$\frac{363}{980}$	$\frac{481}{3920}$	$\frac{3349}{70560}$	$\frac{2761}{132300}$	$\frac{25961}{2561328}$	$\frac{27767}{4268880}$	$\frac{263111}{88339680}$
7	$\frac{761}{2240}$	$\frac{4609}{45360}$	$\frac{3601}{100800}$	$\frac{32891}{2286900}$	$\frac{28271}{4390848}$	$\frac{4712171}{7935847920}$	$\frac{288851}{176679360}$
8	$\frac{7129}{22680}$	$\frac{4861}{56700}$	$\frac{42131}{1524600}$	$\frac{35201}{3430350}$	$\frac{395243}{92756664}$	$\frac{6694151}{13887733860}$	$\frac{12515}{13250952}$
9	$\frac{7381}{25200}$	$\frac{55991}{762300}$	$\frac{44441}{2032800}$	$\frac{485333}{64414350}$	$\frac{420983}{144288144}$	$\frac{8543999}{347193346500}$	$\frac{134159}{235572480}$
10	$\frac{83711}{304920}$	$\frac{58301}{914760}$	$\frac{605453}{34354320}$	$\frac{511073}{90180090}$	$\frac{445007}{216432216}$	$\frac{69114973}{1111018708800}$	$\frac{2424847}{6808044672}$

The general formula for the entries of this table is

$$C_a^{(b)}(1;2) = \frac{a'}{\binom{b+a'}{b}} \sum_{j=a'}^{b+a'} 1/j \qquad \text{(where } a' = 1/a \text{ is an integer).}$$

The decimal C-values for $a' = 1(1)5$ and $b = 1(1)15$ can be found in Vol. 9.

REFERENCES

Coxeter, H.S.M. (1973). *Regular Polytopes.* Dover Publications: New York.

Sobel, M. (1987). Multinomial problems in geometric probability with Dirichlet analysis. Technical Report 239, Stanford University, Stanford, CA.

Sobel, M., Uppuluri, V.R.R. and Frankowski, K. (1977). Vol. 4 of *Selected Tables in Mathematical Statistics.* Published jointly by the Institute of Mathematical Statistics and the American Mathematical Society.

Sobel, M., Uppuluri, V.R.R. and Frankowski, K. (1985). Vol. 9 of *Selected Tables in Mathematical Statistics.* Published jointly by the Institute of Mathematical Statistics and the American Mathematical Society.

8

Probability Inequalities for n-Dimensional Rectangles via Multivariate Majorization

Y.L. Tong[1]

ABSTRACT Inequalities for the probability content $P\left[\bigcap_{j=1}^{n}\{a_{1j} \le X_j \le a_{2j}\}\right]$ are obtained, via concepts of multivariate majorization (which involves the diversity of elements of the $2\times n$ matrix $A = (a_{ij})$). A special case of the general result is that $P\left[\bigcap_{j=1}^{n}\{a_{1j} \le X_j \le a_{2j}\}\right] \le P\left[\bigcap_{j=1}^{n}\{\bar{a}_1 \le X_j \le \bar{a}_2\}\right]$ for $\bar{a}_i = \frac{1}{n}\sum_{j=1}^{n} a_{ij}$ $(i = 1, 2)$. The main theorems apply in most important cases, including the exchangeable normal, t, chi-square and gamma, F, beta, and Dirichlet distributions. The proofs of the inequalities involve a convex combination of an n-dimensional rectangle and its permutation sets.

1 Introduction

This paper concerns probability inequalities (or integral inequalities) for n-dimensional rectangles and their applications in statistics. The mathematical tools used are Schur-concavity and concepts of multivariate majorization (for definitions see Marshall and Olkin (1979), p. 54 and Chapter 15).

Let $\boldsymbol{X} = (X_1, \ldots, X_n)$ be an n-dimensional random variable whose density is $f(\boldsymbol{x})$. It is known that (Tong (1982)) if $f(\boldsymbol{x})$ is a Schur-concave function of $\boldsymbol{x}$ and if D is a region given by $D = \{\boldsymbol{x} \mid \boldsymbol{x} \in \Re^n,\ -d_j \le x_j \le d_j,\ j = 1, \ldots, n\}$, which is an n-dimensional rectangle centered at the origin, then the probability content

$$P[\boldsymbol{X} \in D] = P\left[\bigcap_{j=1}^{n}\{-d_j \le X_j \le d_j\}\right]$$

[1]Georgia Institute of Technology

is a Schur-concave function of $\boldsymbol{d} = (d_1, \ldots, d_n)$. Consequently we have

$$P\left[\bigcap_{j=1}^{n}\{-d_j \le X_j \le d_j\}\right] \le P\left[\bigcap_{j=1}^{n}\{-\bar{d} \le X_j \le \bar{d}\}\right], \qquad \bar{d} = \frac{1}{n}\sum_{j=1}^{n} d_j. \tag{1.1}$$

A natural question to ask is whether similar results hold if D is not centered at the origin. To answer this question we prove two theorems in Section 3 via multivariate majorization. For $n \ge 2$ let

$$A = \begin{pmatrix} \boldsymbol{a}_1 \\ \boldsymbol{a}_2 \end{pmatrix} = \begin{pmatrix} a_{11} \cdots a_{1n} \\ a_{21} \cdots a_{2n} \end{pmatrix}, \qquad B = \begin{pmatrix} \boldsymbol{b}_1 \\ \boldsymbol{b}_2 \end{pmatrix} = \begin{pmatrix} b_{11} \cdots b_{1n} \\ b_{21} \cdots b_{2n} \end{pmatrix}, \tag{1.2}$$

denote two real matrices such that $a_{1j} < a_{2j}$ and $b_{1j} < b_{2j}$ $(j = 1, \ldots, n)$. From Marshall and Olkin (1979), Chapter 15, we say that

Definition 1.1 *(1) A chain majorizes B (in symbols $A \overset{c}{\succ} B$) if there exists an $n \times n$ matrix Q which is the product of finitely many $n \times n$ T-transform matrices (defined in Marshall and Olkin (1979), p. 21), such that $B = AQ$. (2) A majorizes B in a multivariate sense (in symbols $A \overset{m}{\succ} B$) if there exists a doubly stochastic matrix Q such that $B = AQ$. (3) A row-wise majorizes B (in symbols $A \overset{r}{\succ} B$) if $\boldsymbol{a}_1 \succ \boldsymbol{b}_1$ and $\boldsymbol{a}_2 \succ \boldsymbol{b}_2$.*

These concepts imply the following fact: If $A \overset{c}{\succ} B$ or $A \overset{m}{\succ} B$ or $A \overset{r}{\succ} B$, then the components of $\boldsymbol{b}_i$ are simultaneously less diverse than that of $\boldsymbol{a}_i$, given $\sum_{j=1}^{n} a_{ij} = \sum_{j=1}^{n} b_{ij}$ $(i = 1, 2)$. Clearly we have

$$A \overset{c}{\succ} B \Rightarrow A \overset{m}{\succ} B \Rightarrow A \overset{r}{\succ} B, \tag{1.3}$$

and it is known that for $n > 2$ the implications in (1.3) are strict. The statement that $A \overset{r}{\succ} B$ does not imply $A \overset{m}{\succ} B$ can be seen easily; and "$A \overset{m}{\succ} B$ does not imply $A \overset{c}{\succ} B$ for $n > 2$" follows from the fact that a doubly stochastic matrix may not be the finite product of T-transform matrices (Marshall and Olkin (1979), p. 39).

The concepts of multivariate majorization have been found useful for deriving certain types of probability (or integral) inequalities. For example, the results in Rinott (1973), Marshall and Olkin (1979) Chapter 15, and Karlin and Rinott (1981) concern inequalities via the diversity of the elements of a parameter matrix. In this paper we study inequalities for the probability content of an n-dimensional rectangle whose location and shape are determined either by A or by B. That is, denoting

$$\gamma(A) = P\left[\bigcap_{j=1}^{n}\{a_{ij} \le X_j \le a_{2j}\}\right], \tag{1.4}$$

we establish the inequality

$$\gamma(A) \le \gamma(B) \tag{1.5}$$

via multivariate majorization, which yields

$$P\left[\bigcap_{j=1}^{n}\{a_{ij} \le X_j \le a_{2j}\}\right] \le P\left[\bigcap_{j=1}^{n}\{\bar{a}_1 \le X_j \le \bar{a}_2\}\right],$$

$$\bar{a}_i = \frac{1}{n}\sum_{j=1}^{n} a_{ij} \ (i = 1,\ 2) \tag{1.6}$$

as a special case. It is shown in Section 3 that (1.5) holds when $A \overset{c}{\succ} B$ and $f(\boldsymbol{x})$ is any Schur-concave function, and that (1.5) holds when $A \overset{m}{\succ} B$ and $f(\boldsymbol{x})$ is a permutation invariant and log-concave function of $\boldsymbol{x}$ (which then implies that $f(\boldsymbol{x})$ is a Schur-concave function). It is also shown in a counterexample that under the weakest concept of multivariate majorization, the row-wise majorization, such inequalities are no longer possible under useful conditions on $f(\boldsymbol{x})$.

2 Transformations of n-dimensional Rectangles

The method we use to prove the results in Section 3 involves the convex combination of an n-dimensional rectangle and its permutation set(s). For arbitrary but fixed real matrix A let us define

$$S = S(A) = \{\, \boldsymbol{x} \mid \boldsymbol{x} \in \Re^n,\ a_{1j} \le x_j \le a_{2j},\ j = 1, \ldots, n\,\}. \tag{2.1}$$

Let $\boldsymbol{\pi} = (\pi_1, \ldots, \pi_n)$ be a permutation of $(1, 2, \ldots, n)$, and define the permutation set of S by

$$S_{\boldsymbol{\pi}} = \{\, \boldsymbol{x} \mid \boldsymbol{x} \in \Re^n,\ a_{1j} \le x_{\pi_j} \le a_{2j},\ j = 1, \ldots, n\,\}. \tag{2.2}$$

For arbitrary but fixed $\alpha \in [0, 1]$ consider the set

$$\begin{aligned} S_\alpha &= \{\, \boldsymbol{z} \mid \boldsymbol{z} \in \Re^n,\ \boldsymbol{z} = \alpha\boldsymbol{x} + (1-\alpha)\boldsymbol{y} \text{ for } \boldsymbol{x} \in S,\ \boldsymbol{y} \in S_{\boldsymbol{\pi}}\,\} \\ &= \alpha S + (1+\alpha) S_{\boldsymbol{\pi}}. \end{aligned} \tag{2.3}$$

Now let $P_{\boldsymbol{\pi}}$ be the permutation matrix such that $\boldsymbol{\pi} = (1, 2, \ldots n)P_{\boldsymbol{\pi}}$. It can be verified that for every fixed $\alpha \in [0, 1]$, S_α is also a rectangle and is of the form

$$S_\alpha = \{\, \boldsymbol{x} \mid \boldsymbol{x} \in \Re^n,\ a^*_{1j} \le x_j \le a^*_{2j},\ j = 1, \ldots, n\,\}, \tag{2.4}$$

where

$$\begin{pmatrix} \boldsymbol{a}^*_1(\alpha) \\ \boldsymbol{a}^*_2(\alpha) \end{pmatrix} = \begin{pmatrix} a^*_{11}(\alpha) \cdots a^*_{1n}(\alpha) \\ a^*_{21}(\alpha) \cdots a^*_{2n}(\alpha) \end{pmatrix} = \begin{pmatrix} \boldsymbol{a}_1 \\ \boldsymbol{a}_2 \end{pmatrix} \cdot (\alpha I + (1-\alpha)P_{\boldsymbol{\pi}}^{-1}) \tag{2.5}$$

and I is the identity matrix. Such a transformation can be extended to the convex combination of any finite number of permutation matrices. For fixed $K \leq n!$ let $\boldsymbol{\pi}(1), \ldots, \boldsymbol{\pi}(K)$ denote permutations of $(1, 2, \ldots, n)$, and let $P_{\boldsymbol{\pi}(k)}$ satisfy $\boldsymbol{\pi}(k) = (1, 2, \ldots, n)P_{\boldsymbol{\pi}(k)}$ $(k = 1, \ldots, K)$. If we define, for a fixed real vector $\boldsymbol{\alpha} = (\alpha_1, \ldots, \alpha_k)$, $\alpha_k \geq 0$, $\sum_{k=1}^{K} \alpha_k = 1$,

$$S_{\boldsymbol{\alpha}} = \sum_{k=1}^{K} \alpha_k S_{\boldsymbol{\pi}(k)}, \tag{2.6}$$

then we can similarly verify that

$$S_{\boldsymbol{\alpha}} = \{\, \boldsymbol{x} \mid \boldsymbol{x} \in \Re^n,\ a_{1j}^*(\boldsymbol{\alpha}) \leq x_j \leq a_{2j}^*(\boldsymbol{\alpha}),\ j = 1, \ldots, n \,\}, \tag{2.7}$$

where

$$\begin{pmatrix} \boldsymbol{a}_1^*(\boldsymbol{\alpha}) \\ \boldsymbol{a}_2^*(\boldsymbol{\alpha}) \end{pmatrix} = \begin{pmatrix} \boldsymbol{a}_1 \\ \boldsymbol{a}_2 \end{pmatrix} \sum_{k=1}^{K} \alpha_k P_{\boldsymbol{\pi}(k)}^{-1}. \tag{2.8}$$

Note that, since $P_{\boldsymbol{\pi}(k)}^{-1}$ is a permutation matrix for each k, the matrix $\sum_{k=1}^{K} \alpha_k P_{\boldsymbol{\pi}(k)}^{-1}$ is doubly stochastic. For convenience we say that $S_{\boldsymbol{\alpha}}$ in (2.6) is an S-transform of the rectangle S. Also, note that the matrix $\alpha I + (1 - \alpha)P_{\boldsymbol{\pi}}^{-1}$ in (2.5) is a T-transform (Marshall and Olkin (1979), p. 21) if exactly two of the diagonal elements of $P_{\boldsymbol{\pi}}^{-1}$ are zero; in this special case we say that S_α in (2.3) is a T-transform of S. By Definition 1.1, Lemma 2.B.1 in Marshall and Olkin (1979), p. 21, and Birkhoff's Theorem (Marshall and Olkin (1979), p. 19), we can obtain the following result:

Proposition 2.1 *Let $S(A)$, $S(B)$ be two n-dimensional rectangles. If $A \overset{c}{\succ} B$ holds, then $S(B)$ can be derived from $S(A)$ through a finite number of T-transforms. If $A \overset{m}{\succ} B$, then $S(B)$ is an S-transform of $S(A)$.*

Such transformations can be viewed as an averaging process, because the components in $\boldsymbol{b}_i$ are less diverse than those in $\boldsymbol{a}_i$ $(i = 1, 2)$ after the transformation. This is particularly true for $n = 2$ and $\alpha = 1/2$, in which case the transformation carries any rectangle in $\Re^2$ into a square centered on the 45°-line. Thus if an inequality for the probability contents of the sets S and $S_{\boldsymbol{a}}$ can be obtained, then a chain of inequalities will follow through these concepts of multivariate majorization, by repeating this averaging process for a finite number of times.

3 The Main Results

In this section we prove two theorems and some related results. The first theorem (Theorem 3.1) involves chain majorization, its proof depends on the T-transform of an n-dimensional rectangle. The second theorem (Theorem 3.8) involves the concept of $\overset{m}{\succ}$ majorization, and its proof depends on the (more general) S-transform given in (2.6).

Theorem 3.1 *Let $S(A)$, $S(B)$ denote two n-dimensional rectangles, and let $f(\boldsymbol{x}) : \Re^n \to [0, \infty)$ be Lebesgue measurable. If $f(\boldsymbol{x})$ is a Schur-concave function of $\boldsymbol{x}$ and if $A \overset{c}{\succ} B$, then (provided that the integrals exist)*

$$\int_{S(A)} f(\boldsymbol{x})\, d\boldsymbol{x} \le \int_{S(B)} f(\boldsymbol{x})\, d\boldsymbol{x}. \tag{3.1}$$

Proof Without loss of generality we may assume that

$$\begin{aligned} b_{i1} &= \alpha a_{i1} + (1-\alpha) a_{i2}, \\ b_{i2} &= (1-\alpha) a_{i1} + \alpha a_{i2}, \\ b_{ij} &= a_{ij}, \qquad j = 3, \ldots, n \end{aligned}$$

for some $\alpha \in [0, 1]$ for $i = 1, 2$. Therefore it suffices to show that, for every fixed $(x_3, \ldots, x_n) \in \underset{j=3}{\overset{n}{\times}} [a_{1j}, a_{2j}]$, the inequality

$$\int_{a_{11}}^{a_{21}} \int_{a_{12}}^{a_{22}} g(x_1, x_2)\, dx_2\, dx_1 \le \int_{b_{11}}^{b_{21}} \int_{b_{12}}^{b_{22}} g(x_1, x_2)\, dx_2\, dx_1 \tag{3.2}$$

holds, where for notational convenience g stands for $f(x_1, x_2, x_3, \ldots, x_n)$ when $(x_3, \ldots, x_n)$ is kept fixed. Moreover, since g is a non-negative Schur-concave function of (x_1, x_2) we may further assume, without loss of generality, that $\alpha \in [1/2, 1]$ and

$$a_{21} - a_{11} \le a_{22} - a_{12}, \tag{3.3}$$

because all Schur-concave functions are permutation invariant.

It can be verified analytically that (3.3) implies

$$\begin{aligned} & a_{11} + a_{12} = b_{11} + b_{12} \equiv \lambda_1 < a_{21} + a_{12} \equiv \lambda_2 \\ & \le b_{21} + b_{12} \equiv \lambda_3 \le b_{11} + b_{22} \equiv \lambda_4 \le a_{11} + a_{22} \equiv \lambda_5 \\ & < b_{21} + b_{22} = a_{21} + a_{22} \equiv \lambda_6. \end{aligned} \tag{3.4}$$

For each fixed $\lambda \in [\lambda_m, \lambda_{m+1}]$ ($m = 1, \ldots, 5$) we can find the line segment $\ell_\lambda(A)$ ($\ell_\lambda(B)$) of the intersect of the straight line $x_1 + x_2 = \lambda$ and the rectangle $\underset{j=1}{\overset{2}{\times}} [a_{1j}, a_{2j}]$ (the rectangle $\underset{j=1}{\overset{2}{\times}} [b_{1j}, b_{2j}]$). For $* = A, B$ let $\|\ell_\lambda(*)\|$, $\delta_\lambda(*)$ denote, respectively, the length of $\ell_\lambda(*)$ and the distance between its midpoint and $(\lambda/2, \lambda/2)$. Then, by $\alpha \ge 1/2$, the inequality in (3.3), and $\lambda_m \le \lambda \le \lambda_{m+1}$, it can be verified that for each λ and each $m = 1, \ldots, 5$, at least one of the following statements is true:

$$\ell_\lambda(A) \text{ is a subset of } \ell_\lambda(B), \tag{3.5}$$

$$\|\ell_\lambda(A)\| \le \|\ell_\lambda(B)\| \quad \text{and} \quad \delta_\lambda(A) \ge \delta_\lambda(B). \tag{3.6}$$

Now apply the orthogonal transformation

$$v_1 = \frac{x_1 + x_2}{\sqrt{2}}, \qquad v_2 = \frac{x_1 - x_2}{\sqrt{2}},$$

and for $* = A$, B define

$$c_\lambda(*) = \inf\{ (x_1 - x_2)/\sqrt{2} \mid (x_1, x_2) \in \ell_\lambda(*) \},$$
$$d_\lambda(*) = \sup\{ (x_1 - x_2)/\sqrt{2} \mid (x_1, x_2) \in \ell_\lambda(*) \},$$

Then we can write

$$\int_{a_{11}}^{a_{21}} \int_{a_{12}}^{a_{22}} g(x_1, x_2)\, dx_2\, dx_1$$
$$= \int_{\lambda_1/\sqrt{2}}^{\lambda_6/\sqrt{2}} \int_{c_\lambda(A)}^{d_\lambda(A)} g\left(\frac{v_1 + v_2}{\sqrt{2}}, \frac{v_1 - v_2}{\sqrt{2}}\right) dv_2\, dv_1, \tag{3.7}$$

$$\int_{b_{11}}^{b_{21}} \int_{b_{12}}^{b_{22}} g(x_1, x_2)\, dx_2\, dx_1$$
$$= \int_{\lambda_1/\sqrt{2}}^{\lambda_6/\sqrt{2}} \int_{c_\lambda(B)}^{d_\lambda(B)} g\left(\frac{v_1 + v_2}{\sqrt{2}}, \frac{v_1 - v_2}{\sqrt{2}}\right) dv_2\, dv_1. \tag{3.8}$$

For each fixed v_1, if (3.5) holds then the inner integral in (3.7) is less than or equal to that in (3.8); if (3.6) holds then the same statement is true by Lemma 2.1 in Tong (1982). Thus (3.2) follows, and the proof of the theorem is complete. □

A special case of Theorem 3.1 is the following corollary.

Corollary 3.2 *If $f(\boldsymbol{x}) : \Re^n \to [0, \infty)$ is a Schur-concave function of $\boldsymbol{x}$, then*

$$\int_{a_{11}}^{a_{21}} \cdots \int_{a_{1n}}^{a_{2n}} f(\boldsymbol{x})\, d\boldsymbol{x} \le \int_{\bar{a}_1}^{\bar{a}_2} \cdots \int_{\bar{a}_1}^{\bar{a}_2} f(\boldsymbol{x})\, d\boldsymbol{x} \tag{3.9}$$

for $\bar{a}_i = \frac{1}{n} \sum_{j=1}^{n} a_{ij}$ $(i = 1,\ 2)$.

Proof By Definition 1.1 and Theorem 3.1 it suffices to show that for arbitrary but fixed $\boldsymbol{a}_1 = (a_{11}, \ldots, a_{1n})$ and $\boldsymbol{a}_2 = (a_{21}, \ldots, a_{2n})$ there exists a matrix Q, which is the product of at most $2(n-1)$ T-transform matrices, such that

$$\begin{pmatrix} \bar{\boldsymbol{a}}_1 \\ \bar{\boldsymbol{a}}_2 \end{pmatrix} \equiv \begin{pmatrix} \bar{a}_1 \cdots \bar{a}_1 \\ \bar{a}_2 \cdots \bar{a}_2 \end{pmatrix} = \begin{pmatrix} \boldsymbol{a}_1 \\ \boldsymbol{a}_2 \end{pmatrix} Q.$$

Since $\boldsymbol{a}_1 \succ \bar{\boldsymbol{a}}_1$ where "$\succ$" stands for the univariate majorization, by 2.B.1.a of Marshal and Olkin (1979), p. 22, there exists r $(r \le n-1)$ T-transform matrices $T_1, \ldots, T_r$ such that $\bar{\boldsymbol{a}}_1 = \boldsymbol{a}_1 \left(\prod_{k=1}^{r} T_k\right)$. Let $\boldsymbol{c}$ denote the vector $\boldsymbol{a}_2 \left(\prod_{k=1}^{r} T_k\right)$, which obviously majorizes $\bar{\boldsymbol{a}}_2$. Then again there exists s $(s \le$

$n-1$) T-transform matrices $T_{r+1}, \ldots, T_{r+s}$ such that $\bar{\boldsymbol{a}}_2 = \boldsymbol{c}\,(\prod_{k=1}^{s} T_{r+k})$. Let $Q = \prod_{k=1}^{r+s} T_k$; it follows that

$$\begin{pmatrix} \boldsymbol{a}_1 \\ \boldsymbol{a}_2 \end{pmatrix} Q = \begin{pmatrix} \bar{\boldsymbol{a}}_1 \\ \boldsymbol{c} \end{pmatrix} \left(\prod_{k=1}^{s} T_{r+k} \right) = \begin{pmatrix} \bar{\boldsymbol{a}}_1 \\ \bar{\boldsymbol{a}}_2 \end{pmatrix}$$

as to be shown. □

Now let $\boldsymbol{X} = (X_1, \ldots, X_n)$ be an n-dimensional variable, $\boldsymbol{a}_1$, $\boldsymbol{a}_2$, $\boldsymbol{b}_1$, $\boldsymbol{b}_2$, be vectors, and let $\gamma(A)$, $\gamma(B)$ denote the probability contents defined in (1.4). From Theorem 3.1 and Corollary 3.2 it follows immediately that

Corollary 3.3 *If $f(\boldsymbol{x})$ (the density of $\boldsymbol{X}$) is a Schur-concave function of $\boldsymbol{x}$, and if $A \overset{c}{\succ} B$, then $\gamma(A) \le \gamma(B)$. In particular, the inequality*

$$P\left[\bigcap_{j=1}^{n} \{a_{1j} \le X_j \le a_{2j}\}\right] \le P\left[\bigcap_{j=1}^{n} \{\bar{a}_1 \le X_j \le \bar{a}_2\}\right] \tag{3.10}$$

holds for $\bar{a}_i = \frac{1}{n}\sum_{j=1}^{n} a_{ij}$, $i = 1,\ 2$.

Remark 3.4 *Theorem 3.1 immediately yields Theorem 2.1 in Tong (1982) by letting $\boldsymbol{a}_1 = -\boldsymbol{a}_2$. In this special case, since $\boldsymbol{a}_1$ is a sign change of $\boldsymbol{a}_2$, the condition of chain majorization in Definition 1.1 is equivalent to that of $\boldsymbol{a}_2 \succ \boldsymbol{b}_2$.*

The above remark suggests that in certain special cases the condition of chain majorization can be simplified to the univariate version of majorization (which is easier to verify). In the proposition below we state two such cases. The proof is immediate.

Proposition 3.5 *(1) If $\boldsymbol{a}_1 = \lambda \boldsymbol{a}_2$ for some real number $\lambda \ne 0$, then $A \overset{c}{\succ} B$ is satisfied iff $\boldsymbol{a}_1 \succ \boldsymbol{b}_1$ (or equivalently $\boldsymbol{a}_2 \succ \boldsymbol{b}_2$). (2) If $\boldsymbol{a}_1 = (\bar{a}_1, \ldots, \bar{a}_1)$ ($\boldsymbol{a}_2 = (\bar{a}_2, \ldots, \bar{a}_2)$), then $A \overset{c}{\succ} B$ is satisfied iff $\boldsymbol{a}_1 = \boldsymbol{b}_1$ and $\boldsymbol{a}_2 \succ \boldsymbol{b}_2$ ($\boldsymbol{a}_1 \succ \boldsymbol{b}_1$ and $\boldsymbol{a}_2 = \boldsymbol{b}_2$).*

Theorem 3.1 concerns inequalities via the strongest concept of multivariate majorization, the chain majorization. An immediate question is whether similar results also follow under weaker concepts of multivariate majorization. The following example shows that, under the row-wise majorization (the weakest concept), it is no longer possible to obtain useful results.

Example 3.6

Let $f(\boldsymbol{x})$ be any continuous function which is > 0 almost everywhere, and let

$$\begin{pmatrix} \boldsymbol{a}_1 \\ \boldsymbol{a}_2 \end{pmatrix} = \begin{pmatrix} c_1 + \epsilon & c_1 - \epsilon & a_{13} \cdots a_{1n} \\ c_2 + \epsilon & c_2 - \epsilon & a_{23} \cdots a_{2n} \end{pmatrix},$$

$\boldsymbol{b}_1 = (c_1, c_1, a_{13}, \ldots, a_{1n})$, $\boldsymbol{b}_2 = \boldsymbol{a}_2$. Then for $\epsilon = c_2 - c_1 > 0$ and $a_{2j} > a_{1j}$ $(j = 3, \ldots, n)$ we have $A \overset{r}{\succ} B$ and

$$\int_{S(A)} f(\boldsymbol{x})\, d\boldsymbol{x} > 0, \qquad \int_{S(B)} f(\boldsymbol{x})\, d\boldsymbol{x} = 0.$$

□

It is not yet known to the author whether or not the condition on chain majorization "$\overset{c}{\succ}$" in Theorem 3.1 can be weakened to "$\overset{m}{\succ}$". However, a result has been obtained, under this weaker concept of multivariate majorization, by imposing a stronger condition on $f(\boldsymbol{x})$. This result is given below as Theorem 3.8; its proof depends on the S-transform (as defined in (2.6)) and an application of Prékopa's inequality (1971).

The new condition on $f(\boldsymbol{x})$ is stated below.

Condition 3.7 *f is of the form $f(\boldsymbol{x}) = \psi(\phi(\boldsymbol{x}))$, where (1) $\phi(\boldsymbol{x}) : \Re^n \to (-\infty, \infty)$ is a permutation invariant and convex function of $\boldsymbol{x}$, and (2) $\psi(z) : (-\infty, \infty) \to [0, \infty)$ is nonincreasing, differentiable, and $-\psi'(z)$ is a log-concave function of z. Moreover, $\int_{\Re^n} f(\boldsymbol{x})\, d\boldsymbol{x} < \infty$.*

Theorem 3.8 *Let $S(A)$ and $S(B)$ denote two n-dimensional rectangles, and let $f(\boldsymbol{x}) : \Re^n \to [0, \infty)$ satisfy Condition 3.7. If $A \overset{m}{\succ} B$, then*

$$\int_{S(A)} f(\boldsymbol{x})\, d\boldsymbol{x} \le \int_{S(B)} f(\boldsymbol{x})\, d\boldsymbol{x}. \tag{3.11}$$

To prove this theorem, let us define

$$\int_{C} f(\boldsymbol{x})\, d\boldsymbol{x} \equiv H(C) \tag{3.12}$$

and observe the following lemma.

Lemma 3.9 *For arbitrary but fixed K let $P_1, \ldots, P_K$ be permutation matrices, and let $\alpha_k \ge 0$, $\sum_{k=1}^{K} \alpha_k = 1$. If $f(\boldsymbol{x})$ satisfies Condition 3.7, then*

$$H\left(\sum_{k=1}^{K} \alpha_k P_k(C)\right) \ge H(C) \tag{3.13}$$

holds for all convex and bounded sets $C \subset \Re^n$.

Proof Since f is permutation invariant we have $H(P_m(C)) = H(C)$ for each P_m. Thus by Theorem 2 of Prékopa (1971) we have, for $\alpha_m \ge 0$,

$$\sum_{m=1}^{M+1} \alpha_m = 1, \qquad \alpha_{M+1} < 1,$$

$$\begin{aligned}
H\Big(\sum_{m=1}^{M+1} \alpha_m P_m(C)\Big) \\
&= H\Big((1-\alpha_{M+1})\sum_{m=1}^{M} \frac{\alpha_m}{1-\alpha_{M+1}} P_m(C) + \alpha_{M+1} P_{M+1}(C)\Big) \\
&\geq \Big(H\Big(\sum_{m=1}^{M} \frac{\alpha_m}{1-\alpha_{M+1}} P_m(C)\Big)\Big)^{1-\alpha_{M+1}} \cdot \big(H(P_{M+1}(C))\big)^{\alpha_{M+1}} \\
&\geq \big(H(C)\big)^{1-\alpha_{M+1}} \cdot \big(H(C)\big)^{\alpha_{M+1}} = H(C).
\end{aligned}$$

Hence the lemma. □

Proof of Theorem 3.8 Immediate from Proposition 2.1 and Lemma 3.9. □

In certain applications f is of the form $f(\boldsymbol{x}) = \psi(\phi(\boldsymbol{x}))$ only for $\boldsymbol{x}$ in a permutation invariant and convex set E, and zero elsewhere. In this case Theorem 3.8 again applies provided that $S(A)$ is a subset of E (note that if $S(A)$ is in E, then any of its S-transforms must be in E). This observation yields the following corollary.

Corollary 3.10 *Let the density function $f(\boldsymbol{x})$ of $\boldsymbol{X}$ satisfy Condition 3.7 for $\boldsymbol{x} \in \Re^n$; or for $\boldsymbol{x} \in E$ (a permutation invariant and convex set), zero elsewhere, and $S(A)$ a subset of E. If $A \overset{m}{\succ} B$, then the inequality*

$$P\Big[\bigcap_{j=1}^{n}\{a_{ij} \leq X_j \leq a_{2j}\}\Big] \leq P\Big[\bigcap_{j=1}^{n}\{b_{1j} \leq X_j \leq b_{2j}\}\Big] \tag{3.14}$$

holds; as a special consequence,

$$P\Big[\bigcap_{j=1}^{n}\{a_{1j} \leq X_j \leq a_{2j}\}\Big] \leq P\Big[\bigcap_{j=1}^{n}\{\bar{a}_1 \leq X_j \leq \bar{a}_2\}\Big] \tag{3.15}$$

holds for $\bar{a}_i = \frac{1}{n}\sum_{j=1}^{n} a_{ij}$, $i = 1, 2$.

It may be useful to point out that Condition 3.7 is satisfied for a variety of density functions. In particular, we note that

Proposition 3.11 *If $f(\boldsymbol{x})$ is a permutation invariant and log-concave function of $\boldsymbol{x}$ for $\boldsymbol{x} \in \Re^n$, or for $\boldsymbol{x}$ in a permutation invariant and convex set E and zero elsewhere, then the condition on $f(\boldsymbol{x})$ in Corollary 3.10 is satisfied.*

Proof Immediate by choosing $\phi(\boldsymbol{x}) = -\log f(\boldsymbol{x})$ and $\psi(z) = e^{-z}$. □

In the following proposition, we show that the class of functions satisfying Condition 3.7 is in fact a sub-class of Schur-concave functions.

Proposition 3.12 *If $f(\boldsymbol{x})$ satisfies Condition 3.7, or if it satisfies Condition 3.7 for $\boldsymbol{x} \in E$ (a permutation invariant and convex set) and zero elsewhere, then it is a Schur-concave function of $\boldsymbol{x}$.*

Proof First assume that f satisfies Condition 3.7 for all $\boldsymbol{x}$, and let $\boldsymbol{x}$, $\boldsymbol{y}$ be in $\Re^n$ such that $\boldsymbol{x} \succ \boldsymbol{y}$ holds. Without loss of generality it may be assumed that $\boldsymbol{y} = \boldsymbol{x}T$ where T is a T-transform matrix, and that $\boldsymbol{x}$ and $\boldsymbol{y}$ differ in the first two coordinates only. Consequently, we can write $\boldsymbol{y} = \alpha\boldsymbol{x} + (1-\alpha)\boldsymbol{x}^*$ for some $\alpha \in [0,1]$, where $\boldsymbol{x}^*$ is obtained by interchanging the first two coordinates in $\boldsymbol{x}$. It then follows that

$$\phi(\boldsymbol{y}) = \phi\big(\alpha\boldsymbol{x} + (1-\alpha)\boldsymbol{x}^*\big) \leq \alpha\phi(\boldsymbol{x}) + (1-\alpha)\phi(\boldsymbol{x}^*) = \phi(\boldsymbol{x}),$$

and that

$$f(\boldsymbol{y}) = \psi\big(\phi(\boldsymbol{y})\big) \geq \psi\big(\phi(\boldsymbol{x})\big) = f(\boldsymbol{x}). \tag{3.16}$$

If f satisfies Condition 3.7 for $\boldsymbol{x} \in E \neq \Re^n$, and zero elsewhere, the $\boldsymbol{x} \in E$ implies $\boldsymbol{y} \in E$. It can be verified that for either (1) $\boldsymbol{x}, \boldsymbol{y} \in E$, or (2) $\boldsymbol{x} \notin \boldsymbol{E}$, $\boldsymbol{y} \in E$, or (3) $\boldsymbol{x} \notin E$, $\boldsymbol{y} \notin E$, the inequality in (3.16) always holds. □

Thus, when comparing Theorems 3.1 and 3.8, we now see that in Theorem 3.1 an inequality is given for all Schur-concave functions, under a stronger condition of multivariate majorization; while the majorization condition in Theorem 3.8 is weaker, the inequality applies to only a sub-class of Schur-concave functions.

A related interesting question is the following: suppose that $A \overset{m}{\succ} B$ holds, is it always possible to find a finite number of S-transforms S_α given in (2.3) (i.e., by taking the convex combination of only *two* sets at a time) to transform $S(A)$ into $S(B)$? The answer to this question would be in the affirmative if the following conjecture were true.

Conjecture 3.13 *Every doubly stochastic matrix Q can be expressed as the product of a finite number of S-matrices, where each S-matrix is of the form $\alpha I + (1-\alpha)P_k$ for some $\alpha \in [0,1]$ and some permutation matrix P_k (I is the identity matrix).*

If Conjecture 3.13 were true, then it would yield the following statement.

Conjecture 3.14 *Let A, B be two $p \times n$ matrices (p, $n \geq 2$). If $A \overset{m}{\succ} B$, then B can be derived from A by successive applications of a finite number of S-transforms defined in Conjecture 3.13.*

This, if true, would be a multivariate version of the result concerning (univariate) majorization and the T-transform (see, e.g., Marshall and Olkin (1979), Lemma 2.B.1, p. 21). Thus, in the proof of Theorem 3.8, we would not have to depend on the transformation given in (2.6) because then the transformation in (2.3) would be sufficient. It was communicated to me by M. Shaked that Conjecture 3.13 is false. The following is one of the counterexamples he obtained.

Example 3.15

The doubly stochastic matrix

$$Q = \frac{1}{16}\begin{pmatrix} 8 & 8 & 0 \\ 7 & 5 & 4 \\ 1 & 3 & 12 \end{pmatrix}$$

is not a finite product of S-transforms.

This can be justified using a geometric argument; the details are not given here. □

4 Some Applications

In this section we point out some of the important cases (distributions) in which the main results in Section 3 apply. The applications are presented for the purpose of illustration, so, of course, the list is not complete. Note that in each of the applications, the inequality

$$P\left[\bigcap_{j=1}^{n}\{a_{1j} \leq X_j \leq a_{2j}\}\right] \leq P\left[\bigcap_{j=1}^{n}\{\bar{a}_1 \leq X_j \leq \bar{a}_2\}\right]$$

holds as a special case for $\bar{a}_i = \frac{1}{n}\sum_{j=1}^{n} a_{ij}$, $i = 1, 2$.

Application 4.1 *In each of the following cases the joint density function* $f(\boldsymbol{x})$ *(of* $\boldsymbol{X}$*) satisfies the condition in Corollary 3.10. Thus (3.14) holds if* $A \overset{m}{\succ} B$.

1. $X_1, \ldots, X_n$ *are i.i.d. random variables whose common (marginal) density* $h(x)$ *is a log-concave function of* x *for all* x*, or for* x *in some interval* I*, and zero elsewhere.*

   In particular, this applies when $h(x)$ is the density function of a normal distribution, or an exponential, gamma, or a chi-square distribution with degrees of freedom ≥ 2.

2. $\boldsymbol{X}$ *is a multivariate normal variable with equal means, equal variances, and equal correlation coefficients* $\rho \in \big(-1/(n-1), 1\big)$.

3. *As a more general case of (2),* $f(\boldsymbol{x})$ *is a permutation invariant elliptically contoured density function such that* $f(\boldsymbol{x}) = \psi((\boldsymbol{x} - \boldsymbol{\mu}) / \Sigma^{-1}(\boldsymbol{x} - \boldsymbol{\mu}))$*, where* Σ *is a positive definite matrix with equal diagonal elements and equal off-diagonal elements,* $\boldsymbol{\mu} = (\mu, \ldots, \mu)$ *for some real number* μ*, and* ψ *satisfies Condition 3.7.*

4. *$\boldsymbol{X}$ is a multivariate beta or Dirichlet variable with density*

$$f(\boldsymbol{x}) = c\prod_{j=1}^{n} w_j^{\omega-1}\Big(1-\sum_{j=1}^{n} x_j\Big)^d \qquad \textit{for} \sum_{j=1}^{n} x_j \le 1,$$

and zero elsewhere; where $c > 0$, $\omega \ge 1$, and $d > 0$.

This application implies an earlier result of Olkin (1972), which involves probability contents of the for $P\big[\bigcap_{j=1}^{n}\{0 \le X_j \le a_j\}\big]$, as a special case.

5. *Other Dirichlet-type integrals with integrand*

$$f(\boldsymbol{x}) = c\prod_{j=1}^{n} w_j^{\omega-1}\Big(1-\sum_{j=1}^{n} x_j\Big)^d \qquad \textit{for } \boldsymbol{x} \in E,$$

and zero elsewhere; where $c > 0$, $\omega \ge 1$, $d > 0$, and E is a permutation-invariant and bounded convex set.

Note that if $h(x)$ is a log-concave function of x, then it is well-known that $f(\boldsymbol{x})$ is a Schur-concave function of $\boldsymbol{x}$. But the statement in (1) above is stronger because all functions f satisfying Condition 3.7 are Scur-concave (Proposition 3.12). The statement in (1) follows from the inequality

$$\begin{aligned}\log f\big(\alpha\boldsymbol{x} + (1-\alpha)\boldsymbol{y}\big) &= \sum_{j=1}^{n} \log h\big(\alpha x_j + (1-\alpha)y_j\big) \\ &\ge \sum_{j=1}^{n} \big(\alpha \log h(x_j) + (1-\alpha)\log h(y_j)\big) \\ &= \alpha \log f(\boldsymbol{x}) + (1-\alpha)\log f(\boldsymbol{y})\end{aligned}$$

(that is, f is a log-concave function of $\boldsymbol{x}$), and Proposition 3.11. The statement in (2) can be verified easily. The statements in (3), (4), (5) follow from Proposition 3.11 and the fact that $f(\boldsymbol{x})$ is a log-concave function (which can be verified analytically).

Below we state some important examples in which the density function $f(\boldsymbol{x})$ is a Schur-concave function and it may or may not satisfy Condition 3.7.

Application 4.2 *In each of the following cases the joint density $f(\boldsymbol{x})$ is a Schur-concave function of $\boldsymbol{x}$. Hence Theorem 3.1 and Corollaries 3.2 and 3.3 apply.*

1. *$f(\boldsymbol{x})$ is permutation invariant and unimodal.*

2. *$f(\boldsymbol{x})$ is the density of one of the following variables (as discussed in Section 3.3 of Tong (1980)):*

(i) *Multivariate t variable* $\boldsymbol{t} = \boldsymbol{X}/\sqrt{U/\nu}$, *where* $\boldsymbol{X}$ *is defined as in Application 4.1. (2) with means zero, and is independent of* U; U *is a chi-squared variable with* ν *d.f.*

(ii) *Multivariate chi-squared variable* $\boldsymbol{\chi}^2 = (U_1 + U_0, \ldots, U_n + U_0)$, *where* $U_1, \ldots, U_n, U_0$ *are independent chi-squared variables with d.f.'s* $\nu_1 = \cdots = \nu_n \geq 2$ *and* ν_0, *respectively.*

(iii) *Multivariate F-variable* $\boldsymbol{F} = (\nu_0/\nu_1)\boldsymbol{U}/U_0$, *where* $\boldsymbol{U} = (U_1, \ldots, U_n)$ *and* U_0 *are as in (2) (ii).*

Remark 4.3 *The following is an example in which Theorem 3.8 applies and Theorem 3.1 does not apply. Consider the matrices*

$$A = \lambda \begin{pmatrix} 1 & 1 & 0 \\ 2 & 4 & 6 \end{pmatrix}, \qquad B = \lambda \begin{pmatrix} 1 & 0.5 & 0.5 \\ 3 & 4 & 5 \end{pmatrix},$$

where $\lambda > 0$ *is arbitrary but fixed. Let* Φ *denote the c.d.f. of a normal (or an exponential or a gamma or a chi-squared) variable. By Example 15.A.3 of Marshall and Olkin (1979), p. 431, we have* $A \overset{m}{\succ} B$ *but not* $A \overset{c}{\succ} B$. *Hence Theorem 3.1 fails to apply and Theorem 3.8 yields*

$$\begin{aligned}&\big(\Phi(2\lambda) - \Phi(\lambda)\big)\big(\Phi(4\lambda) - \Phi(\lambda)\big)\big(\Phi(6\lambda) - \Phi(0)\big) \\ &\qquad \leq \big(\Phi(3\lambda) - \Phi(\lambda)\big)\big(\Phi(4\lambda) - \Phi(0.5\lambda)\big)\big(\Phi(5\lambda) - \Phi(0.5\lambda)\big).\end{aligned}$$

Application 4.4 *When choosing* $f(\boldsymbol{x}) = c > 0$ *for* $\boldsymbol{x} \in E$ *(and zero elsewhere), where* E *is a convex and permutation-invariant set in* $\Re^n$, *Theorem 3.1 yields the following geometric inequality: If the chain majorization* $A \overset{c}{\succ} B$ *is satisfied, then the value of* $S(A)$ *is less than or equal to that of* $S(B)$. *It in turn yields the algebraic expression*

$$\prod_{j=1}^{n} (a_{2j} - a_{1j}) \leq (\bar{a}_2 - \bar{a}_1)^n$$

for $a_{1j} \leq a_{2j}$ $(j = 1, \ldots, n)$ *and* $\bar{a}_i = \frac{1}{n}\sum_{j=1}^{n} a_{ij}$, $i = 1, 2$.

Finally, we state a result for the scale parameter families. The proof is easy, and is omitted.

Application 4.5 *Let* $X_1, \ldots, X_n$ *be independent random variables with densities* $h_j(x) = h(x/\theta_j)$, $\theta_j > 0$, $j = 1, \ldots, n$. *If* $h(x)$ *is a log-concave function of* x *for all* x *(for* $x \geq 0$ *and zero elsewhere) then for all finite intervals* (a_1, a_2) *(for all finite intervals* $(a_1, a_2) \subset [0, \infty)$*) the probability content* $P\big[\bigcap_{j=1}^{n}\{a_1 \leq X_j \leq a_2\}\big]$ *is a Schur-concave function of* $(\theta_1^{-1}, \ldots, \theta_2^{-1})$.

As a special case, Application 4.5 yields inequalities for the independent normal variables via the diversity of their standard deviations; the details are left to the reader.

Acknowledgments: I am grateful to Professor Ingram Olkin for his encouragement and for his comments on an earlier draft of this paper, and to Professor Moshe Shaked for providing me with the result in Example 3.15. Also, I thank Professor Samuel Karlin for bringing to my attention that Karlin and Rinott (1983) had independently ascertained results similar to our Theorem 3.1 and 3.8 when this paper was put out as a technical report in 1983.

This research was partially supported by NSF grants MCS-8100775 and DMS-8502346.

REFERENCES

Karlin, S. and Rinott, Y. (1981). Entropy inequalities for classes of probability distributions II. The multivariate case. *Adv. Appl. Probability* **13**, 325–351.

Karlin, S. and Rinott, Y. (1983). Comparison of measures, multivariate majorization and applications to statistics. In *Studies in Econometrics, Time Series and Multivariate Analysis*, S. Karlin, T. Amemiya and L. Goodman, eds., Academic Press, New York, 465–489.

Marshall, A.W. and Olkin, I. (1979). *Inequalities: Theory of Majorization and Its Applications.* Academic Press, New York.

Olkin, I. (1972). Monotonicity properties of Dirichlet integrals with applications to the multinomial distributions and the analysis of variance. *Biometrika* **59**, 303–307.

Prékopa, A. (1971). Logarithmic concave measures with applications. *Acta Sci. Math.* **32**, 301–316.

Rinott, Y. (1973). Multivariate majorization and rearrangement inequalities with some applications to probability and statistics. *Israel J. Math.* **15**, 60–67.

Tong, Y.L. (1980). *Probability Inequalities in Multivariate Distributions.* Academic Press, New York.

Tong, Y.L. (1982). Rectangular and elliptical probability inequalities for Schur-concave random variables. *Ann. Statist.* **10**, 637–642.

9

Minimum Majorization Decomposition

Joseph S. Verducci[1]

ABSTRACT A famous theorem of Birkhoff says that any doubly stochastic matrix D can be decomposed into a convex combination of permutation matrices R. The various decompositions correspond to probability distributions on the set of permutations that satisfy the linear constraints $E[R] = D$. This paper illustrates how to decompose D so that the resulting probability distribution is minimal in the sense that it does not majorize any other distribution satisfying these constraints.

Any distribution maximizing a strictly Schur concave function g under these linear constraints will be minimal in the above sense (Joe (1987)). In particular, for D in the relative interior of the convex hull of the permutation matrices, the probability functions p that maximize $g(p) = -\sum_{\pi} p(\pi) \log p(\pi)$, subject to $E[R] = D$, form an exponential family $\mathcal{L}$ with sufficient statistic R.

This paper provides a theorem that characterizes the exponential family $\mathcal{L}$ by a property called quasi-independence. Quasi-independence is defined in terms of the invariance of the product measure over Latin sets. The characterization suggests an algorithm for an explicit minimal decomposition of a doubly stochastic matrix.

1 Introduction

Chapter 2 of Marshall and Olkin (1979) describes the theory of doubly stochastic matrices, a term apparently coined by Feller (1950) for square matrices with nonnegative elements, whose rows and columns all sum to one. Perhaps the most famous theorem (Birkhoff (1947)) about doubly stochastic matrices is that they form the convex hull of the set of all permutation matrices, which are doubly stochastic matrices consisting of only 0's and 1's. Marshall and Olkin review several proofs of Birkhoff's theorem, including explicit methods for decomposing a doubly stochastic matrix into a convex combination of permutation matrices. All of these decompositions express a doubly stochastic matrix D in terms of only a few permutation matrices, the smallest number needed for a general $n \times n$ doubly stochastic

[1]The Ohio State University

matrix being $n^2 - 2n + 2$.

Any decomposition of a doubly stochastic matrix D induces a probability distribution on the set Ω of all permutations of n objects as follows. Let $\pi : \{1, \ldots, n\} \to \{1, \ldots, n\}$ be a general permutation in Ω, and define the $n \times n$ permutation matrix $R(\pi)$ by $[R(\pi)]_{ij} = 1$ if $\pi(i) = j$, and 0 otherwise. Then the decomposition $D = \sum_{\pi \in \Omega} p(\pi) R(\pi)$ makes $p(\cdot)$ a probability mass function on Ω. Traditional decompositions of D concentrate $p(\cdot)$ on a few permutations. Here we are interested in making $p(\cdot)$ as diffuse as possible.

The motivation behind a diffuse decomposition is statistical. Through the principle of maximal entropy, Good (1963) extolls the usefulness of probability models that are as diffuse as possible, subject to constraints that represent knowledge of particular features of the probability distribution. If Π represents a random permutation sampled from Ω according to a probability mass function p, then the doubly stochastic matrix $D = E[R(\Pi)|p]$ has ith row equal to the marginal distribution of the random variable $\Pi(i), i = 1, \ldots, n$. A diffuse decomposition of D thus corresponds to a probability distribution p on Ω that is as diffuse as possible, subject to the fixed marginal constraints

$$P\Big[\Pi(i) = j\Big] = \sum_{\{\pi | \pi(i) = j\}} p(\pi) = D_{ij}, \tag{1.1}$$

for each $j = 1, \ldots, n$ and each $i = 1, \ldots, n$.

The preordering $>$ of majorization [cf. Marshall and Olkin, 1979, p13] on the set $\{p(\cdot)\}$ of all probability mass functions on Ω gives a criterion for relative diffusion; namely, p is more diffuse than q if $q > p$. Let P_D denote the set of probability functions on Ω that satisfy the linear constraints (1.1); by Birkhoff's theorem, P_D is not empty. A probability function $p \in P_D$ is *minimal* with respect to majorization if $p > q$ for any $q \in P_D$ implies $q > p$. A result due to Joe (1987) is that if p maximizes a strictly Schur concave function on P_D, then p is minimal.

In this paper, we focus attention on the problem of finding the minimal p that maximizes the Schur concave function $g(p) = -\sum_{\pi \in \Omega} p(\pi) \log[p(\pi)]$ known as Shannon entropy. Let F be an exponential family with sufficient statistic R, that includes the uniform distribution, on a discrete space. It is easy to show that each member of F is the unique distribution that maximizes Shannon entropy among all distributions with the same value of $E[R]$. Thus if the range of $E[R]$ is restricted to $\mathcal{D}_0 \equiv$ relative interior of convex hull of $\{R(\pi)\}$, then for each $D \in \mathcal{D}_0$ the unique distribution of P_D that maximizes Shannon entropy has the form

$$p(\pi|\theta) = \Psi^{-1}(\theta) \exp\{\mathrm{tr}[\theta R(\pi)]\}, \tag{1.2}$$

where θ is an $n \times n$ real valued matrix, and Ψ is the (Laplace transform) normalizing function that ensures $\sum p(\pi) = 1$.

Using the language and tools of log-linear models, Holland and Silverberg (cf. Silverberg, 1980, unpublished Ph. D. dissertation, Princeton) investigated the exponential family $\mathcal{E}$ of probability functions defined by (1.2). This model has not been applied much, chiefly because of the difficulty in estimating the parameter θ, due to the intractable form of Ψ for large values of n. The proposed decomposition of D solves this problem, because the maximum likelihood estimate of θ corresponds to the unique (cf. Brown, 1986, pp. 148–149) solution of the equation

$$E[R(\pi)|\theta] = D_0, \tag{1.3}$$

whenever D_0, the average of the sample permutation matrices, is in $\mathcal{D}_0$. The proposed decomposition of any $D_0 \in \mathcal{D}_0$ will be given in terms of the solution θ of (1.3).

In the next section, the exponential family $\mathcal{E}$ is characterized by the property that its members are the only probability functions p for which

$$\prod_{\pi \in L} p(\pi) = c \tag{1.4}$$

where L is any subset of Ω whose members form the rows of a Latin Square, and $c > 0$ is a constant that does not depend on L.

In Section 3, the algorithm for decomposing D is developed. First it is shown that θ in (1.2) may be determined from the values $p(\pi|\theta)$ on a set S of Ω, which contains $n(n-1)$ members. Second, this set is embedded into the union of Latin sets L. Third, an initial, near-minimal decomposition is performed. Fourth, iteratively transferring probability across Latin sets increases entropy while staying inside the family P_D, and leads to the minimal decomposition. Exact decompositions are computationally feasible only for small values of n; however, the method may be used to obtain approximate decompositions in the case of larger n.

2 Quasi-independence

Definitions. A *Latin set* L is a subset of Ω with the property that for each i and $j \in \{1, \ldots, n\}$, there is a unique $\pi \in L$ such that $\pi(i) = j$. A probability mass function p on Ω is *quasi-independent* if p satisfies (1.4) for all Latin sets L. The class of all quasi-independent probability functions of Ω is denoted by $\mathcal{Q}$.

Remark A. A Latin set, of necessity, consists of n permutations $\{\pi_k\}$, which if displayed in the $n \times n$ array $[\pi_k(i)]$ form a Latin square.

Remark B. The random variables $\{\Pi(i)|i = 1, \ldots, n\}$ induced by a probability distribution on Ω cannot be independent unless they are all constant. Suppose the Ω is generalized to the space Ω^* of all functions (not just bijections) from $\{1, \ldots, n\}$ into itself. Then the corresponding random variables

$\{\Pi^*(i)|i = 1, \ldots, n\}$ could be independent under some probability p^* on Ω^*, in which case the product of probabilities $p^*(\pi)$ over any Latin set L would not depend on the particular choice of L. Quasi-independence preserves this aspect of independence.

Remark C. One corollary of the characterization $\mathcal{Q} = \mathcal{L}$ proven below is that the above definition of quasi-independence matches that of Bishop, Fienberg and Holland (1975, p. 178) for the n^n contingency table $[P\{\Pi(i) = j_i\}; i, j_i \in \{1, \ldots, n\}]$ with $n^n - n!$ structural zeroes.

Lemma 2.1 $\mathcal{Q} \supset \mathcal{L}$.

Proof For any $p \in \mathcal{L}$ and any Latin set L,

$$\prod_{\pi \in L} p(\pi) = \Psi^{-n}(\theta) e^{\sum \theta_{ij}} = \text{constant}.$$

□

The goal of this section is to prove, conversely, that $\mathcal{L} \supset \mathcal{Q}$, so that the exponential family $\mathcal{L}$ is characterized by the property of quasi-independence. To understand the idea of the proof, one more definition is necessary.

Definitions. Let G be a group and M an additive abelian group. The group M is called a *left G-module* if for each $g \in G$ and $m \in M$ a product $gm \in M$ is defined such that

$$g(m + m') = gm + gm', \tag{2.1}$$
$$gg'(m) = g[g'(m)],$$

and

$$em = m \text{ for the identity element } e \text{ of } G.$$

Similarly, M is called a *right G-module* if a product $mg \in M$ is defined with the right multiplication properties analogous to (2.1). M is called a *G-module* if it is both a left and right G-module. A subgroup N of M is a *G-submodule* if it is closed under left and right multiplication be G. A G-submodule of M is *trivial* if it consists of only the zero element of M or of M itself.

The set Ω is a group under composition $\circ$ of permutations defined by $\pi \circ \sigma(i) = \pi[\sigma(i)]$. Let $\Im$ be the vector space of real valued functions on Ω, and for each $\lambda, \pi \in \Omega$, define left multiplication by $\lambda f(\pi) = f(\lambda \circ \pi)$ so that $\Im$ becomes a left Ω-module. Similarly for each $\sigma, \pi \in \Omega$ define right multiplication by $f\sigma(\pi) = f(\pi \circ \sigma)$, so that $\Im$ is indeed an Ω-module. Let V be the vector subspace of $\Im$ spanned by $\{\log p(\cdot)|p \in \mathcal{L}\}$, and let W be the vector subspace of $\Im$ spanned by $\{\log q(\cdot)|q \in \mathcal{Q}\}$. We shall show that V and W are the same Ω-submodule of $\Im$.

First, however, we need to make precise the relationships between $\mathcal{L}$ and V and between $\mathcal{Q}$ and W, so that we can conclude from $V = W$ that $\mathcal{L} = \mathcal{Q}$. To do this, define $\phi : \Im \to \mathcal{L}$ by $\phi[f(\pi)] = \exp[f(\pi)] / \sum_{\nu \in \Omega} \exp[f(\nu)]$. Then we get the following lemma.

Lemma 2.2 *$\mathcal{L} = \phi(V)$ and $\mathcal{Q} = \phi(\mathcal{W})$.*

Proof $\underline{\mathcal{L} = \phi(V)}$ V is also spanned by $\{[R(\cdot)]_{ij} | i, j \in \{1, \ldots, n\}\}$, so that each $f \in V$ can be expressed as $f(\pi) = tr[\theta R(\pi)]$. Thus $\phi[f(\pi)]$ has exactly the form (1.2).

$\underline{\mathcal{Q} = \phi(\mathcal{W})}$ If $f \in W$, then $f = \sum_k a_k \log q_k = \log[\prod_k (q_k)^{a_k}]$ for some $a_k \in \Re$ and $q_k \in \mathcal{Q}$. Let $c = \sum_{\nu \in \Omega} \exp[f(\nu)]$ so that $\phi(f) = c^{-1} \prod_k (q_k)^{a_k}$ is a probability function $q \in \Im$. Then for any Latin set L,

$$\prod_{\pi \in L} q(\pi) = c^{-n} \prod_{\pi \in L} \prod_k [q_k(\pi)]^{a_k}$$

$$= c^{-n} \prod_k \left[\prod_{\pi \in L} q_k(\pi) \right]^{a_k},$$

which does not depend on L because each $q_k \in \mathcal{Q}$ implies that $\prod_{\pi \in L} q_k(\pi)$ does not depend on L. Thus $\mathcal{Q} \supset \phi(\mathcal{W})$; but for any $q \in \mathcal{Q}$, $q = \phi(\log q) \in \phi(W)$, so that $\phi(W) \supset \mathcal{Q}$. □

In fact, the stronger result $V = \phi^{-1}(\mathcal{L})$ and $W = \phi^{-1}(\mathcal{Q})$ is true, but not needed for our theorem. We will show that V and W are both Ω-modules, that they must have the same decomposition into irreducible Ω-modules, and hence must be equal. Then, by Lemma 2.2, $\mathcal{Q} = \phi(\mathcal{W}) = \phi(\mathcal{V}) = \mathcal{L}$.

Lemma 2.3 *V is an Ω-submodule of $\Im$.*

Proof As vector subspace of $\Im$, V is clearly a subgroup of $\Im$. It only needs to be shown that V is closed under left and right multiplication by permutations. Suppose that $f(\pi) = tr[\theta R(\pi)] \in V$. Then $\lambda f(\pi) = f(\lambda\pi) = tr[\theta R(\lambda\pi)] = tr\{[\theta R(\lambda)] R(\pi)\} \in V$, and similarly $f\sigma(\pi) = f(\pi\sigma) = tr[\theta R(\pi\sigma)] = tr\{[R(\sigma)\theta] R(\pi)\} \in V$, as required. □

Lemma 2.4 *W is an Ω-submodule of $\Im$.*

Proof Again, we only need to show that W is closed under left and right multiplication by permutations. Let L be a Latin set. Then for any $f \in W$, $\sum_{\pi \in L} f(\pi)$ does not depend on the specific choice of L; and for each $\lambda, \sigma \in \Omega$, $L_1 = \lambda L = \{\lambda\pi | \pi \in L\}$ is also a Latin set, as is $L_2 = L\sigma = \{\pi\sigma | \pi \in L\}$. Therefore

$$\sum_{\pi \in L} \lambda f(\pi) = \sum_{\pi \in L} f(\lambda\pi) = \sum_{\pi \in L_1} f(\pi)$$

$$= \text{constant}$$

$$= \sum_{\pi \in L_2} f(\pi) = \sum_{\pi \in L} f(\pi\sigma) = \sum_{\pi \in L} f\sigma(\pi)$$

for any Latin set L. It follows that $q\lambda = \phi(\lambda f)$ and $q\sigma = \phi(f\sigma)$ are $\mathcal{Q}$; moreover, λf is proportional to $q\lambda$ and $f\sigma$ is proportional to $q\sigma$. Thus λf and $f\sigma$ are both in W. □

Remark D. Suppose that people independently rank items in a set $\{1, \ldots, n\}$, with each person's ranking represented by a permutation π. Suppose further that these people come from two distinct populations. Let $q_k(\pi)$ be the probability that a person from population k ($k = 1, 2$) produces the ranking π, and define the *agreement* between q_1 and q_2 as $q_1@q_2(\pi) = [q_1(\pi)q_2(\pi)]/\sum_\nu[q_1(\nu)q_2(\nu)]$. This is the conditional distribution of rankings given that two people from different populations agree on the same ranking. If $q_1, q_2 \in \mathcal{Q}$, Lemma 2.4 implies that their agreement is also in $\mathcal{Q}$. In fact, if M is any Ω-module, then $\phi(M)$ is closed under agreement.
Definition. A G-module M is *irreducible* if its only G-submodules are trivial; otherwise it is *reducible.*

The next lemma, a version of Maschke's theorem (see, for example, Curtis and Reiner, 1962, p.41), has a simple proof in the present setting, where $\Im$ is an inner product vector space rather than just a group.

Lemma 2.5 *Any Ω-submodule A of $\Im$ may be expressed as a vector space direct sum of irreducible Ω-submodules.*

Proof Because $\Im$ is finite dimensional, we need only show that if A is reducible, then it can be written as a direct sum of irreducible Ω-submodules. Specifically, we show that if B is a nontrivial Ω-submodule of A, the orthogonal complement B^c of B in A is an Ω-submodule of A.

Let $< h, f >= \sum_{\nu\in\Omega} h(\nu)f(\nu)$ denote the standard inner product of h and $f \in \Im$. This inner product is both left and right invariant in the sense that for every $\lambda, \sigma \in \Omega$, $< \lambda h, \lambda f >=< h, f >=< h\sigma, f\sigma >$. Thus if $f \in B^c$ and $h \in B$, then $< h, \lambda f >=< \lambda^{-1}h, f >= 0$, since $\lambda^{-1}h \in B$ because B is an Ω-submodule. Likewise, $< h, f\sigma >=< h\sigma^{-1}, f >= 0$, shows that B^c is closed under right multiplication. B^c is therefore an Ω-submodule of A. □

The simplest example of an irreducible Ω-module is the subspace V_0 of $\Im$ consisting of all constant functions on Ω. Let V_1 be the orthogonal complement of V_0 in V. By Lemma 2.5, it is an Ω-submodule. The following example illustrates the construction of irreducible Ω-modules in general, and shows that $V = V_0 \oplus V_1$ is the decomposition of V into irreducible Ω-submodules promised by Maschke's theorem.
Example. V_1 is irreducible.

Proof We show that V_1 is generated by a single function. Let τ_j be the permutation in Ω that transposes j and n; that is, $\tau_j(n) = j$, $\tau_j(j) = n$, and $\tau_j(i) = i$ otherwise. Define f_1 to be the indicator function of the set $\{\pi(n) = n\}$ so that $h_0 = f_1 - n^{-1}$ is in V_1. Any $f \in V$ can then be expressed as

$$f(\pi) = \sum_{i,j} \theta_{j,i}[R(\pi)]_{i,j} = \sum_{i,j} \theta_{j,i}\tau_j[f_1\tau_i](\pi).$$

Since

$$(n!)^{-1}\sum_{\nu\in\Omega} f(\nu) = n^{-1}\sum_{j,i}\theta_{j,i},$$

any $h \in V_1$ can be expressed as

$$f(\pi) - (n!)^{-1}\sum_{\nu\in\Omega} f(\nu) = \sum_{i,j}\theta_{j,i}\{\sigma_j[f_1\sigma_i](\pi) - n^{-1}\} = \sum_{i,j}\theta_{j,i}\sigma_j[h_0\sigma_i](\pi).$$

Thus h_0 generates V_1. Similarly, since $h_1 = f_1 - \tau_1 f_1$ is in V_1 and

$$h_0 = n^{-1}\sum_{j=2}^{n}\tau_j\tau_1 h_1,$$

it follows that h_1 also generates V_1; we mention this because h_1 is the standard generator described in Lemma 2.7 below. We conclude that any Ω-submodule that contains h_1 must contain all of V_1. Together with Lemmas 2.3 and 2.5, this proves that V_1 is an irreducible Ω-module. □

In fact, each irreducible Ω-module on $\Im$ is generated by a single element (see, for example, Curtis and Reiner, 1962, p. 190), a fact that we now need. Including this fact, we need only two non-elementary results, the enumeration of the irreducible Ω-submodules of $\Im$ and the form of a generating function for each of the irreducible Ω-submodules. Fortunately, we require just a few more definitions to discuss these results.

Definition. Let n be a positive integer. A *partition* $z = (z_1, \ldots, z_n)$ of n is a vector with non-negative integer components arranged in nonincreasing order and summing to n. For any fixed partition z, let

ζ be the number of positive components of z;

$$s_k = \sum_{j=1}^{k} z_j,\ k = 1, \ldots, \zeta \text{ with } s_0 = 0;$$

$$Z_k = \{1 + s_{k-1}, \ldots, s_k\},\ k = 1, \ldots, \zeta;$$

and

$f_z \in \Im$ be the indicator function of the event

$$\bigcap_{k=1}^{\zeta}\{\pi(Z_k) = Z_k\}.$$

Remark E. It is helpful to think of the partition z as the triangular array

$$\begin{array}{c} 1,\ldots\ldots\ldots\ldots\ldots\ldots, s_1 \\ s_1+1,\ldots\ldots\ldots\ldots, s_2 \\ \ldots\ldots\ldots\ldots \\ s_{\zeta-1}+1,\ldots\ldots, s_\zeta, \end{array}$$

called an *ordered tableau*, with row k consisting of the elements of Z_k, in order. This visual aid is especially useful for constructing the column stabilizer of z, defined below. (See, for example, Stanley, 1971.)

Remark F. The Ω-module M_z generated by f_z is sometimes called a *Young module.* The structure of Young modules, together with the ordering of majorization on partitions, makes $\{Q(M_z)\}$ an interesting system of probability models, with $Q(M_z)$ including more complicated forms of interaction among the component variables $\Pi(i)$ as z decreases in the sense of majorization.

Proofs of the following two lemmas may be found in James (1978, pp. 6 and 16, respectively).

Lemma 2.6 *There is a one-to-one correspondence $z \to S_z$ between partitions of n and irreducible Ω-modules.*

Definition. For any partition z, let $\zeta_j = \#\{z_i \geq j\}$, and let $Z_j^c = \{s_{k-1} + j | k = 1, \ldots, \zeta_j\}$ be the jth column of the ordered tableau associated with z. Then the *column stabilizer* Γ_z of z is defined by $\Gamma_z = \{\gamma \in \Omega | \gamma(Z_j^c) = Z_j^c, j = 1, \ldots, z_1\}$.

Lemma 2.7 *For each partition z, the irreducible Ω-module S_z is generated by*

$$g_z = \sum_{\gamma \in \Gamma_z} sgn(\gamma)\gamma f_z,$$

where $sgn(\gamma) = 1$ or -1 as γ is even or odd, respectively.

Example. $n = 4$
$z = (2, 2, 0, 0)$. In this case

$$f_z(\pi) = \begin{cases} 1 & \text{if } \{\pi(1), \pi(2)\} = \{1, 2\}; \\ 0 & \text{otherwise.} \end{cases}$$

and

$$g_z(\pi) = \begin{cases} 1 & \text{if } \{\pi(1), \pi(2)\} = \{1, 2\} \text{ or } \{3, 4\} \\ -1 & \text{if } \{\pi(1), \pi(2)\} = \{1, 4\} \text{ or } \{2, 3\} \\ 0 & \text{otherwise.} \end{cases}$$

For Latin sets L_1 and L_2 defined by

$$L_1 = \begin{bmatrix} (1 & 2 & 3 & 4) \\ (2 & 1 & 4 & 3) \\ (3 & 4 & 1 & 2) \\ (4 & 3 & 2 & 1) \end{bmatrix} \quad \text{and} \quad L_2 = \begin{bmatrix} (1 & 4 & 2 & 3) \\ (2 & 3 & 4 & 1) \\ (3 & 2 & 1 & 4) \\ (4 & 1 & 2 & 3) \end{bmatrix}$$

$$\sum_{\pi \in L_1} g_z(\pi) = 4 \neq -4 = \sum_{\pi \in L_2} g_z(\pi)$$

and so $S_z \notin W$.

$z = (2,\ 1,\ 1\ 0)$. In this case

$$f_z(\pi) = \begin{cases} 1 & \text{if } \pi(3) = 3 \text{ and } \pi(4) = 4; \\ 0 & \text{otherwise.} \end{cases}$$

and

$$g_z(\pi) = \begin{cases} 1 & \text{if } [\pi(3),\ \pi(4)] = [3,\ 4] \text{ or } [1,\ 3] \text{ or } [4,\ 1]; \\ -1 & \text{if } [\pi(3),\ \pi(4)] = [4,\ 3] \text{ or } [3,\ 1] \text{ or } [1,\ 4]; \\ 0 & \text{otherwise.} \end{cases}$$

For Latin sets L_3 and L_4 defined by

$$L_3 = \begin{bmatrix} (1 & 2 & 3 & 4) \\ (2 & 3 & 4 & 1) \\ (3 & 4 & 1 & 2) \\ (4 & 1 & 2 & 3) \end{bmatrix} \quad \text{and} \quad L_4 = \begin{bmatrix} (1 & 4 & 2 & 3) \\ (2 & 1 & 4 & 3) \\ (3 & 2 & 1 & 4) \\ (4 & 3 & 2 & 1) \end{bmatrix}$$

$$\sum_{\pi \in L_3} = 2 \neq -2 = \sum_{\pi \in L_4} g_z(\pi)$$

and so $S_z \notin W$.

$z = (1,\ 1,\ 1,\ 1)$. Here

$$g_z(\pi) = \begin{cases} 1 & \text{if } \pi \text{ is even} \\ 0 & \text{if } \pi \text{ is odd.} \end{cases}$$

As for the case $z = (2,\ 2,\ 0,\ 0)$, we again get

$$\sum_{\pi \in L_1} g_z(\pi) = 4 \neq -4 = \sum_{\pi \in L_2} g_z(\pi)$$

and so $S_z \notin W$.

The above example effectively shows that $\mathcal{Q} = \mathcal{L}$ in case $n = 4$. The following theorem gives a formal proof for general n. The key of the proof is to find for each partition $z < (n-2,\ 2,\ 0, \ldots, 0)$ a pair of Latin sets L_1 and L_2 such that $\sum_{L_1} g_z(\pi) \neq \sum_{L_z} g_z(\pi)$. The theorem shows how to construct such pairs for a general partition z.

Theorem 2.8 $\mathcal{Q} = \mathcal{L}$

Proof By Lemmas 2.1 and 2.2, we have to show $V \supset W$; by Lemmas 2.3, 2.4, 2.5, and 2.6, $V \supset W$ if $z < (n-2,\ 2,\ 0,\ \ldots, 0)$ in the sense of majorization implies that $S_z \notin W$; by Lemma 2.7 we just need to show that $g_z \notin W$ for any $z < (n-2,\ 2,\ 0, \ldots, 0)$.

Let τ_z be the transposition of 1 and $z_1 + 1$. Then $\tau_z \in \Gamma_z$ and $\text{sgn}(\tau_z) = -1$, so that

$$g_z(\tau_z \pi) = \sum_{\gamma \in \Gamma_z} \text{sgn}(\gamma) \gamma \tau_z f_z(\pi)$$

$$= \sum_{\nu \in \Gamma_z} \text{-sgn}(\nu)\nu f_z(\pi) \tag{2.2}$$

$$= -g_z(\pi). \tag{2.3}$$

Thus for any Latin set L, $\sum_{\tau_z L} g_z(\pi) = -\sum_L g_z(\pi)$, and it remains only to show that for every $z < (n-2,\ 2,\ 0, \ldots, 0)$, there exists a Latin set L such that $\sum_L g_z(\pi) \neq 0$. In terms of ordered tableaux, we proceed as follows: First construct such a set L for the two cases that correspond to the two basic forms for the bottom of an ordered tableau, and then show how to construct L when successively larger blocks are placed on top of the tableau. □

Case A $z = (z_1,\ z_2,\ 0, \ldots, 0)$ with $z_1 \geq z_2 \geq 2$.

If z_2 is even, let L_{12} be a Latin square formed from Z_2, let L_{22} be a Latin square formed from $\{1, \ldots, z_2\}$, and let L be a Latin square of the form

$$L = \begin{bmatrix} \cdot & L_{12} \\ \cdot & L_{22} \\ \cdot & \cdot \end{bmatrix}.$$

Then $\sum_L g_z(\pi) = 2z_2 > 0$, since any permutation π in the last $n - 2z_2$ rows of L must have $f_z(\pi) = 0$.

If z_2 is odd, the above construction gives $\sum_L g_z(\pi) = 0$, and must thus be modified. In this situation, let L'_{12} be a Latin square formed from $\{1, (n-z_2)+2, \ldots, n\}$, let L''_{12} be the square matrix obtained by replacing the column of L'_{12} with first entry 1 by the column $[(n-z_2)+1,\ 2, \ldots, z_2]$, and let L_1 be the Latin rectangle of the form $L_1 = [\ \cdot\ L'_{12}]$. By construction, each row of L_1 has $g_z(\pi) = 1$, because each $\pi \in L_1$ has the form $\pi = \tau_j \tau_1 \rho$, where $\rho Z_2 = Z_2$, and τ_j is the transposition in the column stabilizer Γ_z that transposes j and $(n - z_2) + j$.

The Latin rectangle L_1 is now completed into a Latin square L whose remaining rows π have $g_z(\pi) = 0$. Let L'_{22} be the Latin square formed from Z_2, let L''_{22} be the square matrix obtained by replacing the column of L'_{22} with first entry $(n-z_2)+1$ by the column $[(n-z_2)+2, \ldots, n,\ 1]$, and let L_2 be the Latin rectangle of the form $L_2 = [\cdot \quad L''_{22}]$. Then any Latin square of the form

$$L = \begin{bmatrix} L_1 \\ L_2 \\ \cdot \end{bmatrix}$$

has $\sum_L g_z(\pi) = z_2$.

Case B $z = (z_1,\ z_2,\ 1,\ 0, \ldots,\ 0)$.

Setting L equal to the Latin set corresponding to the cyclic group generated by $\pi_c = (2, \ldots, n, 1)$ always works in this case. To see this, first note that the only π for which $\gamma f_z(\pi) \neq 0$ for some $\gamma \in \Gamma_z$ must have $\pi(n) \in \{1,\ z_1 + 1,\ n\}$. For π equal to the identity $f_z(\pi) = 1$; for $\pi = \pi_c$,

$\gamma f_z(\pi) = 1$ when γ is the cycle $n \to (z_1 + 1) \to 1 \to n$, which is even. Thus $\sum_L g_z(\pi) \geq 1$ since two out of its three possible non-zero terms are 1.

As mentioned earlier, the above two cases will be used to form the base or bottom of a general ordered partition. In order to prove that larger blocks may be added on top of these bases, we need to show that these additional blocks themselves have the desired property. Thus we consider

Case C (block form) $z = (z_1, \ldots, z_n)$ with $z_1 = \cdots = z_k = n/k$ for $k \geq 3$ and $z_{k+1} = \cdots = z_n = 0$.

We first show that there is a Latin set K of permutations of $\{1, \ldots, k\}$ consisting of only even permutations: If k is odd, the the cyclic group K_0 generated by $(2, \ldots, k, 1)$ constitutes such a set; if k is even, transpose items 1 and $k - 1$ in each of the odd permutations of K_0 to obtain K. Because 1 and $k - 1$ appear only in the even columns of K_0, this K is indeed a Latin set.

Now let L_j be a Latin square formed from $Z_j (j = 1, \ldots, k)$, and let $K = \{\sigma_1, \ldots, \sigma_k\}$ be a Latin set of even permutations of $\{1, \ldots, k\}$. Then the $n \times n$ Latin square L with (i, j) block equal to $L_{\sigma_j(i)}$ satisfies $\sum_L gz(\pi) = n$.

Appending of blocks

We now consider partitions of the form

$$z = \begin{bmatrix} w \\ y \end{bmatrix}$$

where w is a $k \times m$ block, and $y = (y_1, \ldots, y_{(n-km)})$ is a general partition of $n - km$ with $y_1 < k$. We assume that y has the desired property, and we want to show that z does also.

Case 1 $k = 1$.

Let L_{12} be a Latin set on $\{m + 1, \ldots, n\}$ such that $\sum_{L_12} g_y(\sigma) > 0$. If $n - m > m$, restrict L_{12} to an $m \times (n - m)$ Latin rectangle by first deleting any σ with $g_y(\sigma) \leq 0$.

For any permutation σ of $\{m + 1, \ldots, n\}$, let π_σ be its embedding into Ω defined by $\pi_\sigma(i) = i$ for $i = 1, \ldots, m$, and $\pi_\sigma(i) - \sigma(i)$ otherwise, and note that $g_z(\pi_\sigma) = g_y(\sigma)$. Let L_{11} be a Latin square formed $\{1, \ldots, m\}$. If $m > n - m$ restrict L_{11} to an $(n - m) \times m$ Latin rectangle with the entry "m" in each of its first $(n - m)$ columns. Then any completed Latin square of the form

$$L = \begin{bmatrix} L_{11} & L_{12} \\ L_{21} & L_{22} \end{bmatrix}$$

has $\sum_L g_z(\pi) > 0$ because each column (and hence row) of L_{22} has an "m" in it.

Case 2 $k > 1$.

Note that for $k = 2$, case A above guarantees that w has an associated Latin set L_w with the desired property that $\sum_{Lw} g_w(\pi) > 0$, and for $k > 2$

case C above makes the same guarantee. The proof now follows as for case 1, with L_w replacing L_{11}. $\|$

Remark G. The proof of the above theorem shows that if a Latin square is chosen at random according to the uniform distribution on the set of all Latin squares of fixed dimension, then for each partition z, $X_z = \sum_L g_z(\pi)$ has a non-degenerate distribution that is symmetric about 0. Further research on the distribution of X_z should provide some insight into the relative frequency of various types of Latin squares.

3 Application

The fact that the distributions in $\mathcal{L}$ are the only distributions possessing the property of quasi-independence can be used as the basis of an algorithm for a minimal majorization decomposition of a doubly stochastic matrix. The idea is to start with an arbitrary decomposition and then diffuse it by transferring probability mass between pairs of Latin sets. The transfers preserve the decomposition but make the distribution approach the maximal entropy distribution. In fact for small values of n it is possible to achieve quasi-independence to any degree of numerical accuracy.

A key to obtaining the desired minimal decomposition is the fact that the parametric matrix θ in (1.2) may be determined simply by knowing the probabilities of $n(n-1)$ particular permutations. First notice that for identifiability we may set $\theta_{jn} = \theta_{ni} = 0$ for $i, j = 1, \ldots, n$. Let τ_i be the permutation in Ω that transposes i and n; that is, $\tau_i(n) = i$, $\tau_i(i) = n$, and $\tau_i(k) = k$ otherwise. Then by (1.2)

$$p(\pi|\theta) = \Psi^{-1}(\theta) \exp\left[\sum \theta_{\pi(i),i}\right],$$

so that

$$\log\{[p(\pi|\theta)]/[p(\pi \circ \tau_i|\theta)]\} = [\theta_{\pi(i),i} + \theta_{\pi(n),n}] - [\theta_{\pi(i),n} + \theta_{\pi(n),i}].$$

In particular, if $\pi(n) = n$, then

$$\log\{[p(\pi|\theta)]/[p(\pi \circ \tau_i|\theta)]\} = \theta_{\pi(i),i}. \tag{3.1}$$

Let $L^- = \{\mu_1, \ldots, \mu_{n-1}\}$ be a Latin set on the first $n-1$ integers. Define $\nu_j \in \Omega (j = 1, \ldots, n-1)$ by $\nu_j(i) = \mu_j(i)$ for $i = 1, \ldots, n-1$, and $\nu_j(n) = n$. Then (3.1) implies that θ is determined by the probabilities of the $n(n-1)$ permutations in

$$S = \{\nu_j \circ \tau_i | j = 1, \ldots, n-1;\ i = 1, \ldots, n\}.$$

Let q be some arbitrary decomposition of a given doubly stochastic matrix D; that is, q is a probability function on Ω such that $E_q[R(\pi)] = D$.

Let Υ be a collection of Latin sets whose union contains both S and the support of q. We now give a practical method for transferring probability across sets in Υ to achieve a new decomposition p such that $E_p[R(\pi)] = D$ and (1.4) holds for all Latin sets $L \in \Upsilon$. In the case where Υ consists of all Latin sets, it follows from the Theorem and the already cited result of Joe (1987) that p is a minimal decomposition of D.

Let L_1 and L_2 be two Latin sets in Υ. There is a one-to-one correspondence between $L_1 - L_2$ and $L_2 - L_1$, and so if an amount Δ of probability is transferred from each $\pi \in L_1 - L_2$ to a corresponding $\sigma \in L_2 - L_1$, $E_q[R(\pi)]$ will not change. The amount Δ that will balance the product of probabilities on L_1 and L_2 is determined by the equation

$$\prod_{\pi \in L_1 - L_2} [q(\pi) - \Delta] = \prod_{\sigma \in L_2 - L_1} [q(\sigma) + \Delta]. \tag{3.2}$$

For practical purposes the solution

$$\Delta = \left[\prod_{\pi \in L_1 - L_2} q(pi) - \prod_{\sigma \in L_2 - L_1} q(\sigma) \right] \Big/ \left\{ \left[\sum_{\pi \in L_1 - L_2} \prod_{\mu \neq \pi} q(\mu) \right] + \left[\sum_{\sigma \in L_2 - L_1} \prod_{\lambda \neq \sigma} q(\lambda) \right] \right\} \tag{3.3}$$

to the linear approximation of (3.2) suffices to balance the product of probabilities on L_1 and L_2. Note that $\mu \in L_1 - L_2$ and $\lambda \in L_2 - L_1$ in (3.3).

Suppose that in successive iterations the sets L_1 and L_2 are chosen from Υ to maximize and minimize, respectively, the product of the probabilities of the permutations inside them. Then after a suitable number of such transfers the products of the probabilities will be balanced on all latin sets in Υ.

At this point (3.1) may be used to compute the $\theta_{ji\cdot}(i,\ j = 1, \ldots, n-1)$. If n is not too large, (1.2) may be used to compute the probabilities for each $\pi \in \Omega$, and the doubly stochastic matrix $E[R(\pi)|\theta]$ may be computed and compared to the original D to check the accuracy of the approximation.

For moderate and large values of n, the accuracy of the approximation depends heavily on the initial decomposition q and the collection Υ of Latin set over which to balance. Optimal choices of q and Υ remain an open research problem.

References

Birkhoff, G. (1946). "Tres observaciones sobre el algebra lineal." *Univ. Nac, Tucuman Rev. Ser. A* **5**, 147–151, [Math. Rev. **8** (1947) 561].

Bishop, Y., Fienberg, S. and Holland, P. (1975). *Discrete Multivariate Analysis: Theory and Practice.* MIT Press, Cambridge, Ma.

Brown, L.D. (1986). *Fundamentals of Statistical Exponential Families.* Institute of Mathematical Statistics Lecture Notes—Monograph Series, Hayward, CA.

Curtis, C.W. and Reiner, I. (1962). *Representation Theory of Finite Groups and Associative Algebras.* Interscience Publishers, New York.

Feller, W. (1950). *An Introduction to Probability Theory and Its Applications, 1st ed. vol. 1*, John Wiley & Sons, New York.

Good, I.J. (1963). "Maximum entropy for hypothesis formulation, especially for multidimensional contingency tables." *Ann. Math. Statis.* **34**, 911–934.

James, G.D. (1978). *Representation Theory of the Symmetric Group.* Springer-Verlag, New York.

Joe, H. (1987). "Majorization, randomness and dependence for multivariate distributions." *Ann. Prob.* **15**, 1217–1225.

Marshall, A. and Olkin, I. (1979). *Inequalities: Theory of Majorization and Its Applications.* Academic Press, New York

Stanley, R.P. (1971). "Theory and application of plane partitions," Parts 1 and 2. *Stud. Appl. Math.* **50**, 167–188, 259–279.

Part B

Multivariate Analysis and Association

10

The Asymptotic Distribution of Characteristic Roots and Vectors in Multivariate Components of Variance

T.W. Anderson[1]

ABSTRACT The asymptotic distribution of the characteristic roots and vectors of one Wishart matrix in the metric of another as the two degrees of freedom increase in fixed proportion is obtained. In the balanced one-way multivariate analysis of variance these two matrices are the sample effect and error covariance matrices, and the numbers of degrees of freedom are (approximately) proportional to the number of classes. The maximum likelihood estimate of the effect covariance matrix of a given rank depends on the characteristic roots and vectors.

1 Introduction

In the one-way multivariate analysis of variance the covariance matrix of effects and the covariance matrix of errors are to be estimated. In the balanced case with normally distributed effects and errors the maximum likelihood estimators of these matrices are based on the sample "between" and "within" cross-product matrices, which have independent Wishart distributions. The estimators, in fact, depend nontrivially on the ordered characteristic roots and corresponding vectors of the "between" matrix in the metric of the "within" matrix (Anderson (1984a), Schott and Saw (1984), and Anderson, Anderson, and Olkin (1986)). This paper develops the asymptotic distribution of such roots and vectors in a general setting. The two matrices do not need to have Wishart distributions; they are only required to have asymptotic distributions with certain symmetry properties. The asymptotic distribution of the estimates of the population covariance matrices are characterized on this basis. Because Ingram Olkin has been involved in deriving the maximum likelihood estimators in the multivariate components of variance as well as playing an important role in multivariate

[1]Stanford University.

statistical analysis in general, it seems fitting to contribute this paper to a volume in his honor.

Let $\boldsymbol{H}$ and $\boldsymbol{G}$ be random $p \times p$ symmetric matrices with $\boldsymbol{G}$ positive definite (with probability 1). We treat the asymptotic distribution of the roots of $|\boldsymbol{H} - t\boldsymbol{G}| = 0$ and the corresponding vectors $\boldsymbol{y}$ satisfying $(\boldsymbol{H} - t\boldsymbol{G})\boldsymbol{y} = \boldsymbol{0}$ and a normalization condition. When $\boldsymbol{H}$ and $\boldsymbol{G}$ have independent Wishart distributions $W(\boldsymbol{\Phi}, M)$ and $W(\boldsymbol{\Psi}, N)$, an explicit asymptotic distribution is obtained.

As noted above, a particular case of $\boldsymbol{H}$ and $\boldsymbol{G}$ arises in the balanced one-way multivariate analysis of variance with random effects. Suppose

$$\boldsymbol{x}_{\alpha j} = \boldsymbol{\mu} + \boldsymbol{v}_\alpha + \boldsymbol{u}_{\alpha j}, \qquad \alpha = 1, \ldots, n, \quad j = 1, \ldots, k, \tag{1.1}$$

where the unobservable $\boldsymbol{v}_\alpha$'s and $\boldsymbol{u}_{\alpha j}$'s are independently normally distributed with means $\boldsymbol{0}$ and covariance matrices $\mathcal{E}\boldsymbol{v}_\alpha \boldsymbol{v}_\alpha' = \boldsymbol{\Theta}$ and $\mathcal{E}\boldsymbol{u}_{\alpha j}\boldsymbol{u}_{\alpha j}' = \boldsymbol{\Psi}$. Then $\boldsymbol{x}_{\alpha j}$ has the distribution $N(\boldsymbol{\mu}, \boldsymbol{\Psi} + \boldsymbol{\Theta})$, $\sqrt{k}\bar{\boldsymbol{x}}_\alpha = (1/\sqrt{k}) \sum_{j=1}^k \boldsymbol{x}_{\alpha j}$ has the distribution $N(\sqrt{k}\,\boldsymbol{\mu}, \boldsymbol{\Psi} + k\boldsymbol{\Theta})$, and $\sqrt{nk}\,\bar{\boldsymbol{x}} = \sqrt{k/n}\,\sum_{\alpha=1}^n \bar{\boldsymbol{x}}_\alpha$ has the distribution $N(\sqrt{nk}\,\boldsymbol{\mu}, \boldsymbol{\Psi} + k\boldsymbol{\Theta})$. The matrices

$$\boldsymbol{H} = k \sum_{\alpha=1}^{n} (\bar{\boldsymbol{x}}_\alpha - \bar{\boldsymbol{x}})(\bar{\boldsymbol{x}}_\alpha - \bar{\boldsymbol{x}})' \tag{1.2}$$

and

$$\boldsymbol{G} = \sum_{\alpha=1}^{n} \sum_{j=1}^{k} (\boldsymbol{x}_{\alpha j} - \bar{\boldsymbol{x}}_\alpha)(\boldsymbol{x}_{\alpha j} - \bar{\boldsymbol{x}}_\alpha)', \tag{1.3}$$

are distributed independently according to $W(\boldsymbol{\Psi} + k\boldsymbol{\Theta}, n-1)$ and $W(\boldsymbol{\Psi}, n(k-1))$, respectively. This is the case of $\boldsymbol{\Phi} = \boldsymbol{\Psi} + k\boldsymbol{\Theta}$, $M = n - 1$, and $N = n(k-1)$, representing k replications on n categories. The covariance matrix of effects $\boldsymbol{\Theta}$ is positive semidefinite, but not necessarily of full rank.

If $k \to \infty$, $\{1/[n(k-1)]\}\boldsymbol{G} \xrightarrow{\text{p}} \boldsymbol{\Psi}$ and $(1/k)\boldsymbol{H} \xrightarrow{\text{d}} \boldsymbol{H}_\infty$, which has the distribution $W(\boldsymbol{\Theta}, n-1)$. A consistent estimator of $\boldsymbol{\Theta}$ is not available. (In this model $\bar{\boldsymbol{x}}, \boldsymbol{H}$, and $\boldsymbol{G}$ are a sufficient set of statistics.) n the other hand, if $n \to \infty$, then $M = n - 1 \to \infty$, $N = n(k-1) \to \infty$, and $M/N = (n-1)/[n(k-1)] \to 1/(k-1)$, and consistent estimators are available. This latter case is covered in this paper.

In our treatment we permit the roots δ of $|\boldsymbol{\Phi} - \delta\boldsymbol{\Psi}| = 0$ to have arbitrary multiplicities. When $\boldsymbol{H}$ and $\boldsymbol{G}$ have Wishart distributions, the asymptotic distribution of the roots of $|\boldsymbol{H} - t\boldsymbol{G}| = 0$ for this general case has been given by Li, Pillai, and Chang (1970), Chang (1973), Chattopadhyay and Pillai (1973), and Sugiura (1976); asymptotic expansions have also been given. uirhead (1978) has given a survey of the results. However, the asymptotic distribution of the related vectors has been obtained only for the case of the roots of $|\boldsymbol{\Phi} - \delta\boldsymbol{\Psi}| = 0$ being simple, but multiple roots must be considered in the multivariate analysis of variance, for example, when the rank of $\boldsymbol{\Theta}$ is less than $p-1$. (See Sections 2 and 6.) The asymptotic distribution of the roots and vectors of the Wishart matrices is a special case of a more general theory which holds when the matrices $\boldsymbol{H}$ and $\boldsymbol{G}$ have asymptotic normal distributions with certain symmetry properties (Theorem 4.3), but the asymptotic distribution of the roots and vectors can be obtained under even more general conditions (Theorem 4.1 and Section A2). In this generality Amemiya (1986) has considered the asymptotic distribution of the roots, but not the vectors. Tyler (1987) among others has studied the asymptotic normal distributions of symmetric matrices under general conditions.

2 Characteristic Roots and Vectors

Let $d_1 > d_2 > \cdots > d_p > 0$ be the roots of

$$|\bar{\boldsymbol{H}} - d\bar{\boldsymbol{G}}| = 0, \tag{2.1}$$

where $\bar{\boldsymbol{H}} = (1/M)\boldsymbol{H}$ and $\bar{\boldsymbol{G}} = (1/N)\boldsymbol{G}$ with $M \geq p$ and $N \geq p$. (If $\boldsymbol{H}$ and $\boldsymbol{G}$ have densities, the roots of (2.1) exist and are distinct and positive with probability 1.) Let $\boldsymbol{y}_1, \ldots, \boldsymbol{y}_p$ be the solutions to

$$(\bar{\boldsymbol{H}} - d_i\bar{\boldsymbol{G}})\boldsymbol{y} = \boldsymbol{0}, \qquad i = 1, \ldots, p, \tag{2.2}$$

normalized by

$$\boldsymbol{y}'\bar{\boldsymbol{G}}\boldsymbol{y} = \boldsymbol{1}. \tag{2.3}$$

Let $\boldsymbol{Y} = (\boldsymbol{y}_1, \ldots, \boldsymbol{y}_p)$ and $\boldsymbol{D} = \mathrm{diag}(d_1, \ldots, d_p)$, where $\mathrm{diag}(d_1, \ldots, d_p)$ denotes a $p \times p$ diagonal matrix with diagonal elements $d_1, \ldots, d_p$. Then

$$\bar{\boldsymbol{H}}\boldsymbol{Y} = \bar{\boldsymbol{G}}\boldsymbol{Y}\boldsymbol{D}, \qquad \boldsymbol{Y}'\bar{\boldsymbol{G}}\boldsymbol{Y} = \boldsymbol{I}. \tag{2.4}$$

From (2.4) we obtain

$$\boldsymbol{Y}'\bar{\boldsymbol{H}}\boldsymbol{Y} = \boldsymbol{D}. \tag{2.5}$$

Let[2] $\boldsymbol{Z} = \boldsymbol{Y}^{-1}$. Then (2.4) and (2.5) imply

$$\bar{\boldsymbol{H}} = \boldsymbol{Z}'\boldsymbol{D}\boldsymbol{Z}, \qquad \bar{\boldsymbol{G}} = \boldsymbol{Z}'\boldsymbol{Z}. \tag{2.6}$$

The vectors $\boldsymbol{y}_1, \ldots, \boldsymbol{y}_p$ are determined uniquely except for multiplication of each by -1; the rows of $\boldsymbol{Z}$ are determined uniquely except for multiplication by -1. We shall indicate later (Section 5) how this indeterminacy is resolved.

[2]In Anderson, Anderson, and Olkin (1986) $\boldsymbol{Z}$ was defined as $(\boldsymbol{Y}')^{-1}$. We have changed that notation to $\boldsymbol{Y}^{-1}$ in order to agree with Anderson (1951).

Suppose that as $M \to \infty$, $\bar{\boldsymbol{H}} \xrightarrow{\text{p}} \boldsymbol{\Phi}$ and that as $N \to \infty$, $\bar{\boldsymbol{G}} \xrightarrow{\text{p}} \boldsymbol{\Psi}$, where $\boldsymbol{\Psi}$ is positive definite. Let $\delta_1 \geq \cdots \geq \delta_p$ be the roots of

$$|\boldsymbol{\Phi} - \delta \boldsymbol{\Psi}| = 0, \tag{2.7}$$

and let γ_i satisfy

$$(\boldsymbol{\Phi} - \delta_i \boldsymbol{\Psi})\gamma = \mathbf{0}, \tag{2.8}$$

$\gamma_i' \boldsymbol{\Psi} \gamma_i = 1$, and $\gamma_i' \boldsymbol{\Psi} \gamma_j = 0$, $i \neq j$. Let $\boldsymbol{\Delta} = \operatorname{diag}(\delta_1, \ldots, \delta_p)$. Then $\boldsymbol{\Gamma} = (\gamma_1, \ldots, \gamma_p)$ satisfies

$$\boldsymbol{\Phi\Gamma} = \boldsymbol{\Psi\Gamma\Delta}, \quad \boldsymbol{\Gamma}'\boldsymbol{\Psi\Gamma} = \boldsymbol{I}, \quad \boldsymbol{\Gamma}'\boldsymbol{\Phi\Gamma} = \boldsymbol{\Delta}. \tag{2.9}$$

In the case of the balanced one-way components of variance, (2.7) is

$$|\boldsymbol{\Psi} + k\boldsymbol{\Theta} - \delta \boldsymbol{\Psi}| = 0. \tag{2.10}$$

If the rank of $\boldsymbol{\Theta}$ is m, then $\delta_{m+1} = \cdots = \delta_p = 1$. As $n \to \infty$, then $N \to \infty$, $M \to \infty$, $\bar{\boldsymbol{H}} \xrightarrow{\text{p}} \boldsymbol{\Phi}$, $\bar{\boldsymbol{G}} \xrightarrow{\text{p}} \boldsymbol{\Psi}$, and $\boldsymbol{D} \xrightarrow{\text{p}} \boldsymbol{\Delta}$.

We shall suppose that

$$\boldsymbol{\Delta} = \begin{bmatrix} \lambda_1 \boldsymbol{I}_{r_1} & \mathbf{0} & \cdots & \mathbf{0} \\ \mathbf{0} & \lambda_2 \boldsymbol{I}_{r_2} & \cdots & \mathbf{0} \\ \vdots & \vdots & & \vdots \\ \mathbf{0} & \mathbf{0} & \cdots & \lambda_K \boldsymbol{I}_{r_K} \end{bmatrix}, \tag{2.11}$$

where $\lambda_1 > \lambda_2 > \cdots > \lambda_K > 0$; that is, λ_j is a root of multiplicity r_j $(\sum_{j=1}^{K} r_j = p)$, and $\lambda_1, \ldots, \lambda_K$ are distinct. Let

$$\boldsymbol{\Gamma} = (\boldsymbol{\Gamma}_1, \ldots, \boldsymbol{\Gamma}_K) = \begin{bmatrix} \boldsymbol{\Gamma}_{11} & \boldsymbol{\Gamma}_{12} & \cdots & \boldsymbol{\Gamma}_{1K} \\ \boldsymbol{\Gamma}_{21} & \boldsymbol{\Gamma}_{22} & \cdots & \boldsymbol{\Gamma}_{2K} \\ \vdots & \vdots & & \vdots \\ \boldsymbol{\Gamma}_{K1} & \boldsymbol{\Gamma}_{K2} & \cdots & \boldsymbol{\Gamma}_{KK} \end{bmatrix}. \tag{2.12}$$

If $\boldsymbol{\Gamma}$ satisfies (2.9) then $\boldsymbol{\Gamma}^* = (\boldsymbol{\Gamma}_1 \boldsymbol{\Omega}_1, \ldots, \boldsymbol{\Gamma}_K \boldsymbol{\Omega}_K)$, where $\boldsymbol{\Omega}_1, \ldots, \boldsymbol{\Omega}_K$ are orthogonal matrices of order $r_1, \ldots, r_K$, respectively, satisfies (2.9) as well. We shall consider the limiting distribution of $\sqrt{N}\,(\boldsymbol{D} - \boldsymbol{\Delta})$ and $\sqrt{N}\,(\boldsymbol{Z} - \boldsymbol{\Gamma})$ as $M \to \infty$ and $N \to \infty$ in such a way that $M/N \to c$, where $0 < c < \infty$.

3 Reduction of the Problem

First we relate $\boldsymbol{H}$ and $\boldsymbol{G}$ to the canonical forms of $\boldsymbol{\Phi}$ and $\boldsymbol{\Psi}$. Let

$$\boldsymbol{H}^* = \boldsymbol{\Gamma}'\boldsymbol{H\Gamma} = M\boldsymbol{\Gamma}'\boldsymbol{Z}'\boldsymbol{DZ\Gamma}, \tag{3.1}$$

$$\boldsymbol{G}^* = \boldsymbol{\Gamma}'\boldsymbol{G\Gamma} = N\boldsymbol{\Gamma}'\boldsymbol{Z}'\boldsymbol{Z\Gamma}. \tag{3.2}$$

Let $\boldsymbol{Z\Gamma} = \boldsymbol{T}$. Then

$$\frac{1}{M}\boldsymbol{H}^* = \bar{\boldsymbol{H}}^* = \boldsymbol{T}'\boldsymbol{DT}, \quad \frac{1}{N}\boldsymbol{G}^* = \bar{\boldsymbol{G}}^* = \boldsymbol{T}'\boldsymbol{T}. \tag{3.3}$$

If the roots of $|\bar{\boldsymbol{H}}^* - d\bar{\boldsymbol{G}}^*| = 0$ are distinct, then (3.3) determines $\boldsymbol{D}$ and $\boldsymbol{T}$ uniquely if the diagonal elements of $\boldsymbol{D}$ are in descending order and $t_{ii} > 0$ (as long as $t_{ii} \neq 0$).

As $M \to \infty$ and $N \to \infty$, $\bar{\boldsymbol{H}}^* \xrightarrow{\text{p}} \boldsymbol{\Delta}$, $\bar{\boldsymbol{G}}^* \xrightarrow{\text{p}} \boldsymbol{I}$. Let

$$\sqrt{N}(\bar{\boldsymbol{H}}^* - \boldsymbol{\Delta}) = \boldsymbol{U}, \tag{3.4}$$

$$\sqrt{N}(\bar{\boldsymbol{G}}^* - \boldsymbol{I}) = \boldsymbol{V}. \tag{3.5}$$

We suppose that $(\boldsymbol{U}, \boldsymbol{V})$ has a limiting distribution.

If $\boldsymbol{H}^*$ has the distribution $W(\boldsymbol{\Delta}, M)$ (as in the balanced components of variance), $\boldsymbol{U}$ has a limiting normal distribution with mean $\mathbf{0}$ as $M \to \infty$. The functionally independent elements are uncorrelated and asymptotically independent. The ith diagonal element of $\boldsymbol{H}^*$ divided by δ_i has a χ^2-distribution with M degrees of freedom. The limiting distribution of u_{ii} is $N(0, 2\delta_i^2/c)$, and the limiting distribution of $u_{ji} = u_{ij}$ is $N(0, \delta_i\delta_j/c), i \neq j$. If $\boldsymbol{G}^*$ has the distribution $W(\boldsymbol{I}, N), \boldsymbol{V}$ has a limiting normal distribution with mean $\mathbf{0}$ as $N \to \infty$. The functionally independent components are uncorrelated and asymptotically independent. The limiting distribution of v_{ii} is $N(0,2)$, and the limiting distribution of $v_{ji} = v_{ij}$ is $N(0,1), i \neq j$. In the balanced components of variance, $\boldsymbol{U}$ and $\boldsymbol{V}$ are independent.

4 The Asymptotic Distribution of Roots and Vectors in Canonical Form

We shall now study the asymptotic distribution of

$$\boldsymbol{D} = \begin{bmatrix} \boldsymbol{D}_1 & \mathbf{0} & \dots & \mathbf{0} \\ \mathbf{0} & \boldsymbol{D}_2 & \dots & \mathbf{0} \\ \vdots & \vdots & & \vdots \\ \mathbf{0} & \mathbf{0} & \dots & \boldsymbol{D}_K \end{bmatrix}, \quad \boldsymbol{T} = \begin{bmatrix} \boldsymbol{T}_{11} & \boldsymbol{T}_{12} & \dots & \boldsymbol{T}_{1K} \\ \boldsymbol{T}_{21} & \boldsymbol{T}_{22} & \dots & \boldsymbol{T}_{2K} \\ \vdots & \vdots & & \vdots \\ \boldsymbol{T}_{K1} & \boldsymbol{T}_{K2} & \dots & \boldsymbol{T}_{KK} \end{bmatrix}. \tag{4.1}$$

Let

$$\boldsymbol{X}_{gh} = \sqrt{N}\,\boldsymbol{T}_{gh}, \quad g \neq h, \quad g, h = 1, \dots, K, \tag{4.2}$$

$$\boldsymbol{T}_{gg} = \boldsymbol{P}_g\boldsymbol{S}_g\boldsymbol{Q}_g, \qquad g = 1, \dots, K, \tag{4.3}$$

where $\boldsymbol{P}_g$ and $\boldsymbol{Q}_g$ are orthogonal matrices of order r_g and $\boldsymbol{S}_g$ is diagonal with positive diagonal elements, arranged in descending order. (4.3) is the singular value decomposition of $\boldsymbol{T}_{gg}$, $\boldsymbol{T}_{gg}\boldsymbol{T}_{gg}' = \boldsymbol{P}_g\boldsymbol{S}_g^2\boldsymbol{P}_g'$, and $\boldsymbol{T}_{gg}'\boldsymbol{T}_{gg} = \boldsymbol{Q}_g'\boldsymbol{S}_g^2\boldsymbol{Q}_g$. Define

$$\boldsymbol{W}_g = \boldsymbol{P}_g\boldsymbol{Q}_g. \tag{4.4}$$

Note that $\boldsymbol{W}_g$ is uniquely defined, although $\boldsymbol{P}_g$ and $\boldsymbol{Q}_g$ may not be. Define

$$\boldsymbol{X}_{gg} = \sqrt{N}(\boldsymbol{T}_{gg} - \boldsymbol{W}_g) = \sqrt{N}\boldsymbol{P}_g(\boldsymbol{S}_g - \boldsymbol{I}_{r_g})\boldsymbol{Q}_g, \tag{4.5}$$

and

$$\boldsymbol{L}_g = \sqrt{N}(\boldsymbol{D}_g - \lambda_g \boldsymbol{I}_{r_g}), \tag{4.6}$$

which is diagonal. Note that $\boldsymbol{W}_g$ is orthogonal and

$$\boldsymbol{W}_g\boldsymbol{T}'_{gg} = \boldsymbol{P}_g\boldsymbol{S}_g\boldsymbol{P}'_g = \boldsymbol{T}_{gg}\boldsymbol{W}'_g. \tag{4.7}$$

is positive definite if $\boldsymbol{T}_{gg}$ is nonsingular. We want the limiting distribution of $\boldsymbol{X}_{gh},\ g,h = 1,\dots,K,\ \boldsymbol{W}_g,\ g = 1,\dots,K$, and $\boldsymbol{L}_g,\ g = 1,\dots,K$.

Let

$$\boldsymbol{W} = \text{block diag}(\boldsymbol{W}_1, \boldsymbol{W}_2, \dots, \boldsymbol{W}_K), \quad \boldsymbol{L} = \text{block diag}(\boldsymbol{L}_1, \boldsymbol{L}_2, \dots, \boldsymbol{L}_K), \tag{4.8}$$

and define $\boldsymbol{X}$, $\boldsymbol{U}$, and $\boldsymbol{V}$ as composed of blocks $\boldsymbol{X}_{gh}, \boldsymbol{U}_{gh}$, and $\boldsymbol{V}_{gh}$, respectively.

We can write (3.3) as

$$\begin{aligned}\boldsymbol{\Delta} + \frac{1}{\sqrt{N}}\boldsymbol{U} &= \left(\boldsymbol{W} + \frac{1}{\sqrt{N}}\boldsymbol{X}\right)'\left(\boldsymbol{\Delta} + \frac{1}{\sqrt{N}}\boldsymbol{L}\right)\left(\boldsymbol{W} + \frac{1}{\sqrt{N}}\boldsymbol{X}\right)\\ &= \boldsymbol{\Delta} + \frac{1}{\sqrt{N}}(\boldsymbol{X}'\boldsymbol{\Delta}\boldsymbol{W} + \boldsymbol{W}'\boldsymbol{\Delta}\boldsymbol{X} + \boldsymbol{W}'\boldsymbol{L}\boldsymbol{W}) + \frac{1}{N}\boldsymbol{C}, \end{aligned}\tag{4.9}$$

$$\begin{aligned}\boldsymbol{I} + \frac{1}{\sqrt{N}}\boldsymbol{V} &= \left(\boldsymbol{W} + \frac{1}{\sqrt{N}}\boldsymbol{X}\right)'\left(\boldsymbol{W} + \frac{1}{\sqrt{N}}\boldsymbol{X}\right)\\ &= \boldsymbol{I} + \frac{1}{\sqrt{N}}(\boldsymbol{X}'\boldsymbol{W} + \boldsymbol{W}'\boldsymbol{X}) + \frac{1}{N}\boldsymbol{X}'\boldsymbol{X}, \end{aligned}\tag{4.10}$$

where the submatrices of $\boldsymbol{C}$ are products of the submatrices of $\boldsymbol{W}, \boldsymbol{X}, \boldsymbol{\Delta}$, and $\boldsymbol{L}$. Note that $\boldsymbol{C}$ and $\boldsymbol{X}'\boldsymbol{X}$ are $O_p(1)$. The submatrix equations of (4.9) and (4.10) when $(1/\sqrt{N})\ \boldsymbol{C}$ and $(1/\sqrt{N})\ \boldsymbol{X}'\boldsymbol{X}$ are dropped are

$$\begin{aligned}\boldsymbol{U}_{gg} &= \boldsymbol{W}'_g\boldsymbol{L}_g\boldsymbol{W}_g + \lambda_g(\boldsymbol{W}'_g\boldsymbol{X}_{gg} + \boldsymbol{X}'_{gg}\boldsymbol{W}_g)\\ &= \boldsymbol{W}'_g\boldsymbol{L}_g\boldsymbol{W}_g + 2\lambda_g\boldsymbol{W}'_g\boldsymbol{X}_{gg}, \end{aligned}\tag{4.11}$$

$$\boldsymbol{U}_{gh} = \lambda_g\boldsymbol{W}'_g\boldsymbol{X}_{gh} + \lambda_h\boldsymbol{X}'_{hg}\boldsymbol{W}_h, \qquad g \neq h, \tag{4.12}$$

$$\boldsymbol{V}_{gg} = \boldsymbol{W}'_g\boldsymbol{X}_{gg} + \boldsymbol{X}'_{gg}\boldsymbol{W}_g = 2\boldsymbol{W}'_g\boldsymbol{X}_{gg}, \tag{4.13}$$

$$\boldsymbol{V}_{gh} = \boldsymbol{W}'_g\boldsymbol{X}_{gh} + \boldsymbol{X}'_{hg}\boldsymbol{W}_h, \qquad g \neq h. \tag{4.14}$$

From these equations we find

$$\boldsymbol{W}'_g\boldsymbol{L}_g\boldsymbol{W}_g = \boldsymbol{U}_{gg} - \lambda_g\boldsymbol{V}_{gg}, \tag{4.15}$$

$$\boldsymbol{X}_{gg} = \frac{1}{2}\boldsymbol{W}_g\boldsymbol{V}_{gg}, \tag{4.16}$$

$$\boldsymbol{X}_{gh} = \frac{1}{\lambda_g - \lambda_h}\boldsymbol{W}_g(\boldsymbol{U}_{gh} - \lambda_h\boldsymbol{V}_{gh}), \qquad g \neq h. \tag{4.17}$$

Theorem 4.1 *Define $\boldsymbol{\Delta}$ by (2.11), where $\lambda_1 > \lambda_2 > \cdots > \lambda_K$. Suppose $(\bar{\boldsymbol{U}}, \bar{\boldsymbol{V}})$ has a distribution such that with probability 1 (4.15) has a unique solution for orthogonal $\boldsymbol{W}_g$ and diagonal $\boldsymbol{L}_g$ with each diagonal element of $\boldsymbol{W}_g$ positive and the diagonal elements of $\boldsymbol{L}_g$ indexed in descending order. If the limiting distribution of $\sqrt{N}\ (\bar{\boldsymbol{H}}^* - \boldsymbol{\Delta},\ \bar{\boldsymbol{G}}^* - \boldsymbol{I})$ as $N \to \infty$ is the distribution of $(\bar{\boldsymbol{U}}, \bar{\boldsymbol{V}})$, then the limiting distribution of orthogonal $\boldsymbol{W}_g$, diagonal $\boldsymbol{L}_g, \boldsymbol{X}_{gg}$, and $\boldsymbol{X}_{gh}$, $g \neq h$, defined by (4.2) to (4.6), where diagonal $\boldsymbol{D}$ and $\boldsymbol{T}$ satisfy (3.3), is determined from the distribution of $(\bar{\boldsymbol{U}}, \bar{\boldsymbol{V}})$ by (4.15), (4.16), and (4.17).*

Corollary 4.2 *Let $d_1 > d_2 > \cdots > d_p$ be the characteristic roots of $\bar{\boldsymbol{H}}\bar{\boldsymbol{G}}^{-1}$. Under the conditions of Theorem 4.1 the limiting distribution of $\sqrt{N}(d_i - \delta_i)$, $i = 1, \ldots, p$, is the distribution of ℓ_i, $i = 1, \ldots, p$, where ℓ_i, $i = \sum_{g=1}^{h-1} r_g + 1, \ldots, \sum_{g=1}^{h} r_g$, have the distribution of the characteristic roots of $\bar{\boldsymbol{U}}_{hh} - \lambda_h \bar{\boldsymbol{V}}_{hh}$, $h = 1, \ldots, K$.*

Theorem 4.1 was proved by Anderson (1951), Section 6, but was not stated in this generality. The corollary, stated by Amemiya (1986), follows from (4.15), which indicates that the diagonal elements of $\boldsymbol{L}_h$ are the characteristic roots of $\boldsymbol{U}_{hh} - \lambda_h \boldsymbol{V}_{hh}$. The conclusions require only that the sequence of symmetric matrices $\bar{\boldsymbol{H}}^* = \bar{\boldsymbol{H}}^*(N)$ and $\bar{\boldsymbol{G}}^* = \bar{\boldsymbol{G}}^*(N)$ have joint distributions (depending on N) such that $\sqrt{N}(\bar{\boldsymbol{H}}^* - \boldsymbol{\Delta},\ \bar{\boldsymbol{G}}^* - \boldsymbol{I})$ has a limiting distribution that is the distribution of $(\bar{\boldsymbol{U}}, \bar{\boldsymbol{V}})$. As an example, in (1.1) the $\boldsymbol{v}_\alpha$'s can be independently identically distributed according to any multivariate distribution with finite fourth-order moments, and similarly for the $\boldsymbol{u}_{\alpha j}$'s.

The proof of Theorem 4.1 is treated in the Appendix.

A random orthogonal matrix $\boldsymbol{W}$ is said to have the uniform distribution, or its probability measure is said to be the Haar invariant measure, if the distribution or measure of $\boldsymbol{WP}$ is that of $\boldsymbol{W}$ for every (fixed) orthogonal $\boldsymbol{P}$. The distribution of $\boldsymbol{W}$ that results from the conditioning $w_{i1} \geq 0$ or $w_{ii} \geq 0$ we call the conditional Haar invariant distribution. See Anderson (1951) or Anderson (1984b), Section 13.3, for more discussion.

Theorem 4.3 *Suppose the distribution of the symmetric matrices $\bar{\boldsymbol{U}}$ and $\bar{\boldsymbol{V}}$ is normal with means $\boldsymbol{0}$ and with the functionally independent elements of the matrices being (statistically) independent, the variance of a diagonal element of $\bar{\boldsymbol{U}}_{gg}$ being $2\alpha_g^2$, of an off-diagonal element being α_g^2, of an element of $\bar{\boldsymbol{U}}_{gh}$, $g \neq h$, being $\beta_{gh}\ (= \beta_{hg})$, of a diagonal element of $\bar{\boldsymbol{V}}$ being 2 and of an off-diagonal element of $\boldsymbol{V}$ being 1. In the limiting distribution the triples $\boldsymbol{L}_g, \boldsymbol{W}_g$ and $\boldsymbol{X}_{gg}$ and the pairs $\boldsymbol{X}_{gh}, \boldsymbol{X}_{hg}$ are independent. The limiting marginal distribution of the diagonal elements of $\boldsymbol{L}_g$ is*

$$\frac{\exp\left(-\frac{1}{4(\alpha_g^2+\lambda_g^2)} \sum_{i=r_1+\cdots+r_{g-1}+1}^{r_1+\cdots+r_g} \ell_i^2\right) \prod_{i<j}(\ell_i - \ell_j)}{2^{\frac{1}{2}r_g}[2(\alpha_g^2 + \lambda_g^2)]^{r_g(r_g+1)/4} \prod_{i=1}^{r_g} \Gamma[(r_g + 1 - i)/2]} \tag{4.18}$$

for $\ell_{r_1+\cdots+r_g} \geq \cdots \geq \ell_{r_1+\cdots+r_{g-1}+1}$ *and 0 elsewhere.* $\boldsymbol{W}_g$ *is independent of* $\boldsymbol{L}_g$ *and has the Haar invariant distribution conditional on* $w_{ii} > 0$. *The conditional distribution of* $\boldsymbol{X}_{gg}$ *given* $\boldsymbol{L}_g$ *and* $\boldsymbol{W}_g$ *is normal with expected value* $-\frac{1}{2}\left[\lambda_g/(\alpha_g^2+\lambda_g^2)\right]\boldsymbol{L}_g\boldsymbol{W}_g$ *and covariances*

$$\mathit{Cov}\left[(\boldsymbol{x}_i,\boldsymbol{x}_j)|\boldsymbol{L}_g,\boldsymbol{W}_g\right] = \frac{1}{4}\frac{\alpha_g^2}{\alpha_g^2+\lambda_g^2}(\boldsymbol{\delta}_{ij}\boldsymbol{I}+\boldsymbol{w}_j\boldsymbol{w}_i'), \tag{4.19}$$

where $\boldsymbol{X}_{gg} = (\boldsymbol{x}_{r_1+\cdots+r_{g-1}+1},\ldots,\boldsymbol{x}_{r_1+\cdots+r_g})$ *and* $\boldsymbol{W}_g = (\boldsymbol{w}_{r_1+\cdots+r_{g-1}+1},\ldots,\boldsymbol{w}_{r_1+\cdots+r_g})$. *In the conditional distribution of the off-diagonal blocks of* $\boldsymbol{X}$, *the pairs* $\boldsymbol{X}_{gh},\boldsymbol{X}_{hg}$, $g<h$, *are independent normal with means* $\boldsymbol{0}$. *The elements of* $\boldsymbol{X}_{gh}$ $(g\neq h)$ *are independent with mean 0 and variance* $(\beta_{gh}+\lambda_g^2)/(\lambda_g-\lambda_h)^2$. *The covariance of an element* x_{ij} *in* $\boldsymbol{X}_{gh}$ *and* $x_{k\ell}$ *in* $\boldsymbol{X}_{hg}$ *conditional on* $\boldsymbol{W}_g$ *and* $\boldsymbol{W}_h$ *is* $-[(\beta_{gh}+\lambda_g\lambda_h)/(\lambda_g-\lambda_h)^2]w_{i\ell}^{(g)}w_{kj}^{(h)}$.

The proof is considered in the Appendix.

The case of $\boldsymbol{H}^*$ and $\boldsymbol{G}^*$ having distributions $W(\boldsymbol{\Delta},M)$ and $W(\boldsymbol{I},N)$ follows from Theorem 4.3.

Corollary 4.4 *Let* $\boldsymbol{U}$ *and* $\boldsymbol{V}$ *(depending on* N *and* M*) be defined by (3.4) and (3.5) and* $\boldsymbol{L}_g,\boldsymbol{W}_g$, *and* $\boldsymbol{X}_{gh}$ *be defined by (4.2), (4.5), (4.6), and (4.7). Then Theorem 4.3 holds with* $\alpha_g^2=\lambda_g^2/c$ *and* $\beta_{gh}=\lambda_g\lambda_h/c$.

In the balanced one-way multivariate components of variance model, $\lambda_K = 1, c = 1/(k-1), \alpha_K^2 = \lambda_K^2/c = k-1$, and $2(\alpha_K^2+\lambda_K^2) = 2k$. The limiting distribution of $\boldsymbol{L}_K,\boldsymbol{W}_K$ is the same as the limiting distribution of the characteristic roots and vectors of $\sqrt{N}(\boldsymbol{S}-\boldsymbol{I})$, where $\boldsymbol{S}$ is the sample covariance matrix of a sample of N from $N(\boldsymbol{0},k\boldsymbol{I}_{p-m})$. Let $\boldsymbol{A}=(1/\sqrt{2k})(\boldsymbol{U}_{KK}-\boldsymbol{V}_{KK})$ and (diagonal) $\boldsymbol{B}=(1/\sqrt{2k})\boldsymbol{L}_K$. Then

$$\boldsymbol{A} = \boldsymbol{W}_K'\boldsymbol{B}\boldsymbol{W}_K \tag{4.20}$$

has the limiting distribution of $\sqrt{N}(\boldsymbol{S}-\boldsymbol{I})$, where $\boldsymbol{S}$ is the sample covariance matrix of a sample of N from $N(\boldsymbol{0},(1/2)\boldsymbol{I}_{p-m})$. The density of the limiting distribution of $\boldsymbol{A}=\boldsymbol{A}'$ is

$$\text{const.}\ \exp\left(-\frac{1}{2}\text{tr}\boldsymbol{A}^2\right) \tag{4.21}$$

[Anderson (1984b), Section 13.3]. For an arbitrary orthogonal $\boldsymbol{P}$ we have $\boldsymbol{P}'\boldsymbol{A}\boldsymbol{P} \overset{\text{d}}{=} \boldsymbol{A}$. Thus $\boldsymbol{W}_K\boldsymbol{P} \overset{\text{d}}{=} \boldsymbol{W}_K$ except for conditioning $(w_{ii}\geq 0)$. Thus $\boldsymbol{W}_K$ has the Haar invariant measure. The limiting distribution of $b_{m+1},\ldots,b_p$ (the diagonal elements of $\boldsymbol{B}$) has the density

$$\text{const.}\ \exp\left(-\frac{1}{2}\sum_{i=m+1}^{p} b_i^2\right)\prod_{i<j}(b_i-b_j). \tag{4.22}$$

See Anderson (1984b), Section 13.3.

5 The Asymptotic Distribution of Characteristic Roots and Vectors in the General Case

We now consider the roots $d_1, \ldots, d_p$ of (2.1) and the corresponding vectors $\boldsymbol{y}_1, \ldots, \boldsymbol{y}_p$ satisfying (2.2) and (2.3). The matrix $\boldsymbol{Y}$ is

$$\boldsymbol{Y} = \boldsymbol{\Gamma}\boldsymbol{T}^{-1} = \boldsymbol{\Gamma}\left(\boldsymbol{W} + \frac{1}{\sqrt{N}}\boldsymbol{X}\right)^{-1}. \tag{5.1}$$

Let

$$\boldsymbol{Y} = \boldsymbol{R} + \frac{1}{\sqrt{N}}\boldsymbol{F}, \tag{5.2}$$

where

$$\boldsymbol{\Gamma}^{-1}\boldsymbol{R} = \boldsymbol{R}^* = \text{block diag}(\boldsymbol{R}_1^*, \boldsymbol{R}_2^*, \ldots, \boldsymbol{R}_K^*), \tag{5.3}$$

$\boldsymbol{\Gamma}^{-1}\boldsymbol{F} = \boldsymbol{F}^*$, $\boldsymbol{R}_g^*\boldsymbol{R}_g^{*\prime} = \boldsymbol{I}$, $\boldsymbol{R}_g^{*\prime}\boldsymbol{F}_{gg}^* = \boldsymbol{F}_{gg}^{*\prime}\boldsymbol{R}_g^*$, the matrix $\boldsymbol{I} + (1/\sqrt{N})\boldsymbol{R}_g^{*\prime}\boldsymbol{F}_{gg}^*$ is positive definite, and $\boldsymbol{Y}, \boldsymbol{F}$ and $\boldsymbol{F}^*$ are partitioned as $\boldsymbol{T}$. These definitions amount to decomposing $\boldsymbol{T}^{-1}$ as $\boldsymbol{R}^* + (1/\sqrt{N})\boldsymbol{F}^*$. If the indeterminacy of multiplication of each column of $\boldsymbol{Y}$ by ± 1 is resolved so that the diagonal elements of $\boldsymbol{\Gamma}^{-1}\boldsymbol{Y}$ are positive, the matrices $\boldsymbol{R}_g^*$ and $\boldsymbol{F}_{gh}^*$ can be treated as $\boldsymbol{W}_g'$ and $-\boldsymbol{W}_g'\boldsymbol{X}_{gh}\boldsymbol{W}_h'$, respectively. See Anderson (1951), Section 7, for a rigorous treatment.

The limiting distribution of $\boldsymbol{L} = \sqrt{N}(\boldsymbol{D} - \boldsymbol{\Delta})$, $\boldsymbol{R}$, and $\boldsymbol{F}$ is the distribution of $\boldsymbol{L}, \boldsymbol{\Gamma}\boldsymbol{R}^*$, and $\boldsymbol{F}$, which may be described as follows:

The marginal distribution of $\boldsymbol{L}, \boldsymbol{R}^*$ is such that the pair $\boldsymbol{L}_g, \boldsymbol{R}_g^*$ is independent of the other pairs $\boldsymbol{L}_h, \boldsymbol{R}_h^*$; the density of $\boldsymbol{L}_g$ is (4.18) with $\alpha_g^2 = \lambda_g^2/c$; and the distribution of $\boldsymbol{R}_g^*$ is the conditional Haar invariant distribution (conditional on $r_{ii}^* \geq 0$). The conditional distribution of $\boldsymbol{F}$ given $\boldsymbol{L}, \boldsymbol{R}$ is normal. The conditional expectation of $\boldsymbol{F}_{gh}$ is

$$\mathcal{E}(\boldsymbol{F}_{gh}|\boldsymbol{L}, \boldsymbol{R}) = \frac{c}{2(1+c)\lambda_h}\,\boldsymbol{\Gamma}_{gh}\boldsymbol{R}_h^*\boldsymbol{L}_h. \tag{5.4}$$

Let the i-th column of $\boldsymbol{F}$ be $\boldsymbol{f}_i$; the conditional covariance between $\boldsymbol{f}_i$ and $\boldsymbol{f}_j$ for $r_1 + \cdots + r_{g-1} + 1 \leq i, j \leq r_1 + \cdots + r_g$ is

$$\frac{1}{4(1+c)}\boldsymbol{\Gamma}_g(\delta_{ij}\boldsymbol{I} + \boldsymbol{r}_j^{*(g)}\boldsymbol{r}_i^{*(g)}) + \sum_{\substack{h=1\\h\neq g}}^{k} \frac{\lambda_g\lambda_h + c\lambda_h^2}{c(\lambda_g - \lambda_h)^2}\,\boldsymbol{\Gamma}_h\boldsymbol{\Gamma}_h', \tag{5.5}$$

and the conditional covariance between $\boldsymbol{f}_i$ and $\boldsymbol{f}_j$ for $r_1 + \cdots + r_{g-1} + 1 \leq i \leq r_1 + \cdots + r_g$ and $r_1 + \cdots + r_{h-1} + 1 \leq j \leq r_1 + \cdots + r_h$, $g \neq h$, is

$$-\frac{(1+c)\lambda_g\lambda_h}{c(\lambda_g - \lambda_h)^2}\,\boldsymbol{\Gamma}_h\boldsymbol{r}_j^{*(h)}\boldsymbol{r}_i^{*(g)\prime}\boldsymbol{\Gamma}_g', \tag{5.6}$$

where $\boldsymbol{r}_i^{*(g)}$ is the ith column of $\boldsymbol{R}_g^*$.

An important case occurs when all of the roots δ_j are simple except possibly one; that is, $r_1 = \cdots = r_{K-1} = 1$. Then $\boldsymbol{W}_i = \boldsymbol{R}_i^* = 1, i = 1, \ldots, K-1$. Let

$$\boldsymbol{Y} = (\boldsymbol{y}_1, \ldots, \boldsymbol{y}_{K-1}, \boldsymbol{Y}_K), \quad \boldsymbol{\Gamma} = (\gamma_1, \ldots, \gamma_{K-1}, \boldsymbol{\Gamma}_K). \tag{5.7}$$

Then $\boldsymbol{y}_i$ is a consistent estimator of $\gamma_i, i = 1, \ldots, K-1$. The conditional expectation of $\sqrt{N}(\boldsymbol{y}_i - \gamma_i) = \boldsymbol{f}_i$ in the limiting distribution given $\ell_1, \ldots, \ell_p$ and $\boldsymbol{R}_K^*$ is $\{c\ \ell_i/[2(1+c)\lambda_i]\}\gamma_i$. In the conditional limiting distribution of $f_1, \ldots, f_{K-1}$ given $\boldsymbol{L}_K$ and $\boldsymbol{R}_K^*$, the expectation of $\boldsymbol{f}_i$ is $\mathbf{0}$ and the covariances are

$$\begin{aligned} \mathcal{E}(\boldsymbol{f}_i \boldsymbol{f}_i' | \boldsymbol{L}_K, \boldsymbol{R}_K) = & \frac{1}{2(1+c)} \gamma_i \gamma_i' + \sum_{\substack{j=1 \\ j \neq i}}^{K-1} \frac{\lambda_i \lambda_j + c\lambda_j^2}{c(\lambda_i - \lambda_j)^2} \gamma_j \gamma_j' \\ & + \frac{\lambda_i \lambda_K + c\lambda_K^2}{c(\lambda_i - \lambda_K)^2} \boldsymbol{\Gamma}_K \boldsymbol{\Gamma}_K', \quad i = 1, \ldots, K-1, \end{aligned} \tag{5.8}$$

$$\mathcal{E}(\boldsymbol{f}_i \boldsymbol{f}_j' | \boldsymbol{L}_K, \boldsymbol{R}_K) = -\frac{2\lambda_i \lambda_j}{(\lambda_i - \lambda_j)^2} \gamma_j \gamma_i', \quad i \neq j,\ i, j = 1, \ldots, K-1. \tag{5.9}$$

6 The Asymptotic Distribution of the Maximum Likelihood Estimators of the Covariance Matrices

In the multivariate components of variance the two sample matrices $\boldsymbol{H}$ and $\boldsymbol{G}$ have distributions $W(\boldsymbol{\Psi} + k\boldsymbol{\Theta}, M)$ and $W(\boldsymbol{\Psi}, N)$, respectively, where $M = n-1$ and $N = n(k-1)$. Let p^* be the number of roots of (2.1) that are greater than 1, and let $m^* = \min(m, p^*)$, where m is the assumed rank of $\boldsymbol{\Theta}$, which may be less than p. Let

$$\boldsymbol{D} = \begin{pmatrix} \boldsymbol{D}^* & \mathbf{0} \\ \mathbf{0} & \boldsymbol{D}^{**} \end{pmatrix}, \quad \boldsymbol{Z} = \begin{pmatrix} \boldsymbol{Z}^* \\ \boldsymbol{Z}^{**} \end{pmatrix} \tag{6.1}$$

where the order of $\boldsymbol{D}^*$ and number of rows of $\boldsymbol{Z}^*$ is m^*. Then the maximum likelihood estimators of $\boldsymbol{\Theta}$ and $\boldsymbol{\Psi}$ based on the Wishart likelihood are

$$\hat{\boldsymbol{\Theta}} = \frac{1}{k} \boldsymbol{Z}^{*\prime}(\boldsymbol{D}^* - \boldsymbol{I}_{m^*})\boldsymbol{Z}^*, \tag{6.2}$$

$$\hat{\boldsymbol{\Psi}} = \boldsymbol{Z}^{*\prime}\boldsymbol{Z}^* + \boldsymbol{Z}^{**\prime}\left(\frac{n(k-1)}{nk-1}\boldsymbol{I}_{p-m^*} + \frac{n-1}{nk-1}\boldsymbol{D}^{**}\right)\boldsymbol{Z}^{**}, \tag{6.3}$$

Anderson, Anderson, and Olkin (1986) found the maximum likelihood estimators based on the likelihood function of the kn observations; the expressions are the same with $(1/M)\boldsymbol{H} = \big[1/(n-1)\big]\boldsymbol{H}$ replaced by $(1/n)\boldsymbol{H}$. Note that m^* is a random variable.

The number of roots of (2.9) that are greater than 1 is m. Since $\boldsymbol{D}$ is a consistent estimator of $\boldsymbol{\Delta}$ as $n \to \infty$, $\Pr\{p^* \geq m\} \to 1$ and $m^* \xrightarrow{\text{p}} m$. If $m = p$ (that is, $\boldsymbol{\Theta}$ is nonsingular), then $\Pr\{\boldsymbol{Z}^* = \boldsymbol{Z}\} \to 1$ and (6.2) is asymptotically equivalent to $k\hat{\boldsymbol{\Theta}} = \boldsymbol{Z}'(\boldsymbol{D} - \boldsymbol{I})\boldsymbol{Z} = \bar{\boldsymbol{H}} - \bar{\boldsymbol{G}}$ and (6.3) is equivalent to $\hat{\boldsymbol{\Psi}} = \boldsymbol{Z}'\boldsymbol{Z} = \bar{\boldsymbol{G}}$.

A more interesting case is $1 \leq m < p$. Let

$$\boldsymbol{Z} = \begin{bmatrix} \boldsymbol{Z}_1 \\ \boldsymbol{Z}_2 \\ \vdots \\ \boldsymbol{Z}_K \end{bmatrix}, \quad \boldsymbol{\Gamma}^{-1} = \boldsymbol{\Pi} = \begin{bmatrix} \boldsymbol{\Pi}_1 \\ \boldsymbol{\Pi}_2 \\ \vdots \\ \boldsymbol{\Pi}_K \end{bmatrix}. \tag{6.4}$$

Note that $\Pr\left\{\boldsymbol{Z}^* = (\boldsymbol{Z}_1', \boldsymbol{Z}_2', \ldots, \boldsymbol{Z}_{K-1}')'\right\} \longrightarrow 1$. We have

$$\begin{aligned} \boldsymbol{Z} &= \boldsymbol{T}\boldsymbol{\Gamma}^{-1} = \boldsymbol{T}\boldsymbol{\Pi} \\ &= (\boldsymbol{W} + \frac{1}{\sqrt{N}}\boldsymbol{X})\boldsymbol{\Pi} \\ &= \begin{bmatrix} \boldsymbol{W}_1\boldsymbol{\Pi}_1 + \frac{1}{\sqrt{N}}\sum_{h=1}^K \boldsymbol{X}_{1h}\boldsymbol{\Pi}_h \\ \boldsymbol{W}_2\boldsymbol{\Pi}_2 + \frac{1}{\sqrt{N}}\sum_{h=1}^K \boldsymbol{X}_{2h}\boldsymbol{\Pi}_h \\ \vdots \\ \boldsymbol{W}_K\boldsymbol{\Pi}_K + \frac{1}{\sqrt{N}}\sum_{h=1}^K \boldsymbol{X}_{Kh}\boldsymbol{\Pi}_h \end{bmatrix}. \end{aligned} \tag{6.5}$$

With probability approaching 1, $k\hat{\boldsymbol{\Theta}}$ is

$$\begin{aligned} &(\boldsymbol{Z}_1', \ldots, \boldsymbol{Z}_{K-1}') \begin{bmatrix} \boldsymbol{D}_1 - \boldsymbol{I} & \ldots & \boldsymbol{0} \\ \vdots & & \vdots \\ \boldsymbol{0} & \ldots & \boldsymbol{D}_{K-1} - \boldsymbol{I} \end{bmatrix} \begin{bmatrix} \boldsymbol{Z}_1 \\ \vdots \\ \boldsymbol{Z}_{K-1} \end{bmatrix} \\ &= \boldsymbol{Z}'(\boldsymbol{D} - \boldsymbol{I})\boldsymbol{Z}' - \boldsymbol{Z}_K'(\boldsymbol{D}_K - \boldsymbol{I})\boldsymbol{Z}_K' \\ &= \boldsymbol{\Pi}'(\bar{\boldsymbol{H}}^* - \bar{\boldsymbol{G}}^*)\boldsymbol{\Pi} - \frac{1}{\sqrt{N}}\boldsymbol{Z}_K'\boldsymbol{L}_K\boldsymbol{Z}_K' \\ &= k\boldsymbol{\Theta} + \frac{1}{\sqrt{N}}\left[\boldsymbol{\Pi}'(\boldsymbol{U} - \boldsymbol{V})\boldsymbol{\Pi} - \boldsymbol{\Pi}_K'\boldsymbol{W}_K'\boldsymbol{L}_K\boldsymbol{W}_K\boldsymbol{\Pi}_K\right] \\ &\quad + O_p\left(\frac{1}{N}\right). \end{aligned} \tag{6.6}$$

Thus $\sqrt{N}\ (k\hat{\boldsymbol{\Theta}} - k\boldsymbol{\Theta})$ has the limiting distribution of

$$\boldsymbol{\Pi}'(\boldsymbol{U} - \boldsymbol{V})\boldsymbol{\Pi} - \boldsymbol{\Pi}_K'(\boldsymbol{U}_{KK} - \boldsymbol{V}_{KK})\boldsymbol{\Pi}_K = \sum_{\substack{g,h=1 \\ (g,h)\neq(K,K)}}^{K} \boldsymbol{\Pi}_g'(\boldsymbol{U}_{gh} - \boldsymbol{V}_{gh})\boldsymbol{\Pi}_h. \tag{6.7}$$

With probability approaching 1, $\hat{\boldsymbol{\Psi}}$ is

$$
\begin{aligned}
(\boldsymbol{Z}_1', \ldots, \boldsymbol{Z}_{K-1}') & \begin{pmatrix} \boldsymbol{Z}_1 \\ \vdots \\ \boldsymbol{Z}_{K-1} \end{pmatrix} + \boldsymbol{Z}_K' \left(\frac{k-1}{k} \boldsymbol{I} + \frac{1}{k} \boldsymbol{D}_K \right) \boldsymbol{Z}_K \\
&= \boldsymbol{Z}'\boldsymbol{Z} - \frac{1}{k} \boldsymbol{Z}_K' \boldsymbol{Z}_K + \frac{1}{k} \boldsymbol{Z}_K' \boldsymbol{D}_K \boldsymbol{Z}_K \\
&= \boldsymbol{\Pi}' \bar{\boldsymbol{G}}^* \boldsymbol{\Pi} + \frac{1}{k\sqrt{N}} \boldsymbol{Z}_K' \boldsymbol{L}_K \boldsymbol{Z}_K \\
&= \boldsymbol{\Psi} + \frac{1}{\sqrt{N}} \boldsymbol{\Pi}' \boldsymbol{V} \boldsymbol{\Pi} + \boldsymbol{\Pi}_K' \boldsymbol{W}_K' \boldsymbol{L}_K \boldsymbol{W}_K \boldsymbol{\Pi}_K + O_p\left(\frac{1}{N}\right). \qquad (6.8)
\end{aligned}
$$

Thus $\sqrt{N}(\hat{\boldsymbol{\Psi}} - \boldsymbol{\Psi})$ has the limiting distribution of

$$
\boldsymbol{\Pi}'\boldsymbol{V}\boldsymbol{\Pi} + \boldsymbol{\Pi}_K'(\boldsymbol{U}_{KK} - \boldsymbol{V}_{KK})\boldsymbol{\Pi}_K = \sum_{\substack{g,h=1 \\ (g,h)\neq(K,K)}}^{K} \boldsymbol{\Pi}_g' \boldsymbol{V}_{gh} \boldsymbol{\Pi}_h + \boldsymbol{\Pi}_K' \boldsymbol{U}_{KK} \boldsymbol{\Pi}_K. \qquad (6.9)
$$

Thus $\sqrt{N}(\hat{\boldsymbol{\Theta}} - \boldsymbol{\Theta})$ and $\sqrt{N}(\hat{\boldsymbol{\Psi}} - \boldsymbol{\Psi})$ have a limiting normal distribution with mean $\boldsymbol{0}$. The covariances can be calculated from the covariances of $\boldsymbol{U}_{gh}$ and $\boldsymbol{V}_{gh}$ using (6.7) and (6.9).

7 Linear Structural Relations

The model (1.1) can be viewed in another way. The unobserved vector $\boldsymbol{\mu} + \boldsymbol{v}_\alpha$ can be called the "systematic part" and $\boldsymbol{u}_{\alpha j}$ the "error." If the systematic part satisfies $\boldsymbol{B}(\boldsymbol{\mu} + \boldsymbol{v}_\alpha) = \boldsymbol{\beta}$ or equivalently

$$
\boldsymbol{B}\boldsymbol{v}_\alpha = \boldsymbol{0}, \qquad (7.1)
$$

the model is called a linear structural relation. Then

$$
\boldsymbol{B}\boldsymbol{\Theta} = \boldsymbol{0}. \qquad (7.2)
$$

See Anderson (1984a), for example. When the $\boldsymbol{v}_\alpha$'s and $\boldsymbol{u}_\alpha$'s are normally distributed, a maximum likelihood estimator of $\boldsymbol{B}$ is any matrix $\hat{\boldsymbol{B}}$ such that $\hat{\boldsymbol{B}}\hat{\boldsymbol{\Theta}} = \boldsymbol{0}$. Let $\boldsymbol{Y}^{**}$ be made up of the characteristic vectors of $\bar{\boldsymbol{G}}^{-1}\bar{\boldsymbol{H}}$ corresponding to the $p - m^*$ smallest characteristic roots of $\bar{\boldsymbol{G}}^{-1}\bar{\boldsymbol{H}}$; then $\boldsymbol{Y}^{**\prime}$ is a maximum likelihood estimator of $\boldsymbol{B}$.

Let $\boldsymbol{Y}_K$ consist of the last r_K columns of $\boldsymbol{Y} = \boldsymbol{Z}^{-1}$. Then $\Pr\{\boldsymbol{Y}^{**} = \boldsymbol{Y}_K\} \to 1$ as $n \to \infty$. The asymptotic distribution of $\boldsymbol{Y}_K$ is the asymptotic distribution of

$$
\boldsymbol{Y}_K = \begin{bmatrix} \boldsymbol{R}_{1K} \\ \boldsymbol{R}_{2K} \\ \vdots \\ \boldsymbol{R}_{KK} \end{bmatrix} + \frac{1}{\sqrt{n(k-1)}} \begin{bmatrix} \boldsymbol{F}_{1K} \\ \boldsymbol{F}_{2K} \\ \vdots \\ \boldsymbol{F}_{KK} \end{bmatrix}
$$

$$= \boldsymbol{\Gamma}_K \boldsymbol{R}_K^* + \frac{1}{\sqrt{n(k-1)}} \sum_{g=1}^{K} \boldsymbol{\Gamma}_g \boldsymbol{F}_{gK}^*, \tag{7.3}$$

which is asymptotically equivalent to

$$\boldsymbol{\Gamma}_K \boldsymbol{W}_K' - \frac{1}{\sqrt{n(k-1)}} \sum_{g=1}^{K} \boldsymbol{\Gamma}_g \boldsymbol{W}_g' \boldsymbol{X}_{gK} \boldsymbol{W}_K'. \tag{7.4}$$

In (7.1) there is the indeterminacy of multiplying $\boldsymbol{B}$ on the left by an arbitrary nonsingular matrix. If $\boldsymbol{B}$ is normalized by

$$\boldsymbol{B}\boldsymbol{\Psi}\boldsymbol{B}' = \boldsymbol{I}, \tag{7.5}$$

there remains the indeterminacy of multiplication of $\boldsymbol{B}$ on the left by an arbitrary orthogonal matrix. That indeterminacy is reflected in the fact that (7.4) can be multiplied on the right by an arbitrary orthogonal matrix (except for the requirement of $w_{ii} > 0$.)

All of the ambiguity in $\boldsymbol{B}$ can be eliminated by requiring that

$$\boldsymbol{B} = (\boldsymbol{B}^* \ \boldsymbol{I}_{p-m}). \tag{7.6}$$

This is sometimes known as the "errors in variables" model. Then the maximum likelihood estimator of $\boldsymbol{B}'$ is $\hat{\boldsymbol{B}}' = \boldsymbol{Y}_K \boldsymbol{Y}_{KK}^{-1}$, which is treated asymptotically as

$$\begin{aligned}
&\left(\boldsymbol{\Gamma}_K \boldsymbol{W}_K' - \frac{1}{\sqrt{n(k-1)}} \sum_{g=1}^{K} \boldsymbol{\Gamma}_g \boldsymbol{W}_g' \boldsymbol{X}_{gK} \boldsymbol{W}_K' \right) \\
&\quad \times \left(\boldsymbol{\Gamma}_{KK} \boldsymbol{W}_K' - \frac{1}{\sqrt{n(k-1)}} \sum_{g=1}^{K} \boldsymbol{\Gamma}_{Kg} \boldsymbol{W}_g' \boldsymbol{X}_{gK} \boldsymbol{W}_K' \right)^{-1} \\
&= \boldsymbol{\Gamma}_K \boldsymbol{\Gamma}_{KK}^{-1} - \frac{1}{\sqrt{n(k-1)}} \sum_{g=1}^{K} (\boldsymbol{\Gamma}_g - \boldsymbol{\Gamma}_K \boldsymbol{\Gamma}_{KK}^{-1} \boldsymbol{\Gamma}_{Kg}) \boldsymbol{W}_g' \boldsymbol{X}_{gK} \boldsymbol{\Gamma}_{KK}^{-1} \\
&\quad + O_p\left(\frac{1}{n}\right).
\end{aligned} \tag{7.7}$$

The limiting distribution of $\sqrt{n}(\hat{\boldsymbol{B}} - \boldsymbol{B})'$ as $n \to \infty$ is the limiting distribution of

$$\begin{aligned}
&- \sum_{g=1}^{K-1} (\boldsymbol{\Gamma}_g - \boldsymbol{\Gamma}_K \boldsymbol{\Gamma}_{KK}^{-1} \boldsymbol{\Gamma}_{Kg}) \boldsymbol{W}_g' \boldsymbol{X}_{gK} \boldsymbol{\Gamma}_{KK}^{-1} \\
&\quad = - \sum_{g=1}^{K-1} (\boldsymbol{\Gamma}_g - \boldsymbol{\Gamma}_K \boldsymbol{\Gamma}_{KK}^{-1} \boldsymbol{\Gamma}_{Kg}) \frac{1}{\lambda_g - 1} (\boldsymbol{U}_{gK} - \boldsymbol{V}_{gK}) \Gamma_{KK}^{-1}.
\end{aligned} \tag{7.8}$$

We have

$$\begin{aligned}\boldsymbol{U}_{gK} &= \sqrt{n(k-1)}\,\boldsymbol{\Gamma}_g'(\bar{\boldsymbol{H}}-\boldsymbol{\Phi})\boldsymbol{\Gamma}_K \\ &= \sqrt{n(k-1)}\,\boldsymbol{\Gamma}_g'\left[\frac{k}{n-1}\sum_{\alpha=1}^{n}(\boldsymbol{v}_\alpha-\bar{\boldsymbol{v}}+\bar{\boldsymbol{u}}_\alpha-\bar{\boldsymbol{u}})(\boldsymbol{v}_\alpha-\bar{\boldsymbol{v}}+\bar{\boldsymbol{u}}_\alpha-\bar{\boldsymbol{u}})'\right. \\ &\qquad \left. -\boldsymbol{\Phi}\right]\boldsymbol{\Gamma}_K, \end{aligned} \tag{7.9}$$

where $\bar{\boldsymbol{v}} = (1/n)\sum_{\alpha=1}^{n}\boldsymbol{v}_\alpha$, $\bar{\boldsymbol{u}}_\alpha = (1/k)\sum_{j=1}^{k}\boldsymbol{u}_{\alpha j}$, and $\bar{\boldsymbol{u}} = (1/n)\sum_{\alpha=1}^{n}\bar{\boldsymbol{u}}_\alpha$. Because $\boldsymbol{\Gamma}_K = \boldsymbol{B}'\boldsymbol{\Gamma}_{KK}$, it follows from (7.1) and (7.2) that $\boldsymbol{v}_\alpha'\boldsymbol{\Gamma}_K = \boldsymbol{0}$, $\bar{\boldsymbol{v}}'\boldsymbol{\Gamma}_K = \boldsymbol{0}$, and $\boldsymbol{\Theta}\boldsymbol{\Gamma}_K = \boldsymbol{0}$. Thus

$$\begin{aligned}\boldsymbol{U}_{gK} &= \frac{k\sqrt{n(k-1)}}{n-1}\boldsymbol{\Gamma}_g'\sum_{\alpha=1}^{n}(\boldsymbol{v}_\alpha-\bar{\boldsymbol{v}})(\bar{\boldsymbol{u}}_\alpha-\bar{\boldsymbol{u}})'\boldsymbol{\Gamma}_K \\ &\quad + k\sqrt{n(k-1)}\boldsymbol{\Gamma}_g'\left[\frac{1}{n-1}\sum_{\alpha=1}^{n}(\bar{\boldsymbol{u}}_\alpha-\bar{\boldsymbol{u}})(\bar{\boldsymbol{u}}_\alpha-\bar{\boldsymbol{u}})'-\boldsymbol{\Psi}\right]\boldsymbol{\Gamma}_K. \end{aligned} \tag{7.10}$$

The first matrix on the right-hand side of (7.10) has a limiting normal distribution under quite general conditions, in particular, if $\{\boldsymbol{v}_\alpha\}$ and $\{\boldsymbol{u}_{\alpha j}\}$ are independent, $\boldsymbol{v}_\alpha$ are iid with finite second-order moments, and $\boldsymbol{u}_{\alpha j}$ are iid with finite second-order moments. (See Anderson (1987b), for example.) The second matrix on the right-hand side of (7.10) has a limiting normal distribution if the $\boldsymbol{u}_{\alpha j}$ are iid with finite fourth-order moments. Under the above conditions the limiting distribution of (7.8) will depend on the distribution of $\boldsymbol{v}_\alpha$ only through $\boldsymbol{\Theta}$.

Amemiya and Fuller (1984) have given the maximum likelihood estimator of $\boldsymbol{B}$ and its asymptotic distribution when the $\boldsymbol{u}_{\alpha j}$ are normally distributed and when the $\boldsymbol{v}_\alpha$ are iid or when the $\boldsymbol{v}_\alpha$ are nonstochastic and $(1/n)\sum_{\alpha=1}^{n}\boldsymbol{v}_\alpha\boldsymbol{v}_\alpha'$ converges. See also Fuller (1987), Section 4.1.

A1 Proofs of Theorems 4.1 and 4.3

Theorems 4.1 and 4.3 were essentially proved in Anderson (1951), although in that paper the results were stated for a regression model that includes the multivariate analysis of variance with fixed effects. In this appendix we shall sketch the proofs in the notation of the current paper. More details were given in Anderson (1987a).

The proofs of Theorem 4.1 and 4.3 are based on the (nonstochastic) matrices $\boldsymbol{W} = \boldsymbol{W}(N), \boldsymbol{L} = \boldsymbol{L}(N)$ and $\boldsymbol{X} = \boldsymbol{X}(N)$ being (single-valued) functions of $\boldsymbol{U} = \boldsymbol{U}(N)$ and $\boldsymbol{V} = \boldsymbol{V}(N)$ as defined by (3.3) to (4.6). The limit of these functions are $\boldsymbol{W}, \boldsymbol{L}$, and $\boldsymbol{X}$ as defined by (4.15), (4.16), and

(4.17). It is to be shown that these functions are continuous in the sense that if nonstochastic $\boldsymbol{U}(N) \to \boldsymbol{U}_0$ and $\boldsymbol{V}(N) \to \boldsymbol{V}_0$, then $\boldsymbol{W}(N), \boldsymbol{L}(N)$, and $\boldsymbol{X}(N)$ converge to the solutions of (4.15), (4.16), and (4.17) for $\boldsymbol{U} = \boldsymbol{U}_0$ and $\boldsymbol{V} = \boldsymbol{V}_0$, except for a set of probability zero relative to the distribution of $\boldsymbol{U}$ and $\boldsymbol{V}$. Then a theorem of Rubin (stated in Anderson (1951) and proved in Anderson (1963)) applies to the effect that the limiting distribution of (stochastic) $\boldsymbol{W}(N), \boldsymbol{L}(N)$, and $\boldsymbol{X}(N)$ is that defined by (4.15), (4.16), and (4.17) and the distribution of $\boldsymbol{U}$ and $\boldsymbol{V}$.

First we argue that $\boldsymbol{L}_g(N) \to \boldsymbol{L}_g^0$ consisting of the characteristic roots of $\boldsymbol{U}_{gg}^0 - \lambda_g \boldsymbol{V}_{gg}^0$. For $g = 1$, for example, (2.1) is

$$\begin{aligned}
&\left| \boldsymbol{\Delta} + \frac{1}{\sqrt{N}} \boldsymbol{U}(N) - \left(\lambda_1 + \frac{1}{\sqrt{N}} \ell \right) \left[\boldsymbol{I} + \frac{1}{\sqrt{N}} \boldsymbol{V}(N) \right] \right| \\
&= \left| \boldsymbol{\Delta} - \lambda_1 \boldsymbol{I} + \frac{1}{\sqrt{N}} \left[\boldsymbol{U}(N) - \lambda_1 \boldsymbol{V}(N) \right] - \frac{1}{\sqrt{N}} \ell \left[\boldsymbol{I} + \frac{1}{\sqrt{N}} \boldsymbol{V}(N) \right] \right| \\
&= 0. \qquad \text{(A1.1)}
\end{aligned}$$

Since the first r_1 columns of $\boldsymbol{\Delta} - \lambda_1 \boldsymbol{I}$ in the second determinant are zero, we can factor $1/\sqrt{N}$ out of the first r_1 rows of the entire matrix. Then as $N \to \infty$ that determinant converges to

$$\left| \boldsymbol{U}_{11}^0 - \lambda_1 \boldsymbol{V}_{11}^0 - \ell \boldsymbol{I} \right| \cdot \prod_{g=2}^{K} (\lambda_g - \lambda_1)^{r_g}, \qquad \text{(A1.2)}$$

the zeroes of which are the characteristic roots of $\boldsymbol{U}_{11}^0 - \lambda_1 \boldsymbol{V}_{11}^0$.

Since r_1 of the roots of

$$\left| \boldsymbol{\Delta} + \frac{1}{\sqrt{N}} \boldsymbol{U}(N) - d \left[\boldsymbol{I} + \frac{1}{\sqrt{N}} \boldsymbol{V}(N) \right] \right| = 0 \qquad \text{(A1.3)}$$

converge to λ_i, $i = 1, \ldots, p$, then $p - r_1$ of the roots of (A1.1) diverge to ∞. The argument depends on the following proposition, which is a simplification of a proposition given by Bai (1984) correcting Hsu (1941). (See, also, Amemiya (1988).)

Proposition A1.1 *Suppose the zeroes of*

$$P_N(x) = a_p(N) x^p + \cdots + a_0(N), \quad a_p(N) \neq 0, \qquad \text{(A1.4)}$$

are real and ordered as $x_1(N) \geq \cdots \geq x_p(N)$, *and suppose the zeroes of*

$$p(x) = a_r x^r + \cdots + a_0, \quad a_r \neq 0, \qquad \text{(A1.5)}$$

are real and ordered as $x_1 \geq \cdots \geq x_r$. *Suppose further that as* $N \to \infty$

$$a_i(N) \to \begin{cases} a_i, & i = 0, 1, \ldots, r, \\ 0, & i = r+1, \ldots, p, \end{cases} \qquad \text{(A1.6)}$$

$$x_i(N) \to \begin{cases} \infty, & i = 1, \ldots, q, \\ -\infty, & i = q+r+1, \ldots, p. \end{cases} \qquad \text{(A1.7)}$$

Then

$$x_{i+q}(N) \to x_i, \qquad i = 1, \ldots, r. \tag{A1.8}$$

Proof Because $x - x_i(N)$ is a factor of $P_N(x)$ we can define a polynomial of degree r as

$$\begin{aligned} Q_N(x) &= \frac{P_N(x)}{\prod_{i=1}^q \left[1 - \frac{x}{x_i(N)}\right] \prod_{i=q+r+1}^p \left[1 - \frac{x}{x_i(N)}\right]}, \\ &= b_r(N)x^r + \cdots + b_0(N). \end{aligned} \tag{A1.9}$$

Since $|x_i(N)| \to \infty$, $i = 1, \ldots, q$, $q+r+1, \ldots, p$, $x_i(N) \neq 0$ for sufficiently large N. The zeroes of $Q_N(x)$ are $x_{q+1}(N) \geq \cdots \geq x_{q+r}(N)$. Because $P_N(x) \to p(x)$, we have $Q_N(x) \to p(x)$. By continuity of the roots as functions of the coefficients (with leading coefficient different from 0), we obtain (A1.8). □

Now we argue that in (4.9) $(1/\sqrt{N})\boldsymbol{C}(N) \to 0$ and further that in (4.10) $(1/\sqrt{N})\boldsymbol{X}'(N)\boldsymbol{X}(N) \to 0$ by consideration of the characteristic vectors of $\left[\boldsymbol{I} + (1/\sqrt{N})\boldsymbol{V}(N)\right]^{-1} \left[\boldsymbol{\Delta} + (1/\sqrt{N})\boldsymbol{U}(N)\right]$ normalized by $\boldsymbol{y}^{*\prime}\left[\boldsymbol{I} + (1/\sqrt{N})\boldsymbol{V}(N)\right]\boldsymbol{y}^* = 1$. For convenience consider the set of r_1 (column) vectors associated with $\ell_1, \ldots, \ell_{r_1}$, denoted by $(\boldsymbol{Y}_{11}^{*\prime}, \boldsymbol{Y}_{21}^{*\prime}, \ldots, \boldsymbol{Y}_{K1}^{*\prime})'$. They satisfy

$$\begin{aligned} &\frac{1}{\sqrt{N}} \sum_{g=1}^K \left[\boldsymbol{U}_{1g}(N) - \lambda_1 \boldsymbol{V}_{1g}(N)\right]\boldsymbol{Y}_{g1}^*(N) \\ &= \frac{1}{\sqrt{N}} \left[\boldsymbol{Y}_{11}^*(N) + \frac{1}{\sqrt{N}} \sum \boldsymbol{V}_{1g}(N)\boldsymbol{Y}_{g1}^*(N)\right] \boldsymbol{L}_1(N), \end{aligned} \tag{A1.10}$$

$$\begin{aligned} &(\lambda_h - \lambda_1)\boldsymbol{Y}_{h1}^*(N) + \frac{1}{\sqrt{N}} \sum_{g=1}^K \left[\boldsymbol{U}_{hg}(N) - \lambda_1 \boldsymbol{V}_{hg}(N)\right]\boldsymbol{Y}_{g1}^*(N) \\ &= \left[\frac{1}{\sqrt{N}}\boldsymbol{Y}_{h1}^* + \frac{1}{N} \sum_{g=1}^K \boldsymbol{V}_{hg}(N)\boldsymbol{Y}_{g1}^*(N)\right] \boldsymbol{L}_1(N), \quad h = 2, \ldots, K. \end{aligned} \tag{A1.11}$$

(A1.11) shows that $\boldsymbol{Y}_{g1}^*(N) = O(1/\sqrt{N})$, $g \neq 1$. Then the limit of (A1.10) as $N \to \infty$ is

$$(\boldsymbol{U}_{11}^0 - \lambda_1 \boldsymbol{V}_{11}^0)\boldsymbol{Y}_{11}^* = \boldsymbol{Y}_{11}^* \boldsymbol{L}_1^0; \tag{A1.12}$$

hence $\boldsymbol{Y}_{11}^*(N) \to \boldsymbol{W}_1^{0\prime}$. Similarly, $\boldsymbol{Y}_{gh}^*(N) = O(1/\sqrt{N})$, $g \neq h$, and $\boldsymbol{Y}_{gg}^*(N) \to \boldsymbol{W}_g^{0\prime}$. Since $\boldsymbol{T} = \boldsymbol{Y}^{*-1}$, $\boldsymbol{T}_{gh}(N) = O(1/\sqrt{N})$ and $\boldsymbol{T}_{gg}(N) \to \boldsymbol{W}_g^0$. Then $\boldsymbol{X}_{gh}(N) = O(1)$, $g \neq h$. Then (4.10) implies $\boldsymbol{X}_{gg}(N) = O(1)$. Hence, the limits of (4.9) and (4.10) as $N \to \infty$ yield (4.11) to (4.14), and $\boldsymbol{L}_g(N), \boldsymbol{W}_g(N), \boldsymbol{X}_{gh}(N)$ converge to $\boldsymbol{L}_g^0, \boldsymbol{W}_g^0, \boldsymbol{X}_{gh}^0$. Rubin's theorem applies to complete the proof of Theorem 4.1.

For Theorem 4.3 the algebra of Anderson (1951) applies. In that paper the variance of a diagonal term of $\boldsymbol{E}_{ii}$ (which corresponds to $\boldsymbol{U}_{gg}$ here) is $4\lambda_i$ and of an off-diagonal term is $2\lambda_i$ (corresponding to $2\alpha_g^2$ and α_g^2, respectively, here). The variance of a element of an off-diagonal block $\boldsymbol{E}_{ij}$ is $\lambda_i + \lambda_j$ (corresponding to β_{gh} in $\boldsymbol{U}_{gh}$ here). The distribution of $\boldsymbol{U}$ in that paper is the distribution of $\boldsymbol{V}$ here.

In the proof of Theorem 4.1 we have used the fact that (4.5) and (4.7) define $\boldsymbol{X}_{gg}$ and $\boldsymbol{W}_g$ uniquely (for orthogonal $\boldsymbol{W}_g$). The proof of that statement in Anderson (1951) needs to take account of the condition that $\boldsymbol{W}_g \boldsymbol{T}'_{gg} = \boldsymbol{T}_{gg} \boldsymbol{W}'_g$ is positive definite.

A2 Parameters Depending on N

Suppose $\boldsymbol{H}^*$ has the distribution $W(\boldsymbol{\Delta}(N), M)$, where

$$\boldsymbol{\Delta}(N) = \text{block diag}\big(\lambda_1(N)\boldsymbol{I}_{r_1}, \ldots, \lambda_K(N)\boldsymbol{I}_{r_K}\big), \tag{A2.1}$$

$\lambda_1(N) > \lambda_2(N) > \cdots > \lambda_K(N)$, $\lambda_k(N) \to \lambda_k^0$, and $\lambda_1^0 > \cdots > \lambda_K^0$. Note that the ordering and multiplicities do not depend on N. The definitions of the various matrices are unchanged. In (4.9) and (4.10) $\boldsymbol{\Delta}$ is replaced by $\boldsymbol{\Delta}(N)$. In (4.15) and (4.17) λ_g and λ_h are the limiting values λ_g^0 and λ_h^0. The proofs of Theorems 4.1 and 4.3 are based on the applicability of Rubin's theorem, which is justified in Appendix A1. In (A1.1), for example, $\boldsymbol{\Delta}$ and λ_1 depend on N. In (A1.2) λ_1 is the limiting value. In (A1.10) and (A1.11) $\lambda_1(N)$ and $\lambda_h(N)$ are involved, while in (A1.12) $\lambda_1 = \lambda_1^0$ is the limiting value. The proof goes through with these modifications.

Now consider $\boldsymbol{\Phi} = \boldsymbol{\Phi}(N)$ and $\boldsymbol{\Psi} = \boldsymbol{\Psi}(N)$ positive definite such that

$$\boldsymbol{\Phi}(N) \to \boldsymbol{\Phi}_0, \qquad \boldsymbol{\Psi}(N) \to \boldsymbol{\Psi}_0, \tag{A2.2}$$

and such that $\lambda_1(N) > \cdots > \lambda_K(N)$ are the characteristic roots of the matrix $\boldsymbol{\Phi}(N)\boldsymbol{\Psi}^{-1}(N)$ of multiplicity $r_1, \ldots, r_K$ and such that $\lambda_1^0 > \cdots > \lambda_K^0$ are the characteristic roots of $\boldsymbol{\Phi}_0\boldsymbol{\Psi}_0^{-1}$ of multiplicities $r_1, \ldots, r_K$. Then $\lambda_k(N) \to \lambda_k^0$. Let $\boldsymbol{\Gamma}_k(N)$ satisfy

$$\Phi(N)\boldsymbol{\Gamma}_k(N) = \lambda_k(N)\boldsymbol{\Psi}(N)\boldsymbol{\Gamma}_k(N), \tag{A2.3}$$

$$\boldsymbol{\Gamma}'_k(N)\boldsymbol{\Psi}(N)\boldsymbol{\Gamma}_k(N) = \boldsymbol{I}. \tag{A2.4}$$

If $r_k > 1$, some other conditions are needed on $\boldsymbol{\Gamma}_k(N)$ to avoid the indeterminacy of multiplication on the right by an arbitrary orthogonal matrix of order r_k. Let $\boldsymbol{\Gamma}_k^0$ satisfy

$$\boldsymbol{\Psi}_0\boldsymbol{\Gamma}_k^0 = \lambda_k^0\boldsymbol{\Psi}_0\boldsymbol{\Gamma}_k^0, \tag{A2.5}$$

$$\boldsymbol{\Gamma}_k^{0\prime}\boldsymbol{\Psi}_0\boldsymbol{\Gamma}_k^0 = \boldsymbol{I}, \tag{A2.6}$$

and any other conditions for uniqueness. Then

$$\boldsymbol{\Gamma}_k(N) \to \boldsymbol{\Gamma}_k^0. \tag{A2.7}$$

Define

$$\boldsymbol{\Gamma}(N) = \left[\boldsymbol{\Gamma}_1(N), \ldots, \boldsymbol{\Gamma}_K(N)\right]. \tag{A2.8}$$

Then $\boldsymbol{\Gamma}(N)$ is used in Section 3 to define $\boldsymbol{H}^*, \boldsymbol{G}^*$, and $\boldsymbol{T}$. Section 4 is unchanged from the above. In Section 5 $\boldsymbol{\Gamma}$ is $\boldsymbol{\Gamma}^0$. The results hold with these changes.

Anderson (1951) gave details for a different case with population roots and vectors depending on N. It was remarked that if the multiplicities of the roots vary with N one needs

$$\sqrt{N}\,\left|\delta_i(N) - \lambda_k^0\right| \to 0 \tag{A2.9}$$

for such i that $\delta_i(N) \to \lambda_k^0$. Here $\delta_i(N)$ is the i-th root of

$$\left|\boldsymbol{\Phi}(N) - \delta \boldsymbol{\Psi}(N)\right| = 0. \tag{A2.10}$$

Acknowledgments: The author is indebted to Z.D. Bai for pointing out flaws in Anderson (1951) and to Yasuo Amemiya for helpful comments. This paper was completed at the Naval Postgraduate School while the author was a National Research Council Resident Research Associate. Research was supported by National Science Foundation Grant No. DMS 86-03779.

Since this paper was written, attention has been drawn to Eaton and Tyler (1988).

References

Amemiya, Yasuo (1986). Limiting distributions of certain characteristic roots under general conditions and their applications. Technical Report No. 17. Econometric Workshop, Stanford University, June, 1986.

Amemiya, Yasuo (1988). On the convergence of the ordered roots of a sequence of determinantal equations. Preprint No. 88-26, Statistical Laboratory, Iowa State University.

Amemiya, Yasuo, and Wayne A. Fuller (1984). Estimation for the multivariate errors-in-variables model with estimated covariance matrix. *Annals of Statistics* **12**, 497–509.

Anderson, Blair M., T.W. Anderson, and Ingram Olkin (1986). Maximum likelihood estimators and likelihood ratio criteria in multivariate components of variance. *Annals of Statistics* **14**, 405–417.

Anderson, T.W. (1951). The asymptotic distribution of certain characteristic roots and vectors. *Proceedings of the Second Berkeley Symposium on Mathematical Statistics and Probability*, University of California Press, Berkeley, 103–130.

Anderson, T.W. (1963). Asymptotic theory for principal component analysis. *Annals of Mathematical Statistics* **34**, 122–148.

Anderson, T.W. (1984a). Estimating linear statistical relationships. *Annals of Statistics* **12**, 1–45.

Anderson, T.W. (1984b). *An Introduction to Multivariate Statistical Analysis*, Second Edition, John Wiley and Sons, Inc.

Anderson, T.W. (1987a). The asymptotic distribution of characteristic roots and vectors in multivariate components of variance. Technical Report No. 23. Econometric Workshop, Stanford University, March 1987.

Anderson, T.W. (1987b). Linear latent variable models and covariance structures, Technical Report No. 27, Econometric Workshop, Stanford University, October, 1987, *Journal of Econometrics*, in press.

Bai, Z.D. (1984). A note on asymptotic joint distribution of the eigenvalues of a noncentral multivariate F matrix. *Journal of Mathematical Research and Exposition* **19**, 113–118.

Chang, T.C. (1973). On an asymptotic distribution of the characteristic roots of $S_1S_2^{-1}$ when roots are not all distinct. *Annals of the Institute of Statistical Mathematics* **25**, 447–452.

Chattopadhyay, A.K., and Pillai, K.C.S. (1973). Asymptotic expansions for the distributions of characteristic roots when the parameter matrix has several multiple roots. In *Multivariate Analysis III* (P. R. Krishnaiah, Ed.), Academic Press, New York.

Eaton, Morris L. and Tyler, David E. (1988). Wielandt's inequality and its application to the asymptotic distribution of the eigenvalues of a random symmetric matrix. School of Statistics at the University of Minnesota, Technical Report No. 518, September.

Fuller, Wayne A. (1987). *Measurement Error Models*, John Wiley and Sons, Inc.

Hsu, P.L. (1941). On the limiting distribution of roots of certain determinantal equations. *Journal of the London Mathematical Society* **16**, 183–194.

Li, H.C., Pillai, K.C.S., and Chang, T.C. (1970). Asymptotic expansions for distributions of the roots of two matrices from classical and complex Gaussian populations. *Annals of Mathematical Statistics* **41**, 1541–1556.

Muirhead, Robb J. (1978). Latent roots and matrix variates: A review of some asymptotic results. *Annals of Statistics* **6**, 5–33.

Schott, J.R., and J.G. Saw (1984). A multivariate one-way classification model with random effects, *Journal of Multivariate Analysis* **15**, 1–12.

Sugiura, N. (1976). Asymptotic expansions of the distributions of the latent roots and the latent vector of the Wishart and multivariate F matrices. *Journal of Multivariate Analysis* **6**, 500–525.

Tyler, David E. (1987). A distribution-free M-estimator of multivariate scatter. *Annals of Statistics* **15**, 234–251.

11

Univariate and Multivariate Analyses of Variance for Incomplete Factorial Experiments Having Monotone Structure in Treatments

R.P. Bhargava [1]

ABSTRACT In this paper, we give the analysis of variance for an incomplete factorial experiment with three fixed factors with a monotone structure (in the treatments) when the design of the experiment is completely randomized. We consider both the univariate and the multivariate case. For the multivariate case, the analysis is given for the general situation when the data vectors are incomplete and have a monotone sample pattern. (See Bhargava 1962, 1975.) The monotone sample case includes as a special case the complete sample case in which no observations are missing in any vector. A univariate example is given.

1 Introduction

Consider a factorial experiment with three fixed factors A, B and C. Let A, B, C have I levels, J levels and K levels respectively. Thus, in all, there are IJK treatment combinations possible. Let τ be the number of treatment combinations actually used in the experiment. If $\tau <$ IJK, then IJK $- \tau$ treatment combinations are missing and the experiment is said to be an incomplete factorial. Let $a_i b_j c_k$ denote the treatment combination (hereafter, called treatment) composed of the ith level of A, the jth level of B, and the kth of C. Let $a_i b_j c_k \neq \phi$ mean that the treatment $a_i b_j c_k$ is occurring in the experiment, while $a_i b_j c_k = \phi$ means that it is missing.

An incomplete factorial experiment (design) will be called *monotone factorial*, or will be said to have a monotone structure in treatments, if (i) $a_i b_j c_k \neq \phi$ implies $a_{i'} b_{j'} c_{k'} \neq \phi$ for all $i' \leq i, j' \leq j, k' \leq k$ and (ii)

[1]Ontario Institute for Studies in Education, University of Toronto

$a_i b_j c_k = \phi$ implies $a_{i'} b_{j'} c_{k'} = \phi$ for all $i' \geq i, j' \geq j, k' \geq k$. Such a design for the case of two additive factors has been called "triangular" by Ditchburne (1955) and "staircase" by Graybill and Pruitt (1957), and by Bhargava (1967).

A monotone factorial design may be appropriate when the following conditions hold:

(i) the number of treatment combinations (i.e. IJK in our case) is large,

(ii) the resources available to do the experiment are low, and

(iii) the experimenter is able to order the levels of each factor in some decreasing (non-increasing) order of importance and wants to have a more precise comparison between important levels of factors and their interactions than the unimportant ones.

Also, in certain experiments some treatment combinations may not be feasible. For example, consider a pharmacological experiment which investigates the effects of three ingredients (factors) A, B and C with I, J and K levels respectively. It may be known, *a priori*, that the toxic effects of the highest levels of any two of A, B and C may be tolerable but that the highest levels of all three combined may be nearly lethal. In such a situation, the experiment will not have the treatment combination with the highest levels of all three factors. The design will therefore be an incomplete factorial with monotone structure.

The univariate case is considered in Section 2. Some preliminaries and notational conventions introduced at the start of this section are used throughout the paper. Tests of main effects and two- and three-factor interaction are derived, and illustrated by an example.

In Section 3, the multivariate case is considered. Modified likelihood ratio tests of main effects and interactions are presented, along with their approximate null distributions.

In Section 4, some comments are made regarding the use of the monotone factorial design when the experiments have block designs with blocks of unequal sizes.

The generalization to the m-factor case ($m > 3$) is straightforward.

2 The Univariate Case

Some preliminaries and notation.

Let $a_i b_j c_k$ be a treatment included in our experiment, i.e., $a_i b_j c_k \neq \phi$. Let x_{ijkl} denote the lth observation for $a_i b_j c_k, l = 1, \ldots, n_{ijk}$, for the completely randomized design. Let $I(j,k)$ denote the highest level of A occurring with the jth level of B and the kth level of C, (i.e., $a_f b_j c_k \neq \phi$ and $a_{i'} b_{j'} c_{k'} = \phi$ for $f = I(j,k)$, $i' > f, j' \geq j, k' \geq k$.) Define $J(i,k)$ and $K(i,j)$ similarly.

Note $I(1,1) = I$, $J(1,1) = J$ and $K(1,1) = K$. $I(j,k)$ is a non-increasing function of each j and k. Similar remarks hold for $J(i,k)$ and $K(i,j)$.

We use dot notation for totals and (later on) for means. Let

$$I(j,\cdot) = \sum_{k=1}^{K(1,j)} I(j,k), \; I(\cdot,k) = \sum_{j=1}^{J(1,k)} I(j,k),$$

and

$$I(\cdot,\cdot) = \sum_{j=1}^{J} I(j,\cdot) = \sum_{k=1}^{K} I(\cdot,k).$$

Define similarly, $J(i,\cdot)$, $J(\cdot,k)$, $J(\cdot,\cdot)$ and $K(i,\cdot)$, $K(\cdot,j)$, $K(\cdot,\cdot)$. Observe that $J(i,\cdot) = K(i,\cdot)$ is the total number of treatments having the ith level of A, $I(j,\cdot) = K(\cdot,j)$ is the total number of treatments having the jth level of B and $I(\cdot,k) = J(\cdot,k)$ is the total number of treatments having the kth level of C. And $I(\cdot,\cdot) = J(\cdot,\cdot) = K(\cdot,\cdot) = \tau$ is the total number of treatments included in our experiment.

We find the following notation convenient for describing contrasts, conditions on the parameters and transformations on the variables. For any singly indexed array $\{z_i\}$, define

$$\Delta_i z_i = \big(i(i-1)\big)^{-1/2}\Big((i-1)z_i - \sum_{i'=1}^{i-1} z_{i'}\Big),$$

Similarly, for any double indexed array $\{z_{ij}\}$ define

$$\Delta_i z_{ij} = \big(i(i-1)\big)^{-1/2}\Big((i-1)z_{ij} - \sum_{i'=1}^{i-1} z_{i'j}\Big),$$

$$\Delta_j z_{ij} = \big(j(j-1)\big)^{-1/2}\Big((j-1)z_{ij} - \sum_{j'=1}^{j-1} z_{ij'}\Big).$$

Finally, define

$$\Delta_i z_{ijk} = \big(i(i-1)\big)^{-1/2}\Big((i-1)z_{ijk} - \sum_{i'=1}^{i-1} z_{i'jk}\Big),$$

$$\Delta_j z_{ijk} = \big(j(j-1)\big)^{-1/2}\Big((j-1)z_{ijk} - \sum_{j'=1}^{j-1} z_{ij'k}\Big),$$

$$\Delta_k z_{ijk} = \big(k(k-1)\big)^{-1/2}\Big((k-1)z_{ijk} - \sum_{k'=1}^{k-1} z_{ijk'}\Big).$$

Observe that

$$\Delta_i\Delta_j z_{ij} = \Delta_j\Delta_i z_{ij}, \quad \Delta_i\Delta_j z_{ijk} = \Delta_j\Delta_i z_{ijk}$$

and $\Delta_i\Delta_j\Delta_k$ operated in any order on the array $\{z_{ijk}\}$ gives the same result.

The quantities z_i, z_j, z_k, z_{ijk} in the above could be parameters or variables. $V[X]$ will denote the variance of the random variable (r.v.) X.

Model

The fixed effects model for our three factor experiment (with a monotone structure) having a completely randomized design is

$$\begin{aligned} x_{ijkl} = \mu + \alpha_i + \beta_j + \gamma_k + (\alpha\beta)_{ij} + (\alpha\gamma)_{ik} + (\beta\gamma)_{jk} \\ + (\alpha\beta\gamma)_{ijk} + e_{ijkl}, \end{aligned} \tag{2.1}$$

for $i = 1, \ldots, I$; $j = 1, \ldots, J(i, 1)$; $k = 1, \ldots, K(i, j)$; $1 = 1, \ldots, n_{ijk}$, where all the symbols in the model except e_{ijkl} are parameters, e_{ijkl} is $N(0, \sigma^2)$, $\sigma^2 > 0$ is unknown, and the e_{ijkl}'s are independent random variables (r.v.'s). We here consider the balanced case $n_{ijk} = M \geq 2$. The approach goes through when the design is proportionally balanced: $n_{ijk} = n_i^{(1)} n_j^{(2)} n_k^{(3)}$. This is more general than the sub-class numbers given by Ditchburne (1955) for the 2-factor additive model. The model (2.1) has too many parameters and they may be assumed to satisfy the following conditions:

$$\begin{aligned} \sum_{j=1}^{J(i,1)} K(i,j)[\Delta_i(\alpha\beta)_{ij}] = 0, \ 2 \leq i \leq I, \\ \sum_{i=1}^{I(j,1)} K(i,j)[\Delta_j(\alpha\beta)_{ij}] = 0, \ 2 \leq j \leq J, \end{aligned} \tag{2.2}$$

$$\begin{aligned} \sum_{k=1}^{K(i,1)} J(i,k)[\Delta_i(\alpha\gamma)_{ik}] = 0, \ 2 \leq i \leq I, \\ \sum_{i=1}^{I(1,k)} J(i,k)[\Delta_k(\alpha\gamma)_{ik}] = 0, \ 2 \leq k \leq K, \end{aligned} \tag{2.3}$$

$$\begin{aligned} \sum_{k=1}^{K(1,j)} I(j,k)[\Delta_j(\beta\gamma)_{jk}] = 0, \ 2 \leq j \leq J, \\ \sum_{j=1}^{J(1,k)} I(j,k)[\Delta_k(\beta\gamma)_{jk}] = 0, \ 2 \leq k \leq K, \end{aligned} \tag{2.4}$$

$$\sum_{j=1}^{J(i,1)} \sum_{k=1}^{K(i,j)} \Delta_i(\alpha\beta\gamma)_{ijk} = 0, \ 2 \le i \le I,$$
$$\sum_{i=1}^{I(j,1)} \sum_{k=1}^{K(i,j)} \Delta_j(\alpha\beta\gamma)_{ijk} = 0, \ 2 \le j \le J, \tag{2.5}$$
$$\sum_{i=1}^{I(1,k)} \sum_{j=1}^{J(i,k)} \Delta_k(\alpha\beta\gamma)_{ijk} = 0, \ 2 \le k \le K,$$

$$\sum_{k=1}^{K(i,j)} \Delta_j\Delta_i(\alpha\beta\gamma)_{ijk} = 0, \ 2 \le j \le J(i,1), \ 2 \le i \le I,$$
$$\sum_{j=1}^{J(i,k)} \Delta_k\Delta_i(\alpha\beta\gamma)_{ijk} = 0, \ 2 \le k \le K(i,1), \ 2 \le i \le I, \tag{2.6}$$
$$\sum_{i=1}^{I(j,k)} \Delta_k\Delta_j(\alpha\beta\gamma)_{ijk} = 0, \ 2 \le k \le K(1,j), \ 2 \le j \le J.$$

Analysis

Let

$$\bar{x}_{ijk\cdot} = M^{-1}\sum_{l=1}^{M} x_{ijkl}, \qquad 1 \le k \le K(i,j), \ 1 \le j \le J(i,1), \ 1 \le i \le I,$$
$$U_{ijk} = \Delta_i \bar{x}_{ijk\cdot}, \qquad 1 \le k \le K(i,j), \ 1 \le j \le J(i,1), \ 2 \le i \le I,$$
$$V_{ijk} = \Delta_j \bar{x}_{ijk\cdot}, \qquad 1 \le k \le K(i,j), \ 1 \le i \le I(j,1), \ 2 \le j \le J,$$
$$W_{ijk} = \Delta_k \bar{x}_{ijk\cdot}, \qquad 1 \le j \le J(i,k), \ 1 \le i \le I(1,k), \ 2 \le k \le K.$$

The U_{ijk}'s are Helmert transformations on the $\bar{x}_{ijk\cdot}$'s. Hence $\{U_{ijk}, \ 1 \le k \le K(i,j), \ 1 \le j \le J(i,1), \ 2 \le i \le I\}$ is a set of mutually independent normal r.v.'s each with variance equal to $M^{-1}\sigma^2$. Similar remarks hold for V_{ijk} and W_{ijk}.

Define

$$\bar{U}_{i\cdot\cdot} = \sum_{j=1}^{J(i,1)} \sum_{k=1}^{K(i,j)} U_{ijk}/K(i,\cdot), \ i = 2, \ldots, I. \tag{2.7}$$

Now

$$EU_{ijk} = \Delta_i\alpha_i + \Delta_i(\alpha\beta)_{ij} + \Delta_i(\alpha\gamma)_{ik} + \Delta_i(\alpha\beta\gamma)_{ijk}. \tag{2.8}$$

Hence, from (2.2), (2.3) and (2.5),

$$E\bar{U}_{i\cdot\cdot} = \Delta_i\alpha_i.$$

Similarly, we can define

$$\bar{V}_{\cdot j \cdot} = \sum_{i=1}^{I(j,1)} \sum_{k=1}^{K(i,j)} V_{ijk}/I(\cdot, j), \quad \bar{W}_{\cdot\cdot k} = \sum_{i=1}^{I(1,k)} \sum_{j=1}^{J(i,k)} W_{ijk}/I(\cdot, k),$$

and then

$$E\bar{V}_{\cdot j \cdot} = \Delta_j \beta_j, \ j = 2, \ldots, J, \ E\bar{W}_{\cdot\cdot k} = \Delta_k \gamma_k, \ k = 2, \ldots, K\cdot$$

Note that

$$V[\bar{U}_{i\cdot\cdot}] = \left(MJ(i, \cdot)\right)^{-1} \sigma^2,$$
$$V[\bar{V}_{\cdot j \cdot}] = \left(MK(\cdot, j)\right)^{-1} \sigma^2,$$
$$V[\bar{W}_{\cdot\cdot k}] = \left(MI(\cdot, k)\right)^{-1} \sigma^2.$$

The $\{ \bar{U}_{i\cdot\cdot} \mid i = 2, \ldots, I \}$ are the best linear unbiased estimators for the I-1 Helmert contrasts $\Delta_i \alpha_i, \ i = 2, \ldots, I$, that involve only the α_i's. Further, the $\bar{U}_{i\cdot\cdot}$'s are mutually independent. Hence the valid sum of squares (SS) for the main effect A is

$$SS_A = M \sum_{i=2}^{I} J(i, \cdot) \bar{U}_{i\cdot\cdot}^2, \qquad (d.f = I - 1).$$

Similarly,

$$SS_B = M \sum_{j=2}^{J} K(\cdot, j) \bar{V}_{\cdot j \cdot}^2, \qquad (d.f = J - 1),$$

$$SS_C = M \sum_{k=2}^{K} I(\cdot, k) \bar{W}_{\cdot\cdot k}^2, \qquad (d.f = k - 1).$$

It is straightforward to show that

$$EMS_A = E\left[\frac{SS_A}{I-1}\right] = \frac{M}{I-1} \sum_{i=2}^{I} J(i, \cdot)(\Delta_i \alpha_i)^2 + \sigma^2,$$

$$EMS_B = E\left[\frac{SS_B}{J-1}\right] = \frac{M}{J-1} \sum_{j=2}^{J} K(\cdot, j)(\Delta_j \beta_j)^2 + \sigma^2,$$

and

$$EMS_C = E\left[\frac{SS_B}{K-1}\right] = \frac{M}{K-1} \sum_{k=2}^{K} I(\cdot, k)(\Delta_k \gamma_k)^2 + \sigma^2.$$

We now proceed to obtain the SS for AB, AC and BC.

Define,

$$U_{ijk}(j) = \Delta_j U_{ijk} = \big(j(j-1)\big)^{-1/2}\left((j-1)U_{ijk} - \sum_{j'=1}^{j-1} U_{ij'k}\right),$$

$$i = 2,\ldots,I,\ j = 2,\ldots,J(i,1),\ k = 1,\ldots,K(i,j),$$

$$U_{ijk}(k) = \Delta_k U_{ijk}\ i = 2,\ldots,I,\ j = 1,\ldots,J(i,2),\ k = 2,\ldots,K(i,j),$$

$$V_{ijk}(k) = \Delta_k V_{ijk}\ i = 1,\ldots,I,\ j = 2,\ldots,J(i,2),\ k = 2,\ldots,K(i,j).$$

$W_{ijk}(i)$, $W_{ijk}(j)$, and $V_{ijk}(i)$ are defined similarly. Note that

$$U_{ijk}(j) = V_{ijk}(i),\ U_{ijk}(k) = W_{ijk}(i),\ V_{ijk}(k) = W_{ijk}(j).$$

Further, $\{U_{ijk}(j),\ i = 2,\ldots,I,\ j = 2,\ldots,J(i,1),\ k = 1,\ldots,K(i,j)\}$ is a set of mutually independent r.v.'s each having a variance equal to $M^{-1}\sigma^2$. Similar results hold about $U_{ijk}(k)$ and $V_{ijk}(k)$.

Let

$$K(i,j)\bar{U}_{ij\cdot}(j) = \sum_{k=1}^{K(i,j)} U_{ijk}(j),\ J(i,k)\bar{U}_{i\cdot k}(k) = \sum_{j=1}^{J(i,k)} U_{ijk}(k),$$

$$I(j,k)\bar{V}_{\cdot jk}(k) = \sum_{i=1}^{I(j,k)} V_{ijk}(k),$$

and let $\bar{V}_{ij\cdot}(i)$, $\bar{W}_{i\cdot k}(i)$ and $\bar{W}_{\cdot jk}(j)$ be defined similarly. We note that $\bar{U}_{ij\cdot}(j) = \bar{V}_{ij\cdot}(i)$, $\bar{U}_{i\cdot k}(k) = \bar{W}_{i\cdot k}(i)$ and $\bar{V}_{\cdot jk}(k) = \bar{W}_{\cdot jk}(j)$.

From (2.8),

$$EU_{ijk}(j) = E\Delta_j U_{ijk} = \Delta_j EU_{ijk} = \Delta_i\Delta_j(\alpha\beta)_{ij} + \Delta_i\Delta_j(\alpha\beta\gamma)_{ijk}. \quad (2.9)$$

Hence, from (2.6),

$$E\bar{U}_{ij\cdot}(j) = \Delta_i\Delta_j(\alpha\beta)_{ij}.$$

Similarly,

$$E\bar{U}_{i\cdot k}(k) = E\bar{W}_{i\cdot k}(i) = \Delta_i\Delta_k(\alpha\gamma)_{ik},\ i = 2,\ldots,I,\ k = 2,\ldots,K(i,1),$$

$$E\bar{V}_{\cdot jk}(k) = E\bar{W}_{\cdot jk}(j) = \Delta_j\Delta_k(\beta\gamma)_{jk},\ j = 2,\ldots,J,\ k = 2,\ldots,K(1,j).$$

Note that

$$\{\,\bar{U}_{ij\cdot}(j) \mid i = 2,\ldots,I,\ j = 2,\ldots,\ J(i,1)\,\}$$

is a set of mutually independent r.v.'s, and that

$$\bar{U}_{ij\cdot}(j) \sim N(\Delta_i\Delta_j(\alpha\beta)_{ij},\ \big(MK(i,j)\big)^{-1}\sigma^2).$$

The $\bar{U}_{ij\cdot}(j)$'s are the best linear unbiased estimators for those contrasts involving only the $(\alpha\beta)_{ij}$'s (i.e., the parameters relating to the two-factor interactions AB). Similar remarks hold about the $\bar{U}_{i\cdot k}(k)$'s and $\bar{V}_{\cdot jk}(k)$'s.

Now the valid SS's for the interactions AB, AC and BC are as follows:

$$SS_{AB} = M \sum_{i=2}^{I} \sum_{j=2}^{J(i,1)} K(i,j) \bar{U}_{ij\cdot}^2(j),$$

based on $n_{AB} = \sum_{i=1}^{I} \sum_{j=2}^{J(i,1)} 1 = J(\cdot,1) - I - J + 1$, d.f.,

$$SS_{AC} = M \sum_{i=2}^{I} \sum_{k=2}^{K(i,1)} J(i,k) \bar{U}_{i\cdot k}^2(k),$$

based on $n_{AC} = K(\cdot,1) - I - K + 1$ d.f., and

$$SS_{BC} = M \sum_{j=2}^{J} \sum_{k=2}^{K(1,j)} I(j,k) \bar{V}_{\cdot jk}^2(k),$$

based on $n_{BC} = K(1,\cdot) - J - K + 1$, d.f.

It can easily be shown that

$$E(MS_{AB}) = E(SS_{AB}/n_{AB}) = M(n_{AB})^{-I} \sum_{i=2}^{I} \sum_{j=2}^{J(i,1)} K(i,j)(\Delta_i \Delta_j (\alpha\beta)_{ij})^2 + \sigma^2,$$

$$E(MS_{AC}) = E(SS_{AC}/n_{AC}) = M(n_{AC})^{-I} \sum_{i=2}^{I} \sum_{k=2}^{K(i,1)} J(i,k)(\Delta_i \Delta_k (\alpha\gamma)_{ik})^2 + \sigma^2,$$

and

$$E(MS_{BC}) = (SS_{BC}/n_{BC}) = M(n_{BC})^{-1} \sum_{j=2}^{J} \sum_{k=2}^{K(1,j)} I(j,k)(\Delta_j \Delta_k (\beta\gamma)_{jk})^2 + \sigma^2.$$

Finally, we proceed to obtain the valid SS for the interaction ABC. Define

$$U_{ijk}(j,k) = \Delta_k U_{ijk}(j) = \Delta_k \Delta_j U_{ijk} = \Delta_k \Delta_j \Delta_i \bar{x}_{ijk},$$
$$i = 2, \ldots, I, \; j = 2, \ldots, J(i,2), \; k = 2, \ldots, K(i,j).$$

Note that

$$U_{ijk}(j,k) = U_{ijk}(k,j) = V_{ijk}(i,k) = V_{ijk}(k,i) = W_{ijk}(i,j) = W_{ijk}(j,i).$$

From (2.9),

$$EU_{ijk}(j,k) = E\Delta_k U_{ijk}(j) = \Delta_k EU_{ijk}(j) = \Delta_i \Delta_j \Delta_k (\alpha\beta\gamma)_{ijk}.$$

Note that

$$\{ U_{ijk}(j,k) \mid i = 2,\ldots,I,\ J = 2,\ldots,J(i,2),\ k = 2,\ldots,K(i,j) \}$$

is a set of mutually independent r.v.'s and that

$$U_{ijk}(j,k) \sim N(\Delta_i\Delta_j\Delta_k(\alpha\beta\gamma)_{ijk},\ M^{-1}\sigma^2)$$

Thus

$$SS_{ABC} = M\sum_{i=2}^{I}\sum_{j=2}^{J(i,2)}\sum_{k=2}^{K(i,j)} U_{ijk}^2(j,k),$$

is based on n_{ABC} d.f., where

$$n_{ABC} = \sum_{i=2}^{I}\sum_{j=2}^{J(i,2)}\sum_{k=1}^{K(i,j)} 1 = \tau - I(\cdot,1) - J(1,\cdot) - K(\cdot,1) + I + J + K - 1.$$

Let $MS_{ABC} = SS_{ABC}/n_{ABC}$. Then

$$E(MS_{ABC}) = M(n_{ABC})^{-1}\sum_{i=2}^{I}\sum_{j=2}^{J(i,2)}\sum_{k=2}^{K(i,j)} (\Delta_i\Delta_j\Delta_k(\alpha\beta\gamma)_{ijk})^2 + \sigma^2.$$

The error SS based on $n_e = (M-1)\tau$ d.f. is

$$SS_e = \sum_{i=1}^{I}\sum_{j=1}^{J(i,1)}\sum_{k=1}^{K(i,j)}\sum_{l=1}^{M}(x_{ijkl} - \bar{x}_{ijk\cdot})^2.$$

Let $MS_e = SS_e/n_e$. Then

$$E(MS_e) = E[((M-1)\tau)^{-1} SS_e] = \sigma^2.$$

MS_A, MS_B, MS_C, MS_{AB}, MS_{AC}, MS_{BC} and MS_{ABC} are each tested against MS_e by an F test with the appropriate d.f.

COMMENTS

SS_{AB}, SS_{AC} and SS_{ABC} each is independent of SS_A; SS_{AB}, SS_{BC} and SS_{ABC} each is independent of SS_B; SS_{AC}, SS_{BC} and SS_{ABC} each is independent of SS_C; SS_{ABC} is independent of each of SS_{AB}, SS_{AC} and SS_{BC}. SS_e is independent of all other SS's. Otherwise, the SS's are dependent. Thus SS_A, SS_B and SS_C are dependent; and SS_{AB}, SS_{AC} and SS_{BC} are dependent. The factorial design is non-orthogonal.

Example

The data for our example are a subset of the data given in Table 5.14 on page 129 of Guenther (1964). This data came from a factorial experiment with 3 factors. Factor A (Wheat Varieties) had 4 levels, Factor B (Kind of Fertilizer) had 3 levels and Factor C (Plowing Methods) had 2 levels. The design of the experiment was completely randomized, and each treatment was replicated 3 times.

The data for our example (see Table 1) are obtained by omitting all the observations for (a) the 3rd replicate, (b) the 4th level of A (so that there are only 3 levels of A) and (c) the treatment $a_3b_3c_2$. Consequently, $\tau = 17$, $M = 2$, $I = 3$, $J = 3$, $K = 2$, $a_3b_3c_2 = \phi$, and $a_ib_jc_k \neq \phi$ otherwise.

TABLE 1. Data for the Example

		Kind of fertilizer					
		1		2		3	
Plowing methods		1	2	1	2	1	2
	1	52	48	38	32	44	20
		44	44	38	18	29	8
		43	51	21	13	27	1
	2	57	48	62	43	32	26
Wheat		69	51	43	25	13	20
varieties		53	42	39	34	27	11
	3	65	58	64	37	27	—
		76	36	46	51	50	—
		79	73	63	48	50	—

TABLE 2. Values of $\bar{x}_{ijk\cdot}$

	$j=1$		$j=2$		$j=3$	
$k=$	1	2	1	2	1	2
$i=1$	48	46	38	25	36.5	14
2	63	49.5	52.5	34	22.5	23
3	70.5	47	55	44	38.5	—

From Table 1, we first find $SS_e = 1767$. Since SS_e has $n_e = (M-1)\tau = 17$ degrees of freedom, it follows that $MS_e = 103.94$.

Next, from Table 2, we obtain Table 3 which gives values of the U_{ijk}'s. Recall that $U_{ijk} = \Delta_i \bar{x}_{ijk\cdot}$, $i \geq 2$.

We wish to compute

$$SS_A = M \sum_{i=2}^{I} J(i,\cdot)\bar{U}_{i\cdot\cdot}^2.$$

TABLE 3. Values of U_{ijk}.

	$j=1$		$j=2$		$j=3$			
$k=$	1	2	1	2	1	2	Sum	Mean $\bar{U}_{i\cdot\cdot}$
$i=2$	$\frac{15}{\sqrt{2}}$	$\frac{3.5}{\sqrt{2}}$	$\frac{14.5}{\sqrt{2}}$	$\frac{9}{\sqrt{2}}$	$\frac{-14}{\sqrt{2}}$	$\frac{9}{\sqrt{2}}$	$\frac{37}{\sqrt{2}}$	$\frac{37}{\sqrt{2}}\frac{1}{6}$
3	$\frac{30}{\sqrt{6}}$	$\frac{-1.5}{\sqrt{6}}$	$\frac{19.5}{\sqrt{6}}$	$\frac{29}{\sqrt{6}}$	$\frac{18}{\sqrt{6}}$	—	$\frac{95}{\sqrt{6}}$	$\frac{95}{\sqrt{6}}\frac{1}{5}$

In our case $M=2,\ I=3,\ J(2,\cdot)=6,\ J(3,\cdot)=5$ and the values of $\bar{U}_{i\cdot\cdot}$ are given in the last column of Table 3. Thus

$$SS_A = 2\left[6\left(\frac{37}{\sqrt{2}}\frac{1}{6}\right)^2 + 5\left(\frac{95}{\sqrt{6}}\frac{1}{5}\right)^2\right]$$
$$= 829.82$$

with d.f. $n_A = 2$. Similarly SS_B and SS_C can be calculated. In fact, $SS_B = 3960$, with $n_B = 2$ d.f. and $SS_C = 1339.03$, with $n_C = 1$ d.f.

Now we obtain SS_{AB}. For this purpose, from Table 3 we obtain the $U_{ijk}(j)$'s which are given below in Table 4. Note that $U_{ijk}(j) = \Delta_j U_{ijk},\ i \geq 2,\ j \geq 2$.

TABLE 4. Values of $U_{ijk}(j)$

	$j=2$		$j=3$	
$k=$	1	2	1	2
$i=2$	$\frac{-0.5}{\sqrt{2}\sqrt{2}}$	$\frac{5.5}{\sqrt{2}\sqrt{2}}$	$\frac{57.5}{\sqrt{2}\sqrt{6}}$	$\frac{5.5}{\sqrt{2}\sqrt{6}}$
3	$\frac{-10.5}{\sqrt{6}\sqrt{2}}$	$\frac{30.5}{\sqrt{6}\sqrt{2}}$	$\frac{-13.5}{\sqrt{6}\sqrt{6}}$	—

From Table 4,

$$\bar{U}_{22\cdot}(2) = \frac{5}{\sqrt{2}\sqrt{2}}\frac{1}{2}, \qquad \bar{U}_{23\cdot}(3) = \frac{63}{\sqrt{2}\sqrt{6}}\frac{1}{2},$$

$$\bar{U}_{32\cdot}(2) = \frac{20}{\sqrt{6}\sqrt{2}}\frac{1}{2}, \qquad U_{33\cdot}(3) = U_{331}(3) = -\frac{13.5}{\sqrt{6}\sqrt{6}}.$$

Note that $K(i,j) = 2$ for $i \neq 3,\ j \neq 3,\ K(3,3) = 1$, and that $J(2,1) = J(3,1) = 3$. In our case,

$$SS_{AB} = 2\sum_{i=2}^{3}\sum_{j=2}^{J(i,1)} K(i,j)\bar{U}^2_{ij\cdot}(j)$$

$$= 2\left[\left(\frac{5}{\sqrt{2}\sqrt{2}}\right)^2 \Big/ 2 + \left(\frac{63}{\sqrt{2}\sqrt{6}}\right)^2 \Big/ 2 + \left(\frac{20}{\sqrt{6}\sqrt{2}}\right)^2 \Big/ 2 + \left(\frac{13.5}{\sqrt{6}\sqrt{6}}\right)^2\right]$$
$$= 380.46$$

with d.f. $n_{AB} = 4$. Similarly, $SS_{AC} = 46.33$, with $n_{AC} = 4$ d.f., and $SS_{BC} = 2.79$, with $n_{BC} = 2$ d.f.

Now we obtain SS_{ABC}. For this purpose, from Table 4 we obtain $U_{ijk}(j,k)$, $i \geq 2$, $j \geq 2$, $k = 2$ which are given in Table 5 below. Note that

$$U_{ijk}(j,k) = \Delta_k U_{ijk}(j), \ i \geq 2, \ j \geq 2, \ k \geq 2.$$

Since k assumes only the value 2, $U_{ijk}(j,k) = U_{ij2}(j,2)$.

TABLE 5. Values of $U_{ij2}(j,2)$

	$j =$	
	2	3
$i = 2$	$\frac{6.0}{\sqrt{2}\sqrt{2}\sqrt{2}}$	$\frac{-52.0}{\sqrt{2}\sqrt{6}\sqrt{2}}$
3	$\frac{41.0}{\sqrt{6}\sqrt{2}\sqrt{2}}$	—

Here, $J(2,2) = 3$ and $J(3,2) = 2$. In our case,

$$SS_{ABC} = 2\sum_{i=2}^{3}\sum_{j=2}^{J(i,2)} U_{ij2}^2(j,2)$$
$$= 2\left[\frac{36}{8} + \frac{52^2}{24} + \frac{41^2}{24}\right] = 374.42$$

with $n_{ABC} = 3$ degrees of freedom. All the main effects and interactions are to be tested against $MS_e = 103.94$.

3 The Multivariate Case

Preliminaries

Let the p (≥ 2) response variables be $x^{(1)}, x^{(2)}, \ldots, x^{(p)}$. Let $x_{ijkl}^{(t)}$ denote the observations on the variable $x^{(t)}$ for the lth experimental unit assigned to $a_i b_j c_k$, $l = 1, \ldots, M_t$, $M_t \geq 2$, $t = 1, \ldots, p$. Let the design of the experiment be completely randomized, and let the structure of the treatments be monotone factorial for each of the p variables taken singly. Further, let the design be monotone in respect of the variables, i.e., (i) $x_{ijkl}^{(t)}$ missing implies that $x_{ijkl'}^{(t')}$ is also missing for $t' \geq t$, $l' \geq l$, and (ii) $x_{ijkl}^{(t)}$ not missing implies that $x_{ijkl'}^{(t')}$ is also not missing for $t' \leq t$, $l' \leq l$. A multivariate sample possessing this property has been termed a monotone sample by Bhargava (1962, 1975). Let r_t be the number of treatment combinations used for the experiment in $x^{(t)}$ and let R_t denote the set of these r_t treatments. Then the design being monotone in variables implies that $R_t \supset R_{t'}$ and $r_t \geq r_{t'}$, if $t \leq t'$, and that $M_1 \geq M_2 \geq \cdots \geq M_p \geq 2$. The complete sample case, viz, $M_1 = M_2 = \cdots = M_p \geq 2$, is a particular case of this. To sum up the above conditions, $x_{ijkl}^{(t)}$ missing implies $x_{i'j'k'l'}^{(t')}$ is also missing for $t' \geq t,\ i' \geq i,\ j' \geq j,\ k' \geq k,\ l' \geq l$.

The completely randomized design in this case means that there are $M_1 r_1$ experimental units. M_1 experimental units are assigned at random to each of the r_1 treatments of the set R_1. $x^{(1)}$ is measured on these $M_1 r_1$ units. Let $a_i b_j c_k \varepsilon R_2$. From the M_1 experimental units assigned to $a_i b_j c_k$ on which $x^{(1)}$ is measured, a subset of M_2 experimental units, $M_2 \leq M_1$ is chosen at random and on these $x^{(2)}$ is measured. This is done for each of $a_i b_j c_k \varepsilon R_2$. This is continued until $x^{(p)}$ is measured for the M_p experimental units of each $a_i b_j c_k \varepsilon R_p$. Note $R_t \supset R_{t'}$ and $r_t \geq r_{t'}$ if $t \leq t'$. After relabeling l, the observations will have the structure described above, namely, $x_{ijkl}^{(t)} = \phi$ implies $x_{i'j'k'l'}^{(t')} = \phi$ for $t' \geq t,\ i' \geq i,\ j' \geq j,\ k' \geq k,\ l' \geq l$.

We briefly outline the analysis of this design. Our notations remain the same as described in section 2 except that now a superscript t is added to the variables, to the parameters and to the indexing variables that give the range of the levels of the factors. The joint distribution of the p variables is assumed to be multivariate normal with the covariance matrix $\Sigma = (\sigma_{ij}),\ i,j = 1,\ldots,p,\ \Sigma$ positive definite and unknown.

Model

Our model is

$$x_{ijkl}^{(t)} = \mu^{(t)} + \alpha_i^{(t)} + \beta_j^{(t)} + \gamma_k^{(t)} + (\alpha\beta)_{ij}^{(t)} + (\alpha\gamma)_{ik}^{(t)} + (\beta\gamma)_{jk}^{(t)} + (\alpha\beta\gamma)_{ijk}^{(t)} + e_{ijkl}^{(t)},$$
$$t = 1,\ldots,p, \quad i = 1,\ldots,I^{(t)}(1,1), \quad j = 1,\ldots,J^{(t)}(i,1),$$
$$k = 1,\ldots,K^{(t)}(i,j), \quad l = 1,\ldots,M^{(t)}. \tag{3.1}$$

Note that

$$\operatorname{cov}(x_{ijkl}^{(t)},\ x_{i'j'k'l'}^{(t')}) = \begin{cases} \sigma_{tt'} & \text{if } i = i',\ j = j',\ k = k',\ l = l' \\ 0 & \text{otherwise} \end{cases}. \tag{3.2}$$

Also

$$\begin{aligned} I &= I^{(1)}(1,1), & J &= J^{(1)}(1,1), & K &= K^{(1)}(1,1), \\ I^{(t)}(1,1) &\geq I^{(t')}(1,1), & J^{(t)}(i,1) &\geq J^{(t')}(i,1), & K^{(t)}(i,j) &\geq K^{(t')}(i,j), \end{aligned} \tag{3.3}$$

if $t' \geq t$.

Let (2.2) to (2.6) hold, with superscript $t,\ t = 1,\ldots,p$. Note that $I^{(t)}(\cdot,\cdot) = J^t(\cdot,\cdot) = K^t(\cdot,\cdot)$ is the number of treatments having observations on $x^{(t)}$.

Testing the Hypotheses Relating to the Main Effects of A, B and C.

We now proceed to test hypotheses relating to the main effects of A, B and C, which we respectively denote by H_A, H_B and H_C. Similar notation will be followed for interactions. Now H_A is

$$H_A\colon\ \Delta_i\alpha_i^{(t)}=0,\ t=1,\dots,p,\ i=2,\dots,I^{(t)}(1,1).$$

The alternative hypothesis to H_A is that H_A is not true; that is, at least one $\Delta_i\alpha_i^{(t)}\neq 0$. H_B and H_C can be defined similarly. Define

$$\begin{aligned}\underset{\sim}{x}_{ijkl}(s)&\equiv(x_{ijkl}^{(1)},\ x_{ijkl}^{(2)},\dots,x_{ijkl}^{(s)})',\ s\le p,\\ M_t\bar{x}_{ijk[M_t]}^{(t')}&=\sum_{l=1}^{M_t}x_{ijkl}^{(t')},\quad t'\le t,\\ \underset{\sim}{\bar{x}}_{ijk[M_t]}(s)&\equiv(\bar{x}_{ijk[M_t]}^{(1)},\bar{x}_{ijk[M_t]}^{(2)},\dots,\bar{x}_{ijk[M_t]}^{(s)})',\ s\le t\le p,\end{aligned}\tag{3.4}$$

and

$$S_s^{(t)}=\begin{cases}\sum_{i=1}^{I^{(t)}(1,1)}\sum_{j=1}^{J^{(t)}(i,1)}\sum_{k=1}^{K^{(t)}(i,j)}\sum_{l=1}^{M_t}\big[\big(\underset{\sim}{x}_{ijkl}(s)-\underset{\sim}{\bar{x}}_{ijk[M_t]}(s)\big)\\ \quad\times\big(\underset{\sim}{x}_{ijk[M_t]}(s)-\underset{\sim}{\bar{x}}_{ijk[M_t]}(s)\big)'\big],\quad 1\le s\le t,\\ 1,\quad s=0.\end{cases}\tag{3.5}$$

Let

$$U_{ijk[M_t]}^{(t')}\equiv\Delta_i\bar{x}_{ijk[M_t]}^{(t')},\ t'\le t$$

for those values of i, j, k for which $\Delta_i\bar{x}_{ijk[M_t]}^{(t')}$ exists, viz, for $i=2,\dots,I^{(t)}(1,1)$, $j=1,\dots,J^{(t)}(i,1)$, $k=1,\dots,K^{(t)}(i,j)$.

Similarly, define

$$V_{ijk[M_t]}^{(t')}\equiv\Delta_j\bar{x}_{ijk[M_t]}^{(t')},\quad W_{ijk[M_t]}^{(t')}\equiv\Delta_k\bar{x}_{ijk[M_t]}^{(t')},\ t'\le t,$$

for the values of i, j, k for which they exist.

Let

$$\bar{U}_{i\cdot\cdot[M_t]}^{(t')}=\sum_{j=1}^{J^{(t)}(i,1)}\sum_{k=1}^{K^{(t)}(i,j)}U_{ijk[M_t]}^{(t')}/K^{(t)}(i,\cdot),\ t'\le t,\ i=2,\dots,I^{(t)}(1,1),$$

where $K^{(t)}(i,\cdot)=\sum_{j=1}^{J^{(t)}(i,1)}K^{(t)}(i,j)$, and

$$\underset{\sim}{\bar{U}}_{i\cdot\cdot[M_t]}(s)\equiv\begin{cases}(\bar{U}_{i\cdot\cdot[M_t]}^{(1)},\bar{U}_{i\cdot\cdot[M_t]}^{(2)},\dots,\bar{U}_{i\cdot\cdot[M_t]}^{(s)})' & \text{for } 1\le s\le t,\\ 0 & \text{for } s=0.\end{cases}$$

Let

$$M_tK^{(t)}(\cdot,\cdot)\hat{\sigma}_t^2 = |S_t^{(t)}| \div |S_{t-1}^{(t)}|,\ t = 1,\ldots,p, \tag{3.6}$$

where $S_s^{(t)}$ is defined in (3.5), and

$$M_tK^{(t)}(\cdot,\cdot)\hat{\hat{\sigma}}_t^2(A) = \frac{\left|S_t^{(t)} + M_t \sum_{i=2}^{I^{(t)}(1,1)} K^{(t)}(i,\cdot)\bar{\underset{\sim}{U}}_{i\cdot\cdot[M_t]}(t)\bar{\underset{\sim}{U}}'_{i\cdot\cdot[M_t]}(t)\right|}{\left|S_{t-1}^{(t)} + M_t \sum_{i=2}^{I^{(t)}(1,1)} K^{(t)}(i,\cdot)\bar{\underset{\sim}{U}}_{i\cdot\cdot[M_t]}(t-1)\bar{\underset{\sim}{U}}'_{i\cdot\cdot[M_t]}(t-1)\right|}$$

for $t = 1,\ldots,p$. Finally, let

$$Z_t(A) = \frac{\hat{\sigma}_t^2}{\hat{\hat{\sigma}}_t^2(A)}, \quad t = 1,\ldots,p.$$

The likelihood ratio (L.R) statistic to test H_A (see Bhargava (1962, 1975)) is

$$Z(A) = \prod_{t=1}^{p} \left(Z_t(A)\right)^{M_tK^{(t)}(\cdot,\cdot)/2}.$$

Let $Y_{m,n}$ denote a beta random variable with parameters $m > 0$ and $n > 0$; i.e. its density is

$$f(x) = \begin{cases} x^{m-1}(1-x)^{n-1}/B(m,n) & \text{for } 0 \le x \le 1, \\ 0, & \text{otherwise} \end{cases}.$$

From Bhargava (1962, 1975) under H_A, $Z_1(A)$, $Z_2(A),\ldots,Z_p(A)$ are independent beta random variables with

$$Z_t(A) \sim Y_{a_t/2,\ b_t/2},\ t = 1,\ldots,p,$$

where $a_t = (M_t - 1)K^{(t)}(\cdot,\cdot) - (t-1)$ and $b_t = I^{(t)}(1,1) - 1$.

Let

$$\lambda''(A) = \prod_{t=1}^{p} \left(Z_t(A)\right)^{(a_t - l_t)/2},$$

where the l_t's are fixed numbers such that $0 < a_t - l_t$. The asymptotic c.d.f of $M(A) \equiv -2ln\lambda''(A)$ under H_A can be obtained by using the result of Bhargava (1962), who followed the method of Box (1949), and is as follows:

$$\begin{aligned} \Pr\{M(A) \le M_0(A)\} &= \Pr\{\rho M(A) \le \rho M_0(A)\} \\ &= (1 - w_2)\Pr\{\chi_f^2 \le \rho M_0\} + w_2 \Pr\{\chi_{f+4}^2 \le \rho M_0\} \\ &\quad + 0(n^{-3}), \end{aligned} \tag{3.7}$$

where

$$f = \sum_{t=1}^{p} b_t, \quad n = \sum_{t=1}^{p} a_t, \quad \rho = 1 + (2f)^{-1}\sum_{t=1}^{p} b_t \frac{2l_t + b_t - 2}{a_t - l_t} > 0,$$

$$2h_t = (1-\rho)a_t + \rho l_t,\ w_2 = \sum_{t=1}^{p} \left(b_t \frac{6h_t(2h_t + b_t - 2) + (b_t - 2)(b_t - 1)}{6\rho^2(a_t - l_t)^2} \right)$$

and χ_a^2 denotes a chi-square random variable with a degrees of freedom.

If $M_0(A)$ is the observed value of $M(A)$ for our experiment, then H_A is rejected at the α level of significance if $\Pr\{M(A) \geq M_0(A)\} \leq \alpha$.

If we put $-l_t = K^t(\cdot,\cdot) + (t-1)$ in $\lambda''(A)$, we get the likelihood ratio (L.R.) test. If we put $l_t = 0$, we get the modified L.R. test based on d.f., rather than the sample size, as suggested by Bartlett (1937) in a different context. The L.R. test and the modified L.R. test, in general, are different when the multivariate observations form a monotone sample pattern.

Approximations of higher order can be obtained by using Theorem A.2 (Appendix) in Bhargava (1962).

The hypotheses H_B and H_C are

$$H_B\colon\ \Delta_j \beta_j^{(t)} = 0,\ t = 1, \ldots, p,\ j = 2, \ldots, J^{(t)}(1,1)$$

and

$$H_C\colon\ \Delta_k \gamma_k^{(t)} = 0,\ t = 1, \ldots, p,\ k = 2, \ldots, K^{(t)}(1,1).$$

The alternative hypotheses for H_B and H_C respectively are H_B not true and H_C not true.

H_B and H_C can be tested similarly by defining

$$\bar{V}^{(t')}_{\cdot j \cdot [M_t]} = \sum_{i=1}^{I^{(t)}(j,1)} \sum_{k=1}^{K^{(t)}(i,j)} V^{(t')}_{ijk[M_t]} / K^{(t)}(\cdot, j),\ t' \leq t,$$

and

$$\bar{W}^{(t')}_{\cdot\cdot k[M_t]} = \sum_{i=1}^{I^{(t)}(1,k)} \sum_{j=1}^{J^{(t)}(i,k)} W^{(t')}_{ijk[M_t]} / I^{(t)}(\cdot, k),\ t' \leq t$$

and proceeding similarly.

Testing the Hypotheses Relating to the Interactions AB, AC and BC

We now proceed to test the hypotheses H_{AB}, H_{AC} and H_{BC}. H_{AB} is

$$H_{AB}\colon\ \Delta_i \Delta_j (\alpha\beta)_{ij}^{(t)} = 0,\ t = 1, \ldots, p,\ i = 2, \ldots, I^{(t)}(1,1),\ j = 2, \ldots, J^{(t)}(i,1).$$

The alternative to H_{AB} is that H_{AB} is not true; i.e., at least one $\Delta_i \Delta_j (\alpha\beta)^{(t)} ij \neq 0$. H_{AC} and H_{BC} are defined similarly.

Define

$$P^{(t')}_{ijk[M_t]} = \Delta_j U^{(t')}_{ijk[M_t]} = \Delta_j \Delta_i \bar{x}^{(t')}_{ijk[M_t]},\ t' \leq t,$$
$$Q^{(t')}_{ijk[M_t]} = \Delta_k U^{(t')}_{ijk[M_t]} = \Delta_k \Delta_i \bar{x}^{(t')}_{ijk[M_t]},\ t' \leq t,$$

and

$$R^{(t')}_{ijk[M_t]} = \Delta_k V^{(t')}_{ijk[M_t]} = \Delta_k \Delta_j \bar{x}^{(t')}_{ijk[M_t]}, \; t' \le t,$$

for those values of i, j, k for which they exist.

Also define

$$H^{(t')}_{ijk[M_t]} = \Delta_k \Delta_j \Delta_i \bar{x}^{(t')}_{ijk[M_t]}, \; t' \le t, \tag{3.8}$$

for those values of i, j k for which they exist.

Let

$$\bar{P}^{(t')}_{ij\cdot[M_t]} = \sum_{k=1}^{K^{(t)}(i,j)} P^{(t')}_{ijk[M_t]} / K^{(t)}(i,j), \; t' \le t$$
$$\text{for } i = 2, \ldots, I^{(t)}(1,1), j = 2, \ldots, J^{(t)}(i,1).$$

Similarly, $\bar{Q}^{(t')}_{i\cdot k[M_t]}$ and $\bar{R}^{(t')}_{\cdot jk[M_t]}$ are defined.

Let

$$\bar{P}_{ij\cdot[M_t]}(s) \equiv \begin{cases} \left(\bar{P}^{(1)}_{ij\cdot[M_t]}, \bar{P}^{(2)}_{ij\cdot[M_t]}, \ldots, \bar{P}^{(s)}_{ij\cdot[M_t]}\right)', & \text{for } 1 \le s \le t, \\ 1, & \text{for } s = 0. \end{cases}$$

Define

$$M_t K^{(t)}(\cdot,\cdot) \hat{\hat{\sigma}}^2_t(AB) = \frac{\left| S^{(t)}_t + M_t \sum_{i=2}^{I^{(t)}(1,1)} \sum_{j=2}^{J^{(t)}(i,1)} K^{(t)}_{ij} \left(\underset{\sim}{\bar{P}}_{ij\cdot[M_t]}(t) \underset{\sim}{\bar{P}}'_{ij\cdot[M_t]}(t) \right) \right|}{\left| S^{(t)}_{t-1} + M_t \sum_{i=2}^{I^{(t)}(1,1)} \sum_{j=2}^{J^{(t)}(i,1)} K^{(t)}_{ij} \left(\underset{\sim}{\bar{P}}_{ij\cdot[M_t]}(t-1) \underset{\sim}{\bar{P}}'_{ij\cdot[M_t]}(t-1) \right) \right|}$$

and

$$Z_t(AB) \equiv \frac{\hat{\sigma}^2_t}{\hat{\hat{\sigma}}^2_t(AB)}, \; t = 1, \ldots, p,$$

where $\hat{\sigma}^2_t$ is given by (3.6).

The L.R. statistic to test H_{AB} (see Bhargava (1962, 1975)) is

$$Z(AB) = \prod_{t=1}^{p} \left(Z_t(AB) \right)^{M_t K^{(t)}(\cdot,\cdot)/2}$$

From Bhargava (1962, 1975), under H_{AB}, $Z_1(AB)$, $Z_2(AB), \ldots, Z_p(AB)$ are independent beta random variables with

$$Z_t(AB) \sim Y_{a_t/2, b_t/2}, \qquad t = 1, \ldots, p,$$

where $a_t = (M_t - 1)K^{(t)}(\cdot,\cdot) - (t-1)$ and $b_t = I^{(t)}(\cdot,1) - I^{(t)}(1,1) - J^{(t)}(1,1) + 1$.

Let $\lambda''(AB) = \prod_{t=1}^{p} \left(Z_t(AB)\right)^{(a_t - l_t)/2}$, where the l_t's are fixed numbers such that $0 < a_t - l_t$. The asymptotic c.d.f. of $M(AB) = -2ln\lambda''(AB)$ under H_{AB} can be obtained by using the result given in (3.7).

Testing of the hypotheses H_{AC} and H_{BC} can similarly be done by defining $\bar{\underset{\sim}{Q}}_{i\cdot k[M_t]}(s)$ and $\bar{\underset{\sim}{R}}_{\cdot jk[M_t]}(s)$.

TESTING THE HYPOTHESIS RELATING TO THE INTERACTION ABC

We now proceed to test the hypothesis H_{ABC}. H_{ABC} is

$$H_{ABC}\colon\ \Delta_i \Delta_j \Delta_k (\alpha\beta\gamma)_{ijk}^{(t)} = 0,$$

$t = 1, \ldots, p,\ i = 2, \ldots, I^{(t)}(1,1),\ j = 2, \ldots, J^{(t)}(i,1),\ k = 2, \ldots, K^{(t)}(i,j)$. The alternative to H_{ABC} is that H_{ABC} is not true.

Using (3.8), define

$$\underset{\sim}{H}_{ijk[M_t]}(s) \equiv \begin{cases} \left(H_{ijk[M_t]}^{(1)}, H_{ijk[M_t]}^{(2)}, \ldots, H_{ijk[M_t]}^{(s)}\right)' & \text{for } 1 \le s \le t \\ 0 & \text{for } s = 0. \end{cases}$$

Let

$$G_{ijk}(s) \equiv \begin{cases} \underset{\sim}{H}_{ijk[M_t]}(s) \underset{\sim}{H}'_{ijk[M_t]}(s) & \text{for } 1 \le s \le t, \\ 0 & \text{for } s = 0, \end{cases}$$

and

$$M_t K_{(\cdot,\cdot)}^{(t)} \hat{\hat{\sigma}}_t^2(ABC) = \frac{\left| S_t^{(t)} + M_t \sum_{i=2}^{I^{(t)}(1,1)} \sum_{j=2}^{J^{(t)}(i,1)} \sum_{k=2}^{K^{(t)}(i,j)} G_{ijk}(t) \right|}{\left| S_{t-1}^{(t)} + M_t \sum_{i=2}^{I^{(t)}(1,1)} \sum_{j=2}^{J^{(t)}(i,1)} \sum_{k=2}^{K^{(t)}(i,j)} G_{ijk}(t-1) \right|},$$

$t = 1, \ldots, p$. Finally, let

$$Z_t(ABC) \equiv \frac{\hat{\sigma}_t^2}{\hat{\hat{\sigma}}_t^2(ABC)}, \qquad t = 1, \ldots, p,$$

where $\hat{\sigma}_t^2$ is given by (3.6).

The L.R. statistic to test H_{ABC} (See Bhargava (1962), (1975)) is

$$Z(ABC) = \prod_{t=1}^{p} \left(Z_t(ABC)\right)^{M_t K^{(t)}(\cdot,\cdot)/2}.$$

From Bhargava (1962, 1975), under H_{ABC}, $Z_1(ABC)$, $Z_2(ABC)$, $\ldots$, $Z_p(ABC)$ are independent beta random variables with

$$Z_t(ABC) \sim Y_{\frac{a_t}{2}, \frac{b_t}{2}}, t = 1, \ldots, p,$$

where

$$a_t = (M_t - 1)K^{(t)}(\cdot,\cdot) - (t-1)$$

and

$$b_t = K^{(t)}(\cdot,\cdot) - I^{(t)}(\cdot,1) - J^{(t)}(1,\cdot) - K^{(t)}(\cdot,1) + I^{(t)}(1,1) + J^{(t)}(1,1) + K^{(t)}(1,1) - 1.$$

Let $\lambda''(ABC) = \prod_{t=1}^{p} \left(Z_t(ABC)\right)^{(a_t-l_t)/2}$, where the l_t's are fixed numbers such that $0 < a_t - l_t$. The asymptotic c.d.f. of $M(ABC) \equiv -2\ln\lambda''(ABC)$ under H_{ABC} can be obtained by using the result given in (3.7).

4 Final Comments

Incomplete factorial designs with monotone structure can be used very naturally in experiments having block designs with blocks of unequal sizes. In medical experiments litter sizes are generally unequal. If we want to use all the experimental material, then a monotone factorial design is one choice. However, due to lack of space here, this interesting further work will be presented elsewhere.

Acknowledgments: I am grateful to Professor D.F. Burrill and Samprasad for their assistance. I also wish to thank Professor Leon Jay Gleser for going through the paper thoroughly and making suggestions. I also wish to express my appreciation to Betty Gick and Teena Seele who typed the manuscript.

This work was supported in part by National Science and Engineering Research Council of Canada under grant number A9263.

References

Bartlett, M.S. (1937). "Properties of sufficiency and statistical tests." *Proc. Roy. Soc. London Series A*, 160, 268–282.

Bhargava, R.P. (1962). "Multivariate tests of hypotheses with incomplete data." Technical Report No. 3, Applied Mathematics and Statistics Laboratories, Stanford University, Stanford, California.

Bhargava, R.P. (1967). "Analysis of variance for Scheffé's mixed model for stair-case design." *Sankhyā, Series A*, 29, Part 4, 391–398.

Bhargava, R.P. (1975). "Some one-sample hypothesis testing problems when there is a monotone sample from a multivariate normal population." *Ann. Inst. Math.*, 27, No. 2, 327–339.

Box, G.E.P. (1949). "A general distribution theory for a class of likelihood ratio criteria." *Biometrika* 36, 317–346.

Ditchburne, Nell (1955). "A method of analysis for a double classification arranged in a triangular table." *Biometrics* II, 453–464.

Graybill, Franklin A. and Pruitt William E. (1958). "The staircase design: theory." *Ann. Math. Statist.*, 29, 523–533.

Guenther, William G. (1964). *Analysis of Variance.* Prentice-Hall, N.Y.

12

The Limiting Distribution of the Rank Correlation Coefficient R_g

Rudy A. Gideon[1]
Michael J. Prentice[2]
Ronald Pyke[3]

ABSTRACT A new correlation coefficient, R_g, based on ranks and greatest deviation was defined in Gideon and Hollister (1987). In there the exact distributions were obtained by enumeration for small sample sizes, and by computer simulations for larger sample sizes. In this note, it is shown that the asymptotic distribution of $n^{1/2}R_g$ is $N(0,1)$ when the variables are independent and n is the sample size. This limit is derived by restating the definition of R_g in terms of a rank measure and then using a limit theorem on set-indexed empirical processes which appears in Pyke (1985). The limiting distribution can be compared to the critical values for large samples given in Figure 2 of Gideon and Hollister (1987). Methods for deriving the limiting distribution under fixed and contiguous alternatives are also described.

1 Introduction

In Gideon and Hollister (1987), a new rank correlation coefficient, R_g, is defined that is more resistant to outliers than classical coefficients. Critical values for tests based on R_g for sample sizes $n = 2, 3, \ldots, 100$ were provided. In this paper, the limiting distribution of R_g is obtained so that the new robust correlation procedures may be used in all cases. Using the notation of Gideon and Hollister (1987), the correlation coefficient R_g is defined as follows. Let $\boldsymbol{p}$ denote any permutation of the first n positive integers and let $\boldsymbol{\varepsilon}$ denote the particular "reverse" permutation, $(n, n-1, \ldots, 2, 1)$. The symbol $\circ$ denotes the cyclic group operation, $[\cdot]$ the greatest integer function, and $\mathbf{1}$ the indicator function. Let (X_k, Y_k), $k = 1, \ldots, n$ be a random

[1]University of Montana, Missoula, Montana, USA
[2]University of Edinburgh, Edinburgh, Scotland, UK
[3]University of Washington, Seattle, Washington, USA

sample from an absolutely continuous bivariate distribution H, and denote the corresponding order statistics by $-\infty < X_{n1} < \cdots < X_{nn} < +\infty$ and $-\infty < Y_{n1} < \cdots < Y_{nn} < +\infty$. If r_i is the y rank of that y value which is paired with the i-th smallest x value, then $(X_{n1}, Y_{nr_1}), \ldots, (X_{nn}, Y_{nr_n})$ are the data recorded by increasing x values. Now for $\boldsymbol{p} = (r_1, r_2, \ldots, r_n)$, define

$$d_i(\boldsymbol{p}) = \sum_{j=1}^{i} 1(r_j > i), \quad d_i(\boldsymbol{\varepsilon} \circ \boldsymbol{p}) = \sum_{j=1}^{i} 1(n+1-r_j > i), \quad d(\boldsymbol{p}) = \max_{1 \le i \le n} d_i(\boldsymbol{p}).$$

The new correlation coefficient is then defined by

$$R_g = \{d(\boldsymbol{\varepsilon} \circ \boldsymbol{p}) - d(\boldsymbol{p})\}/[n/2].$$

For notational convenience during the derivation of the limiting distribution, it is assumed that the sample size n is even; the modifications needed when n is odd are straightforward. Thus, the normalized coefficient becomes

$$n^{1/2} R_g = 2n^{-1/2}\{d(\boldsymbol{\varepsilon} \circ \boldsymbol{p}) - d(\boldsymbol{p})\}. \tag{1.1}$$

In terms of this definition, one sees that R_g may be described as the maximum deviation between sums of forward and backward ranks. However, for purposes of this paper, it is important to observe that the ranks in R_g may be viewed as random measures of particular sets, and this set-indexed approach enables one to obtain the asymptotic null distribution rather directly. It also indicates the large family of correlation coefficients that may be considered. To obtain the set-indexed representation, define the following two families of Borel sets on the unit square $I^2 = [0,1] \times [0,1]$,

$$\begin{aligned} \mathcal{A} &= \{A_t : 0 \le t \le 1\} \quad \text{where } A_t = \{(x,y) \in I^2 : y > t, x \le t\}, \\ \mathcal{B} &= \{B_t : 0 \le t \le 1\} \quad \text{where } B_t = \{(x,y) \in I^2 : y < 1-t, x \le t\}. \end{aligned} \tag{1.2}$$

Since R_g is distribution free with respect to the class of absolutely continuous distribution functions, without loss of generality we assume that the marginal distributions of the X_k and Y_k are both uniform on the unit interval I. The problem addressed here is that of finding the limiting distribution of $n^{1/2} R_g$ under the null hypothesis of independence in which the joint distribution H is the uniform distribution in I^2; that is, for all Borel sets $C \in I^2$, $H(C) = |C|$, where $|\cdot|$ denotes Lebesgue measure.

For the x-ordered data (X_{nj}, Y_{nr_j}), $j = 1, 2, \ldots$, n, the rank measure R_n is defined on the Borel sets in I^2 (cf. Pyke (1985)) by

$$R_n(C) = n^{-1}\#\{(X_k, Y_k) : X_k = X_{ni},\ Y_k = Y_{nj} \ \text{ for some } (i,j) \in (n+1)C\} \tag{1.3}$$

in which $\#$ denotes the cardinal number. Thus, R_n is the probability random measure set function that assigns equal measure of $1/n$ to each of the

n points $(i/(n+1), r_i/(n+1)), 1 \leq i \leq n$. Notice that by the definition of ranks, this rank measure has (discrete) uniform marginals. It is easily checked that $d_i(\boldsymbol{p}) = nR_n(A_{i/(n+1)})$, so that

$$d(\boldsymbol{p}) = \max_{1\leq i\leq n} nR_n(A_{i/(n+1)}) = n \sup_{0\leq t\leq 1} R_n(A_t).$$

Similarly, $d_i(\boldsymbol{\varepsilon} \circ \boldsymbol{p}) = nR_n(B_{i/(n+1)})$ and

$$d(\boldsymbol{\varepsilon} \circ \boldsymbol{p}) = \max_{1\leq i\leq n} nR_n(B_{i/(n+1)}) = n \sup_{0\leq t\leq 1} R_n(B_t).$$

For the last equality in each case, observe that $R_n(B_t)$ and $R_n(A_t)$ change values only at $t = i/(n+1)$, $i = 1, \ldots, n$. It now follows that, for n even, (1.1) may be written as

$$R_g = 2\{\sup_{0\leq t\leq 1} R_n(B_t) - \sup_{0\leq t\leq 1} R_n(A_t)\}. \tag{1.4}$$

See Figure 1 for an illustration based on the original YMCA data reported in Gideon and Hollister (1987).

2 The Limiting Null Distribution

Before deriving the limiting distribution, we record some properties of the exact distributions. First of all, $d_i(\boldsymbol{p})$ is a hypergeometric random variable (r. v.); specifically, if Hyper (N, k, m) denotes a hypergeometric r.v. with population size N, sample size m, and k individuals of the type which is being counted, then $d_i(\boldsymbol{p})$ and $d_i(\boldsymbol{\varepsilon} \circ \boldsymbol{p})$ are Hyper $(n, n-i, i)$ under the null hypothesis. To see this, note that the r.v. $d_i(\boldsymbol{p})$ counts the number of ranks in the first i positions of $\boldsymbol{p}$ which exceed i and there are $n-i$ such possibilities. Thus i ranks are samples, $n-i$ ranks are classified as a "success" and the total population size is n. For the null distribution all permutations are equally likely and the conclusion follows. By the same argument, it follows that for any rectangle $A = B \times C$, $B, C \subset I$, $nR_n(A)$ is a Hyper (n, b, c) r.v. where b and c are the number of integers in $(n+1)B$ and $(n+1)C$, respectively. In the above, $d_i(\boldsymbol{p}) = nR_n(A_{i/(n+1)})$ and $A_t = [0, t) \times [t, 1]$.

FIGURE 1. Set-indexed Representation of R_g Illustrated for the YMCA Data of Gideon and Hollister (1987): (i, r_i) is plotted at $(i/17, r_i/17), i = 1, 2, \ldots, 16$ for $(r_1, \ldots, r_{16}) = (14, 11, 16, 2, 12, 13, 7, 9, 10, 3, 8, 1, 15, 6, 4, 5)$

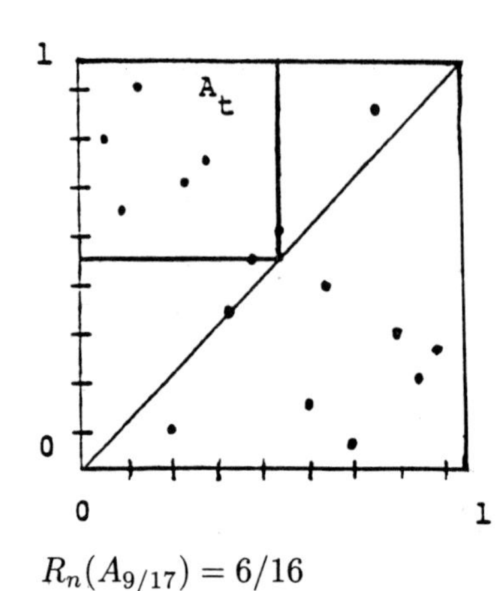

$$\begin{aligned} A_t &= \{(x,y) \in I^2 : y > t, x \leq t\} \\ d(\mathbf{p}) &= \max_{1 \leq i \leq n} d_i(\mathbf{p}) \\ &= n \sup_{0 \leq t \leq 1} R_n(A_t) \\ &= 16 R_{16}(A_{9/17}) = 6 \end{aligned}$$

$R_n(A_{9/17}) = 6/16$

$$\begin{aligned} B_t &= \{(x,y) \in I^2 : y < 1 - t, x \leq \\ d(\varepsilon \circ \mathbf{p}) &= \max_{1 \leq i \leq n} d_i(\varepsilon \circ \mathbf{p}) \\ &= n \sup_{0 \leq t \leq 1} R_n(B_t) \\ &= 16 R_{16}(B_{12/17}) = 3 \end{aligned}$$

$R_n(B_{12/17}) = 3/16$

$$R_g = \frac{2}{n}\{d(\varepsilon \circ \mathbf{p}) - d(\mathbf{p})\} = (3 - 6)/8 = -3/8$$

It then follows that

$$E(d_i(\boldsymbol{p})) = E(nR_n(A_{i/n+1})) = i(1 - i/n)$$

and

$$\operatorname{var}(d_i(\boldsymbol{p})) = \operatorname{var}(nR_n(A_{i/n+1})) = i^2(n-i)^2/n^2(n-1).$$

Similar equations hold for $d_i(\boldsymbol{\varepsilon} \circ \boldsymbol{p})$ and $nR_n(B_{i/n+1})$. The covariances are

$$\operatorname{cov}(d_i(\boldsymbol{p}), d_k(\boldsymbol{p})) = i^2(n-k)^2/n^2(n-1), \quad 0 \le i \le k \le n.$$

To show this, write $\boldsymbol{p} = (r_1, \ldots, r_n)$. Note that $d_n(\boldsymbol{p}) \equiv 0$. Set $d_0(\boldsymbol{p}) \equiv 0$. For $i \le k$,

$$\begin{aligned} \operatorname{cov}(d_i(\boldsymbol{p}), d_k(\boldsymbol{p})) &= \operatorname{cov}\left\{ \sum_{j=1}^{i} 1(i < r_j), \sum_{\ell=1}^{k} 1(k < r_\ell) \right\} \\ &= \sum_{j=1}^{i} \sum_{\ell=1}^{k} \operatorname{cov}\{1(i < r_j), 1(k < r_\ell)\}. \end{aligned}$$

Clearly,

$$\operatorname{cov}(1(i < r_j), 1(k < r_\ell)) = P(r_j > i \text{ and } r_\ell > k) - P(r_j > i)\ P(r_\ell > k).$$

Since $P(r_j > i) = (n-i)/n$ and

$$\begin{aligned} P(r_j > i, r_\ell > k) &= P(r_j > i | r_\ell > k) P(r_\ell > k) \\ &= \begin{cases} (n-k)/n & \text{for } j = \ell \\ \{(n-i-1)/(n-1)\}\{(n-k)/n\} & \text{for } j \ne \ell. \end{cases} \end{aligned}$$

Thus

$$\operatorname{cov}(1(i < r_\ell), 1(k < r_\ell)) = \frac{n-k}{n} - \frac{n-i}{n}\frac{n-k}{n} = \frac{(n-k)i}{n^2}$$

and for $j \ne \ell$,

$$\operatorname{cov}(1(i < r_j), 1(k < r_\ell)) = \frac{n-i-1}{n-1}\frac{n-k}{n} - \frac{n-i}{n}\frac{n-k}{n} = \frac{-i(n-k)}{n^2(n-1)}.$$

Thus for $0 \le i \le k \le n$,

$$\begin{aligned} \operatorname{cov}(d_i(\boldsymbol{p}), d_k(\boldsymbol{p})) &= \sum_{j=1}^{i} \sum_{j \ne \ell = 1}^{k} \frac{(-i)(n-k)}{n^2(n-1)} + \sum_{j=1}^{i} \frac{i(n-k)}{n^2} \\ &= \frac{(-i)(n-k)}{n^2(n-1)} i(k-1) + \frac{i(n-k)}{n^2} i = \frac{i^2(n-k)^2}{n^2(n-1)}. \end{aligned}$$

as desired. Again, the result depends only on the rectangular nature of the sets A_t, so that similar results could be stated for general rectangles.

From the above, the covariance structure of the limiting process can be suggested: If i, k and n increase while $i/n \to t_1$ and $k/n \to t_2$ for $0 < t_1 \le t_2 < 1$, then the limiting covariance is $t_1^2(1-t_2)^2$.

In order to study the asymptotic behavior of $n^{1/2}R_g$, we describe this normalized rank coefficient in terms of the normalized rank processes introduced in Pyke (1985). For any collection C of Borel sets in I^2, define the rank process $S_n = \{S_n(C) : C \in \mathcal{C}\}$ by

$$S_n(C) = n^{1/2}\{R_n(C) - |C|\}, \quad C \in \mathcal{C},$$

where R_n is the rank measure defined in (1.3). From (1.4), it follows that

$$n^{1/2}R_g = 2\sup_{\mathcal{B}}\{S_n(B) + n^{1/2}|B|\} - 2\sup_{\mathcal{A}}\{S_n(A) + n^{1/2}|A|\}. \tag{2.1}$$

By means of this representation of R_g as a function of the set-indexed rank process S_n, its limiting normality can be derived from the weak convergence of S_n that was established in Pyke (1985). To this end, rewrite (2.1) as

$$\begin{aligned} n^{1/2}R_g = 2\{S_n(B_{1/2}) - S_n(A_{1/2})\} + 2\sup_{\mathcal{B}}\{S_n(B) - S_n(B_{1/2}) - n^{1/2}(1/4 - |B|)\} \\ -2\sup_{\mathcal{A}}\{S_n(A) - S_n(A_{1/2}) - n^{1/2}(1/4 - |A|)\}. \end{aligned} \tag{2.2}$$

Theorem 2.1 of Pyke (1985) states that under certain assumptions on the index family $\mathcal{C}$, there exists a probability space on which equivalent versions of the S_n-processes and a Gaussian process S can be defined for which $\sup_{\mathcal{C}} |S_n(C) - S(C)|$ converges to zero as $n \to \infty$. The assumptions on $\mathcal{C}$ for this form of weak convergence can be straightforwardly checked to be satisfied for both $\mathcal{A}$ and $\mathcal{B}$ of this paper, and hence for $\mathcal{C} = \mathcal{A} \cup \mathcal{B}$. Both families inherit the metric structure of $[0,1]$ and the rectangular shape of the sets permits Assumption II of Pyke (1985) to hold. The limiting process, S, is a mean zero Gaussian process with covariance given by

$$\operatorname{cov}(S(A), S(B)) = |A \cap B| + |A||B| - \int_I |A \cap (B_{1y} \times I)|\, dy - \int_I |A \cap (I \times B_{2x})| dx$$

for any $A, B \in \mathcal{C}$, where the sections B_{1y} and B_{2x} are defined by $B_{1y} = \{x \in I : (x,y) \in B\}$, $B_{2x} = \{y \in I : (x,y) \in B\}$. It is easily calculated that when $\mathcal{C} = \mathcal{B}$, this covariance reduces to

$$\operatorname{cov}(S(B_s), S(B_t)) = s^2(1-t)^2, \quad 0 \le s \le t \le 1, \tag{2.3}$$

which agrees with (1.4). It is also true that S is continuous over $\mathcal{C}$ where continuity is with respect to the Lebesgue symmetric-difference pseudometric, $d(A, B) = |A \Delta B|$.

The limiting null distribution of $n^{1/2}R_g$ can now be obtained from (2.1) since it can be shown that for the versions of S_n and S in Pyke (1985), both of the suprema converge to zero. The argument is the same for both; we consider only the term involving B. Observe first that $|B_t| \leq 1/4 = |B_{1/2}|$ for all $B \in \mathcal{B}$. The deterministic function $n^{1/2}(1/4 - |B|)$ is therefore non-negative and diverges to $+\infty$ except at $B_{1/2}$, where it is zero for every n. The supremum is clearly non-negative; try $B = B_{1/2}$. Since S_n converges to S uniformly over $\mathcal{B}$, it suffices to show that

$$s_n \equiv \sup_{\mathcal{B}}\{S(B) - S(B_{1/2}) - n^{1/2}(1/4 - |B|)\} \to 0. \tag{2.4}$$

Since S is continuous over $\mathcal{B}$ and $\mathcal{B} = \{B_t : t \in I\}$, S is uniformly bounded, by L say. Also, for any given $\varepsilon > 0$, there exists a $\delta > 0$ such that

$$\sup\{S(B_t) - S(B_{1/2}) : |t - 1/2| < \delta\} < \varepsilon.$$

Consequently, for all n,

$$s_n \leq \varepsilon + \sup_{|t-1/2|\leq\delta} \{2L - n^{1/2}(1/4 - t(1-t))\}^+ \to \varepsilon.$$

where x^+ denotes the non-negative part of x. Since this is true for all $\varepsilon > 0$, it follows that $s_n \to 0$.

By a similar argument, the term in (2.2) that involves the supremum over $\mathcal{A}$ also converges to zero. Consequently, it follows that the right-hand side of (2.2) converges to

$$Z \equiv 2\{S(B_{1/2}) - S(A_{1/2})\} = 4S(B_{1/2}),$$

since $S([0,1/2] \times I) = 0$ and $A_{1/2} \cup B_{1/2} = [0, \frac{1}{2}] \times I$. However, Z is a mean zero normal random variable whose variance, by (2.3) is 1. This completes the proof of

Theorem 2.1 *Under the uniform null hypothesis, $n^{1/2}R_g$ is asymptotically distributed as a $N(0,1)$ random variable.*

3 The Limiting Distribution under Alternatives

In order to be able to evaluate the power or efficiency of procedures based on R_g, it is necessary to know the distribution of R_g under alternatives as well as under the null hypothesis. For moderate sample sizes this can best be done by simulation. For large sample sizes the asymptotic distribution is needed. For this reason, we outline below methods for obtaining the limiting distributions of R_g under both fixed and contiguous alternatives to the null-hypothesis assumption of independence.

Consider first the case of a fixed alternative, H. We assume that both marginals of H are continuous, and so, without loss of generality, we take the marginals to be uniform on $[0, 1]$, even though H itself is not uniform on I^2. In this situation, the rank process is

$$S_n(C) = n^{1/2}\{R_n(C) - H(C)\}, \qquad C \in \mathcal{C}. \tag{3.1}$$

As for the null-hypothesis case above, these limiting results for R_g depend on knowing the weak convergence of the rank process. Since the results for non-uniform (non-independence) cases require lengthy proofs, we outline here only the steps needed to obtain the results for R_g *once one has the results for the rank process.* The fuller study of the non-uniform rank process is left for a later study; in what follows, the needed results for the rank process are stated as assumptions.

As for (2.1), one may write

$$n^{1/2}R_g = 2\sup_{\mathcal{B}}\{S_n(B) + n^{1/2}H(B)\} - 2\sup_{\mathcal{A}}\{S_n(A) + n^{1/2}H(A)\}. \tag{3.2}$$

Set

$$a = \sup_{\mathcal{A}} H(A), \quad b = \sup_{\mathcal{B}} H(B)$$

and

$$\mathcal{A}^* = \{A \in \mathcal{A} : H(A) = a\}, \quad \mathcal{B}^* = \{B \in \mathcal{B} : H(B) = b\}.$$

Clearly $0 < a,\ b < 1/2$ since, for example, $H(A_t) \leq t \wedge (1-t)$ for all $t \in I$. Suppose that $\mathcal{A}^*$ and $\mathcal{B}^*$ are finite, say $\mathcal{A}^* = \{A_i^* : 1 \leq i \leq k\}$ and $\mathcal{B}^* = \{B_j^* : 1 \leq j \leq m\}$. The limiting distributions of the terms on the right hand side of (3.2) are determined by the members of $\mathcal{A}^*$ and $\mathcal{B}^*$. Consider the term involving the supremum over $\mathcal{B}$. Partition $\mathcal{B}$ into a finite union of sets, say $\mathcal{B} = \cup_1^m \mathcal{B}_j$, such that each $\mathcal{B}_j$ contains exactly one element of $\mathcal{B}^*$ and that element is an interior point. Then write

$$\begin{aligned}&\sup_{\mathcal{B}}\{S_n(B) + n^{1/2}H(B)\}\\ &\quad = \max_j\{S_n(B_j^*) + \sup_{\mathcal{B}_j}\{S_n(B) - S_n(B_j^*) - n^{1/2}(b - H(B))\}\} + n^{1/2}b.\end{aligned}$$

By a similar argument to that used to prove (2.4), one may show that the middle term involving the supremum over $\mathcal{B}_j$ converges to zero. This fact, together with a similar one for $\mathcal{A}$, enables one to deduce from (3.1) that

$$n^{1/2}(R_g - b + a) \xrightarrow{L} 2 \max_{1\leq j\leq m} S(B_j^*) - 2 \max_{1\leq i\leq k} S(A_i^*),$$

provided only that on some probability space there exists a process S and equivalent versions of the rank processes S_n for which $\|S_n - S\|_{\mathcal{A}\cup\mathcal{B}} \to 0$, where $\|f\|_M$ is the sup-norm, $\|f\|_M = \sup\{|f(x)| : x \in M\}$. This establishes the limiting distribution under H. In summary,

Theorem 3.1 *For a fixed alternative H with uniform marginals, if $a = \sup\{H(A) : A \in \mathcal{A}\}$ and $b = \sup\{H(B) : B \in \mathcal{B}\}$ are attained on finite subfamilies $\mathcal{A}^*$ and $\mathcal{B}^*$, respectively, and if S_n converges in distribution to S in the sense that versions exists for which $\|S_n - S\|_{\mathcal{A}\cup\mathcal{B}} \to 0$, then $n^{1/2}(R_g - b + a)$ converges in law to*

$$2 \sup_{\mathcal{B}^*} S(B) - 2 \sup_{\mathcal{A}^*} S(A).$$

When H is uniform on I^2, so that one is in the null-hypothesis case, $\mathcal{A}^* = \{A_{1/2}\}$ and $\mathcal{B}^* = \{B_{1/2}\}$. Thus Theorem 3.1 is consistent with Theorem 2.1.

To obtain the limiting distribution under a sequence of contiguous alternatives, say $\{H^{(n)}\}$, one first needs to have a convergence result for the rank processes that is in some sense uniform over alternatives. To this end, suppose $\mathcal{H}$ is a family of distributions H on I^2 that have uniform marginals. Interpret the rank process in (3.1) as indexed by both a family $\mathcal{C}$ of sets and such a family $\mathcal{H}$ of distributions. That is, view the rank process as

$$S_n(C, H) = n^{1/2}\{R_n(C) - H(C)\}, \qquad (C, H) \in \mathcal{C} \times \mathcal{H}. \tag{3.3}$$

Moreover, when the true distribution of (X_i, Y_i) is H, it is possible to construct equivalent observations that are functions of random variables that are uniform on I^2. In this way, one can simultaneously construct all of the rank processes embodied in (3.3) on the one (null-hypothesis) probability space. Suppose it is then possible to show the convergence-in-law of the process S_n to a limiting process, say $S = \{S(C, H) : C \in \mathcal{C}, H \in \mathcal{H}\}$, in such a way that one may assume without loss of generality that $\|S_n - S\|_{\mathcal{C}\times\mathcal{H}} \to 0$. Suppose $\mathcal{C} = \mathcal{A} \cup \mathcal{B}$ and $\mathcal{H}$ is a family containing the specified contiguous sequence of alternatives, $\{H^{(n)}\}$. If the sequence converges to the uniform null hypothesis at a rate of $n^{-1/2}$ so that

$$\|n^{1/2}(H^{(n)}(\cdot) - |\cdot|) - \nu(\cdot)\|_{\mathcal{C}} \to 0$$

for some bounded set function ν, then arguments similar to those applied to (1.4) can be used to show that

$$n^{1/2} R_g \xrightarrow{L} 2\{S(B_{1/2}) + \nu(B_{1/2}) - S(A_{1/2}) - \nu(A_{1/2})\}.$$

The major step in deriving such a result is again that of establishing the weak convergence of the basic rank process S_n. This will be the focus of later research, but it should be remarked here that conditions on $\mathcal{H}$ will permit large families of alternatives and hence of contiguous sequences.

References

Gideon, R.A. and Hollister, R.A. (1987). A rank correlation coefficient resistant to outliers. *Journal of the American Statistical Assoc.* **82**, 656–666.

Pyke, Ronald (1985). Opportunities for set-indexed empirical and quantile processes in inference. *Bulletin International Statistical Institute* **51,** Book #25.2, 1–11.

13

Mean and Variance of Sample Size in Multivariate Heteroscedastic Method

Hiroto Hyakutake[1]
Minoru Siotani[2]

ABSTRACT For statistical inference on several mean vectors when population covariance matrices are different, the heteroscedastic method is employed to overcome difficulties under a two-stage sampling scheme. Total sample size for each sample is thus a random variable. Both exact and approximate upper and lower bounds for the mean and variance of the sample size are given. Tables are computed for some special cases of these bounds in order to have some information on their numerical behavior.

1 Introduction

Consider a statistical inference on mean vectors $\mu_1, \ldots, \mu_k$ of k p-variate normal populations $N_p(\mu_i, \Sigma_i)$, $i = 1, \ldots, k$, where μ_i and Σ_i are unknown, and Σ_i are different. The heteroscedastic method was introduced by Dudewicz and Bishop (1979) to treat the inference in this situation, and is essentially based on the two-stage sampling scheme devised by Stein (1945) for the univariate case and by Chatterjee (1959) for the multivariate case. That is, we first take a sample of fixed size N_0 from each of the k populations and compute

$$\bar{x}^{(i)} = N_0^{-1} \sum_{r=1}^{N_0} x_r^{(i)}, \qquad \nu S_i = \sum_{r=1}^{N_0} (x_r^{(i)} - \bar{x}^{(i)})(x_r^{(i)} - \bar{x}^{(i)})' \tag{1.1}$$

for $i = 1, \ldots, k$, where $\nu = N_0 - 1$. Then for each i, we define the total sample size by

$$N_i = \max\bigl(N_0 + p^2, (c\,\mathrm{tr}(TS_i)) + 1\bigr), \tag{1.2}$$

---

[1]Hiroshima University
[2]Science University of Tokyo. Partially supported by Grant-in-Aid for Scientific Research of the Ministry of Education under Contract Number 321-6009-61530017.

where c is a given positive constant, T is a given $p \times p$ positive definite (p.d.) matrix, and $[a]$ denotes the largest integer less than a real number a. Then we take $N_i - N_0$ additional observations from the ith population and define the basic random vector $z_i : p \times 1$ $(i = 1, \ldots, k)$ using whole observations and using some matrices satisfying certain conditions. For details of the heteroscedastic method, see Dudewicz and Bishop (1979) or Siotani, Hayakawa, and Fujikoshi (1985).

We note that the p.d. matrix T is chosen to as to reflect the relative importance of deviations in the means of components of the original variate x. If weights are assumed to be equal, we may take $T = I_p$. The constant c is usually chosen so as to meet the requirement in the problem under investigation; see, for example, Hyakutake and Siotani (1987).

In this paper, we give formulae for upper and lower bounds on $E(N_i)$ and $E(N_i^2)$, from which we may obtain an approximate variance or standard deviation of N_i. The bounds are given both by infinite series of zonal polynomials and by asymptotic expansions. Tables of the upper and lower bounds on the mean and standard deviation of N_i are computed for some special cases in order to have some information on the numerical behavior of the bounds.

It is clear that the evaluation for each i is the same, so that we shall drop the suffix i hereafter.

2 Inequalities for $E(N)$ and $E(N^2)$ or $\mathrm{Var}(N)$

Let N be the random variable representing the total sample size defined by (1.2). (Recall that we are suppressing the subscript i.) Then N takes values $N_0 + p^2 + r$, $r = 0, 1, 2, \ldots$, and we have

$$E(N^i) = \sum_{n=N_0+p^2}^{\infty} n^i \Pr(N = n), \qquad i = 1, 2. \tag{2.1}$$

It is obvious that

$$\Pr(N = N_0 + p^2) = \Pr\bigl(c\,\mathrm{tr}(TS) < N_0 + p^2\bigr) \tag{2.2}$$

and for $n > N_0 + p^2$

$$\Pr(N = n) = \Pr\bigl(n - 1 < c\,\mathrm{tr}(TS) < n\bigr), \tag{2.3}$$

where $\nu S = V$ is distributed according to the Wishart distribution with $\nu = N_0 - 1$ degrees of freedom (d.f.) and covariance matrix Σ, i.e., $W_p(\nu, \Sigma)$. Thus we can write $E(N^i)$ as

$$E(N^i) = (N_0 + p^2)^i \Pr\left(\mathrm{tr}(TV) < \frac{\nu}{c}(N_0 + p^2)\right)$$

$$+ \sum_{n=N_0+p^2+1}^{\infty} n^i \Pr\left(\frac{\nu}{c}(n-1) < \mathrm{tr}(TV) < \frac{\nu}{c}n\right)$$

$$= (N_0+p^2)^i \int_0^{(\nu/c)(N_0+p^2)} f(u)\,du$$

$$+ \sum_{n=N_0+p^2+1}^{\infty} n^i \int_{(\nu/c)(n-1)}^{(\nu/c)n} f(u)\,du \tag{2.4}$$

where $f(u)$ is the probability density function (p.d.f.) of

$$U = \mathrm{tr}(TV) = \mathrm{tr}(W), \qquad W = T^{1/2}VT^{1/2}. \tag{2.5}$$

Since

$$\left(\frac{\nu}{c}n\right)^i \int_{(\nu/c)(n-1)}^{(\nu/c)n} f(u)\,du > \int_{(\nu/c)(n-1)}^{(\nu/c)n} u^i f(u)\,du, \quad i=1,2 \tag{2.6}$$

$$\left(\frac{\nu}{c}n\right) \int_{(\nu/c)(n-1)}^{(\nu/c)n} f(u)\,du < \int_{(\nu/c)(n-1)}^{(\nu/c)n} uf(u)\,du + \left(\frac{\nu}{c}\right) \int_{(\nu/c)(n-1)}^{(\nu/c)n} f(u)\,du, \tag{2.7}$$

$$\left(\frac{\nu}{c}n\right)^2 \int_{(\nu/c)(n-1)}^{(\nu/c)n} f(u)\,du < \int_{(\nu/c)(n-1)}^{(\nu/c)n} u^2 f(u)\,du + 2\left(\frac{\nu}{c}\right) \int_{(\nu/c)(n-1)}^{(\nu/c)n} uf(u)\,du + \left(\frac{\nu}{c}\right)^2 \int_{(\nu/c)(n-1)}^{(\nu/c)n} f(u)\,du, \tag{2.8}$$

$E(N^i)$, $i = 1, 2$, satisfy the following inequalities.

$$L_1 < E(N) < U_1, \tag{2.9}$$

$$L_2 < E(N^2) < U_2, \tag{2.10}$$

where

$$L_i = (N_0+p^2)^i \int_0^a f(u)\,du + \left(\frac{c}{\nu}\right)^i \int_a^{\infty} u^i f(u)\,du, \ i=1,2, \tag{2.11}$$

$$U_1 = (N_0+p^2) \int_0^a f(u)\,du + \left(\frac{c}{\nu}\right) \int_a^{\infty} uf(u)\,du + \int_a^{\infty} f(u)\,du, \tag{2.12}$$

$$U_2 = (N_0+p^2)^2 \int_0^a f(u)\,du + \left(\frac{c}{\nu}\right)^2 \int_a^{\infty} u^2 f(u)\,du + 2\left(\frac{c}{\nu}\right) \int_a^{\infty} uf(u)\,du + \int_a^{\infty} f(u)\,du, \tag{2.13}$$

and $a = (\nu/c)(N_0 + p^2)$. We observe that

$$U_1 - L_1 = \int_a^\infty f(u) < 1, \tag{2.14}$$

$$U_2 - L_2 = 2\left(\frac{c}{\nu}\right)\int_a^\infty uf(u)\,du + \int_a^\infty f(u)\,du. \tag{2.15}$$

It follows from (2.14) that $E(N)$ can be evaluated almost exactly. For the variance of N, we obtain

$$L_2 - U_1^2 < \mathrm{Var}(N) < U_2 - L_1^2. \tag{2.16}$$

To evaluate those upper and lower bounds, we need to know the distribution of the random variable U defined by (2.5).

3 Distribution of $U = \mathrm{tr}(W)$

The exact distribution of U has been given by Khatri and Srivastava (1971) in the following form. Since W is distributed according to $W_p(\nu, \Sigma_0)$, where $\Sigma_0 = T^{1/2}\Sigma T^{1/2}$, the cumulative distribution function (c.d.f.) of U is given by

$$F_U(u) = \Pr(U \le u) = \left|\frac{1}{\lambda}\Sigma_0\right|^{-\nu/2} \sum_{k=0}^\infty \frac{1}{k!} G_{p\nu+2k}\left(\frac{u}{\lambda}\right) \sum_\kappa \left(\frac{\nu}{2}\right)_\kappa C_\kappa(I_p - \lambda\Sigma_0^{-1}), \tag{3.1}$$

and the p.d.f. is then

$$f_U(u) = \left|\frac{1}{\lambda}\Sigma_0\right|^{-\nu/2} \sum_{k=0}^\infty \frac{1}{k!} g_{p\nu+2k,\lambda}\left(\frac{u}{\lambda}\right) \sum_\kappa \left(\frac{\nu}{2}\right)_\kappa C_\kappa(I_p - \lambda\Sigma_0^{-1}), \tag{3.2}$$

where $G_f(x)$ is the c.d.f. of the χ^2-distribution with f d.f., $g_{f,\lambda}(x)$ is

$$g_{f,\lambda}(x) = (2\lambda)^{-f/2}\left(\Gamma\left(\frac{f}{2}\right)\right)^{-1} x^{f/2-1}\exp\left(-\frac{x}{2\lambda}\right), \tag{3.3}$$

the symbol $\sum_\kappa$ denotes the summation over all partitions $\kappa = (k_1, k_2, \ldots, k_p)$, $k_1 \ge k_2 \ge \cdots k_p \ge 0$, $k_1 + k_2 + \cdots k_p = k$ of the integer k, $C_\kappa(A)$ is the zonal polynomial corresponding to κ,

$$(a)_\kappa = \prod_{i=1}^p \Gamma\left(a + k_i - \frac{i-1}{2}\right) \Big/ \Gamma\left(a - \frac{i-1}{2}\right),$$

and $0 < \lambda < \infty$ is arbitrary, but is chosen so that the series (3.1) and (3.2) converge rapidly. Khatri and Srivastava (1971) suggest the choice

$\lambda = 2\lambda_1\lambda_p/(\lambda_1 + \lambda_p)$, where λ_1 and λ_p are the largest and smallest latent roots of Σ_0, respectively.

When $p = 2$, the distribution of U reduces to

$$F_U(u) = \sum_{k=0}^{\infty} b_k\left(\theta, \frac{\nu}{2}\right) G_{2\nu+4k}\left(\frac{u}{\lambda}\right), \tag{3.4}$$

where $\lambda = 2\lambda_1\lambda_2/(\lambda_1 + \lambda_2)$, and $b_k(\theta, \nu/2)$ is the kth probability of the negative binomial distribution with parameter $\nu/2$, and $\theta = 4\lambda_1\lambda_2/(\lambda_1 + \lambda_2)^2$, i.e.,

$$b_k\left(\theta, \frac{\nu}{2}\right) = \binom{k + \nu/2 - 1}{k} \theta^{\nu/2}(1-\theta)^k. \tag{3.5}$$

The distribution function (3.4) was given by Muirhead (1982; problem 3.12, p. 115). Another special case is obtained for $T = I_p$ and $\Sigma = \sigma^2 I_p$, i.e., $\Sigma_0 = \sigma^2 I_p$;

$$F_U(u) = G_{p\nu}(u/\sigma^2), \tag{3.6}$$

in which we use $\lambda = 2\lambda_1\lambda_p/(\lambda_1 + \lambda_p) = \sigma^2$.

We next consider the approximations to $F_U(u)$ and $f_U(u)$ obtained from the asymptotic expansions for them. Note that the asymptotic distribution of $U_0 = (1/\nu)\,\mathrm{tr}(W)$ as $v \to \infty$ is easily found to be normal with mean $\mathrm{tr}(\Sigma_0)$ and variance $(1/v)\,\mathrm{tr}(\Sigma_0^2)$ by applying Theorem 2.7.3 in Siotani, Hayakawa and Fujikoshi (1985), which was given in the class by Professor I. Olkin in 1964.

Fujikoshi (1970) gave an asymptotic expansion for the distribution of the standardized variable

$$\zeta = \frac{\sqrt{\nu}}{\tau}(U_0 - \mathrm{tr}(\Sigma_0)) = \frac{1}{\tau\sqrt{\nu}}(U - \nu\,\mathrm{tr}(\Sigma_0)), \tag{3.7}$$

where $\tau^2 = 2\,\mathrm{tr}(\Sigma_0^2)$, whose limiting distribution is obviously $N(0,1)$. Fujikoshi's result is

$$\begin{aligned} F_\zeta(z) &= \Pr(\zeta \le z) \\ &= \Phi(z) - \frac{4}{3\sqrt{\nu}}\frac{1}{\tau^3}s_3\Phi^{(3)}(z) + \frac{2}{\nu\tau^4}\left(s_4\Phi^{(4)}(z) + \frac{4}{9\tau^2}s_3^2\Phi^{(6)}(z)\right) \\ &\quad - \frac{8}{3\nu\sqrt{\nu}}\frac{1}{\tau^5}\left(\frac{6}{5}s_5\Phi^{(5)}(z) + \frac{1}{\tau^2}s_3s_4\Phi^{(7)}(z) + \frac{4}{27\tau^4}s_3^3\Phi^{(9)}(z)\right) \\ &\quad + O(\nu^{-2}), \end{aligned} \tag{3.8}$$

where $s_r = \mathrm{tr}(\Sigma_0^r) = \sum_{i=1}^p \lambda_i^r$ (λ_i's are latent roots of Σ_0), and $\Phi^{(r)}(z)$ is the $\mathrm{r^{th}}$ derivative of the standard normal c.d.f. $\Phi(z)$. We note that

$$\begin{aligned} F_U(u) = \Pr(U \le u) &= \Pr\left(\frac{1}{\tau\sqrt{\nu}}(U - \nu s_1) \le \frac{1}{\tau\sqrt{\nu}}(u - \nu s_1)\right) \\ &= \Pr(\zeta \le z_u) = F_\zeta(z_u), \end{aligned} \tag{3.9}$$

where $z_u = (u - \nu s_1)/\tau\sqrt{\nu}$, and that

$$f_U(u) = \frac{d}{dz_u} F_\zeta(z_u) \cdot \frac{dz_u}{du} = h_\zeta(z_u)/\tau\sqrt{\nu}, \tag{3.10}$$

$h_\zeta(z_u)$ being the p.d.f. of ζ at $z = z_u$.

4 Evaluation of Upper and Lower Bounds on $E(N)$

Noting that

$$\int_a^\infty u^i g_{p\nu+2k,\lambda}(u)\,du = (2\lambda)^i \frac{\Gamma\big((p\nu+2k+2i)/2\big)}{\Gamma\big((p\nu+2k)/2\big)} \left(1 - G_{p\nu+2k+2i}\left(\frac{a}{\lambda}\right)\right), \tag{4.1}$$

we obtain from (2.9)–(2.16) the following result.

Theorem 4.1 *The upper and lower bounds U_1 and L_1 on $E(N)$ are given by*

$$L_1 = \left|\frac{1}{\lambda}\Sigma_0\right|^{-\nu/2} \sum_{k=0}^{\infty} \frac{1}{k!} Q_{p,k} \sum_{k} \left(\frac{\nu}{2}\right)_k C_k(I_p - \lambda\Sigma_0^{-1}), \tag{4.2}$$

$$U_1 = L_1 + 1 - F_U(a), \tag{4.3}$$

where

$$Q_{p,k} = (N_0 + p^2) G_{p\nu+2k}\left(\frac{a}{\lambda}\right) + \frac{\lambda c}{\nu}(p\nu + 2k)\left(1 - G_{p\nu+2k+2}\left(\frac{a}{\lambda}\right)\right). \tag{4.4}$$

For the cases $p = 2$ and $\Sigma = \sigma^2 I_p$, $T = I_p$, the bounds reduce to the following forms, which can also be obtained from (3.4)

Corollary 4.2 *When $p = 2$, the bounds on $E(N)$ become*

$$L_1 = \sum_{k=0}^{\infty} b_k\left(\theta, \frac{\nu}{2}\right) Q_{2,k}, \qquad U_1 = L_1 + 1 - F_U(a), \tag{4.5}$$

where $F_U(a)$ is given by (3.4) with $u = a$.

Corollary 4.3 *When $\Sigma = \sigma^2 I_p$ and $T = I_p$, we have*

$$L_1 = Q_{p,0}, \qquad U_1 = L_1 + 1 - G_{p\nu}(a/\sigma^2). \tag{4.6}$$

Next we consider the user of the asymptotic expansion formula given by (3.7). Owing to (3.8) and (3.9), the integrals involved in the bounds (2.11) and (2.12) are converted into integrals with respect to the p.d.f. of ζ as follows:

$$\int_0^a f_U(u)\,du = F_U(a) = F_\zeta(z_a), \tag{4.7}$$

where $z_a = (a - \nu s_1)/\tau\sqrt{\nu}$,

$$\int_a^\infty f_U(u)\,du = 1 - F_\zeta(z_a), \tag{4.8}$$

$$\int_a^\infty u f_U(u)\,du = \tau\sqrt{\nu}\int_{z_a}^\infty z h_\zeta(z)\,dz + \nu s_1\big(1 - F_\zeta(z_a)\big). \tag{4.9}$$

To evaluate the integral on the right-hand side of (4.9), note that

$$z\Phi^{(r)}(z) = -\Phi^{(r+1)}(z) - (r-1)\Phi^{(r-1)}(z), \qquad r = 1, 2, \ldots, \tag{4.10}$$

and hence $zh_\zeta(z)$ can be written as

$$\begin{aligned} zh_\zeta(z) = {} & -\Phi^{(2)}(z) + \frac{4}{3\sqrt{\nu}}\frac{1}{\tau^3}s_3\big(\Phi^{(5)}(z) + 3\Phi^{(3)}(z)\big) \\ & - \frac{2}{\nu}\frac{1}{\tau^4}\left(\frac{4}{9\tau^2}s_3^2\Phi^{(8)}(z) + \left(s_4 + \frac{8}{3\tau^2}s_3^2\right)\Phi^{(6)}(z) + 4s_4\Phi^{(4)}(z)\right) \\ & + \frac{8}{3\nu\sqrt{\nu}}\frac{1}{\tau^5}\left[\frac{4}{27\tau^4}s_3^3\Phi^{(11)}(z) + \left(\frac{1}{\tau^2}s_3s_4 + \frac{4}{3\tau^4}s_3^3\right)\Phi^{(9)}(z)\right. \\ & \left. + \left(\frac{6}{5}s_5 + \frac{7}{\tau^2}s_3s_4\right)\Phi^{(7)}(z) + 6s_5\Phi^{(5)}(z)\right] \\ & + O(\nu^{-2}). \end{aligned} \tag{4.11}$$

Since

$$\int_{z_a}^\infty \Phi^{(r+1)}(z)\,dz = \begin{cases} 1 - \Phi(z_a) & \text{for } r = 0, \\ -\Phi^{(r)}(z_a) & \text{for } r = 1, 2, \ldots, \end{cases} \tag{4.12}$$

we have

$$\begin{aligned} \int_{z_a}^\infty zh(z)\,dz = {} & \Phi^{(1)}(z_a) + \frac{4}{3\sqrt{\nu}}\frac{1}{\tau^3}s_3\big(\Phi^{(4)}(z_a) + 3\Phi^{(2)}(z_a)\big) \\ & + \frac{2}{\nu\tau^4}\left(\frac{4}{9\tau^2}s_3^2\Phi^{(7)}(z_a) + \left(s_4 + \frac{8}{3\tau^3}s_3^2\right)\Phi^{(5)}(z_a) + 4s_4\Phi^{(3)}(z_a)\right) \\ & - \frac{8}{3\nu\sqrt{\nu}}\frac{1}{\tau^5}\left[\frac{4}{27\tau^4}s_3^3\Phi^{(10)}(z_a) + \left(\frac{1}{\tau^2}s_3s_4 + \frac{4}{3\tau^4}s_3^3\right)\Phi^{(8)}(z_a)\right. \\ & \left. + \left(\frac{6}{5}s_5 + \frac{7}{\tau^2}s_3s_4\right)\Phi^{(6)}(z_a) + 6s_5\Phi^{(5)}(z_a)\right] \\ & + O(\nu^{-2}). \end{aligned} \tag{4.13}$$

Substituting (4.7)–(4.9) with (4.13) into (2.11) and (2.12), we obtain approximations of the upper and lower bounds on $E(N)$.

5 Upper and Lower Bounds on $E(N^2)$

To evaluate the upper and lower bounds U_2 and L_2 on $E(N^2)$, we observe from (2.11) and (2.13) that we need no additional computation except for $\int_a^\infty u^2 f_U(u)\,du$.

Theorem 5.1 *The exact upper and lower bounds U_2 and L_2 are given by*

$$L_2 = \left|\frac{1}{\lambda}\Sigma_0\right|^{-\nu/2} \sum_{k=0}^{\infty} \frac{1}{k!} R_{p,k} \sum_{k} \left(\frac{\nu}{2}\right)_k C_k(I_p - \lambda\Sigma_0^{-1}), \tag{5.1}$$

where

$$\begin{aligned} R_{p,k} &= (N_0 + p^2)^2 G_{p\nu+2k}\left(\frac{a}{\lambda}\right) \\ &\quad + \frac{\lambda^2 c^2}{\nu}(p\nu + 2k)(p\nu + 2k + 2)\left(1 - G_{p\nu+2k+4}\left(\frac{a}{\lambda}\right)\right), \end{aligned} \tag{5.2}$$

and

$$U_2 = L_2 + L_1 + U_1 - 2(N_0 + p^2)F_U(a), \tag{5.3}$$

where L_1 and U_1 are given by (4.2) and (4.3), respectively.

Corollary 5.2 *For $p = 2$, the bounds in Theorem 5.1 become*

$$L_2 = \sum_{k=0}^{\infty} b_k\left(\theta, \frac{\nu}{2}\right) R_{2,2k}, \qquad U_2 = L_2 + L_1 + U_1 - 2(N_0 + p^2)F_U(a), \tag{5.4}$$

where L_1, U_1 are given by (4.5) and $F_U(a)$ is given by (3.4) with $u = a$.

Corollary 5.3 *When $\Sigma = \sigma^2 I_p$ and $T = I_p$, the bounds on $E(N^2)$ are U_2 and L_2 in Theorem 5.1 for $k = 0$ and $\lambda = \sigma^2$, i.e.,*

$$L_2 = R_{p,0}, \qquad U_2 = L_2 + L_1 + U_1 - 2(N_0 + p^2)G_{p\nu}(a/\sigma^2), \tag{5.5}$$

where L_1 and U_1 are given in (4.6).

For approximate bounds on $E(N^2)$ based on the asymptotic expansion formula (3.8), note that

$$\begin{aligned} \int_{z_a}^{\infty} u^2 f_U(u)\,du &= \int_{z_a}^{\infty} (\tau\sqrt{\nu}z + \nu s_1)^2 h_\zeta(z)\,dz \\ &= \tau^2\nu \int_{z_a}^{\infty} z^2 h_\zeta(z)\,dz \\ &\quad + 2\tau\nu\sqrt{\nu}s_1 \int_{z_a}^{\infty} z h_\zeta(z)\,dz + \nu^2 s_1^2(1 - F_\zeta(z_a)). \end{aligned} \tag{5.6}$$

Hence we have only to consider the evaluation of the first integral in (5.6). Since

$$z^2\Phi^{(r)}(z) = \Phi^{(r+2)}(z) + (2r-1)\Phi^{(r)}(z) + (r-1)(r-2)\Phi^{(r-2)}(z), \quad (5.7)$$

we obtain the following result by computing in a similar manner to the case of $E(N)$:

$$\int_{z_a}^{\infty} z^2 h_\zeta(z)\,dz$$

$$= 1 - \Phi_a - \Phi_a^{(2)} + \frac{4}{3\sqrt{\nu}}\frac{1}{\tau^3}s_3\left(\Phi_a^{(5)} + 7\Phi_a^{(3)} + 6\Phi_a^{(1)}\right) \quad (5.8)$$

$$- \frac{2}{\nu\tau^4}\left[\frac{4}{9\tau^2}s_3^2\Phi_a^{(8)} + \left(s_4 + \frac{52}{9\tau^2}s_3^2\right)\Phi_a^{(6)} + \left(9s_4 + \frac{40}{3\tau^2}s_3^2\right)\Phi_a^{(4)} + 12s_4\Phi_a^{(2)}\right]$$

$$+ \frac{8}{3\nu\sqrt{\nu}}\frac{1}{\tau^5}\left[\frac{4}{27\tau^4}s_3^3\Phi_a^{(11)} + \left(\frac{76}{27\tau^4}s_3^3 + \frac{1}{\tau^2}s_3s_4\right)\Phi_a^{(9)} + \left(\frac{32}{3\tau^4}s_3^3 + \frac{15}{\tau^2}s_3s_4 + \frac{6}{5}s_5\right)\Phi_a^{(7)} + \left(\frac{42}{\tau^2}s_3s_4 + \frac{66}{5}s_5\right)\Phi_a^{(5)} + 24s_5\Phi_a^{(3)}\right]$$

$$+ O(\nu^{-2}), \quad (5.9)$$

where $\Phi_a^{(r)} \equiv \Phi^{(r)}(z_a)$, $r = 0, 1, \ldots, 11$.

6 Tables for Some Special Cases

To have some information on the numerical behavior of the mean and standard deviation of N, tables of the upper and lower bounds on them are computed for

$p = 3;\quad N_0 = 11(10)51,\quad \nu = N_0 - 1 = 10(10)50;$
$c = 10(5)50;$
$\Sigma_0 : \Sigma_{01} = (0.5)I_3,\quad \Sigma_{02} = I_3,\quad \Sigma_{03} = 2I_3,\quad \Sigma_{04} = \mathrm{diag}(1, 0.5, 0.2),$
$\Sigma_{05} = \mathrm{diag}(3, 1, 0.5).$

Tabulated values are L_1 and U_1 such that $L_1 < E(N) < U_1$, L_{SD} and U_{SD} such that

$$L_{\mathrm{SD}} < \sqrt{\mathrm{Var}(N)} = \text{standard deviation of } N < U_{\mathrm{SD}}, \quad (6.1)$$

where

$$L_{\mathrm{SD}} = \begin{cases} (L_2 - U_1^2)^{1/2} & \text{if } L_2 - U_1^2 > 0, \\ 0 & \text{if } L_2 - U_1^2 \le 0; \end{cases} \quad (6.2)$$

$$U_{\text{SD}} = (U_2 - L_1^2)^{1/2} \tag{6.3}$$

and L_2, U_2 are given in (2.10) and in Section 5.

Table 1 is for the case of $\Sigma_0 = \Sigma_{0i}$, $i = 1, 2, 3$ given above, in which case the computations are based on the formulae (4.6) and (5.5). For the χ^2-distribution with m d.f., $(2\chi_m^2)^{1/2} - (2m-1)^{1/2}$ is approximately distributed according to the standard normal distribution $N(0, 1)$ for large m, say for $m > 40$. Hence we observe from (4.4) and (5.2) with $k = 0$ that, if

$$|d| \equiv |(2a/\sigma^2)^{1/2} - (2p\nu - 1)^{1/2}| > t \tag{6.4}$$

where t is a sufficiently large positive constant, $G_{p\nu+2r}(s/\sigma^2)$, $r = 0, 1, 2$ become very close to one or negligibly small depending on whether d in (6.4) is positive or negative, and we have approximately

$$\begin{aligned} L_1 &\simeq (N_0 + p^2), \quad U_1 \simeq L_1, \text{ if } d > 0; \\ L_1 &\simeq pc\sigma^2, \quad U_1 = L_1 + 1, \text{ if } d < 0 \end{aligned} \tag{6.5}$$

and

$$\begin{aligned} L_2 &\simeq (N_0 + p^2)^2, \quad U_2 \simeq L_2, \text{ if } d > 0; \\ L_2 &\simeq pc^2\sigma^4(p\nu + 2)/\nu, \quad U_2 \simeq L_2 + L_1 + U_1, \text{ if } d < 0. \end{aligned} \tag{6.6}$$

The value of the critical point t depends, of course, on $(N_0 + p^2)^i$, $i = 1, 2$, $pc\sigma^2$, $pc^2\sigma^4(p\nu + 2)/\nu$. For the range of parameters in Table 1, the value of t is chosen to be 5, for which

$$G_{p\nu+2r}(a/\sigma^2) > 0.99999971, \quad r = 0, 1, 2, \text{ if } d > 0,$$

$$G_{p\nu+2r}(a/\sigma^2) < 0.00000029, \quad r = 0, 1, 2, \text{ if } d < 0.$$

Table 2 is for the case of $\Sigma_0 = \Sigma_{04}$, Σ_{05}, and is calculated based on the asymptotic formulae (3.8), (4.13), and (5.9). The formulae for L_i, U_i, $i = 1, 2$ in this case are written as

$$L_1 = (N_0 + p^2)F_\zeta(z_a) + \left(\frac{c}{\nu}\right)\left(\tau\sqrt{\nu}\int_{z_a}^{\infty} zh_\zeta(z)\,dz + \nu s_1(1 - F_\zeta(z_a))\right), \tag{6.7}$$

$$U_1 = L_1(1 - F_\zeta(z_a)), \tag{6.8}$$

$$L_2 = (N_0 + p^2)^2 F_\zeta(z_a) + c^2\left(\frac{\tau^2}{\nu}\int_{z_a}^{\infty} z^2 h(z)\,dz + \frac{2\tau s_1}{\sqrt{\nu}}\int_{z_a}^{\infty} zh_\zeta(z)\,dz + s_1^2(1 - F_\zeta(z_a))\right), \tag{6.9}$$

$$U_2 = L_2 + L_1 + U_1 - 2(N_0 + p^2)F_\zeta(z_a). \tag{6.10}$$

It was observed, by comparing $F_\zeta(z_a)$ with $G_{p\nu}(a/\sigma^2)$ for the case $\Sigma_0 = \sigma^2 I_p$ (see (3.6)), that the asymptotic expansion formulae did not give good approximations to the exact values, especially for the case of larger absolute values of z_a, say $|z_a| > 2.5$; in some cases, $F_\zeta(z_a)$ became greater than one or became negative when diagonal elements of $\Sigma_0 = \text{diag}(\sigma_1^2, \sigma_2^2, \sigma_3^2)$ were large, although the absolute amounts of those violations were small. Fortunately, numerical comparisons stated above showed that if we put $F_\zeta(z_a) = 1$ or 0 in those cases, effects of this modification on the final results were pretty small, and the tabulated values preserved the accuracy at least up to one decimal place.

Acknowledgments: We would like to thank the editor and the associate editor for their useful suggestions.

REFERENCES

Chatterjee, S.K. (1959). On an extension of Stein's two-sample procedure to the multi-normal problem. *Calcutta Statist. Assoc. Bull.* **8** 121–148.

Dudewicz, E.J. and Bishop, T.A. (1979). The heteroscedastic method. *Optimizing Methods in Statistics* (J.S. Rustagi, ed.) 183–203, Academic Press, New York.

Fujikoshi, Y. (1970). Asymptotic expansions of the distributions of test statistics in multivariate analysis. *J. Sci. Hiroshima Univ. Ser. A-I* **34** 73–144.

Hyakutake, H. and Siotani, M. (1987). The multivariate heteroscedastic method: Distributions of statistics and an application. *Amer. J. of Math. and Manage. Sci.* **7** 89–111.

Khatri, C.G. and Srivastava, M.S. (1971). On exact non-null distributions of likelihood ratio criteria for sphericity test and equality of two covariance matrices. *Sankhyā A* **33** 201–206.

Muirhead, R.J. (1982). *Aspects of Multivariate Statistical Theory.* Wiley, New York.

Siotani, M., Hyakawa, T. and Fujikoshi, Y. (1985). *Modern Multivariate Statistical Analysis: A Graduate Course and Handbook.* American Sciences Press, Ohio.

Stein, C. (1945). A two-sample test for a linear hypothesis whose power is independent of the variance. *Ann. Math. Statist.* **16** 243–258.

TABLE 1. Upper and lower bounds L_1, U_1, and L_{SD}, U_{SD} when $\Sigma_0 = \Sigma_{01}, \Sigma_{02}, \Sigma_{03}$; $p = 3$

	$\Sigma_{01} = (0.5)I_3$																			
	$N_0 = 11$ $(\nu = 10)$				$N_0 = 21$ $(\nu = 20)$				$N_0 = 31$ $(\nu = 30)$				$N_0 = 41$ $(\nu = 40)$				$N_0 = 51$ $(\nu = 50)$			
c	L_1	U_1	L_{SD}	U_{SD}	L_1	U_1	L_{SD}	U_{SD}	L_1	U_1	L_{SD}	U_{SD}	L_1	U_1	L_{SD}	U_{SD}	L_1	U_1	L_{SD}	U_{SD}
10	20.25	20.36	0	2.42	30.00	31.00	0	1.41	40.00	40.00	0	0.00	50.00	50.00	0	0.00	60.00	60.00	0	0.00
15	23.69	24.33	0	7.35	30.09	30.13	0	1.78	40.00	40.00	0	0.00	50.00	50.00	0	0.00	60.00	60.00	0	0.00
20	30.21	31.12	0	10.63	32.18	32.65	0	6.71	40.05	40.07	0	1.28	50.00	50.00	0	0.00	60.00	60.00	0	0.00
25	37.54	38.52	4.20	12.95	37.87	38.74	0	10.40	41.26	41.57	0	5.89	50.02	50.03	0	0.93	60.00	60.00	0	0.04
30	45.01	46.00	6.62	15.01	45.05	46.03	0	12.47	45.77	46.53	0	10.28	50.71	50.90	0	4.93	60.01	60.01	0	0.67
35	52.50	53.50	8.81	17.02	52.51	53.50	0	14.05	52.59	53.54	0	12.68	54.07	54.70	0	9.83	60.39	60.50	0	1.62
40	60.00	61.00	10.91	19.00	60.00	61.00	0	15.52	60.00	61.00	0	14.15	60.25	61.16	0	12.86	62.76	63.25	0	9.06
45	67.50	68.50	12.95	20.97	67.50	68.50	3.99	16.97	67.50	68.50	0	15.40	67.52	68.51	0	14.47	68.09	68.92	0	12.82
50	75.00	76.00	14.96	22.93	75.00	76.00	6.04	18.40	75.00	76.00	0	16.61	75.00	76.00	0	15.65	75.08	76.04	0	14.80
	$\Sigma_{02} = I_3$																			
	$N_0 = 11$ $(\nu = 10)$				$N_0 = 21$ $(\nu = 20)$				$N_0 = 31$ $(\nu = 30)$				$N_0 = 41$ $(\nu = 40)$				$N_0 = 51$ $(\nu = 50)$			
c	L_1	U_1	L_{SD}	U_{SD}	L_1	U_1	L_{SD}	U_{SD}	L_1	U_1	L_{SD}	U_{SD}	L_1	U_1	L_{SD}	U_{SD}	L_1	U_1	L_{SD}	U_{SD}
10	30.21	31.12	0	10.63	32.18	32.65	0	6.71	40.05	40.07	0	1.13	50.00	50.00	0	0.00	60.00	60.00	0	0.00
15	45.01	46.00	6.63	15.02	45.05	46.03	0	12.47	46.01	46.77	0	9.64	50.71	50.90	0	4.93	60.01	60.01	0	0.67
20	60.00	61.00	10.91	19.00	60.00	61.00	0	15.52	60.00	61.00	0	14.15	60.25	61.16	0	12.86	62.76	63.25	0	9.06
25	75.00	76.00	14.97	22.93	75.00	76.00	6.04	18.40	75.00	76.00	0	16.61	75.00	76.00	0	15.56	75.08	76.04	0	14.80
30	90.00	91.00	18.95	23.24	90.00	91.00	9.43	21.24	90.00	91.00	0	19.00	90.00	91.00	0	17.78	90.00	91.00	0	17.00
35	105.00	106.00	22.89	30.76	105.00	106.00	12.51	24.05	105.00	106.00	5.83	21.35	105.00	106.00	0	19.87	105.00	106.00	0	18.92
40	120.00	121.00	26.81	34.66	120.00	121.00	15.46	26.85	120.00	121.00	8.89	23.69	120.00	121.00	0	21.93	120.00	121.00	0	20.81
45	135.00	136.00	30.72	38.55	135.00	136.00	18.34	29.64	135.00	136.00	11.58	26.00	135.00	136.00	5.72	23.97	135.00	136.00	0	22.67
50	150.00	151.00	34.63	42.44	150.00	151.00	21.19	32.42	150.00	151.00	14.11	28.30	150.00	151.00	8.60	26.00	150.00	151.00	0	24.52

Table 1 continued

	$\Sigma_{03} = 2I_3$																			
	$N_0 = 11$ $(\nu = 10)$				$N_0 = 21$ $(\nu = 20)$				$N_0 = 31$ $(\nu = 30)$				$N_0 = 41$ $(\nu = 40)$				$N_0 = 51$ $(\nu = 50)$			
c	L_1	U_1	L_{SD}	U_{SD}	L_1	U_1	L_{SD}	U_{SD}	L_1	U_1	L_{SD}	U_{SD}	L_1	U_1	L_{SD}	U_{SD}	L_1	U_1	L_{SD}	U_{SD}
10	60.00	61.00	10.91	19.00	60.00	61.00	0	15.52	60.00	60.99	0	14.17	60.25	61.16	0	13.57	62.76	63.25	0	9.06
15	90.00	91.00	18.95	26.85	90.00	91.00	9.43	21.24	90.00	91.00	0	19.00	90.00	91.00	0	17.78	90.00	91.00	0	17.00
20	120.00	121.00	26.81	34.66	120.00	121.00	15.46	26.85	120.00	121.00	8.89	23.69	120.00	121.00	0	21.93	120.00	121.00	0	20.81
25	150.00	151.00	34.63	42.44	150.00	151.00	21.19	32.42	150.00	151.00	14.11	28.30	150.00	151.00	8.60	26.00	150.00	151.00	0	24.52
30	180.00	181.00	42.41	50.21	180.00	181.00	26.81	37.96	180.00	181.00	18.95	32.88	180.00	181.00	13.38	30.02	180.00	181.00	8.43	28.16
35	210.00	211.00	50.19	57.97	210.00	211.00	32.39	43.49	210.00	211.00	23.64	37.43	210.00	211.00	17.72	34.00	210.00	211.00	12.92	31.76
40	240.00	241.00	57.96	65.73	240.00	241.00	37.93	49.00	240.00	241.00	28.27	41.96	240.00	241.00	21.89	37.96	240.00	241.00	16.94	35.34
45	270.00	271.00	65.72	73.49	270.00	271.00	43.46	54.51	270.00	271.00	32.85	46.49	270.00	271.00	25.96	41.90	270.00	271.00	20.76	38.90
50	300.00	301.00	73.48	81.25	300.00	301.00	48.98	60.01	300.00	301.00	37.40	51.00	300.00	301.00	29.98	45.84	300.00	301.00	24.47	42.44

TABLE 2. Upper and lower bounds L_1, U_1, and L_{SD}, U_{SD} when $\Sigma_0 = \Sigma_{04}, \Sigma_{05}$; $p = 3$

$\Sigma_{04} = \text{diag}(1, .05, 0.2)$

	$N_0 = 11$ ($\nu = 10$)				$N_0 = 21$ ($\nu = 20$)				$N_0 = 31$ ($\nu = 30$)				$N_0 = 41$ ($\nu = 40$)				$N_0 = 51$ ($\nu = 50$)			
c	L_1	U_1	L_{SD}	U_{SD}	L_1	U_1	L_{SD}	U_{SD}	L_1	U_1	L_{SD}	U_{SD}	L_1	U_1	L_{SD}	U_{SD}	L_1	U_1	L_{SD}	U_{SD}
10	20.96	21.21	0	4.24	30.00	30.00	0	0.34	40.00	40.00	0	0.00	50.00	50.00	0	0.00	60.00	60.00	0	0.00
15	26.32	27.07	2.08	9.37	30.70	30.89	0	4.16	40.00	40.01	0	0.47	50.00	50.00	0	0.00	60.00	60.00	0	0.00
20	33.34	34.27	9.39	14.73	35.17	35.86	0	9.30	40.56	40.71	0	4.10	50.01	50.01	0	0.57	60.00	60.00	0	0.00
25	42.53	43.52	8.62	15.65	41.68	42.62	8.98	15.48	44.27	44.88	0	9.34	50.45	50.57	0	4.01	60.01	60.01	0	0.65
30	51.01	52.01	11.29	18.25	51.03	52.01	3.58	14.74	51.31	52.21	0	12.84	53.53	54.06	0	9.29	60.37	60.47	0	3.87
35	59.50	60.50	13.91	20.82	59.51	60.50	6.14	16.66	59.54	60.53	0	14.91	60.02	60.88	0	13.14	62.93	63.39	0	9.13
40	68.00	69.00	16.54	23.40	67.99	68.99	8.36	18.54	68.01	69.00	0.68	16.55	68.08	69.05	0	15.31	68.81	69.65	0	13.50
45	76.50	77.50	19.11	25.95	76.50	77.50	10.35	20.37	76.50	77.50	4.48	18.11	76.51	77.51	0	16.85	76.64	77.60	0	15.75
50	85.00	86.00	21.66	28.48	85.00	86.00	12.27	22.20	85.00	86.00	6.61	19.64	85.00	86.00	0	18.22	85.02	86.01	0	17.27

$\Sigma_{05} = \text{diag}(3, 1, .05)$

	$N_0 = 11$ ($\nu = 10$)				$N_0 = 21$ ($\nu = 20$)				$N_0 = 31$ ($\nu = 30$)				$N_0 = 41$ ($\nu = 40$)				$N_0 = 51$ ($\nu = 50$)			
c	L_1	U_1	L_{SD}	U_{SD}	L_1	U_1	L_{SD}	U_{SD}	L_1	U_1	L_{SD}	U_{SD}	L_1	U_1	L_{SD}	U_{SD}	L_1	U_1	L_{SD}	U_{SD}
10	45.02	46.01	10.65	17.15	45.14	46.09	3.31	13.64	46.25	46.96	0	10.77	51.09	57.33	0	5.89	60.05	60.06	0	1.52
15	67.50	68.50	18.03	24.44	67.50	68.50	9.71	19.14	67.53	68.52	4.16	16.99	67.62	68.58	0	15.58	68.51	69.29	0	13.59
20	90.00	91.00	25.28	31.64	90.00	91.00	15.13	24.31	90.00	91.00	9.64	21.33	90.00	91.00	4.89	19.64	90.01	91.01	0	18.55
25	112.50	113.50	32.48	38.82	112.50	113.50	20.36	29.44	112.50	113.50	14.18	25.56	112.50	113.50	9.71	23.37	112.50	113.50	5.50	21.96
30	135.00	136.00	39.67	46.00	135.00	136.00	25.52	34.55	135.00	136.00	18.55	29.77	135.00	136.00	13.79	27.06	135.00	136.00	9.90	25.30
35	157.50	158.50	46.85	53.17	157.50	158.50	30.65	39.64	157.50	158.50	22.83	33.96	157.50	158.50	17.66	30.72	157.50	158.50	13.65	28.61
40	180.00	181.00	54.03	60.34	180.00	181.00	35.76	44.73	180.00	181.00	27.06	38.14	180.00	181.00	21.42	34.37	180.00	181.00	17.18	31.89
45	202.50	203.50	61.20	67.51	202.50	203.50	40.86	49.82	202.50	203.50	31.27	42.31	202.50	203.50	25.14	38.00	202.50	203.50	20.60	35.16
50	225.00	226.00	68.37	74.67	225.00	226.00	45.95	54.90	225.00	226.00	35.46	46.47	225.00	226.00	28.81	41.62	225.00	226.00	23.96	38.42

14

A Comparative Study of Cluster Analysis and MANCOVA in the Analysis of Mathematics Achievement Data

J.E. Lockley[1]

ABSTRACT This article is based upon the author's Ph.D. dissertation (Lockley, 1970), which was one of the first attempts to use cluster analytic techniques in the analysis of educational data. It is published here to illustrate an alternative approach to the analyses of covariance of student achievement utilizing large numbers of related social, demographic, environmental and educational covariates which are widely applied in educational research. In this alternative approach, covariates relating to school environment and teaching method are used to identify clusters of school-community-teaching environments. Such environments are then compared for their effects on student achievement using analysis of variance (with adjustments only for covariates that reflect students' individual prior abilities and achievements). If successful, such an approach both increases statistical power and provides insight into the effects of school environments on student achievement.
The data used in this article comes from junior high schools that participated in a pioneering research project conducted at Stanford University in the late 1960's by the National Longitudinal Study of Mathematical Abilities (NLSMA). Evidence in the data supports the existence of 3 or 4 school-environment clusters, and suggests that such clusters (or the schools themselves) do make a difference in the mathematics achievement of junior high school students.

1 Introduction

Educational researchers are often confronted with the problem of making sense out of a large number of measurements made on each of a set of objects. The measurements are thought to be related, and to reflect an un-

[1]Mountain View College

derlying partition of the population of all such objects that has descriptive or substantiative meaning for research. One approach to this problem is that of cluster analysis, where heterogeneous data vectors are partitioned into smaller, more homogeneous sub-groups called clusters.

Cluster analysis shares with classification analysis the assumption that the data vectors belong to one of G underlying statistical populations. However, in classification analysis the number and identities of such populations are known in advance, and the problem is to assign a given data vector to its appropriate population. In cluster analysis, there is no prespecified number of groups, nor is there necessarily a clear idea of what objects should constitute a group. Often, clustering is done without using measurements at all, but instead is based on some index of similarity between the objects under study. Whether measurements or indices of similarity are used, the hope is that the resulting classifications (clusters) reflect an underlying structure which exists apart from the particular data (or similarity index) used to construct the clusters. In positing and searching for such structures, cluster analysis resembles scaling methods such as factor analysis and principal components analysis. However, unlike such methods, the structures sought by cluster analysis are often qualitative (categorical), rather than quantitative.

Clustering techniques tend to require considerable calculation. Hence, their development has paralleled the growth of ever more powerful computers and related computer software. Numerous techniques for clustering have been proposed. All depend upon: (1) the choice of variables used to describe the objects being clustered, (2) choice of a proximity index (or distance) that measures the similarity or dissimilarity between pairs of objects, and (3) the criterion function that assesses "goodness-of-fit" of the clustering technique. Subjective judgment enters into all of these choices, and also into the decision as to how to trade off the number G of clusters against the criterion of goodness-of-fit. Often, the interpretation of the clusters in terms of the measurements used to create them is used as a guide to evaluate whether achieved clusters are reasonable (i.e. reflect an underlying structure).

However, there are also external ways to evaluate clusters. Frequently, the variables used to describe the objects to be clustered are selected from a larger collection of variables known to be related to one another. For example, in behavioral research one may be interested in "environment" or "background" as an underlying qualitative structure, and will have a wealth of possible measures that reflect this concept. If one collection of such variables is used to form the clusters, and such clusters can be reasonably interpreted in terms of a different collection of variables that might have been used, then one has some confidence that these clusters reflect some stable underlying structure. (One must be careful that the two collections of variables are not too highly correlated with one another.)

Still another check is provided by using different clustering methods on the same set of variables (and objects). If the resulting clusters are in close agreement, one is less inclined to regard them as artifacts of a particular algorithm. This is particularly true if the clustering methods used are based on different proximity indices and criteria and/or use different search algorithms. For example, one could use both hierarchical and non-hierarchical algorithms (Johnson and Wichern, 1988; Chapter 12) to construct clusters.

Finally, in behavioral research the structure sought by a clustering method may be thought to affect or influence certain performance variables (e.g. performance on certain achievement tests). If this effect is demonstrated statistically (by an analysis of variance), one has evidence that the clusters have operational meaning (at least in terms of predicting performance). An important advantage of using clustering on covariates as a precursor to an analysis of performance differences, as opposed to a regression analysis using the explanatory variables (covariates) as predictors, is a reduction in dimensionality (number of predictors) of the regression model, and hence an increase in statistical power to detect such differences.

The goal of the present paper is to apply such an approach via clustering to data collected in a pioneering educational research project conducted at Stanford University in the late 1960's by the National Longitudinal Study of Mathematical Abilities (NLSMA). The objects grouped into clusters are junior high schools. The variables used to cluster these schools are background (school-community) variables. Further insight on this clustering is provided by variables reflecting the training and background of teachers in these schools, and the educational methods emphasized. Four different clustering algorithms are used, including one of hierarchical type and three of non-hierarchical type (including one k-means method). The number (3 or 4) of clusters, and the assignments of schools to clusters, produced by these algorithms are very similar. Using the 3-cluster partition of schools produced by the hierarchical algorithm, an identification of the clusters in terms of the teacher-training variables yields meaningful interpretations, suggesting that a stable underlying partition of the schools has been identified.

Finally, statistical analysis of student mathematical achievement data shows that the 3 clusters of junior high schools are not equally effective in mathematical teaching, even after adjustments are made (MANCOVA) for differences in aptitude and initial understanding of mathematical concepts by the students. These results suggest that background-teacher-school "gestalts" do make a difference in the mathematics achievement of junior high school students.

2 The Clustering Techniques Used

The clustering techniques used in this study were among the first proposed in the literature: HICLUS (Johnson, 1967), ISODATA (Ball and Hall, 1965,

1967), F-R (Friedman and Rubin, 1967), and S-K (Singleton and Kautz, 1965). HICLUS is a hierarchical clustering method, while the remaining three algorithms are non-hierarchical (Johnson and Wichern, 1988). The non-hierarchical methods differ primarily in the search algorithms used to construct optimal clusters, and in the way they choose the number, G, of clusters to be formed.

We here give a brief exposition that may be helpful in understanding these methods. More detailed descriptions can be found in the previously cited articles introducing the methods, and in the present author's dissertation (Lockley, 1970).

Suppose that p variables are measured on each of N objects. If x_{ij} is the jth measurement on the ith object, then $x_i = (x_{i1}, x_{i2}, \ldots, x_{ip})'$ is the vector of measurements for object $i, i = 1, \ldots, N$. The vectors $x_1, x_2, \ldots, x_N$ may be presented as points in p-dimensional space. Let

$$d_{ij} = \left((x_i - x_j)'(x_i - x_j)\right)^{\frac{1}{2}}$$

be the Euclidean distance between the points x_i and x_j.

If the vectors x_i are grouped in G clusters, we can compute the between-cluster scatter matrix B and the within-cluster scatter matrix W familiar from the multivariate analysis of variance (MANOVA). The total scatter matrix T, which is the usual matrix of sums of squares and cross products of deviations of the x_i about their mean $\bar{x} = N^{-1} \sum_{i=1}^{N} x_i$, is fixed regardless of which (and how many) clusters are used. Further, for every choice of clusters

$$T = B + W.$$

A good clustering procedure will produce clusters such that objects within clusters are more homogeneous (closer in terms of the distance d_{ij}) than objects between clusters. This may be accomplished by minimizing the matrix W, or (since $T = B + W$ is fixed) by maximizing the matrix B. Since W is a matrix, some scalar function $h(W)$ of W is usually minimized. The S-K procedure seeks to minimize the trace, $tr(W)$, of W. The ISODATA procedure uses a mean-square-error criterion which is equivalent to minimizing $tr(W)$. The F-R procedure can use $tr(W)$, a criterion equivalent to the determinant $|W|$ of W, or a criterion equivalent to $-tr(W^{-1})$. (For the latter two criteria, the number G of clusters cannot exceed $N - p$.)

A desirable trait of clustering is to produce a small number of homogeneous clusters that would provide as much information about the data as could be obtained using N clusters. Hence, a value as close to zero as possible is sought for the statistic $h(W)$, while keeping G, the number of clusters, small. The optimal number of clusters is reached as a compromise between the desire to produce a small number of clusters and the desire to make clusters as homogeneous as possible (a goal trivially achieved by using N clusters).

Assuming that the criterion used by the F-R method is $tr(W)$, the three non-hierarchical clustering methods used in this paper (ISODATA, S-K, F-R) differ primarily in the search algorithm used to optimize the criterion for a fixed number of clusters, and in the way the number G of clusters is chosen. Since the search algorithms used stop when a local minimum of the criterion is achieved, they may yield different results (for fixed G) if the criterion function has several local minima. (A global minimum is not necessarily achieved by any of the search algorithms.) In the F-R algorithm one object is selected at a time, and the effect on the criterion of moving that object from its present cluster to a new cluster is evaluated. The object is moved only if the criterion function is decreased, and is then moved to the cluster for which the decrease in the criterion is greatest. Several passes are made over all objects until no further moves are possible. This procedure is followed for each of several choices of the number G of clusters, and the number G of clusters is chosen from a graph of the optimal criterion value for each G versus the value of G. The algorithm of the S-K procedure is similar, except that an up-and-down method is used to select G. (The move from G to $G-1$ is accomplished by collapsing two clusters into one, and from G to $G+1$ by using one object as the nucleus of a new cluster.) The ISODATA algorithm is a k-mean clustering procedure (Johnson and Wichern, 1988), with the choice of G made graphically.

Hierarchical clustering procedures such as HICLUS (Johnson, 1967) can use similarity or difference measures between objects in place of the distances d_{ij} between data points. For the present application, however, the distance d_{ij} between x_i and x_j was used as a measure of similarity between objects i and j. HICLUS can form clusters by single linkage (MIN) or complete linkage (MAX); see Johnson and Wichern (1988) for a categorization of linkage methods for hierarchical clustering. HICLUS uses a measure of average linkage to graphically choose the number of clusters. In the present paper, the following variant of this average linkage approach was used together with the MAX option of HICLUS to choose the number of clusters:

Johnson (1968b) defines an experimental cluster statistic $z(g)$ which measures the tightness of a cluster g. If the data is at least interval scaled, or if the data is replaced by ranks, then the statistic can be treated as a standardized score (a z-score). The cluster statistic is denoted here by $Z_j(g)$, which measures the tightness of a cluster g at each stage j of clustering. The larger the value of $Z_j(g)$, the more homogeneous is the cluster g. Johnson did not discuss a method for determining the optimal number of clusters. The optimal number of clusters is obtained here by using a weighted average of the Johnson statistic developed by the present author. We define

$$\bar{Z}_j = \frac{\sum_{g=1}^{G_j} N_j(g) Z_j(g)}{\sum_{g=1}^{G_j} N_j(g)} ,$$

where G_j is the number of clusters at the jth stage which contain more than one object, and $N_j(g)$ is the number of objects in cluster g at stage j. Note that $O_j = N - \sum_{g=1}^{G_j} N_j(g)$ is the number of clusters at the jth stage containing only one object (called outliers). Thus, $\bar{Z}_j$ is a measure of tightness of the set of clusters at stage j which are not outliers. If two sets of clusters have the same value $\bar{Z}_j$, then the one with the smaller dissimilarity (see Johnson, 1968b) is chosen as the more desirable.

Table 1 summarizes the clustering techniques used in this paper in terms of procedure, proximity measure, criterion function, goal and computer restrictions (at the time of the analysis).

TABLE 1. Summary of Concepts Necessary for Applying Clustering Procedures

Procedure	Proximity Measure (Input)	Criterion Function	Restriction
T_1 Friedman-Rubin			Up to 200 objects with 30 variables per object in up to 25 clusters
A	Raw data	$\max[\det(B+W)/\det(W)]$	
B	scaled	$\max[\text{Trace } W^{-1}B]$	
C	internally	$\min[\text{Trace } W]$	
T_2 Ball-Hall ISODATA	Raw or scaled data	minimize MSE (equivalent to min of Trace W)	Up to 1000 objects with 50 variables per object in up to 50 clusters
T_3 Singleton-Kautz	Raw or scaled data	$\min[\text{Trace } W]$	Same as ISO-DATA
T_4 Hierarchical Clustering HICLUS MAX	Similarities dissimilarities	clusters optimally compact in terms of $d_M(x_i, x_j)$	Up to 100 objects.

3 The Mathematics Achievement Data Sets

The data sets used in this paper were collected by the National Longitudinal Study of Mathematical Abilities, NLSMA. NLSMA was a team research effort, funded by the National Science Foundation, and administered by the School Mathematics Study Group (SMSG) at Stanford University. Consultants from mathematics, education, psychology, and statistics contributed. During the Fall of 1962, schools that agreed to participate in the NLSMA five-year mathematics achievement study gave an initial battery of achievement inventories to students in the fourth, seventh and tenth grades in both conventional and new mathematics curricula. The design stressed three features: (1) A long-term longitudinal and cross-sectional study of student groups, (2) The study of mathematical abilities of students on several grade levels over varying times and, (3) Data collection on students' mathematics achievement in grades 4 through 12 for mathematics curriculum revision and modification.

Over 112,000 students in 1,500 schools from 39 states participated in the study. This paper focuses attention on NLSMA's Y-Population of 197 junior high schools. A random sample of thirty schools was selected. These schools had a total of 2,995 students. The sets of measurements taken on each school are divided into the following two main groups:

A. Student Test Variables Mathematical and psychological scales were administered to the students during the fall, and mathematical scales only were administered in the spring, for each of the five years.

B. Non-Test Variables School-community and teacher scales provided information on individual schools in school districts and the communities served by the schools. The teacher scales included information on the educational background of the teachers, and measured their attitudes toward teaching mathematics.

Three batteries of tests designed by the NLSMA staff were administered to the students:

(i) *The Ideas and Preference Inventory* measured attitudes toward mathematics.

(ii) *The Mathematics Inventory* measured achievement in specific mathematical topics.

(iii) *The Reasoning Inventory* contained sub-tests of the Lorge-Thorndike Verbal and Nonverbal Intelligence Tests.

Twelve school-community non-test variables were selected to cluster the junior high schools: "average daily attendance", "residential description", "median yearly income of parents", "teachers' starting salary", "teachers' salary index", "innovations", "use of mathematics supervisors" (persons employed to spend at least 50% of the time in the development of math curriculum), "heavy use of SMSG" (use of the materials by at least 50% of the students in grades 4 through 12), "heavy use of other experimental mathematics" (materials from the University of Maryland, University of Illinois Committee on School Mathematics, and Greater Cleveland), "in-service training", "mathematics class size", and "academic class size", respectively. School means on each of the school-community variables were used to cluster the objects (schools).

4 Results of the Cluster Analysis

PRELIMINARY ANALYSIS: PRINCIPAL COMPONENTS

A principal component analysis was performed on the 12 school-community variables to obtain an idea of the underlying dimensionality of these variables, and thus to suggest guidelines for the number of clusters to use. The first five eigenvalues (2.98, 2.53, 1.31, 1.15, 0.99) of the correlation matrix account for 75% of the total variance. The eigenvectors corresponding to these eigenvalues are presented in Table 2. Each component (factor) is bipolar; coefficients less than 0.25 in absolute value are not reported. The components may be described in terms of the school community variables as: "school characteristics", "district professional expenditures", "family socio-economic status", "number of innovations used" and "heavy use of SMSG materials", respectively.

TABLE 2. Eigenvectors corresponding to the five largest eigenvalues of the school-community correlation matrix

Variables	1	2	3	4	5
S_1		.38			.25
S_2		−.27	.69		
S_3	−.35	.30	−.43		
S_4		.50	.27		
S_5		.43	.30		
S_6	.27			.35	
S_7	.26	.36		.28	−.27
S_8					−.83
S_9	.30			.59	
S_{10}	.31		−.27		−.26
S_{11}	.46			−.29	
S_{12}	.45			−.41	

Dimensions	1	2	3	4	5	6	7	8	9	10	11	12
Percentage of Variance	25	46	57	67	75	82	87	92	96	98	99	100

S_1	Average Daily Attendance	S_2	Residential Description
S_3	Parent's Yearly Income	S_4	Teacher's Starting Salary
S_5	Teacher Salary Index	S_6	Innovation
S_7	Mathematics Supervisor	S_8	SMSG
S_9	Experimental Mathematics	S_{10}	In-Service Training
S_{11}	Mathematics Class Size	S_{12}	Academic Class Size

RESULTS

The optimal number of clusters produced by the three non-hierarchical clustering methods were determined by the algorithms used by these methods. Three clusters were chosen as optimal by F-R and S-K; four clusters were selected by ISODATA.

The linkages produced by the hierarchical clustering method HICLUS are presented in Figure 1 using the data from Table 3. If only the clusterings with three or less outlying schools are considered, then the first 16 stages in Table 3 are eliminated. The maximum value of 4.12 for the clustering measure $\bar{Z}_j$ occurs at $j = 24$ with six clusters and only one outlying school. The next value is 4.01 which occurs at $j = 21$, a clustering with nine clusters, three of which are outlying schools. At $j = 27$, $\bar{Z}_j$ attains its third largest value of 3.99. This clustering has the advantage of having only three clusters and no outlying schools, and thus was chosen as the preferred set of clusters for the HICLUS algorithm.

FIGURE 1. Johnson's max. method HICLUS clusterings

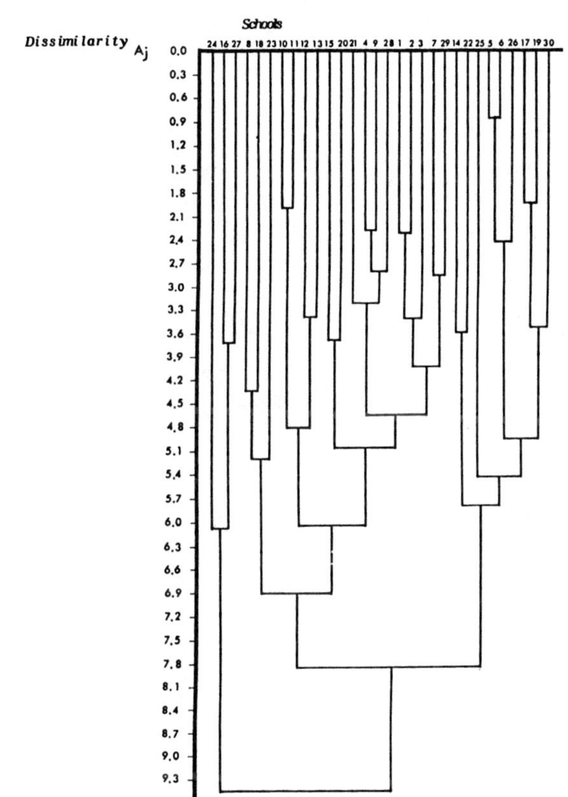

Table 4 gives the memberships of the clusters chosen by the four clustering methods. (For purposes of comparison, memberships in clusters chosen by F-R and S-K are also given for $G = 4$ clusters.) It is apparent that the clusters are quite similar. For example, if the two smallest clusters produced by ISODATA are combined into one cluster, the only disagreement between HICLUS and ISODATA is the placement of school 30. Similarly, F-R and S-K agree exactly for $G = 4$ clusters, and for $G = 3$ clusters disagree only in the placement of school 13. However, the S-K and HICLUS three-group clusters differ in the placement of six schools: 10, 11, 12, 13, 15 and 16. (But these five schools, excluding school 16, do remain together.)

In Figures 2 and 3, the schools are plotted in two dimensions using a multidimensional scaling algorithm (MDSCAL; see Johnson and Wichern, 1988, pp. 572–578) with the cluster identities supplied by HICLUS (Figure 2) and S-K, $G = 3$ clusters (Figure 3) superimposed on the data points. Other graphical representations and comparisons of the clusters obtained by the four clustering algorithms can be found in the author's Ph.D. dissertation (op. cit.). Only the HICLUS grouping of 3 clusters is used in the rest of this paper.

FIGURE 2. HICLUS max grouping imposed upon MDSCAL configuration (stress 0.477)

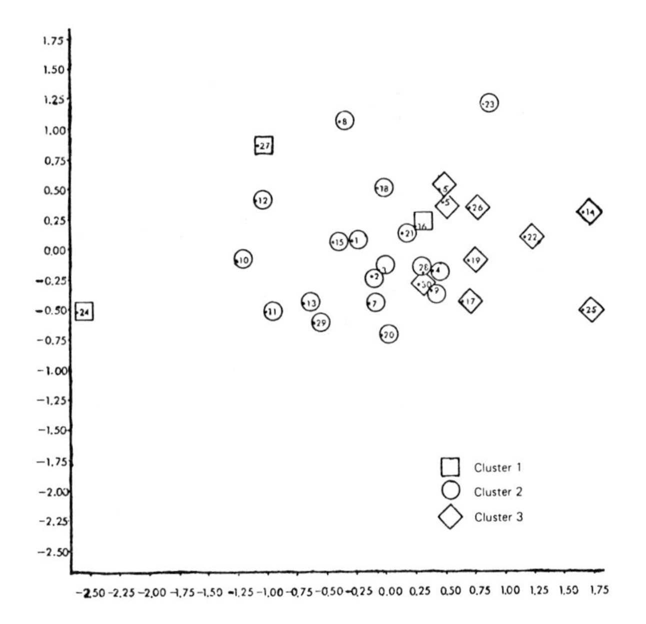

FIGURE 3. S-K grouping imposed upon MDSCAL configuration (stress 0.477)

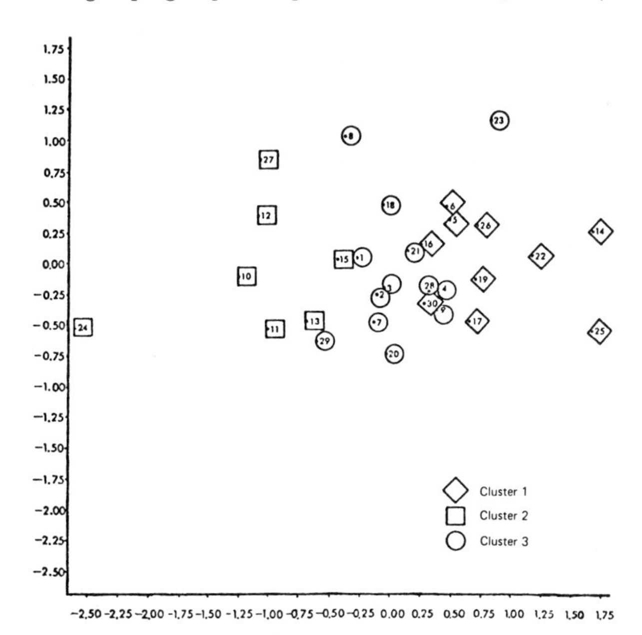

INTERPRETATION OF THE HICLUS CLUSTERS USING SCHOOL-COMMUNITY VARIABLES

In Table 8 a statistical description of the school-community variables is given for the clusters created by HICLUS.

The schools in Cluster 1 tend to be smaller than the schools in the other clusters, to be in communities with low cost housing and lower yearly family income, and (perhaps consequently) to pay lower salaries (and lower starting salaries) to teachers. Two of the three schools in this cluster have math supervisors, with more than 75% of the teachers involved in in-service training.

Cluster 2 tends to have larger schools than those in the other two clusters and also tends to have larger classes. Family incomes are comparable, but somewhat larger, than Cluster 1. Teacher salaries for this cluster are larger than Cluster 1 and comparable to Cluster 3. Schools in this cluster are unlikely to have math supervisors, and highly likely to use innovation in teaching. Like Cluster 1 there are large fractions of teachers involved in in-service training.

The most significant aspects of Cluster 3 are its higher family incomes (which however do not seem to yield highest starting salaries or pay for teachers), and relatively low use of innovation and experimental mathematics programs.

From this description, one might hypothesize that Cluster 1 schools tend to provide training experience for teachers unable (because of lack of experience) to find more desirable employment. The large size and higher pay of Cluster 2 schools (relative to income in the community), and the use of innovative methods, suggest that such schools are recipients of more state aid than schools in the other two clusters, perhaps because the teaching environment is less desirable (heavily urban). Cluster 3 schools may provide the most desirable teaching environment.

INTERPRETATION USING TEACHER VARIABLES

The Teacher Opinion Questionnaire developed by Price (1962), and refined by SMSG, measures teachers' attitudes toward teaching and provides teacher demographic data. A statistical summary of data from this questionnaire by cluster is given in Table 9.

Table 9 confirms our hypothesis that teachers in Cluster 1 tend to be the least experienced. Interestingly enough, they tend to have the strongest background in mathematics in college, and the strongest theoretical and authoritarian orientation to teaching. Schools in this cluster are also more likely to have a majority of female teachers.

From our previous description of schools in Cluster 3 one might hypothesize that schools in this cluster attract the most experienced teachers. However, Table 9 shows that this honor goes to Cluster 2, which also has a larger proportion of teachers whose undergraduate major was not mathematics, and who express the greatest need for social approval. Schools in this cluster may attract teachers interested in innovation in teaching methods and the improvement of society.

The most striking aspects of Cluster 3 schools are (1) the large proportion of schools having a majority of male teachers, (2) a higher proportion of teachers expressing only moderate involvement in teaching or concern for students, and (3) its inclusion of the only schools emphasizing rote learning of mathematics (3 of 9). Teachers in the schools in this cluster are nonauthoritarian, and use a creative approach to teaching. (They may also be more interested in teaching as a means of employment.)

5 Analysis of Students' Mathematical Achievement

Do the clusters formed in our analysis influence students' mathematical achievement? To answer this question, school means on three student mathematics achievement scales were analyzed as dependent variables by a multivariate analysis of covariance:

$$\begin{aligned} y_1 &= \text{Numbers-Whole}, \\ y_2 &= \text{Algebra-Sentences}, \\ y_3 &= \text{Conversion}. \end{aligned}$$

School means for the following 6 variables (scales) were used as covariates, to account for prior ability and achievement of the students:

x_1 = Lorge-Thorndike Verbal, x_2 = Lorge-Thorndike Nonverbal,
x_3 = Rationals-Computation, x_4 = Rationals-Noncomputation,
x_5 = Whole-Numbers, x_6 = Geometry.

The x variables (covariates) were administered during the fall of the first year. The y variables (criterion variables) were administered during the spring of the third year of testing. The criterion variables are used to measure changes in the students' mathematics achievement over the three year period at the junior high school. It is important to note that the x variables need not be direct causal agents of the y variables but may merely reflect characteristics of the environment that also influence the y variables.

Multivariate analysis of covariance

The general linear model for the one-way analysis of variance design for each dependent variable may be expressed as:

$$y = \mu + \beta_0 + \beta_1 x_1 + \beta_2 x_2 + \cdots + \beta_p x_p + \varepsilon$$

where μ denotes the overall mean effect, β_0 is the main class effect due to the treatment, and ε is the residual or error which represents discrepancies between the observed vector and the vector sum of the general mean and treatment effect.

Homogeneity of regression equations assumes that the regression equations across groups are identical, whereas homogeneity of covariances assumes that the error covariance matrix is the same for each group. The term $\mu + \beta_0$ is referred to as the design part of the model. The term $\beta_1 x_1 + \beta_2 x_2 + \cdots + \beta_p x_p$ measures the effect of the covariates, variables measured by but not under the control of the researcher. The covariate model may be written as

$$\tilde{y} = y - (\beta_1 x_1 + \beta_2 x_2 + \cdots + \beta_p x_p) = \mu + \beta_0 + \varepsilon,$$

and multivariate analysis of covariance could be thought of as a multivariate analysis of variance which is performed on y assuming that the x's are known.

The multivariate analysis of covariance involves the testing of the following null hypotheses:

H_0 Equality of treatment effects. The constant terms in the regression equations are the same across groups.

H_1 Parallelism of regression lines. The vectors of beta-weights are the same across groups.

H_2 Homogeneity of covariances. There are no differences between the error covariance matrices across groups.

Data for this study includes $G = 3$ clusters (treatments), $q = 3$ variates, and $p = 6$ covariates. The predicted criterion score of each student in the kth cluster ($k = 1, 2, \ldots, G$) can be thought of as the sum of the terms $\beta_0^{(k)}$ and $\left(\beta_1^{(k)} x_1 + \beta_2^{(k)} x_2 + \cdots + \beta_p^{(k)} x_p\right)$. The first term characterizes the cluster to which the student is assigned, and the second term is proportional to the students' entering score. The assumption of parallelism of the regression planes means that the vector of regression weights for the covariates is the same across clusters.

The hypothesis H_0 states that the differences of the performances of the students in the different clusters can be attributed entirely to their competencies as measured by their scores on the six pretests. And after allowance for this, the effect of the clusters is the same. Hence, we are testing the hypothesis,

$$H_0 : \beta_0^{(1)} = \beta_0^{(2)} = \beta_0^{(3)}.$$

Testing homogeneity of regression

The likelihood ratio test described by Bock (1966) to test the hypothesis H_1 of homogeneity of regression planes is used in this study. In doing so, we obtained a chi-square value of 29.90 with 36 degrees of freedom ($\chi^2_{36} = 28.7, p = .80$) which is clearly not significant. Hence, the data does not contradict the hypothesis H_1.

Univariate F tests for homogeneity of regression appear in Table 10. The F values, with 12 and 2974 degrees of freedom, reported for y_1, y_2, and y_3 are .71, .93, and .79 respectively. All of these F values correspond to $p > .50$. Therefore, none of the differences between the vectors of regression weights are significant and hence, the planes may be considered parallel. It follows that the relationship between each of the variates measured separately and the set of covariates is constant across clusters. Hence, a single plane may be fitted to all of the data to test treatment effects if the assumption of equal dispersion matrices is satisfied.

Testing homogeneity of covariance matrices

When testing for homogeneity of dispersion, after the y values have been adjusted for the x values, there is a likelihood ratio test available which is a multivariate extension of Bartlett's χ^2 approximation for homogeneity of variance. The chi-square value obtained for this test is 20.89 with 12 degrees of freedom. Since a chi-square value of 21.03 corresponds to a p value of .05, we may say that the assumption H_2 of a common within-group covariance matrix is not rejected at $p < .05$ level. As a result of these tests we have satisfied the conditions for applying the multivariate analysis of covariance.

The multiple correlation coefficient, R^2 for totals, is reported in Table 10. We obtain values of 0.40 for y_3-conversion, 0.43 for y_2-algebra sentences, and 0.48 for y_1-numbers-whole. R^2 is the proportion of the variance of the y variables which is explained by the x variables in the regression equation. The value, $1 - R^2$, is the proportion of variation in the criterion which must be ascribed to other sources.

6 Test of the MANCOVA Hypothesis H_0

The means and standard deviations for the three clusters are presented in Table 11. The Rao F transformation is used as the criterion for testing the hypothesis H_0, following covariance adjustment. An F value of 9.14 with 6 and 5,968 degrees of freedom has a p-value, $p < .01$. Hence, the mathematics achievement of the students across the three clusters cannot be considered the same after adjustments have been made for differences in aptitude and initial understanding of mathematical concepts.

The next stage of the analysis includes the analysis of the differences on each of the three variates using all six covariates to determine which particular variable was most significant on an individual basis. The results of the univariate tests after adjustments for the covariates have been made is displayed in Table 3. This table shows that the most significant variable is y_3=conversion, followed by y_2=algebra sentences. The variate y_1=number whole did not discriminate between the groups. These differences between the adjusted means cannot be explained by the competencies of the students, but must be attributed to the effect of the cluster to which the student was assigned.

TABLE 3. F-values for differences between clusters on each variate

	$F_{2,2986}$	p
Y_1 Numbers Whole	0.86	.42
Y_2 Algebra Sentences	4.07	.02
Y_3 Conversation	20.43	.01

The adjusted cluster means for each of the three criterion variables are displayed in Table 4. The students in Cluster 1 have the lowest adjusted means on variables y_1=numbers whole and y_2=algebra sentences. Cluster

2 has the highest student achievers on y_1=numbers whole and the lowest achievers on y_3=conversion. Finally, Cluster 3 has the highest adjusted means over variables y_2=algebra sentences and y_3=conversion.

It must be concluded that even after adjustments are made for the covariates, the cluster differences on the scale y_1=numbers whole are not significant. However, the students' differences across clusters on y_2=algebra sentences and y_3=conversion are significant.

Hence, the mathematics achievement of the students across the clusters cannot be considered the same after adjustments have been made for differences in aptitude and initial understanding of mathematical concepts. We conclude that the clusters may not be considered equally effective.

TABLE 4. The adjusted means for the three variates

	Cluster I	Cluster II	Cluster IIII
Y_1 Numbers Whole	4.13	4.32	4.30
Y_2 Algebra Sentences	2.45	2.81	2.84
Y_3 Conversion	6.04	5.53	6.20

TABLE 5. The error dispersion matrix adjusted for the covariate

	Y_1	Y_2	Y_3
Y_1	2.50	0.73	0.96
Y_2	0.73	1.96	0.93
Y_3	0.96	0.93	6.87

The standard errors of estimate

$s_{\epsilon 1} = 1.58 = \sqrt{2.5044}$ $(R_1^2 = .48,\ s_{y1} = 2.17)$

$s_{\epsilon 2} = 1.40 = \sqrt{1.9556}$ $(R_2^2 = .43,\ s_{y2} = 1.81)$

$s_{\epsilon 3} = 2.62 = \sqrt{6.8678}$ $(R_3^2 = .41,\ s_{y3} = 3.35)$

The error dispersion matrix adjusted for the six covariates is presented in Table 5 along with the standard error of estimate, s_ε. The multiple correlation coefficient R is directly related to s_ε by the relation

$$s_{\varepsilon i} = s_{yi}\sqrt{1 - R^2} \qquad \text{for } i = 1, 2, 3$$

where s_{yi} is the standard deviation of the ith criterion variable (variate). When predicting the mathematics achievement for students similar to those in this study on scales y_1, y_2 and y_3, the observed achievement will be within $\pm 1.58, \pm 1.40$, and ± 2.26 respectively of the adjusted mean about sixty-eight percent of the time.

TABLE 6. Data for HICLUS max clustering (Figure 1)

Stage j	Dissimilarity $A_j^{(1)}$	$\bar{Z}_j^{(2)}$	$O_j^{(3)}$	$G^{(4)}$
1	0.88	3.92	28	29
2	1.95	3.47	26	28
3	1.97	3.54	24	27
4	2.28	3.28	22	26
5	2.31	3.08	20	25
6	2.43	3.44	19	24
7	2.85	3.61	18	23
8	2.87	3.41	16	22
9	3.24	3.58	15	21
10	3.40	3.40	13	20
11	3.43	3.44	12	19
12	3.54	3.44	11	18
13	3.60	3.31	9	17
14	3.70	3.14	7	16
15	3.74	3.02	5	15
16	4.04	3.36	5	14
17	4.34	3.18	3	13
18	4.66	3.70	3	12
19	4.80	3.87	3	11
20	4.96	3.89	3	10
21	5.08	4.01	3	9
22	5.19	3.90	2	8
23	5.43	3.85	1	7
24	5.82	4.12	1	6
25	6.12	3.92	1	5
26	6.14	3.84	0	4
27	6.90	3.99	0	3
28	7.83	2.75	0	2
29	9.45	Strong Clustering	0	1

(1) Dissimilarity measure (Euclidean distance between two clusters/objects joined at stage j).

(2) Measure of clustering excluding outlying schools at stage j.

(3) Number of outlying schools (clusters of size 1) at stage j.

(4) Total number of clusters at stage j.

TABLE 7. Cluster membership for optimal clustering

ISODATA, $G = 4$ clusters:
- Cluster 1: Schools 1, 2, 3, 4, 7, 8, 9, 10, 11, 12, 13, 15, 18, 20, 21, 23, 28, 29, 30
- Cluster 2: School 24
- Cluster 3: Schools 5, 6, 14, 17, 19, 22, 25, 26
- Cluster 4: Schools 16, 27

HICLUS MAX, $G = 3$ clusters:
- Cluster 1: Schools 16, 24, 27
- Cluster 2: Schools 1, 2, 3, 4, 7, 8, 9, 10, 11, 12, 13, 15, 18, 20, 21, 23, 28, 29
- Cluster 3: Schools 5, 6, 14, 17, 19, 22, 25, 26, 30

F-R procedure based on trace W

$G = 4$ clusters:
- Cluster 1: Schools 15, 20, 24, 27
- Cluster 2: Schools 5, 6, 14, 16, 17, 19, 22, 25, 26, 30
- Cluster 3: Schools 10, 11, 12, 13
- Cluster 4: Schools 1, 2, 3, 4, 7, 8, 9, 18, 21, 23, 28, 29

$G = 3$ clusters:
- Cluster 1: Schools 10, 11, 12, 15, 24, 27
- Cluster 2: Schools 5, 6, 14, 16, 17, 19, 22, 25, 26, 30
- Cluster 3: Schools 1, 2, 3, 4, 7, 8, 9, 13, 18, 20, 21, 23, 28, 29

S-K procedure

$G = 4$ clusters:
- Cluster 1: Schools 5, 6, 14, 16, 17, 19, 22, 25, 26, 30
- Cluster 2: Schools 10, 11, 12, 13
- Cluster 3: Schools 15, 20, 24, 27
- Cluster 4: Schools 1, 2, 3, 4, 7, 8, 9, 18, 21, 23, 28, 29

$G = 3$ clusters:
- Cluster 1: Schools 5, 6, 14, 16, 17, 19, 22, 25, 26, 30
- Cluster 2: Schools 10, 11, 12, 13, 15, 24, 27
- Cluster 3: Schools 1, 2, 3, 4, 7, 8, 9, 18, 20, 21, 23, 28, 29

TABLE 8. Statistical description of HICLUS clusters using school-community variables

	Cluster I Lower-Middle		Cluster II Middle-Middle		Cluster III Upper-Middle	
	Schools	**Number of Schools**	**Schools**	**Number of Schools**	**Schools**	**Number of Schools**
Variable 1: Average Daily Attendance. Nos. of Students.						
Under 700	24,27	2	1,28	2	26,30	2
701–1400	16	1	2,3,8,9,10,11,12,15,21,29	10	5,6,14,17,19,22,25	7
Over 1400		0	4,7,13,18,20,23	6		0
		3		18		9
	Mean = 527 Range = 766-213 = 553		Mean = 1188 Range = 1665-630 = 735		Mean = 701 Range = 820-350 = 470	
Variable 2: Residential Description.						
Expensive		0		0		0
Moderately priced		0	2,3,4,7,9,11,13,20,21,23,28,29	12	14,17,19,22,25,26,30	7
Low Cost	16,24,27	3	1,8,10,12,15,18	6	5,6	2
		3		18		9
Variable 3: Medain Yearly Income of Parents.						
Less than $5,000	24	1	10,12,15	3	6	1
$5,000–9,999.99	16,27	2	1,2,3,4,7,8,11,13,18,20,21,23,29	13	5,14,17,19,22,26,30	7
$10,000 and Above		0	9,28	2	25	1
		3		18		9
	Cluster Median = $5,750		Cluster Median = $6,750		Cluster Median = $8,750	
Variable 4: Teacher's Starting Salaries.						
$3,500–3,999.99	24,27	2		0		0
$4,000–4,499.99	16	1	10,11	2		0
$4,500–4,999.99		0	3,12,13,15,18,20,28,29	8	17,22,25,26,30	5
$5,000–5,499.99		0	1,2,4,7,8,9,21,23	8	5,6,14,19	4
		3		18		9
	Median Starting Salary = $3,750		Median Starting Salary = $5,000		Median Starting Salary = $4,750	

Table 8 continued

	Cluster I Lower-Middle		Cluster II Middle-Middle		Cluster III Upper-Middle	
	Schools	**Number of Schools**	**Schools**	**Number of Schools**	**Schools**	**Number of Schools**
Variable 5: Teacher Salary Index.						
1.26–1.46	16,24,27	3	12,13,15	3	22	1
1.46–1.66		0	1,2,4,7,9,10,11,18,20,21,29	11	14,17,19,25,30	5
1.66–1.86		0	3,8,23,28	4	5,6,26	3
		3		18		9
Variable 6: Innovation Points for Innovation.						
0 Methods used		0		0	14	1
1–2 Methods used	16	1	20	1	17,19,26	3
3–4 Methods used	24,27	2	7,9,10,11,15,21,23	7	5,6,22,30	4
5+ Methods used		0	1,2,3,4,8,12,13,18,28,29	10	25	1
		3		18		9
Variable 7: Mathematics Supervisor. Math Supervisor Employed.						
Yes	24,27	2	3,15,20	3	5,6,14,22,25,26	6
No	16	1	1,2,4,7,8,9,10,11,12,13,18,21,23,28,29	15	17,19,30	3
		3		18		9
Variable 8: Heavy Use of SMSG Grades.						
0 Grades	16	1	4,7,9,15,23,28	6	5,6,22,26,30	5
1–3 Grades	27	1	10,11,18,20,21	5		0
4–6 Grades	24	1	1,2,3,29	4	14,17,19,25	4
7–9 Grades		0	8,12,13	3		0
		3		18		9

Table 8 continued

	Cluster I Lower-Middle		Cluster II Middle-Middle		Cluster III Upper-Middle	
	Schools	**Number of Schools**	**Schools**	**Number of Schools**	**Schools**	**Number of Schools**
Variable 9: Heavy Use of Other Experimental Mathematics Programs. Total Grades.						
0 Grades	16,24	2	1,2,3,4,7,8,9,15,18,20,21,23,28,29	14	5,6,14,17,19,22,25,26,30	9
1–3 Grades	27	1	12,13	2		0
4–6 Grades		0	10,11	2		0
7–9 Grades		0		0		0
		3		18		9
Variable 10: In-Service Training Experience. Percentage of Teachers.						
Under 25%		0	23	1	14,22	2
25–50%		0	4,21	2	26	1
51–75%	16	1	8,9,12,13,18,28	6	5,6,19,25	4
Over 75%	24,27	2	1,2,3,7,10,11,15,20,29	9	17,30	2
		3		18		9
Variable 11: Mathematics Class Size.						
25–29		0		0	25	1
30–34	16,27	2	18,20,28	3	5,6,14,17,19,22,26,30	8
35–39		0	1,2,3,4,8,9,10,11,12,13,15,21,23	13		0
40 or more	24	1	7,29	2		0
		3		18		9
Variable 12: Academic Class Size (Other than Mathematics).						
25–29		0		0	25	1
30–34	16,27	2	18,28	2	5,6,14,17,19,22,26	7
35–39		0	1,2,3,4,7,8,9,10,11,12,13,15,21,23	14	30	1
40 or more	24	1	20,29	2		0
		3		18		9

TABLE 9. Statistical description of HICLUS clusters using teacher variables

	Cluster I Lower-Middle		Cluster II Middle-Middle		Cluster III Upper-Middle	
	Schools	**Number of Schools**	**Schools**	**Number of Schools**	**Schools**	**Number of Schools**
Variable 1: Number of Years Teaching.						
1–6	24,27	2	1,10,12,13,15,20,21	7	5,6,14,17,30	5
7–12	16	1	2,4,7,9,11,23,28,29	8	19,25	2
13 or more		0	3,8,18	3	22,26	2
		3		18		9
Variable 2: Highest Academic Degree.						
B.A.	16,24,27	3	1,2,3,7,9,10,11,12,13,15,20,21,23,29	14	5,6,14,17,19,22,25,30	8
M.A.		0	3,8,18,28	4	26	1
Ed.D. or Ph.D.		0		0		0
		3		18		9
Variable 3: Number of Academic Credits Beyond B.A.						
0–19	24	1	9,10,13,20,21	5	6,17,22,30	4
20–39	16,27	2	1,2,4,7,11,12,15,18,23,29	10	5,14,19,25,26	5
40 or more		0	3,8,28	3		0
		3		18		9
Variable 4: Number of College Mathematics Credits, Beginning with Calculus.						
0–3		0	2,28,29	3	22	1
4–9	16	1	1,3,4,7,10,12,13,15,18,20,23	11	5,6,14,17,19,25,26,30	8
10 or more	24,27	2	8,9,11,21	4		0
		3		18		9

Table 9 continued

	Cluster I Lower-Middle		Cluster II Middle-Middle		Cluster III Upper-Middle	
	Schools	**Number of Schools**	**Schools**	**Number of Schools**	**Schools**	**Number of Schools**
Variable 5: Number of Credits in Methods of Teaching Mathematics.						
0–3	27	1	2,3,15,20,23,29	6	6,19,22,25,26,30	6
4–9	16,24	2	1,4,7,9,10,11,12,13,18,21,28	11	5,14,17	3
10 or more		0	8	1		0
		3		18		9
Variable 6: Number of Regular, Inservice, or Extension Courses Taken Since Last Degree.						
0–3		0	13,20,29	3	5,22,25,30	4
4–6	24,27	2	1,2,3,4,7,9,10,11,12,15,18,21,28	13	6,14,17,19,26	5
7 or more	16	1	8,23	2		0
		3		18		9
Variable 7: Number of Other Types of Mathematics Preparation Since Last Degree.						
1 or 2		0	2,9	2	5	1
3 or 4	16,24,27	3	1,3,4,7,10,11,12,13,15,18,20,21,23,28,29	15	6,14,17,19,22,25,26,30	8
5		0	8	1		0
		3		18		9
Variable 8: Sex.						
More Males	27	1	2,4,7,8,10,11,15,18,21,28,29	11	5,6,14,17,19,25,26	7
More Females	16,24	2	1,3,9,12,13,20,23	7	22,30	2
		3		18		9

Table 9 continued

	Cluster I Lower-Middle		Cluster II Middle-Middle		Cluster III Upper-Middle	
	Schools	**Number of Schools**	**Schools**	**Number of Schools**	**Schools**	**Number of Schools**
Variable 9: Age						
20–25	24	1	20	1	30	1
26–30		0	1,10,12,13,21	5	6,14	2
31–35	27	1	15	1	5,19,25	3
36 and Over	16	1	2,3,4,7,8,9,11,18,23,28,29	11	17,22,26	3
		3		18		9
Variable 10: Undergraduate Major Mathematics?						
Yes	16,24,27	3	1,4,7,8,9,10,11,13,15,18,20,21	12	6,14,17,19,25,26,30	7
No		0	2,3,12,23,28,29	6	5,22	2
		3		18		9
Variable 11: Degree of Theoretical Orientation.						
Moderate		0	1,2,3,4,8,9,20,29	8	5,14,19,26	4
Strong	16,24,27	3	7,10,11,12,13,15,18,21,23,28	10	6,17,22,25,30	5
		3		18		9
Variable 12: Concern for Students.						
Moderate		0	1,9,20,21	4	5,14,17,26,30	5
Great	16,24,27	3	2,3,4,7,8,10,11,12,13,15,18,23,28,29	14	6,19,22,25	4
		3		18		9

Table 9 continued

	Cluster I Lower-Middle		Cluster II Middle-Middle		Cluster III Upper-Middle	
	Schools	**Number of Schools**	**Schools**	**Number of Schools**	**Schools**	**Number of Schools**
Variable 13: Involvement in Teaching.						
Moderate	24	1	2,3,7,9,12,13,23,28,29	9	5,14,19,30	4
Great	16,27	2	1,4,8,10,11,15,18,20,21	9	6,17,22,25,26	5
		3		18		9
Variable 14: Orientation.						
Authoritarian	24	1	1,11,12,13,23,28	6	22,25,30	3
Non-Authoritarian	16,27	2	2,3,4,7,8,9,10,15,18,20,21,29	12	5,6,14,17,19,26	6
		3		18		9
Variable 15: Attitude Toward Mathematics.						
Dislikes		0		0		0
Likes	All Schools	3	All Schools	18	All Schools	9
		3		18		9
Variable 16: Creative vs. Rote View of Learning Mathematics.						
Rote		0		0	2,25,30	3
Creative	All Schools	3	All Schools	18	6,14,17,19,22,26	6
		3		18		9
Variable 17: Teachers' Need for Social Approval.						
Moderate Need	24	1	1,2,3,9,20,29	6	5,19,26,30	4
Great Need	16,27	2	4,7,8,10,11,12,13,15,18,21,23,28	12	6,14,17,22,25	5
		3		18		9

TABLE 10. Raw score regression weights

Variates	Cluster	Covariates x_1	x_2	x_3	x_4	x_5	x_6	R^2	R	F Test on R	(df)	Significance of Regression 6, 2974 df	Homogeneity of Regression 12, 1974 df
Y_1 Number-Whole	I ($N = 119$)	.14	.02	.21	.04	.32	.18	.53	.72	20.62*	(6, 112)		
	II ($N = 1898$)	.08	.04	.19	.19	.22	.07	.50	.71	314.13*	(6, 1891)		
	III ($N = 978$)	.07	.04	.16	.21	.22	.05	.42	.65	118.40*	(6, 971)		
	Pooled $N = 2995$	.08	.04	.18	.19	.22	.07	.47	.69	451.51*	(6, 2988)	$F = 465.15$*	$F = 0.71$ (N.S)
	Total $N = 2995$	.08	.04	.18	.19	.22	.07	.48	.70	465.77*	(6, 2988)		
Y_2 Algebra-Sentences	I	.04	.05	.19	.24	−.12	.01	.36	.60	10.50*	(6, 112)		
	II	.05	.04	.11	.17	.13	.13	.43	.66	240.14*	(6, 1891)		
	III	.05	.04	.10	.17	.17	.12	.40	.64	109.31*	(6, 971)	$F = 373.59$*	$F = 0.93$ (N.S)
	Pooled	.05	.04	.11	.17	.14	.12	.42	.65	365.12*	(6, 2988)		
	Total	.05	.04	.11	.18	.13	.12	.43	.65	372.98	(6, 2988)		
Y_3 Conversion	I	.16	.03	.06	.41	.23	.38	.43	.65	14.31*	(6, 112)		
	II	.08	.03	.21	.37	.36	.28	.41	.64	219.50*	(6, 1891)		
	III	.06	.05	.31	.31	.36	.32	.36	.60	92.78*	(6, 971)	$F = 346.68$*	$F = 0.79$ (N.S)
	Pooled	.07	.04	.23	.35	.40	.30	.40	.63	324.94*	(6, 2988)		
	Total	.08	.04	.20	.35	.37	.28	.40	.63	342.76*	(6, 2988)		

* Significant at .01 level

TABLE 11. Means and standard deviations for each cluster and the total group

Student Test Variable	Total Possible Score	"Lower Average Group" Cluster I $N_1 = 119$		"Average Group" Cluster II $N_2 = 1898$		"Upper Average Group" Cluster III $N_3 = 978$		Total Group $N = 2995$	
Covariates		**Mean**	**SD**	**Mean**	**SD**	**Mean**	**SD**	**Mean**	**SD**
Lorge-Thorndike Verbal — X_1	40	20.78	5.90	20.59	6.39	22.7	5.97	21.31	6.23
Lorge-Thorndike Nonverbal — X_2	58	33.42	7.99	33.20	9.81	37.22	8.34	34.53	9.28
Rationals-Computation — X_3	6	2.76	1.31	3.56	1.47	3.56	1.37	3.53	1.43
Rationals-Noncomputation — X_4	11	3.45	2.08	4.22	2.26	4.77	2.17	4.37	2.22
Whole Number — X_5	9	5.69	1.57	5.58	1.68	6.03	1.53	5.73	1.63
Geometry — X_6	4	1.00	0.98	1.19	1.02	1.25	1.01	1.20	1.02
Variates									
Number-Whole — Y_1	8	3.72	2.02	4.16	2.22	4.68	2.08	4.31	2.17
Algebra-Sentences — Y_2	6	2.13	1.63	2.67	1.82	3.14	1.81	2.80	1.81
Conversion — Y_3	12	5.41	3.25	5.33	3.38	6.67	3.30	5.77	3.35

References

Anderson, T.W. (1958). *An Introduction to Multivariate Statistical Analysis.* New York, John Wiley and Sons.

Ball, G. (1965). Data Analysis in the Social Sciences, What About the Details? *Proceedings of the 1965 Fall Joint Computer Conference*, 27-I.

Ball, G. and Hall, D. (1965). ISODATA, A Novel Method of Data Analysis and Pattern Recognition. Technical Report, Stanford Research Institute, Menlo Park, CA.

Ball, G. and Hall, D. (1967). PROMENADE, An On-line Pattern Recognition System. Technical Report, Stanford Research Institute, Menlo Park, CA.

Bartlett, M.S. (1937). Properties of Sufficiency and Statistical Tests. *Proceedings of the Royal Society*, A **160**, 268–282.

Bartlett, M.S. (1947). Multivariate Analysis. *Journal of the Royal Statistical Society*, Series B, 176–197.

Bock, R. (1966). Contributions of Multivariate Experimental Design to Educational Research. *Handbook of Multivariate Experimental Psychology* (ed. R. Cattell), Rand McNally and Co., Chicago, Ill., 820–840.

Bock, R. and Haggard, E. (1968). The Use of Multivariate Analysis of Variance in Behavioral Research. *Handbook of Measurement and Assessment in Behavioral Sciences*, (ed. D. Whitla), Addison and Wesley, Mass., 100-142.

Bonner, R. (1964). On Some Clustering Techniques. *IBM Journal*, Jan., 22–32.

Box, G.E.P. (1949). A Generalized Distribution Theory for a Class of Likelihood Criteria. *Biometrika*, **36**, 317–346.

Cooley, W. and Lohnes, P. (1962). *Multivariate Procedures for the Behavioral Sciences.* New York, John Wiley and Sons.

Dixon, W. (editor) (1967). *Biomedical Computer Programs.* University of California Press, Berkeley and Los Angeles.

Edwards, A. and Cavelli-Sforza (1965). A Method for Cluster Analysis. *Biometrics*, **21**, 362–375.

Fisher, R.A. (1936). The Use of Multiple Measurements in Taxonomic Problems. *Annals of Eugenics*, **7**, 179–188.

Friedman, H. and Rubin, J. (1967) On Some Invariant Criteria for Grouping Data. *Journal of American Statistical Association*, **62**, 1159–1177.

Friedman, H. and Rubin, J. (1967) A Cluster Analysis and Taxonomy System for Grouping and Classifying Data. IBM Corporation, New York Scientific Center, New York.

Hotelling, H. (1936). Relations Between Two Sets of Variates, *Biometrika*, **28**, 321–377.

Johnson, Richard A. and Wichern, Dean W. (1988). *Applied Multivariate Statistical Analysis.* (2nd edit.), Prentice Hall, Englewood Cliffs, New Jersey.

Johnson, S. (1967). Hierarchical Clustering Schemes. *Psychometrika*, **32**, 241–254.

Johnson, S. (1968a). How to Use HICLUS-A Hierarchical Cluster Analytic Program, Version 1s, IBM 360 Fortran IV Level H. Revised by Campus Facility, Stanford Computation Center, Stanford University, Stanford, CA.

Johnson, S. (1968b). A Simple Cluster Statistic. Abstract, Bell Telephone Laboratories, Inc., Murray Hill, New Jersey.

Lockley, J.E. (1970). *A Comparative Study of Some Cluster Analytic Techniques with Application to the Mathematics Achievement in Junior High Schools.* Unpublished Dissertation, Stanford University, (Ingram Olkin, Principal Adviser), May 1970. (Abstracted in *Dissertation Abstracts International*, XXXI, Number 8, 1971.)

McQuitty, L. and Clark, J. (1968). Clusters From Iterative, Intercolumnar Correlational Analysis. *Educational and Psychological Measurement*, **28**, 211–238.

Morrison, D. (1967). *Multivariate Statistical Methods.* McGraw-Hill, New York.

Price, S. (1962). The Teacher Opinion Questionnaire: A Methodological Investigation of Attitude Measurement. Unpublished Honors Thesis, Radcliffe College.

Singleton, R. and Kautz, W. (1965). Technical Report, Stanford Research Institute, Menlo Park, California.

Singleton, R. and Kautz, W. (1966). Technical Report, Stanford Research Institute, Menlo Park, California.

Smith, H., Gnanadesikan, R. and Hughes, J. (1962). "Multivariate Analysis of Variance (MANOVA)." *Biometrics*, **54**, 22–41.

Sokal, R. and Sneath, P. (1963). *Principles of Numerical Taxonomy.* W.H. Freeman and Co., San Francisco, California.

Sokal, R. and Michener, C. (1957). A Quantitative Approach to a Problem in Classification. *Evolution*, **11**, 130–162.

Tryon, R. (1967). Person Clusters on Intellectual Abilities and on MMPI Attributes. *Multivariate Behavioral Research*, **2**, 5–34.

Wallace, D. (1968). Cluster Analysis. *International Encyclopedia of the Social Sciences*, Crowell Collier, New York, 519–524.

Ward, J. and Hook, M. (1963). Application of an Hierarchical Grouping Procedure to a Problem of Grouping Profiles. *Educational and Psychological Measurement*, **23**, 69–81.

Wilks, S. (1932). Certain Generalizations in the Analysis of Variance. *Biometrika*, **24**, 471–494.

Wilks, S. (1959). *Mathematical Statistics.* John Wiley and Sons, Inc., New York.

Wilson, J. and Carry, L. (1969). Homogeneity of Regression — Its Rationale Computation and Use. *American Educational Research Journal*, **VI**, 80–90.

15

Bayesian Inference in Factor Analysis

S. James Press[1]
K. Shigemasu[2]

ABSTRACT We propose a new method for analyzing factor analysis models using a Bayesian approach. Normal theory is used for the sampling distribution, and we adopt a model with a full disturbance covariance matrix. Using vague and natural conjugate priors for the parameters, we find that the marginal posterior distribution of the factor scores is approximately a matrix T-distribution, in large samples. This explicit result permits simple interval estimation and hypothesis testing of the factor scores. Explicit point estimators of the factor score elements, in large samples, are obtained as means of the respective marginal posterior distributions. Factor loadings are estimated as joint modes (with the factor scores), or alternatively as means or modes of the distribution of the factor loadings conditional upon the estimated factor scores. Disturbance variances and covariances are estimated conditional upon the estimated factor scores and factor loadings.

1 Introduction

This paper proposes a new method for analyzing factor analysis models using a Bayesian point of view. We use normal theory for the sampling distribution, and vague and natural conjugate theory for the prior distributions. We adopt a general disturbance covariance matrix whose prior mean is diagonal. We show that in large samples, for a variety of prior distributions for the factor scores (including vague and normal priors), the marginal posterior distribution of the factor scores is approximately matrix T. As a result, we are able to make both point and interval estimates of the factor scores, thereby improving upon most earlier research. For comparison, we give below a brief review of earlier research in this area.

An early formulation (see Press 1972, 1982) of a Bayesian factor analysis model used the Wishart distribution for the sample covariance matrix, and a vague prior distribution for the parameters. Only implicit numerical

---

[1]University of California, Riverside
[2]Tokyo Institute of Technology

solutions could be obtained from this model. Kaufman and Press (1973a,b) proposed a new formulation of that model in terms of a model with more factors than observations (a characteristic the model shared with that proposed by Guttman, 1953), but the prior on most of the factors was centered at zero. This work was developed further in Kaufman and Press (1976), who showed that the posterior distribution of the factor loading matrix was truncated multivariate normal in large samples.

Martin and McDonald (1975) approached the factor analysis problem looking for a Bayesian solution to Heywood cases. They adopted a diagonal disturbance covariance matrix and used a Jeffreys type vague prior for the elements. They proposed finding posterior joint modal estimators of the factor loading and disturbance covariance matrices, and obtained an implicit numerical solution. A point estimate of the factor loading matrix was also obtained.

Wong (1980) addressed the factor analysis problem from the empirical Bayes point of view, adopting normal priors for the factor loadings. He suggested use of the EM algorithm (see Dempster, Laird, and Rubin, 1977) to find a posterior mode for the factor loading matrix, but an explicit algorithm was not obtained.

Lee (1981) adopted a heirarchial Bayesian approach to confirmatory factor analysis, starting from the assumption that the free parameters in the factor loading matrix were exchangeable and normally distributed. The disturbance covariance matrix was assumed to be diagonal. Joint modal estimates were found of the factor loading matrix, the variances of the disturbances, and the covariance matrix of the factors. This modal solution was implicit and numerical. A point estimate of the factor loading matrix was obtained.

Euverman and Vermulst (1983) studied the Bayesian factor analysis model with a diagonal disturbance covariance matrix, and a preassigned number of factors. A numerical computer routine was described for implicitly finding the posterior joint mode of the factor loadings and error variances.

Mayekawa (1985) studied the Bayesian factor analysis problem examining factor scores as well as factor loadings and error (specific) variances. The factor loadings were assumed to be normal, a priori. The author used the EM algorithm to find point estimates of the parameters as marginal modes of the posterior distributions. Unfortunately, however, there were no proofs about convergence of the EM algorithm used.

Shigemasu (1986) used a natural conjugate prior Bayesian approach to the factor analysis model and found implicit numerical solutions for the factor loading matrix and specific variances.

Akaike (1987) suggested that the AIC criterion could be used to select the appropriate number of factors to use in a Bayesian model (see also Press, 1982). He was motivated by the desire to deal with the problem of frequent occurrence of improper solutions in maximum likelihood factor analysis caused by overparameterization of the model. The introduction of prior information in our model directly addresses this issue. By minimizing the AIC in this, or other factor analysis models, results can be used to test hypotheses about the appropriate number of factors.

In contrast to the earlier research on Bayesian factor analysis which focused upon point estimation, in this paper we also develop methods for obtaining large sample interval estimators of factor scores, factor loadings, and specific variances. Consequently, standard Bayesian hypothesis testing methods (see, e.g., Press, 1982; 1989) can be used for testing hypotheses about all of the fundamental quantities (apart from the number of factors) in the model. Because we develop exact (large sample) posterior distributions for these quantities, level curves, or contours, of the posterior distributions can be studied for sensitivity around the point estimators by examining the steepness and shape of the level curves. (Earlier research in which only point estimators were proposed has not suggested simple methods for studying estimator sensitivity.) Finally, our development yields *explicit* analytical results for the distributions of the quantities of interest (as well as some general implicit solutions), whereas most earlier work focused only on *implicit* numerical solutions of matrix equations.

The paper is constructed so that the basic model we are adopting is set out in Section 2. The procedures for estimating factor scores and loadings, and disturbance variances and covariances, are discussed in Sections 3, 4, and 5, respectively. The paper concludes in Section 6 with a numerical illustration of the procedures.

2 Model

In this section we develop the basic factor analysis model. We first define the likelihood function. Then we introduce prior distributions on the parameters and calculate the joint posterior density of the parameters. Finally we find the marginal posterior densities for the parameters.

Likelihood function

Define p-variate observation vectors, $(x_1, \ldots, x_N) \equiv X'$ on N subjects. The means are assumed to have been subtracted out, so that $E(X') = 0$. The prime denotes transposed matrix. The traditional factor analysis model is

$$\underset{(p \times 1)}{x_j} = \underset{(p \times m)}{\Lambda} \; \underset{(m \times 1)}{f_j} + \underset{(p \times 1)}{\varepsilon_j}, \qquad m < p, \tag{2.1}$$

for $j = 1, \ldots, N$, where Λ denotes a matrix of constants called the factor loading matrix; f_j denotes the factor score vector for subject j; $F' \equiv (f_1, \ldots, f_N)$. The ε_j's are assumed to be mutually uncorrelated and normally distributed as $N(0, \Psi)$, for Ψ a symmetric positive definite matrix, i.e., $\Psi > 0$. Note that Ψ is not assumed to be diagonal (but note from (2.5b) that $E(\Psi)$ is diagonal).

We assume that (Λ, F, Ψ) are unobserved and fixed quantities, and we assume that we can write the probability law of x_j as

$$\mathcal{L}(x_j|\Lambda, f_j, \Psi) = N(\Lambda f_j, \Psi), \tag{2.2}$$

where $\mathcal{L}(\cdot)$ denotes probability law. Equivalently, if "$\propto$" denotes proportionality, the likelihood for (Λ, F, Ψ) is

$$p(X|\Lambda, F, \Psi) \propto |\Psi|^{-N/2} \exp((-1/2)tr\Psi^{-1}(X - F\Lambda')'(X - F\Lambda')). \tag{2.3}$$

We will use $p(\cdot)$ generically to denote "density"; the p's will be distinguished by their arguments. This should not cause confusion. The proportionality constant in (2.3) is numerical, depending only upon (p, N) and not upon (Λ, F, Ψ).

PRIORS

We use a generalized natural conjugate family (see Press (1982)) of prior distributions for (Λ, Ψ). We take as prior density for the unobservables (to represent our state of uncertainty)

$$p(\Lambda, F, \Psi) = p(\Lambda \mid \Psi)p(\Psi)p(F), \tag{2.4}$$

where

$$p(\Lambda \mid \Psi) \propto |\Psi|^{-m/2} \exp\left\{(-1/2)\,\mathrm{tr}(\Lambda - \Lambda_0)H(\Lambda - \Lambda_0)'\Psi^{-1}\right\}, \tag{2.5a}$$

$$p(\Psi) \propto |\Psi|^{-\nu/2} \exp\left\{(-1/2)\,\mathrm{tr}\,\Psi^{-1}B\right\}, \tag{2.5b}$$

with B a diagonal matrix and $H > 0$. (Choices for the prior density $p(F)$ of F will be discussed in Section 3.) Thus, Ψ^{-1} follows a Wishart distribution, (ν, B) are hyperparameters to be assessed; Λ conditional on Ψ has elements which are jointly normally distributed, and (Λ_0, H) are hyperparameters to be assessed. Note that $E(\Psi \mid B)$ is diagonal, to represent traditional views of the factor model containing "common" and "specific" factors. Also note that if $\Lambda \equiv (\lambda_1, \ldots, \lambda_m)$, $\lambda \equiv \mathrm{vec}(\Lambda) = (\lambda_1', \ldots, \lambda_m')'$, then $\mathrm{var}(\lambda|\Psi) = \Psi \otimes H^{-1}$, $\mathrm{var}(\lambda) = (E\Psi) \otimes H^{-1}$, and $\mathrm{cov}[(\lambda_i, \lambda_j) \mid \Psi] = \Psi_{ij}H^{-1}$. Moreover, we will often take $H = n_0 I$, for some preassigned scalar n_0. These interpretations of the hyperparameters will simplify assessment.

JOINT POSTERIOR

Combining (2.3)–(2.5), the joint posterior density of the parameters becomes

$$p(\Lambda, F, \Psi \mid X) \propto p(F)|\Psi|^{-(N+m+\nu)/2} \exp\left\{(-1/2)\,\mathrm{tr}[\Psi^{-1}G]\right\}, \tag{2.6}$$

where $G \equiv (X - F\Lambda')'(X - F\Lambda') + (\Lambda - \Lambda_0)H(\Lambda - \Lambda_0)' + B$.

Marginal posteriors

Integrating with respect to Ψ, and using properties of the Inverted Wishart density, gives the marginal posterior density of (Λ, F):

$$p(\Lambda, F \mid X) \propto p(F)|G|^{-(N+m+\nu-p-1)/2}. \tag{2.7}$$

We next want to integrate (2.7) with respect to Λ, to obtain the marginal posterior density of F. We accomplish this by factoring G into a form which makes it transparent that in terms of Λ, the density is proportional to a matrix T-density. Thus, completing the square in Λ in the G function defined in (2.6), (2.7) may be rewritten as

$$p(\Lambda, F \mid X) \propto \frac{p(F)}{|R_F + (\Lambda - \Lambda_F)Q_F(\Lambda - \Lambda_F)'|^{\gamma/2}}, \tag{2.8}$$

where

$$\begin{aligned}
Q_F &\equiv H + F'F, \\
R_F &\equiv X'X + B + \Lambda_0 H \Lambda_0' - (X'F + \Lambda_0 H)Q_F^{-1}(X'F + \Lambda_0 H)', && (2.9) \\
\Lambda_F &\equiv (X'F + \Lambda_0 H)(H + F'F)^{-1}, && (2.10) \\
\gamma &\equiv N + m + \nu - p - 1. && (2.11)
\end{aligned}$$

(2.8) is readily integrated with respect to Λ (by using the normalizing constant of the matrix T-distribution) to give the marginal posterior density of F,

$$p(F \mid X) \propto \frac{p(F)}{|R_F|^{(\gamma-m)/2}|Q_F|^{p/2}}. \tag{2.12}$$

After some algebra, the marginal posterior density of F in (2.12) may be rewritten in the form

$$p(F \mid X) \propto \frac{p(F)|H + F'F|^{(\gamma-m-p)/2}}{\left|A + (F - \hat{F})'(I_N - XW^{-1}X')(F - \hat{F})\right|^{(\gamma-m)/2}}, \tag{2.13}$$

where

$$\begin{aligned}
\hat{F} &\equiv (I_N - XW^{-1}X')^{-1}XW^{-1}\Lambda_0 H \\
&= \left(I_N - X(X'X - W)^{-1}X'\right)XW^{-1}\Lambda_0 H, && (2.14) \\
W &\equiv X'X + B + \Lambda_0 H \Lambda_0', && (2.15) \\
A &\equiv H - H'\Lambda_0' W^{-1}\Lambda_0 H \\
&\quad - (H'\Lambda_0' W^{-1}X')(I_N - XW^{-1}X')(XW^{-1}\Lambda_0 H). && (2.16)
\end{aligned}$$

Note 1. In (2.14) the second representation of $\hat{F}$ is more convenient for numerical computation than the first, because we need only invert a matrix of order p, instead of one of order N.

Note 2. In (2.14), the quantity $N^{-1}(X'X)$ is the sample covariance matrix of the observed data (since the data are assumed to have mean zero). If the data are scaled to have a variance of unity, $N^{-1}(X'X)$ denotes the data correlation matrix.

Note 3. In (2.16), $H = H'$, but we have left H' to preserve the symmetry of the representation.

3 Estimation of Factor Scores

Now examine (2.13). There are several cases of immediate interest. We take factor scores of subsets to be independent, a priori, so we can think of $p(F)$ as $p(f_1)p(f_2)\dots p(f_N)$.

Historical data assessment of F

Suppose that, on the basis of historical data that is very similar to the current data set, we can assess $p(F)$. We can then evaluate $p(F \mid X)$ numerically from (2.13) to construct point estimators, and we can make interval estimates from the cdf of $p(F \mid X)$.

Vague prior estimation of F

Suppose instead that we are uninformed about F, a priori, and we accordingly adopt the vague prior

$$p(F) \propto \text{constant}. \tag{3.1}$$

Then (2.13) becomes

$$p(F \mid X) \propto \frac{|H + F'F|^{(\gamma-m-p)/2}}{\left|A + (F - \hat{F})'(I_N - XW^{-1}X')(F - \hat{F})\right|^{(\gamma-m)/2}}. \tag{3.2}$$

Again, interval estimates of F can be made numerically from the cdf of $p(F \mid X)$, and point estimators can also be obtained numerically from (3.2). Such numerical evaluations are treated in Press and Davis (1987).

Large sample estimation of F

We note that

$$\frac{F'F}{N} = \frac{1}{N}\sum_{1}^{N} f_j f_j'.$$

If we assume (without loss of generality) that $E(f_j) = 0$, $\text{var}(f_j) = I$, then for large N, by the law of large numbers,

$$\frac{F'F}{N} \approx I.$$

Thus, for large N, $|H+F'F| \approx |H+NI|$, a term which can be incorporated into the proportionality constant in (2.13), because it no longer depends on F. (2.13) may now be rewritten, for large N, as

$$p(F \mid X) \propto \frac{p(F)}{\left|A + (F - \hat{F})'(I_N - XW^{-1}X')(F - \hat{F})\right|^{(\gamma - m)/2}},$$

where $\hat{F}$ is defined by (2.14).

Suppose $p(F) \propto$ constant. Then, $F \mid X$ follows a matrix T-distribution with density

$$p(F \mid X) \propto \left|A + (F - \hat{F})'(I_N - XW^{-1}X')(F - \hat{F})\right|^{(\gamma - m)/2}. \tag{3.3}$$

Alternately, suppose $\mathcal{L}(f_j) = N(0, 1)$, and the f_j's are mutually independent. Then,

$$p(F) \propto \exp\left(-\frac{1}{2}\,\text{tr}\, F'F\right).$$

For large N, since $F'F \approx NI$, $p(F)$ can be incorporated into the proportionality constant in (2.13) to yield (3.3). The same argument applies to any prior density for F which depends upon $F'F$.

In summary, we conclude that for large N, and for a wide variety of important priors for F (a vague prior, or for any prior which depends on F only through $F'F$), the marginal posterior density of F, given the observed data vectors, is approximately matrix T, as given in (3.3), centered at $\hat{F}$. In particular, $E(F \mid X) \approx \hat{F}$, for large N.

Large sample estimation of f_j

Since $(F \mid X)$ is approximately distributed as matrix T, $(f_j \mid X)$ is distributed as multivariate t (see, e.g., Theorem 6.2.4, in Press, 1982, p. 140). In particular, the marginal posterior density for the factor score vector of subject N is given by

$$p(f_N \mid X) \propto \left(\frac{1}{P_{22\cdot 1}} + (f_N - \hat{f}_N)'A^{-1}(f_N - \hat{f}_N)\right)^{-(\gamma - m - N + 1)/2}, \tag{3.4}$$

where $\hat{f}_N$ is the N-th row of $\hat{F}$ in (2.14), and $P_{22\cdot 1} = P_{22} - P_{21}P_{11}^{-1}P_{12}$ is obtained from

$$\underset{(N\times N)}{P} \equiv I_N - XW^{-1}X' \equiv \begin{pmatrix} P_{11} & P_{12} \\ P_{21} & P_{22} \end{pmatrix}, \qquad P_{11} : (N-1) \times (N-1).$$

By reindexing the subjects, (3.4) gives the posterior density of the factor score vector for any of the N subjects. (3.4) can readily be placed into the canonical form

$$p(f_N \mid X) \propto \left(1 + (f_N - \hat{f}_N)' \left(\frac{A}{P_{22\cdot 1}}\right)^{-1} (f_N - \hat{f}_N)\right)^{-(\delta+m)/2}, \quad (3.5)$$

where $\delta \equiv m + \nu - p - 2m$.

Large sample estimation of elements of f_j.

Now suppose we wish to make posterior probability statements about a particular element of $f_N \equiv (f_{kN})$, $k = 1, \ldots, m$, say f_{1N}. We use the posterior density of a Student t-variate obtained as the marginal of the multivariate t-density in (3.5) (see e.g., Press (1982), p. 137). It is given by

$$p(f_{1N} \mid X) \propto \left(1 + \left(\frac{f_{1N} - \hat{f}_{1N}}{\sigma_1}\right)^2\right)^{-(\delta+1)/2}, \quad (3.6)$$

where σ_1^2 is the $(1,1)$ element of

$$\frac{A}{P_{22\cdot 1}} \equiv \begin{pmatrix} \sigma_1^2 & \Sigma_{12} \\ \Sigma_{21} & \Sigma_{22} \end{pmatrix}.$$

$\hat{f}_{1N}$ is of course the $(1, N)$ element of $\hat{F}'$. From (3.6) we can make posterior probability statements about any factor score for any subject; i.e., we can obtain credibility (confidence) intervals for any factor score.

4 Estimation of the Factor Loading Matrix

We now return to the joint posterior density of (Λ, F), given in (2.8). One method of estimating Λ would be to integrate F out of (2.8) to obtain the marginal posterior density of Λ. Then, some measure of location of the distribution could be used as a point estimator of Λ. Unfortunately, while the integration can be carried out, the resulting marginal density is extremely complicated, and it does not seem possible to obtain a mean or mode of the distribution for any realistic prior densities for F, except numerically. The result is

$$P(\Lambda \mid F) \propto |P_\Lambda|^{-\gamma/2} |\Lambda' P_\Lambda^{-1} \Lambda|^{-N/2} |Z|^{-(\gamma-m)/2},$$

where

$$\underset{(p\times p)}{P_\Lambda} \equiv B + (\Lambda - \Lambda_0) H (\Lambda - \Lambda_0)',$$

and

$$Z \equiv I_N + XP_\Lambda^{-1}X' - (XP_\Lambda^{-1}\Lambda)(\Lambda' P_\Lambda^{-1}\Lambda)^{-1}(\Lambda' P_\Lambda^{-1}\Lambda).$$

Since this distribution is so complicated, we will alternatively estimate Λ for given $F = \hat{F}$. First note from (2.8) that

$$p(\Lambda \mid F, X) \propto |R_F + (\Lambda - \Lambda_F)Q_F(\Lambda - \Lambda_F)'|^{-\gamma/2}. \tag{4.1}$$

That is, the conditional distribution of Λ for pre-specified F is matrix T. Our point estimator of Λ is $E(\Lambda \mid \hat{F}, X)$, or

$$\hat{\Lambda} \equiv \Lambda_{\hat{F}} = (X'\hat{F} + \Lambda_0 H)(H + \hat{F}'\hat{F})^{-1}. \tag{4.2}$$

Any scalar element of $\hat{\Lambda}$, conditional on $(\hat{F}, X)$, follows a general Student t-distribution, analogous to the general univariate marginal Student t-density in (3.6), which corresponds to the matrix T-density in (3.3).

We note also that $\hat{\Lambda}$ in (4.2) is both a mean and a modal estimator of the joint distribution of $(\Lambda, F \mid X)$ under a vague prior for F (see (2.8) and (2.10)). This follows from the unimodality and the symmetry of the density in (2.8). Thus, in this case $(\hat{F}, \hat{\Lambda})$ is a joint modal estimator of (F, Λ).

5 Estimation of the Disturbance Covariance Matrix

The disturbance covariance matrix, Ψ, is estimated conditional upon $(\Lambda, F) = (\hat{\Lambda}, \hat{F})$. The joint posterior density of $(\Lambda, F, \Psi \mid X)$ is given in (2.6). The conditional density of $(\Psi \mid \Lambda, F, X)$ is obtained by dividing (2.6) by (2.7) and setting $G = \hat{G}$ ($\hat{G}$ depends only upon the data). The result is

$$p(\Psi \mid \hat{\Lambda}, \hat{F}, X) \propto \frac{\exp\left((-1/2)\operatorname{tr}\Psi^{-1}\hat{G}\right)}{|\Psi|^{(N+mp+\nu)/2}} \tag{5.1}$$

where

$$\hat{G} = (X - \hat{F}\hat{\Lambda}')'(X - \hat{F}\hat{\Lambda}') + (\hat{\Lambda} - \Lambda_0)H(\hat{\Lambda} - \Lambda_0)' + B. \tag{5.2}$$

That is, the posterior conditional distribution of Ψ given $(\hat{\Lambda}, \hat{F}, X)$ is inverted Wishart. A point estimator of Ψ is given by $\hat{\Psi} = E(\Psi \mid \hat{\Lambda}, \hat{F}, X)$. Equation (5.2.4), page 119, in Press (1982) gives

$$\hat{\Psi} = \frac{\hat{G}}{N + m + \nu - 2p - 2}, \tag{5.3}$$

with $\hat{G}$ given in (5.2).

6 Example

We have extracted some data from an illustrative example in Kendall (1980), and have analyzed this data from a Bayesian viewpoint using our model. There are 48 applicants for a certain job, and they have been scored on 15 variables regarding their acceptability. They are:

(1) Form of letter of application
(2) Appearance
(3) Academic ability
(4) Likeability
(5) Self-confidence
(6) Lucidity
(7) Honesty
(8) Salesmanship
(9) Experience
(10) Drive
(11) Ambition
(12) Grasp
(13) Potential
(14) Keenness to join
(15) Suitabilty

The raw scores of the applicants on these 15 variables, measured on the same scale, are presented in Table 1. The question is, is there an underlying subset of factors that explain the variation observed in the scores? If so, then each applicant could be compared more easily. The correlation matrix for the 15 variables is given in Table 2. (Note: we assume the sample size of 48 is large enough to estimate the mean well enough for it to be ignored after subtracting it out.)

TABLE 1. Raw scores of 48 applicants scaled on 15 variables

Person	1	2	3	4	5	6	7	8	9	10	11	12	13	14	15
1	6	7	2	5	8	7	8	8	3	8	9	7	5	7	10
2	9	10	5	8	10	9	9	10	5	9	9	8	8	8	10
3	7	8	3	6	9	8	9	7	4	9	9	8	6	8	10
4	5	6	8	5	6	5	9	2	8	4	5	8	7	6	5
5	6	8	8	8	4	5	9	3	8	5	5	8	8	7	7
6	7	7	7	6	8	7	10	5	9	6	5	8	6	6	6
7	9	9	8	8	8	8	8	8	10	8	10	8	9	8	10
8	9	9	9	8	9	9	8	8	10	9	10	9	9	9	10
9	9	9	7	8	8	8	8	5	9	8	10	9	9	9	10
10	4	7	10	2	10	10	7	10	3	10	10	10	9	3	10
11	4	7	10	0	10	8	3	9	5	9	10	8	10	2	5
12	4	7	10	4	10	10	7	8	2	8	8	10	10	3	7
13	6	9	8	10	5	4	9	4	4	4	5	4	7	6	8
14	8	9	8	9	6	3	8	2	5	2	6	6	7	5	6
15	4	8	8	7	5	4	10	2	7	5	3	6	6	4	6
16	6	9	6	7	8	9	8	9	8	8	7	6	8	6	10
17	8	7	7	7	9	5	8	6	6	7	8	6	6	7	8
18	6	8	8	4	8	8	6	4	3	3	6	7	2	6	4
19	6	7	8	4	7	8	5	4	4	2	6	8	3	5	4
20	4	8	7	8	8	9	10	5	2	6	7	9	8	8	9
21	3	8	6	8	8	8	10	5	3	6	7	8	8	5	8
22	9	8	7	8	9	10	10	10	3	10	8	10	8	10	8
23	7	10	7	9	9	9	10	10	3	9	9	10	9	10	8
24	9	8	7	10	8	10	10	10	2	9	7	9	9	10	8
25	6	9	7	7	4	5	9	3	2	4	4	4	4	5	4
26	7	8	7	8	5	4	8	2	3	4	5	6	5	5	6
27	2	10	7	9	8	9	10	5	3	5	6	7	6	4	5
28	6	3	5	3	5	3	5	0	0	3	3	0	0	5	0
29	4	3	4	3	3	0	0	0	0	4	4	0	0	5	0
30	4	6	5	6	9	4	10	3	1	3	3	2	2	7	3
31	5	5	4	7	8	4	10	3	2	5	5	3	4	8	3
32	3	3	5	7	7	9	10	3	2	5	3	7	5	5	2
33	2	3	5	7	7	9	10	3	2	2	3	6	4	5	2
34	3	4	6	4	3	3	8	1	1	3	3	3	2	5	2
35	6	7	4	3	3	0	9	0	1	0	2	3	1	5	3
36	9	8	5	5	6	6	8	2	2	2	4	5	6	6	3
37	4	9	6	4	10	8	8	9	1	3	9	7	5	3	2
38	4	9	6	6	9	9	7	9	1	2	10	8	5	5	2
39	10	6	9	10	9	10	10	10	10	10	8	10	10	10	10
40	10	6	9	10	9	10	10	10	10	10	10	10	10	10	10
41	10	7	8	0	2	1	2	0	10	2	0	3	0	0	10
42	10	3	8	0	2	1	2	0	10	2	0	3	0	0	10
43	3	4	9	8	2	4	5	3	6	2	1	3	3	3	8
44	7	7	7	6	9	8	8	6	8	8	10	8	8	6	5
45	9	6	10	9	7	7	10	2	1	5	5	7	8	4	5
46	9	8	10	10	7	9	10	3	1	5	7	9	9	4	4
47	0	7	10	3	5	0	10	0	0	2	2	0	0	0	0
48	0	6	10	1	5	0	10	0	0	2	2	0	0	0	0

TABLE 2. Correlation matrix of variables 1 through 15

	1	2	3	4	5	6	7	8	9	10	11	12	13	14	15
1	1.00	.24	.04	.31	.09	.23	−.11	.27	.55	.35	.28	.34	.37	.47	.59
2		1.00	.12	.38	.43	.37	.35	.48	.14	.34	.55	.51	.51	.28	.38
3			1.00	.00	.00	.08	−.03	.05	.27	.09	.04	.20	.29	−.32	.14
4				1.00	.30	.48	.65	.35	.14	.39	.35	.50	.61	.69	.33
5					1.00	.81	.41	.82	.02	.70	.84	.72	.67	.48	.25
6						1.00	.36	.83	.15	.70	.76	.88	.78	.53	.42
7							1.00	.23	−.16	.28	.21	.39	.42	.45	.00
8								1.00	.23	.81	.86	.77	.73	.55	.55
9									1.00	.34	.20	.30	.35	.21	.69
10										1.00	.78	.71	.79	.61	.62
11											1.00	.78	.77	.55	.43
12												1.00	.88	.55	.53
13													1.00	.54	.57
14														1.00	.40
15															1.00

Now we postulate a model with 4 factors. This choice is based upon our having carried out a principal components analysis and our having found that 4 factors accounted for 81.5% of the variance. This is therefore our first guess, a conclusion that might be modified if we were to do hypothesis testing to see how well a 4-factor model fit the data. Based upon underlying theory we constructed the prior factor loading matrix

$$\Lambda_0 = \begin{array}{r} 1\\2\\3\\4\\5\\6\\7\\8\\9\\10\\11\\12\\13\\14\\15 \end{array} \begin{pmatrix} 0 & 0 & .7 & 0\\ 0 & 0 & 0 & 0\\ 0 & .7 & 0 & 0\\ 0 & 0 & 0 & .7\\ .7 & 0 & 0 & 0\\ .7 & 0 & 0 & 0\\ 0 & 0 & 0 & .7\\ .7 & 0 & 0 & 0\\ 0 & 0 & .7 & 0\\ .7 & 0 & 0 & 0\\ .7 & 0 & 0 & 0\\ .7 & 0 & 0 & 0\\ .7 & 0 & 0 & 0\\ 0 & 0 & 0 & 0\\ 0 & .7 & 0 & 0 \end{pmatrix}.$$

The hyperparameter H was assessed as $H = 10I_{15}$. The prior distribution for Ψ was assessed with $B = 0.2I_{15}$, and $\nu = 33$. Note that when our observational data is augmented by proper prior information, as in this example, the identification-of-parameters problem of classical factor analysis disappears.

The factor scores, factor loadings, and disturbance variances and covariances may now be estimated from (2.14), (4.2), and (5.3), respectively. Results are given in Tables 3, 4, and 5.

Note that since we used standardized scores, the elements in Table 4 may be interpreted as correlations. It may be noted from Table 5 that most off-diagonal elements of the estimated disturbance matrix $\hat{\Psi}$ are very small relative to the diagonal elements (the variances). That is, $\hat{\Psi}$ is approximately diagonal. Tables 3, 4, and 5 give the Bayesian point estimates for (F, Λ, Ψ). We next obtain two-tailed 95% credibility intervals for the 48th subject's factor scores, and for the last (15th) row of the factor loading matrix.

The factor scores for subject 48 are given in the last row of the matrix in Table 3 as

$$(-2.155,\ 2.031,\ -2.527,\ -0.749).$$

Now calculate two-tailed credibility intervals at the 95% level from (3.6) and find the intervals

$$\begin{aligned}&[-2.5748,\ -1.7352],\\&[0.9393,\ 3.1227],\\&[-3.1657,\ -1.8883],\\&[-1.5287,\ 0.0307].\end{aligned}$$

The factor loadings for row 15 of the factor loading matrix are obtained from Table 4 as

$$(0.128,\ -0.014,\ 0.677,\ 0.011).$$

Now calculate 95% two-tailed credibility intervals from the marginals of (4.1), just as we obtained the result in (3.6) from (3.3). Results for the last row factor loadings are

$$\begin{aligned}&[0.0629,\ 0.1931],\\&[-0.0625,\ 0.0345],\\&[0.5997,\ 0.7543],\\&[-0.0560,\ 0.0780].\end{aligned}$$

Hypotheses about the elements of (F, Λ, Ψ) may be tested using the associated marginal posterior densities. These are quite simple, being Student t, Student t for given F, and Inverted Wishart, respectively. For example, note that the confidence intervals for the second and fourth factor loadings corresponding to the last row of Table 4 both include the origin. A commonly used Bayesian hypothesis testing procedure suggests that we should therefore conclude that we cannot reject the hypotheses that these two factor loadings are zero.

TABLE 3. Bayes estimates of factor scores

1	0.728	−3.541	0.404	−0.301
2	1.475	−1.455	1.226	0.735
3	1.019	−2.855	0.728	0.231
4	−0.286	0.645	0.226	−0.021
5	−0.390	0.645	0.691	0.735
6	0.264	−0.055	0.869	0.511
7	1.187	0.645	1.940	0.455
8	1.475	1.331	1.940	0.455
9	0.874	−0.055	1.799	0.455
10	1.881	2.031	0.047	−1.337
11	1.550	2.031	−0.381	−2.954
12	1.550	2.031	−0.526	−0.833
13	−0.588	0.645	0.263	1.239
14	−0.619	0.645	0.474	0.707
15	−0.706	0.645	0.047	0.763
16	0.903	−0.755	1.123	0.203
17	0.424	−0.055	0.907	0.203
18	−0.193	0.645	−0.460	−1.106
19	−0.215	0.645	−0.315	−1.386
20	0.730	−0.055	−0.240	1.015
21	0.597	−0.755	−0.413	1.015
22	1.593	0.055	0.648	1.015
23	1.591	−0.055	0.296	1.267
24	1.363	−0.055	0.507	1.519
25	−0.931	−0.055	−0.601	0.483
26	−0.700	−0.055	0.005	0.455
27	0.327	−0.055	−1.024	1.267
28	−1.822	−1.455	−1.461	−1.638
29	−2.045	−2.155	−1.818	−3.038
30	−0.976	−1.455	−1.250	0.511
31	−0.586	−2.155	−0.925	0.763
32	−0.150	−1.455	−1.419	0.763
33	−0.491	−1.455	−1.597	0.763
34	−1.601	−0.755	−1.564	−0.553
35	−2.195	−2.155	−0.893	−0.525
36	−0.698	−1.455	−0.216	−0.301
37	0.676	−0.755	−1.391	−0.553
38	0.720	−0.755	−1.391	−0.329
39	1.723	1.331	2.119	1.519
40	1.861	1.331	2.119	1.519
41	−2.281	0.645	2.119	−3.234
42	−2.707	0.645	2.119	−3.794
43	−1.643	1.331	0.019	−0.378
44	1.088	−0.055	0.583	−0.049
45	−0.006	2.031	−0.070	1.267
46	0.523	2.031	−0.216	1.519
47	−2.155	2.031	−2.528	−0.245
48	−2.155	2.031	−2.527	−0.749

TABLE 4. Bayes estimates of factor loadings

1	−0.045	−0.065	0.711	0.029
2	0.240	0.047	0.094	0.175
3	0.000	0.726	0.000	0.000
4	−0.010	−0.010	0.152	0.702
5	0.775	−0.050	−0.192	−0.028
6	0.722	−0.013	−0.061	0.031
7	0.012	0.010	−0.152	0.722
8	0.754	−0.049	0.017	−0.090
9	−0.082	0.080	0.737	−0.040
10	0.652	−0.021	0.117	−0.021
11	0.771	−0.046	−0.027	−0.100
12	0.658	0.058	0.052	0.067
13	0.592	0.120	0.095	0.142
14	0.244	−0.264	0.221	0.311
15	0.128	−0.014	0.677	0.011

TABLE 5. Bayes estimates of the disturbance covariance matrix

.1352	.0039	−.0002	.0170	.0070	.0058	−.0168	.0131	−.0718	−.0172	.0138	.0036	−.0001	.0400	−.0616
	.3168	.0001	−.0019	.0092	−.0382	.0123	.0174	−.0281	−.0281	.0534	.0078	−.0009	−.0297	.0245
		.0019	.0000	−.0001	.0000	.0000	−.0001	.0002	−.0001	−.0001	.0002	.0003	−.0007	.0000
			.0640	−.0266	.0124	−.0622	.0009	−.0119	−.0075	.0087	−.0005	.0126	.0241	−.0047
				.0722	−.0065	.0265	−.0004	.0066	−.0152	.0069	−.0255	−.0294	−.0091	−.0141
					.0730	−.0123	−.0044	−.0024	−.0377	−.0347	.0248	−.0126	−.0132	−.0036
						.0641	−.0012	.0118	.0075	−.0090	.0007	−.0122	−.0232	.0047
							.0641	−.0052	.0001	−.0015	−.0267	−.0258	.0035	.0184
								.0967	−.0016	−.0015	.0023	.0015	−.0125	−.0230
									.0990	−.0089	−.0372	.0016	.0252	.0191
										.0623	−.0158	−.0064	.0107	−.0124
											.0692	.0129	−.0056	−.0058
												.0612	−.0109	−.0012
													.1501	−.0268
														.0864

Acknowledgments: The authors would like to acknowledge the assistance of Masanori Ickikawa in computing the numerical results in Section 6. We are also grateful to an anonymous referee for helpful comments on related research on which we report in Section 1.

REFERENCES

Akaike, H. (1987). "Factor analysis and AIC." *Psychometrika*, Vol. 52, No. 3, 317–32.

Dempster, A.P., Laird, N.M., Rubin, D.B. (1977). "Maximum likelihood from incomplete data via the EM algorithm." *Jour Roy Statist Soc (B)*, 39, 1–38.

Euverman, T.J. and Vermulst, A.A. (1983). *Bayesian Factor Analysis.* Department of Social Sciences, State University of Groningen, the Netherlands, Nov 3, 1983.

Guttman, L. (1953). "Image theory for the structure of quantitative variates." *Psychometrika*, vol 18, no 4, 277–96.

Kaufman, G.M. and Press, S.J. (1973a). "Bayesian factor analysis." Report No 73t-25, *Bull. Inst. of Math Statist.*, Vol 2, No 2, Issue No 7, March, 1973.

Kaufman, G.M. and Press, S.J. (1973b). "Bayesian factor analysis." Report No. 7322, Center for Mathematical Studies in Business and Economics, University of Chicago, April, 1973.

Kaufman, G.M. and Press, S.J. (1976). "Bayesian factor analysis." Working Paper no 413, Department of Commerce and Business Administration, University of British Columbia, Vancouver, B.C. Canada, September, 1983.

Kendall, M. (1980). *Multivariate Analysis.* Second edition, Charles Griffin Pub, 34.

Lee, S.Y. (1981). "A Bayesian approach to confirmatory factor analysis." *Psychometrika*, Vol 46, no 2, June, 1981, 153–59.

Martin, J.K. and McDonald, R.P. (1975). "Bayesian estimation in unrestricted factor analysis: A treatment for Heywood cases." *Psychometrika*, Vol. 40, No. 4, 505–17.

Mayekawa, S. (1985). "Bayesian factor analysis." ONR Tech Rept No 85-3, School of Education, The University of Iowa, Iowa City, Iowa.

Press, S. James (1972). *Applied Multivariate Analysis.* New York: Holt, Rinehart and Winston, Inc.

Press, S. James (1982). *Applied Multivariate Analysis: Using Bayesian and Frequentist Methods of Inference.* Robert E. Krieger Publishing Co., Malabar, Florida.

Press, S. James (1989). *Bayesian Statistics: Principles, Models, and Applications.* New York, John Wiley and Sons, Inc.

Press, S. James and Davis, W. (1987). "Asymptotics for the ratio of multiple t-densities." in *Contributions to the Theory and Application of Statistics*, ed. Alan Gelfand, New York, Academic Press, 155–77.

Shigemasu, K. (1986). "A Bayesian estimation procedure with informative prior in factor analysis model." *Tokyo Institute of Technology*, Jinmon-Ronso, No 11, 67–77.

Wong, G. (1980). "A Bayesian approach to factor analysis." Program in Statistical Technical Report, No 80-9, Educational Testing Service, Princeton, N.J., August, 1980.

16

Computational Aspects of Association for Bivariate Discrete Distributions

Allan R. Sampson[1]
Lyn R. Whitaker[2]

ABSTRACT For bivariate discrete probability distributions, P, on the $M \times N$ lattice, various aspects for checking association (Esary, Proschan and Walkup (1967)) are considered. A new algorithm is given for verifying whether or not P is associated. The efficiency of this algorithm is obtained and compared to the efficiency of a simple algorithm based on the definition of association. When $M = N = 5$, for example, the new algorithm requires less than 3% of the computations required for the simple algorithm.
In obtaining these results a new set function Q is constructed from P, on all upper sets in the lattice. In order to construct the algorithm to check association, we define a computationally important set of extreme points and consider related combinatorics.

1 Introduction

The conceptual ideas developed for modern notions of positive or monotone dependence have proved very useful in reliability theory, simultaneous inference, and, what may be generally termed, nonparametric concepts of multivariate dependence. While these notions have been considered in the context of discrete multivariate random variables, their application to contingency table analyses have been limited. In part, this is due to the lack of suitable algorithms to check for the presence of these notions. The purpose of this paper is to obtain some new results concerning the notion of association and based upon these results, to develop a computationally efficient algorithm to show whether or not a discrete bivariate distribution is associated.

Of the many notions of positive dependence, one of the most useful is

[1]Department of Mathematics and Statistics, University of Pittsburgh, Pittsburgh, PA 15260.
[2]Operations Research, U.S. Naval Postgraduate School, Monteray, CA 93943

association (Esary, Proschan and Walkup (1967)), where $X_1, \ldots, X_p$ are said to be associated random variables if

$$\operatorname{cov}[f(X_1, \ldots, X_p), g(X_1, \ldots, X_p)] \geq 0 \tag{1.1}$$

for all $f : \Re^p \to \Re^1$ and $g : \Re^p \to \Re^1$ which are nondecreasing in each argument and have finite variances. Let $U \subseteq \Re^p$ be an upper set if $\boldsymbol{s} \in U$ and $\boldsymbol{s} \leq \boldsymbol{t}$ imply $\boldsymbol{t} \in U$ where $\boldsymbol{s} \leq \boldsymbol{t}$ is componentwise ordering $s_i \leq t_i$ $i = 1, \ldots, p$. Then an equivalent formulation of (1.1) is

$$\begin{aligned}\text{Prob}[(X_1, \ldots, X_p) \in U \cap V] \geq {} & \text{Prob}[(X_1, \ldots, X_p) \in U] \\ & \times \text{Prob}[(X_1, \ldots, X_p) \in V]\end{aligned} \tag{1.2}$$

for all upper sets $U, V \subseteq \Re^p$.

Association has been studied extensively and applied in a variety of settings, e.g., Barlow and Proschan (1981), and Tong (1980) . More recently Schriever (1987) considers a partial ordering based upon association and examines the applications of association to observed bivariate data sets. When the specification of the random variables $X_1, \ldots, X_p$ is done structurally (for example, $X_1, \ldots, X_p$ are increasing functions of independent random variables) association can be established in a number of ways. If on the other hand, only the joint distribution of $X_1, \ldots, X_p$ is known, then the usual approaches are to either use theorems specific to a particular family of distributions, for example, Pitt (1982), or to use sufficient conditions, for example, TP_2 (see Barlow and Proschan (1981)).

However, for a distribution described numerically, such as observed contingency table probabilities, association can only be established by essentially checking all the conditions of equation (1.2). To build a straightforward algorithm to check all conditions in (1.2) for discrete random variables is computationally expensive. The number of comparisons for such an algorithm is on the order of κ^2, where κ is the number of all possible upper sets generated by componentwise partial ordering on the support set of $(X_1, \ldots, X_p)$. For example, if X_1, X_2 take values in an $M \times N$ lattice, the number of possible upper sets is $\binom{M+N}{M}$. For three dimensional lattices, the results are more complex. Sampson and Whitaker (1988) show, for example, that the number of upper sets in a $4 \times 4 \times 4$ lattice is 232,848.

The application of positive dependence notions to contingency tables is relatively recent. Agresti (1984) and Grove (1984) discuss some contingency table concepts in light of positive dependence. Nguyen and Sampson (1987) and Krishnaiah, Rao, and Subramanyan (1987) test for positive quadrant dependence (Lehmann (1966)) based on contingency table data. Sampson and Whitaker (1989) obtain maximum likelihood estimators of the underlying probabilities when one contingency table is assumed to be stochastically larger than a second contingency table. Gleser and Moore (1985) discuss applications of certain other allied notions of positive dependence to contingency tables.

In Section 2 we obtain some new theoretical results concerning association for arbitrary random variables. Specifically, we define a new set function Q, study its properties, and show how association can be established employing Q. We give an algorithm to compute Q in Section 3, where X_1, X_2 take values in an $M \times N$ lattice. The computational complexities of this algorithm are considered in Section 4.

2 An Alternate Characterization of Association

In this section we allow $X_1, \ldots, X_p$ to be arbitrary random variables. We denote the class of all upper sets in $\Re^p$ by $\mathcal{U}_{\Re^p}$, and abbreviate $\text{Prob}[X_1, \ldots, X_p \in A]$ as $P_{\mathbf{X}}(A)$ or when there is no ambiguity as $P(A)$, so that (1.2) can be expressed as

$$P(V \cap W) \geq P(V)P(W) \quad \forall\, V, W \in \mathcal{U}_{\Re^p}. \tag{2.1}$$

By noting that the class of upper sets is closed under intersections we can re-express (2.1) as

$$P(U) \geq \sup_{\{V,W \in \mathcal{U}_{\Re^p} : V \cap W = U\}} P(V)P(W) \quad \forall\, U \in \mathcal{U}_{\Re^p} \tag{2.2}$$

We now define the set function $Q(U)$ on $\mathcal{U}_{\Re^p}$ to be the right-hand side of (2.2) and thus we see that $X_1, \ldots, X_p$ are associated if, and only if

$$P(U) \geq Q(U) \quad \forall\, U \in \mathcal{U}_{\Re^p}. \tag{2.3}$$

We return later to the issue of whether Q actually defines a probability measure, but first we consider some properties of Q.

Observe that, whether or not $X_1, \ldots, X_p$ are associated, $Q(U) \geq P(U)P(\Re^p) = P(U)$. This fact and (2.3) lead immediately to the following slightly stronger representation of association.

Theorem 2.1 *A necessary and sufficient condition that the random variables $X_1, \ldots, X_p$ be associated is*

$$P(U) = Q(U) \quad \forall\, U \in \mathcal{U}_{\Re^p}. \tag{2.4}$$

Additional properties of the set function Q are summarized as follows.

Theorem 2.2 *(i) If $U, V \in \mathcal{U}_{\Re^p}$, $U \subseteq V$, then $Q(U) \leq Q(V)$.*

(ii) $Q(\Re^p) = 1$.

(iii) $Q(\emptyset) = 0$.

(iv) If $U = \{(x_1, \ldots, x_p) : x_k \geq x_0\}$, $k = 1, \ldots, p$ then $Q(U) = P(U)$.

Proof (i) Let $U_1, U_2 \in \mathcal{U}_{\Re^p}$ satisfy $U_1 \cap U_2 = U \subseteq V$. Define $V_1 = U_1 \cup V$, $V_2 = U_2 \cup V$ so that $V_1 \cap V_2 = V$, and $U_1 \subseteq V_1$, $U_2 \subseteq V_2$. The class of upper sets is closed under union, therefore $V_1, V_2 \in \mathcal{U}_{\Re^p}$. Thus

$$\begin{aligned} Q(U) &= \sup_{\{U_1, U_2 \in \mathcal{U}_{\Re^p} : U_1 \cap U_2 = U\}} P(U_1)P(U_2) \\ &\leq \sup_{\{V_1, V_2 \in \mathcal{U}_{\Re^p} : V_1 \cap V_2 = V\}} P(V_1)P(V_2) = Q(V). \end{aligned}$$

(iv) Without loss of generality, let $U = [x_1, \infty) \times \Re^{p-1}$. If $U_1 \cap U_2 = U$, then either $U_1 = U$ or $U_2 = U$. Thus

$$\begin{aligned} Q(U) &= \sup_{\{U_1, U_2 \in \mathcal{U}_{\Re^p} : U_1 \cap U_2 = U\}} P(U_1)P(U_2) \\ &= \sup_{\{W \in \mathcal{U}_{\Re^p} : U \subseteq W\}} P(U)P(W) = P(U). \end{aligned}$$

Parts (ii) and (iii) are obvious. □

Although the results presented in this section hold for general random vectors $\boldsymbol{X}$, we will be primarily interested in bivariate random vectors (X_1, X_2) with a finite support set. Without loss of generality we assume that (X_1, X_2) take values in a subset of the two-dimensional lattice $L = \{1, \ldots, M\} \times \{1, \ldots, N\}$. Let $\mathcal{U}_L = \{U \cap L : U \in \mathcal{U}_{\Re^p}\}$. Thus, $\mathcal{U}_{\Re^p}$ can be replaced by $\mathcal{U}_L$ in (2.1), in the definition of Q, and in Theorem 2.1. The class $\mathcal{U}_L$ is the class of all upper sets generated by componentwise partial ordering on the set L.

To systematically describe upper sets in $\mathcal{U}_L$ we use notation developed by Sampson and Whitaker (1988). There is a one-to-one correspondence between upper sets $U \in \mathcal{U}_L$ and the nondecreasing sequence of integers:

$$0 \leq u_1 \leq \cdots \leq u_M \leq N,$$

where $u_k = \#\{(x_1, x_2) \in \mathcal{U}_L : x_1 = k\}$ for $k = 1, \ldots, M$. The value u_k can be thought of as the "depth" of the upper set U above the kth lattice element in the one dimensional lattice $\{1, \ldots, M\}$. Thus for a 3×3 lattice, the upper set $\{(2,3), (3,1), (3,2), (3,3)\}$ has the representation $(0, 1, 3)$. Let $U, V, W \in \mathcal{U}_L$ have corresponding array representations $\boldsymbol{u} = (u_1, \ldots, u_M)$, $\boldsymbol{v} = (v_1, \ldots, v_M)$ and $\boldsymbol{w} = (w_1, \ldots, w_M)$. Then $V \cap W = U$ is equivalent to $u_i = min(v_i, w_i)$, $1 \leq i \leq M$, and $V \subseteq W$ is equivalent to $v_i \leq w_i$, $1 \leq i \leq M$.

For illustrative purposes and for further use in Section 4, we compute Q for the following bivariate distribution on the 3×3 lattice: $P\big((1,1)\big) = P\big((3,3)\big) = 15/64$, $P\big((1,3)\big) = P\big((3,1)\big) = 8/64$, $P\big((2,2)\big) = 18/64$, and

$P((x_1, x_2)) = 0$, otherwise. (Esary, Proschan and Walkup (1967) give this example as a distribution which is positively quadrant dependent, but not associated.) The set function Q is tabulated in Table 1 for each upper set $U \in \mathcal{U}_L$.

TABLE 1. The set function Q computed for the bivariate distribution,
$P((1,1)) = P((3,3)) = 15/64$, $P((1,3)) = P((3,1)) = 8/64$, $P((2,2)) = 18/64$, $P((x_1, x_2)) = 0$ otherwise.

U	P(U)	Q(U)
(0,0,0)	0	0
(0,0,1)	.2344	.2344
(0,0,2)	.2344	.2344
(0,0,3)	.3594	.3594
(0,1,1)	.2344	.2344
(0,1,2)	.2344	.2498*
(0,1,3)	.3594	.3594
(0,2,2)	.5156	.5156
(0,2,3)	.6406	.6406
(0,3,3)	.6406	.6406
(1,1,1)	.3594	.3594
(1,1,2)	.3594	.3594
(1,1,3)	.4844	.4844
(1,2,2)	.6406	.6406
(1,2,3)	.7656	.7651
(1,3,3)	.7656	.7656
(2,2,2)	.6406	.6406
(2,2,3)	.7656	.7656
(2,3,3)	.7656	.7656
(3,3,3)	.7656	.7656

* There is one upper set $(0,1,2)$ for which $Q(U) > P(U)$ thus X_1, X_2 are not associated.

3 Computation of the Set Function Q

In this section we develop the algorithm based on Theorem 2.1 for checking association of the two dimensional random vectors $\boldsymbol{X} = (X_1, X_2)$ with support in an $M \times N$ lattice L. The thrust of the algorithm is to reduce the number of pairs of upper sets needed to calculate $Q(U)$.

For an upper set U, we are interested in those extreme points $(\boldsymbol{X} \neq (M, N))$ which do not lie on the lower and left boundaries of the lattice L, namely we consider the extreme points (x_1, x_2) of U such that $x_1 \neq 1$ or $x_2 \neq 1$ or $(x_1, x_2) \neq (M, N)$ except in the special case that $U = \{(M, N)\}$.

We call this set of extreme points *the outer vertices* of U; they form the "lower corners" of the upper set.

The outer vertices are easily identified from the array representation $(u_1, \ldots, u_M)$ of an upper set, U. The points $(i, N - u_i + 1)$ $i = 1, \ldots, M$ define the lower boundary of U. The outer vertices are defined to be those points $(i, N - u_i + 1)$, $i = 2, \ldots, M$ for which $u_{i-1} < u_i < N$. For example, for the upper set $(0, 1, 3, 4, 4)$ the outer vertices are $(2, 5)$ $(3, 3)$ and $(4, 2)$. (See Figure 1). For the upper set $(0, \ldots, 0, 1)$ there is exactly one outer vertex, (M, N).

Let $\{\boldsymbol{p}_1 = (p_{11}, p_{12}), \ldots, \boldsymbol{p}_m = (p_{m1}, p_{m2})\}$ be the outer vertices of the upper set U. We find $Q(U)$ by searching over 2^m pairs of upper sets defined by subsets of the outer vertices. Let $\{\boldsymbol{q}_1 = (q_{11}, q_{12}), \ldots, \boldsymbol{q}_k = (q_{k1}, q_{k2})\}$, $k = 0, \ldots, m$ be a subset of outer vertices of U; $k = 0$ corresponds to $\emptyset$. For $k \geq 1$, we assume that this subset is ordered so that $q_{11} < \cdots < q_{k1}$. Then the two corresponding upper sets $V(\boldsymbol{q}_1, \ldots, \boldsymbol{q}_k)$ and $W(\boldsymbol{q}_1, \ldots, \boldsymbol{q}_k)$, denoted V and W where there is no ambiguity, are constructed so that they are the largest upper sets in $\mathcal{U}_L$ which alternate tracing U between $\boldsymbol{q}_0, \ldots, \boldsymbol{q}_{k+1}$ where $\boldsymbol{q}_0 = (1, N)$ and $\boldsymbol{q}_{k+1} = (M, 1)$. (The intuitive notion of "tracing" is formally defined after the proof of Theorem 3.1.) For example, consider again the upper set $(0, 1, 3, 4, 4)$ with outer vertices $\boldsymbol{p}_1 = (2, 5)$, $\boldsymbol{p}_2 = (3, 3)$ and $\boldsymbol{p}_3 = (4, 2)$. Suppose $k = 2$, with $\boldsymbol{q}_1 = \boldsymbol{p}_1$ and $\boldsymbol{q}_2 = \boldsymbol{p}_2$. The upper set $V(\boldsymbol{q}_1, \boldsymbol{q}_2) = (0, 3, 3, 4, 4)$ traces U from the upper left corner of the lattice $(1, 5)$ to $\boldsymbol{q}_1$ and from $\boldsymbol{q}_2$ to the lower right corner of the lattice $(5, 1)$, whereas $W(\boldsymbol{q}_1, \boldsymbol{q}_2) = (1, 1, 5, 5, 5)$ traces U between $\boldsymbol{q}_1$ and $\boldsymbol{q}_2$. Where V and W do not trace U, they are chosen to be as large as possible for upper sets in $\mathcal{U}_L$. Note that because V and W alternate tracing U, $V \cap W = U$. Specifically we define V and W in terms of their array representations $\boldsymbol{v} = (v_1, \ldots, v_M)$ and $\boldsymbol{w} = (w_1, \ldots, w_M)$ as follows. Suppose that k is even. Between the outer vertices for which V traces U

$$v_j = u_j \quad \text{and} \quad w_j = u_{q_{2l+1,1}} \tag{3.1}$$

for $q_{2l,1} \leq j < q_{2l+1,1}$, $l = 0, \ldots, k/2 - 1$; between the outer vertices for which W traces U

$$v_j = u_{q_{2l+2,1}} \quad \text{and} \quad w_j = u_j \tag{3.2}$$

for $q_{2l+1,1} \leq j < q_{2l+2,1}$ for $l = 0, \ldots, k/2 - 1$; between the last outer vertex and $\boldsymbol{q}_{k+1}$, $v_j = u_j$ and $w_j = M$, for $q_{k,1} \leq j \leq q_{k+1,1}$. In the case that k is odd, V and W are constructed similarly. Note that if U has no outer vertices ($m = 0$) or if $k = 0$, then $V(\emptyset) = U$ and $W(\emptyset) = L$.

We now show that $Q(U)$ need not be constructed by searching over all pairs of upper sets S, T for which $S \cap T = U$. Rather, we need only consider pairs of upper sets of the form V and W. The next result states that for any $S, T \in \mathcal{U}_L$ such that $S \cap T = U$, either S traces U or T traces U between any two consecutive outer vertices, and between the end points $(M, 1)$, $(1, N)$ and the adjacent outer vertices.

FIGURE 1. The Upper Set $(0, 1, 3, 4, 4)$ and Its Outer Vertices. (Outer Vertices are denoted by $\boxdot$)

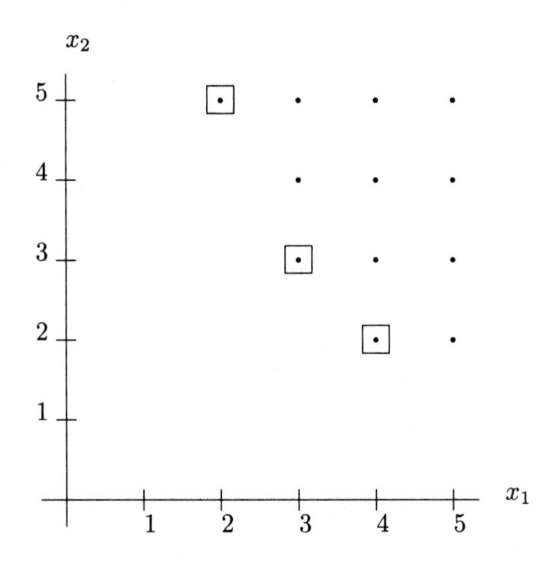

Theorem 3.1 *Let* S, T, $U \in \mathcal{U}_L$, $S \cap T = U$. *Then for* $\mathbf{p}_i, \mathbf{p}_{i+1}$, $i = 0, \ldots, m-1$, *either*

$$s_j = u_j \quad p_{i,1} \leq j \leq p_{i+1,1} - 1 \tag{3.3}$$

or

$$t_j = u_j \quad p_{i,1} \leq j \leq p_{i+1,1} - 1, \tag{3.4}$$

where $\mathbf{p}_0 = (1, N)$ *and* $\mathbf{p}_{m+1} = (M, 1)$; *for* $\mathbf{p}_m$, $\mathbf{p}_{m+1}$ *either*

$$s_j = u_j \qquad p_{m,1} \leq j \leq p_{m+1,1} \tag{3.5}$$

or

$$t_j = u_j \qquad p_{m,1} \leq j \leq p_{m+1,1}. \tag{3.6}$$

Proof Let $a = p_{i,1}$ and $b = p_{i+1,1}$, $i = 0, \ldots, m-1$ then $u_a = \cdots = u_{b-1} < u_b$. Note that the array representations $\boldsymbol{s} = (s_1, \ldots, s_M)$ of S and $\boldsymbol{t} = (t_1, \ldots, t_M)$ of T satisfy $s_1 \leq \cdots \leq s_R$ and $t_1 \leq \cdots \leq t_M$. Further $S \cap T = U$ so that $u_i = \min(s_i, t_i)$, $i = 1, \ldots, M$. Without loss of generality, suppose $s_{b-1} = u_{b-1}$. Then $u_{b-1} = u_a \leq s_a$, so that $s_a = \cdots = s_{b-1} = u_{b-1}$. Thus $s_j = u_j$ for $j = a, \ldots, b-1$. The case that $i = m$ follows similarly. $\square$

When (3.3) or (3.5) holds we say that S is *tracing* U between $\boldsymbol{p}_i$ and $\boldsymbol{p}_{i+1}$, and when (3.4) or (3.6) holds, we say that T is *tracing* U between $\boldsymbol{p}_i$ and $\boldsymbol{p}_{i+1}$.

In light of the previous theorem, there exists a subset $\{\boldsymbol{q}_1, \ldots, \boldsymbol{q}_k\}$, $k = 0, \ldots, m$ (when $k = 0$ the subset is taken to be $\emptyset$) of the outer vertices $\{\boldsymbol{p}_1, \ldots, \boldsymbol{p}_m\}$ so that S and T alternate tracing U between successive points of $\boldsymbol{q}_0, \ldots, \boldsymbol{q}_{k+1}$. From the array representation of $V(\boldsymbol{q}_1, \ldots, \boldsymbol{q}_k)$ and $W(\boldsymbol{q}_1, \ldots, \boldsymbol{q}_k)$ it is clear that $v_j \geq s_j$ and $w_j \geq t_j$ for $j = 1, \ldots, M$. Thus if we take S to be the upper set which traces U between $\boldsymbol{q}_0$ and $\boldsymbol{q}_1$ then $S \subseteq V(\boldsymbol{q}_1, \ldots, \boldsymbol{q}_k)$ and $T \subseteq W(\boldsymbol{q}_1, \ldots, \boldsymbol{q}_k)$ so that

$$P(S)P(T) \leq P\big(V(\boldsymbol{q}_1, \ldots, \boldsymbol{q}_k)\big)P\big(W(\boldsymbol{q}_1, \ldots, \boldsymbol{q}_k)\big). \tag{3.7}$$

This provides the basis of the algorithm for computing $Q(U)$, summarized in the following theorem.

Theorem 3.2 *For any* $U \in \mathcal{U}_L$,

$$Q(U) = max\{P\big(V(\boldsymbol{q}_1, \ldots, \boldsymbol{q}_k)\big)P\big(W(\boldsymbol{q}_1, \ldots, \boldsymbol{q}_k)\big)\} \tag{3.8}$$

where the maximum is taken over all possible subsets $\{\boldsymbol{q}_1, \ldots, \boldsymbol{q}_k\}$ $k = 0, \ldots, m$ *of outer vertices* $\{\boldsymbol{p}_1, \ldots, \boldsymbol{p}_m\}$ *of* U.

We now consider the question posed in Section 2: is Q a p.m.f? To answer this, we examine Q when P exhibits a weaker form of positive dependence than association. Let $R_{\boldsymbol{x}} = \{\boldsymbol{y} \in L : y_1 \geq x_1, y_2 \geq x_2 \text{ where } \boldsymbol{x} = (x_1, x_2)\}$ be an upper rectangle in L. Then a p.m.f. P is said to be positive quadrant dependent (PQD) (see Lehmann (1966)) if

$$P(R_{\boldsymbol{x}}) \geq P(R_{(x_1,1)})P(R_{(1,x_2)}) \quad \forall \boldsymbol{x} \in L. \tag{3.9}$$

Clearly $R_{\boldsymbol{x}} \in \mathcal{U}_L$. Note that upper rectangles of the form $R_{(x_1,1)}$, $1 \leq x_1 \leq M$, or $R_{(1,x_2)}$, $1 \leq x_2 \leq N$, have no outer vertices; furthermore, by Theorem 2.2 (iv)

$$Q(R_{\boldsymbol{x}}) = P(R_{\boldsymbol{x}})P(L) = P(R_{\boldsymbol{x}}), \quad \text{where} \quad \boldsymbol{x} = (x_1, 1) \text{ or } (1, x_2).$$

The remaining upper rectangles $R_{\boldsymbol{x}}$ where $x_1 > 1$ and $x_2 > 1$ have exactly one outer vertex $\boldsymbol{p}_1 = \boldsymbol{x}$. By definition $V(\boldsymbol{p}_1) = R_{(x_1,1)}$ and $W(\boldsymbol{p}_1) = R_{(1,x_2)}$. Thus by Theorem 3.2 for upper rectangles with one outer vertex $\boldsymbol{x}$

$$Q(R_{\boldsymbol{x}}) = max\{P(R_{\boldsymbol{x}}), P(R_{(x_1,1)})P(R_{(1,x_2)})\}. \tag{3.10}$$

This yields the following results

Lemma 3.3 *If* P *is* PQD *then* $Q(R_{\boldsymbol{x}}) = P(R_{\boldsymbol{x}}) \quad \forall \boldsymbol{x} \in L$.

Theorem 3.4 *There exists a* P *such that* Q *is not a p.m.f.*

Proof Let P be a p.m.f. on L. Assume P is PQD but not associated, and suppose that Q is a p.m.f. Then by Lemma 3.3, $P(R_{\boldsymbol{x}}) = Q(R_{\boldsymbol{x}}) \quad \forall \boldsymbol{x} \in L$. Since Q is assumed to be a p.m.f it is uniquely specified by the probabilities it assigns to upper rectangles. Thus $P = Q$. In particular, $P(U) = Q(U) \quad \forall\, U \in \mathcal{U}_L$, which by Theorem 2.1 implies that P is associated, thereby yielding a contradiction. □

An example of P which is PQD but not associated is given in Section 2. Thus Q tabulated in Table 1 is not a p.m.f.

4 Computational Complexity

A "definitional approach" to checking whether or not a pair of random variables on L is associated uses either (2.1) or Theorem 2.1. To use this approach, we would tabulate the probabilities of all upper sets in $\mathcal{U}_L$. Then for each upper set $U \in \mathcal{U}_L$, $P(U)$ would be compared to $Q(U)$. Equivalently $P(V \cap W)$ could be compared to $P(V)P(W)$ for all possible choices of V, $W \in \mathcal{U}_L$. Therefore, the total number of required comparisons for this "definitional approach" is $(\#\mathcal{U}_L - 1)(\#\mathcal{U}_L)/2$, where $\#A$ denotes the cardinality of a set A. As was noted in Section 1, $\#\mathcal{U}_L = \binom{M+N}{M}$, so that the required number of comparisons is $\{\binom{M+N}{N} - 1\}\binom{M+N}{N}/2$.

However, we suggest an approach for checking association which uses outer vertices and Theorem 3.2, thereby, as we show, substantially reducing the total number of comparisons. This proposed "outer vertices" approach again computes $P(U)$ for all $U \in \mathcal{U}_L$ and then compares $P(U)$ to $P(V)P(W)$ for all 2^m pairs of upper sets $V(\boldsymbol{q}_1, \ldots, \boldsymbol{q}_k)$ and $W(\boldsymbol{q}_1, \ldots, \boldsymbol{q}_k)$, where $\boldsymbol{q}_1, \ldots, \boldsymbol{q}_k$ are chosen from the m outer vertices $\boldsymbol{p}_1, \ldots, \boldsymbol{p}_m$ of U. Clearly calculating $Q(U)$ using Theorem 3.2 requires use of fewer than the $\{\binom{M+N}{M} - 1\}\binom{M+N}{M}/2$ comparisons required by the "definitional approach". In fact, assuming $M \leq N$ the required number of comparisons is

$$\sum_{m=0}^{M-1} 2^m K_m \tag{4.1}$$

where K_m is the number of upper sets with m outer vertices. To numerically evaluate the reduction in the number of comparisons requires computing K_m.

The number, K_m, of upper sets with m outer vertices can be found combinatorally by considering the array representation $(u_1, \ldots, u_M)$ for an upper set U. Suppose first that $m > 0$. Every upper set U with m outer vertices can be characterized in a one-to-one fashion by two sets of characteristics:

(i) $1 < p_{11} < \cdots < p_{m1}$, and an integer $Z > p_{m1}$, where $Z = i$, if there exists an i such that $u_i = N$, and $Z = M + 1$, otherwise; and

(ii) $0 \leq u_1 < u_{p_{11}} < \cdots < u_{p_{m1}} \leq N-1$.

For instance, for the upper set $(0,1,3,4,4)$ depicted in Figure 1, $m=3$, and the two sets of characteristics become: (i) $p_{11}=2$, $p_{21}=3$, $p_{31}=4$, $Z=6$; and (ii) $u_1=0$, $u_2=1$, $u_3=3$, $u_4=4$. For the case $m=0$, the sets of characteristics become:

(i′) Z, where $Z=i$, if there exists an i such that $u_i=N$, and $Z=M+1$, otherwise; and

(ii′) $0 \leq u_1 \leq N-1$.

There is, however, one upper set with $m=0$, which is not accounted for in this fashion, and that is $(N,\ldots,N)=L$.

These correspondences lead to the following useful lemma.

Lemma 4.1 *The number of upper sets in an $M \times N$ lattice with m outer vertices is $\binom{M}{m+1}\binom{N}{m+1}$, if $m>0$, and $1+MN$, if $m=0$.*

Obviously the maximum number of possible outer vertices is $min(M-1, N-1)$. There is an interesting probabilistic interpretation of Lemma 4.1. Suppose we sample upper sets uniformly from all $\binom{M+N}{M}$ possible upper sets in the lattice L. For any random upper set U, define the random variable Y by $Y=0$, if $U=L$, and $Y=m+1$, if $U \neq L$, where m is the number of outer vertices of U. Then Y has a hypergeometric distribution with parameters $M, N, M+N$ (following the notation of Johnson and Kotz (1969,p.143)).

Combining Lemma 4.1 with (4.1) we have shown that the number of required comparisons C using the "outer vertices approach" is

$$C = 1 + \sum_{m=0}^{\min(M-1,N-1)} 2^m \binom{M}{m+1}\binom{N}{m+1}. \tag{4.2}$$

Observe that $C = \frac{1}{2} + \binom{M+N}{M}E(2^Y)/2$, so that the ratio of the number of comparisons of the "outer vertices approach" to the "definitional approach" is

$$\frac{1/2 + \binom{M+N}{M}E(2^Y)/2}{\binom{M+N}{M}\left(\binom{M+N}{M}-1\right)/2} \tag{4.3}$$

which is essentially $E(2^Y)/\binom{M+N}{M}$.

Table 2 provides values of $E(2^Y)/\binom{M+N}{M}$ for various choices of M and N. For an 8×8 or dual contingency table the "outer vertices" approach uses approximately $1/600$ of the comparisons that the "definitional approach" uses. Clearly the resultant computational savings are substantial.

TABLE 2. Ratio of Comparisons "outer vertices approach" to "definitional approach"

M	N	Ratio	M	N	Ratio
2	2	0.3611	5	7	0.0064
2	3	0.1900	6	7	0.0049
3	3	0.1575	7	7	0.0041
2	4	0.1156	2	8	0.0316
3	4	0.0784	3	8	0.0120
4	4	0.0655	4	8	0.0060
2	5	0.0771	5	8	0.0036
3	5	0.0440	6	8	0.0025
4	5	0.0316	7	8	0.0019
5	5	0.0265	8	8	0.0016
2	6	0.0548	2	9	0.0251
3	6	0.0269	3	9	0.0086
4	6	0.0168	4	9	0.0038
5	6	0.0125	5	9	0.0021
6	6	0.0105	6	9	0.0013
2	7	0.0409	7	9	0.0009
3	7	0.0176	8	9	0.0007
4	7	0.0097	9	9	0.0006

Acknowledgments: The research of A. Sampson is supported by the Air Force Office of Scientific Research under Contract AFOSR-84-0113. The research of L. Whitaker is supported in part by the Air Force Office of Scientific Research under Contract AFOSR-84-0159.

References

Agresti, A. (1984). *Analysis of Ordinal Categorical Data*, John Wiley, New York.

Barlow, R.E. and Proschan, F. (1981). *Statistical Theory of Reliability and Life Testing.* To Begin With, Silver Spring, MD.

Esary, J.D., Proschan, F., and Walkup, D.W. (1967). Association of random variables, with applications. *Annals of Mathematical Statistics*, **38**, 1466-1474.

Gleser, L. and Moore, D.S. (1985). The effect of positive dependence on chi-squared tests for categorical data. *Journal of the Royal Statistical Society. Ser. B* **47**, 459–465.

Grove, M.D. (1984). Positive association in a two-way contingency table: likelihood ratio tests. *Communications Statist. Theor. Math.*, **13**, 931-945.

Johnson, N.L. and Kotz, S. (1969). *Discrete Distributions.* John Wiley, New York.

Krishnaiah, P.R. Rao, M.B. and Subramanyan, C.K. (1987). A structure theorem on bivariate positive quadrant dependent distributions and tests for independence in two-way contingency tables. *Journal of Multivariate Analysis*, **23**, 93-118.

Lehmann, E.L. (1966). Some concepts of dependence. *Annals of Mathematical Statistics*, **37**, 1137-1153.

Nguyen, T.T. and Sampson, A.R. (1987). Testing for positive quadrant dependence in ordinal contingency tables. *Naval Research Logistics Quarterly*, **34**, 859-878.

Pitt, L. (1982). Positively correlated random variables are associated. *Annals of Probability*, **10**, 496-499.

Sampson, A.R. and Whitaker, L.R. (1988). Positive dependence, upper sets, and multidimensional partitions. *Mathematics of Operations Research*, **13**, 254–264.

Sampson, A.R. and Whitaker, L.R. (1989). Estimation of multivariate distributions under stochastic ordering. To appear in the *Journal of the American Statistical Association.*

Schriever, B.F. (1987), An ordering for positive dependence. *Annals of Statistics*, **15**, 1208-1214.

Tong, Y.L. (1980). *Probability Inequalities*, Academic Press, New York.

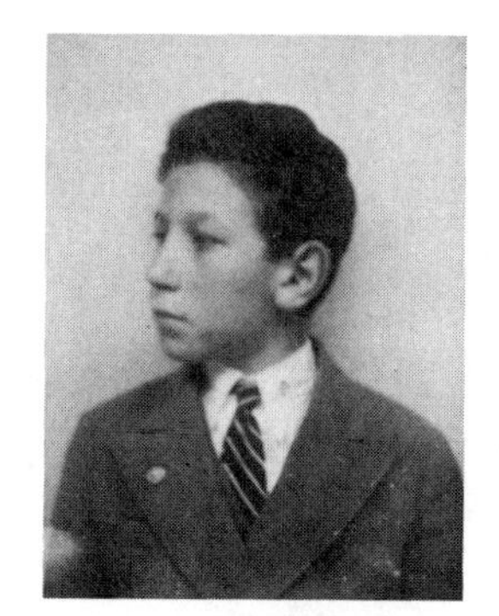
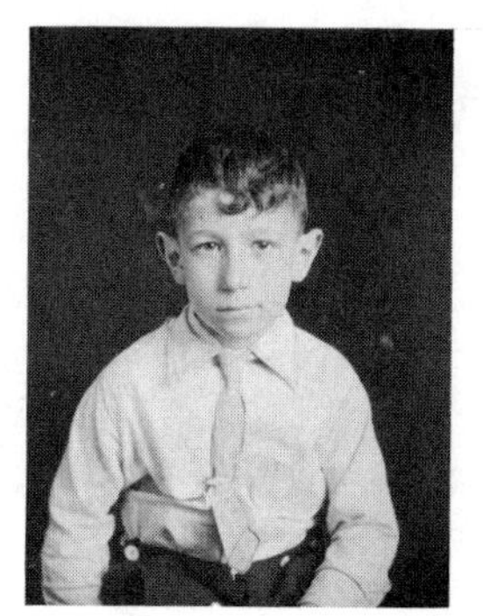

Ingram Olkin, in his youth.

Graduation from DeWitt Clinton High School (1941).

Clockwise from top left: At MIT in Air Force meteorology training; on his wedding day with Anita (May 19, 1945); and with his first child, Vivian (1950).

Clockwise from top left: With (left to right) Vivian, Julia, and Rhoda at Santa Barbara in 1978; at his daughter Rhoda's wedding (July 1984); and with his grandson, Noah (July 1987).

On his birthday with granddaughter Leah (July 1987).

At his daughter Julia's wedding on May 19, 1985 (also his and Anita's 40th wedding anniversary) with (left to right) son-in-law Sim, Vivian, Anita, son-in-law Juan, Julia, Rhoda, and son-in-law Mike.

Part C

Linear and Nonlinear Models, Ranking and Selection, Design

17

A Comparison of the Performances of Procedures for Selecting the Normal Population Having the Largest Mean when the Variances are Known and Equal

Robert E. Bechhofer[1]
David M. Goldsman[2]

ABSTRACT We study the performance characteristics of procedures for selecting the normal population which has the largest mean when the variances are known and equal. The procedures studied are the single-stage procedure of Bechhofer, the closed two-stage procedure of Tamhane and Bechhofer, the open sequential procedure of Bechhofer, Kiefer, and Sobel and a truncated version of that procedure by Bechhofer and Goldsman, the closed multi-stage procedure with elimination of Paulson and improved closed versions of that procedure by Fabian and by Hartmann. The performance characteristics studied are the achieved probability of a correct selection, the expected number of stages required to terminate experimentation, and the expected total number of observations required to terminate experimentation. Except for the single-stage procedure, all performance characteristics are estimated by Monte Carlo sampling. Based on these results, recommendations are made concerning which procedure to use in different circumstances.

1 Introduction and Summary

Over the years considerable research effort has been devoted to devising procedures for selecting the normal population which has the largest population mean. Bechhofer (1954), adopting the so-called *indifference-*

---

[1]School of Operations Research and Industrial Engineering, Cornell University, Ithaca, NY 14853
[2]School of Industrial and Systems Engineering, Georgia Institute of Technology, Atlanta, GA 30332

zone approach, developed a *single-stage* procedure for the case of common *known* variance. Following that paper, several authors, also adopting the indifference-zone approach, have proposed different procedures for dealing with this same problem. Among these is a closed two-stage procedure, an open sequential procedure, and several closed multi-stage procedures. All of these procedures guarantee the same indifference-zone probability requirement but each, at the time that it was introduced, had a different virtue and in some sense represented an improvement over the procedures proposed earlier. The purpose of the present article is to compare certain performance characteristics of these procedures—specifically their achieved probability of a correct selection, and for sequential and multi-stage procedures, their expected number of *stages* required to terminate experimentation and their expected *total number of observations* required to terminate experimentation.

In addition to the single-stage procedure of Bechhofer, the specific procedures that we study are the closed two-stage procedure of Tamhane and Bechhofer (1977,1979); the open sequential procedure of Bechhofer, Kiefer, and Sobel (1968) and the truncated version of that procedure by Bechhofer and Goldsman (1987); the closed multi-stage procedure of Paulson (1964) and the improved version of that procedure by Fabian (1974) with further improvements by Hartmann (1988). The numerical estimates of the performance characteristics for the Tamhane-Bechhofer, Paulson, Fabian, and Hartmann procedures are new, and have not been reported in detail elsewhere; all of our results for these procedures were obtained by carefully executed Monte Carlo sampling experiments. Based on these results we assess the virtues and drawbacks of the procedures, and discuss the alternatives available to the experimenter. Our findings should be of assistance to experimenters who wish to decide which of these procedures is appropriate to use in a particular real-life setting.

We formulate the selection problem in Section A2. The procedures under consideration will be described and their attributes reviewed in Section A3. Numerical estimates of the performance characteristics studied are given in Section A4. We discuss these results in Section A5. In Section A6 we state our conclusions.

A2 Statement of the Problem

We assume that we have $k \geq 2$ normal populations $\Pi_1, \Pi_2, \ldots, \Pi_k$ with unknown population means μ_i $(1 \leq i \leq k)$ and a *common known* variance σ^2. The ordered values of the μ_i are denoted by $\mu_{[1]} \leq \mu_{[2]} \leq \cdots \leq \mu_{[k]}$. We further assume that the values of the $\mu_{[j]}$ $(1 \leq j \leq k)$ are unknown, and that the pairing of the $\mu_{[j]}$ with the Π_i $(1 \leq i, j \leq k)$ is completely unknown. The objective of the experiment, and the associated probability requirement, are stated in (2.1) and (2.2) below.

$$\textbf{Goal:} \quad \textbf{To select the population associated with } \mu_{[k]}. \tag{2.1}$$

If the population selected by a procedure is indeed the population associated with $\mu_{[k]}$, then we say that a *correct selection* (CS) has been made. We limit consideration to procedures that guarantee the following *indifference-zone* requirement on the $P\{\text{CS}\}$:

Probability requirement

$$P\{\text{CS}\} \geq P^* \qquad \text{whenever} \quad \mu_{[k]} - \mu_{[k-1]} \geq \delta^*. \tag{2.2}$$

The quantities $\{\delta^*, P^*\}$ with $0 < \delta^* < \infty$, $1/k < P^* < 1$ are specified prior to the start of experimentation. All of the aforementioned procedures have been shown to guarantee (2.2).

The present investigation is limited to the following representative cases: $k = 4$, $\delta^* = 0.2$, $P^* = 0.75, 0.90, 0.95, 0.99$, and three configurations of the population means, namely, (i) equally-spaced δ^*-apart (ES(δ^*)), i.e., $\mu_{[i+1]} - \mu_{[i]} = \delta^*$ $(1 \leq i \leq k-1)$, (ii) least favorable (LF), i.e., $\mu_{[1]} = \mu_{[k-1]} = \mu_{[k]} - \delta^*$ and (iii) equal means (EM), i.e., $\mu_{[1]} = \mu_{[k]}$. The extent to which our findings can be extrapolated to additional $(k; \delta^*, P^*)$ and other configurations will be indicated.

3 Procedures Considered

In order to make the present article self-contained, we describe and critique in this section the procedures under consideration. The reader is referred to the cited articles for more detailed information concerning these procedures and their performance characteristics.

Single-stage procedure of Bechhofer

For this procedure (referred to herein as B) a common number n of independent observations X_{ij} $(1 \leq i \leq k,\ 1 \leq j \leq n)$ is taken in a *single stage* from each Π_i $(1 \leq i \leq k)$, and the k sample means $\bar{x}_i = \sum_{j=1}^{n} x_{ij}/n$ $(1 \leq i \leq k)$ are calculated. Let $\bar{x}_{[k]} = \max\{\bar{x}_1, \ldots, \bar{x}_k\}$. The experimenter then selects the population that yielded $\bar{x}_{[k]}$ as the one associated with $\mu_{[k]}$. Here $n = [(\sigma c_{k,P^*}/\delta^*)^2]^+$ where c_{k,P^*} is a constant which is given (as $\sqrt{N}\lambda$) for $k = 2(1)10$ and selected P^* in Table I of Bechhofer (1954), and $[b]^+$ is the smallest integer greater than or equal to b.

The single-stage procedure can be quite conservative, i.e., if, unknown to the experimenter, the population means are in a very favorable configuration (e.g., widely spaced), then the actual *achieved* probability of a correct selection when $\mu_{[k]} - \mu_{[k-1]} = \delta^*$ may exceed the specified P^* by a considerable amount. The experimenter thus has taken a larger number of observations than would have been required had the true configuration been known, and had that configuration then been taken into account when choosing n. The fact that observations are taken in a *single* stage does not permit the experimenter to gain information concerning the configuration of the population means as sampling progresses, as would be the case if observations were taken in (say) *several* stages. In particular, no information can be obtained as to whether or not any populations are likely to be contenders or non-contenders for "best." It is for this reason that *multi-stage* procedures were introduced.

Two-stage procedure of Tamhane and Bechhofer

Tamhane and Bechhofer (1977,1979) proposed a two-stage selection procedure. At the end of the first stage, this procedure (referred to herein as T-B) screens out (permanently eliminates) populations indicated as being non-contending for best, and takes observations in the second stage only from those populations indicated as being contending. The *cumulative* sample mean based on first- and second-stage observations is calculated for populations which enter the second stage, and the experimenter selects the population that yielded the largest of these cumulative means. Thus this procedure can capitalize on favorable configurations of the population means permitting a reduction in the expected total number of observations required to terminate sampling, and in extreme situations leads to termination after the first stage. Implementation of the T-B procedure involves the use of three predetermined constants $(\hat{n}_1, \hat{n}_2, \hat{h})$ where $\hat{n}_1$, $\hat{n}_2$ are positive integers and $\hat{h} > 0$. The source of these constants is given below.

We now formally describe the T-B procedure. A common number $\hat{n}_1$ of independent observations $X_{ij}^{(1)}$ $(1 \leq j \leq \hat{n}_1)$ is taken in the first stage from Π_i $(1 \leq i \leq k)$ and the k first-stage sample means $\bar{x}_i^{(1)} = \sum_{j=1}^{\hat{n}_1} x_{ij}^{(1)} / \hat{n}_1$ $(1 \leq i \leq k)$ are calculated. Let $\bar{x}_{[k]}^{(1)} = \max\{\bar{x}_1^{(1)}, \ldots, \bar{x}_k^{(1)}\}$. Determine the subset I of $\{1, 2, \ldots, k\}$ where $I = \{\, i \mid \bar{x}_i^{(1)} \geq \bar{x}_{[k]}^{(1)} - \hat{h} \,\}$, and let Π_I denote the associated subset of $\{\Pi_1, \Pi_2, \ldots, \Pi_k\}$. If Π_I consists of one population, stop sampling and select the population yielding $\bar{x}_{[k]}^{(1)}$ as the one associated with $\mu_{[k]}$. If Π_I consists of more than one population, proceed to the second stage and take $\hat{n}_2$ additional observations $X_{ij}^{(2)}$ $(1 \leq j \leq \hat{n}_2)$ from each population in Π_I. Compute the *cumulative* sample means $\bar{x}_i = \left(\sum_{j=1}^{\hat{n}_1} x_{ij}^{(1)} + \sum_{j=1}^{\hat{n}_2} x_{ij}^{(2)}\right) / (\hat{n}_1 + \hat{n}_2)$ for $i \in I$. The experimenter then selects the population that yielded $\max\{\, \bar{x}_i \mid i \in I \,\}$ as the one associated with $\mu_{[k]}$.

For given k and specified $\{\delta^*, P^*\}$ the procedure depends on the three constants $(\hat{n}_1, \hat{n}_2, \hat{h})$. Here $\hat{n}_1 = [(\sigma\hat{c}_1/\delta^*)^2]^+$, $\hat{n}_2 = [(\sigma\hat{c}_2/\delta^*)^2]^+$, and $\hat{h} = \hat{d}\delta^*/\hat{c}_1$ where $(\hat{c}_1, \hat{c}_2, \hat{d})$ are found for $k = 2$ in Table I (E) of Tamhane and

Bechhofer (1979) and for $k = 3(1)10$, 12, 15, 25 and selected P^* in Table II of that article. The choice $(\hat{n}_1, \hat{n}_2, \hat{h})$ guarantees (2.2) and minimizes $\max_{\underset{\sim}{\mu}} E\{\text{Total number of observations}\}$ where $\underset{\sim}{\mu} = (\mu_1, \mu_2, \ldots, \mu_k)$. This *minimax* two-stage procedure has the property that $E\{\text{Total number of observations}\} < kn$, uniformly in $\underset{\sim}{\mu}$ for all $\{\delta^*, P^*\}$, where n is the number of observations per population required by the corresponding single-stage procedure.

Multi-stage procedures offer the potential of further improvements over the two-stage procedure. We next describe such procedures.

Sequential procedure of Bechhofer, Kiefer, and Sobel, and a truncated version

Bechhofer, Kiefer, and Sobel (1968), Section 12.6.1.1, proposed a fully sequential open selection procedure. For this procedure (referred to herein as B-K-S) independent vector-observations consisting of one observation from each of the k populations are taken one vector-at-a-time until a stopping rule calls for termination of sampling. The procedure can capitalize on favorable configurations of the population means, thus leading to early termination of sampling. In fact, if $\mu_{[k]} - \mu_{[k-1]} \gg \sigma$, then with high probability sampling will stop after a single stage (requiring therefore only a total of k scalar observations).

Let x_{ij} $(1 \leq i \leq k,\ j = 1, 2, \ldots)$ denote the jth observation from Π_i. At stage m of experimentation $(m = 1, 2, \ldots)$, observe the random vector $(X_{1m}, X_{2m}, \ldots, X_{km})$. Let $y_{im} = \sum_{j=1}^{m} x_{ij}$ $(1 \leq i \leq k)$, and denote the ordered values of the y_{im} by $y_{[1]m} < \cdots < y_{[k]m}$. Calculate $z_m = \sum_{i=1}^{k-1} \exp\{-\delta^*(y_{[k]m} - y_{[i]m})/\sigma^2\}$. Stop sampling when, for the first time, $z_m \leq (1 - P^*)/P^*$. Let N, a random variable, denote the value of m at termination. After stopping, the experimenter selects the population that yielded $y_{[k]N}$ as the one associated with $\mu_{[k]}$.

This procedure has the drawback that if $\mu_{[k]} - \mu_{[1]}$ is small, then the distribution of N is highly skewed to the right, and large values of N can occur with sizable probability. (See Table I in Bechhofer-Goldsman (1987).) Also, the variance of N can be very large. To avoid these undesirable effects, B-G (1987) studied a *truncated* version of the B-K-S procedure (referred to herein as $(\text{B-K-S})_T$); truncation is possible here because, when the population means are in the LF-configuration, the achieved $P\{\text{CS}\}$ is greater than P^*. For this version sampling is stopped when, for the first time, *either* $z_m \leq (1 - P^*)/P^*$ *or* $m = n_0$, whichever occurs first. Let N denote the value of m at termination, and let $n_0(k; \delta^*, P^*) = n_0$ (say) be predetermined as the smallest integer that will guarantee (2.2) using $(\text{B-K-S})_T$. After stopping, the experimenter selects the population that yielded $y_{[k]N}$ as the one associated with $\mu_{[k]}$. The truncated procedure retains all of the virtues of the original procedure, and in addition reduces $E\{N\}$ and

$\text{Var}\{N\}$. Values of n_0 are given for $k = 2, 3, 4, 5$ in Tables II, III, IV, V, respectively, of B-G (1987) and in Tables I, II, III, IV of B-G (1989); each table of B-G (1987) gives values for $P^* = 0.75, 0.90, 0.95, 0.99$ with $\delta^* = 0.2(0.2)0.8$, while each table of B-G (1989) gives values for $P^* = 0.75, 0.90, 0.95, 0.99$ with $\delta^* = 0.3(0.2)0.7$.

Neither B-K-S nor $(\text{B-K-S})_T$ permits *elimination* of populations indicated as being non-contending for best. Such elimination-type procedures are considered below. (T-B, which was considered in the foregoing, is also such a procedure.)

Sequential procedure of Paulson, and the Fabian and Hartmann improvements

The B-K-S procedure has two drawbacks: i) it is *open*, i.e., before experimentation starts it is not possible to give a finite upper bound on the number of stages required to terminate experimentation, and ii) it does not eliminate from sampling populations which, based on the earlier stages, would appear to be out of contention for selection as "best." The following *closed* sequential procedure of Paulson (1964) (referred to herein as P), which employs *elimination* of populations, overcomes both of these drawbacks. For this procedure independent vector-observations are taken, one vector at a time, until a stopping rule calls for termination of sampling; each vector consists of one observation from each population *not eliminated at an earlier stage.* Paulson's procedure is actually a family of procedures which depends on a constant λ $(0 \leq \lambda \leq \delta^*/2)$ which is chosen by the experimenter. Paulson recommended the choice $\lambda = \delta^*/4$. However, the choice $\lambda = \delta^*/2$ has the desirable property that the maximum possible *number of stages* to termination of the experiment is *minimized* for given k and specified $\{\delta^*, P^*\}$.

Let x_{ij} $(1 \leq i \leq k,\ j = 1, 2, \ldots)$ denote the jth observation from Π_i. At stage m $(m = 1, 2, \ldots)$ let $y_{im} = \sum_{j=1}^{m} x_{ij}$ $(1 \leq i \leq k)$. Denote the ordered values of the y_{im} for the R populations *still retained at stage* m by $y_{[1]m} < \cdots < y_{[R]m}$. Here $R \geq 2$ is a random variable. Let $a_\lambda = (\sigma^2/(\delta^* - \lambda)) \ln((k-1)/(1-P^*))$ and let W_λ denote the largest integer $< a_\lambda/\lambda$. Start sampling by taking one observation from each of the k populations. Eliminate from further sampling and consideration any population Π_i for which $a_\lambda - \lambda < y_{[k]1} - y_{i1}$. If all but one population is eliminated after the first stage, then stop sampling and select the remaining population as the one associated with $\mu_{[k]}$. Otherwise, proceed to the second stage and take one observation from each population *not yet eliminated.* In general, at stage m $(2 \leq m \leq W_\lambda)$ take one observation from each population *not eliminated after the* $(m-1)$*st stage*, and then eliminate from further consideration any remaining population Π_i for which $a_\lambda - m\lambda < y_{[R]m} - y_{im}$. If all but one population is eliminated after the

mth stage, stop sampling and select the remaining population as the one associated with $\mu_{[k]}$; otherwise proceed to the $(m+1)$st stage. If more than one population remains after stage W_λ, take one additional observation from each of these populations. Then select the population with the largest sum of the $(W_\lambda + 1)$ observations as the one associated with $\mu_{[k]}$.

The Fabian (1974) improvement of the P-procedure (referred to herein as P-F) is obtained by replacing the constant $\big((k-1)/(1-P^*)\big)$ by the *smaller* constants $\big((k-1)/2(1-P^*)\big)$ for $\lambda = \delta^*/2$ and by $1/\theta$ for $\lambda = \delta^*/4$ where θ solves $(1-P^*)/(k-1) = \theta - (1/2)\theta^{4/3}$.

Hartmann (1988) improved the P-F procedure (and thus the P-procedure). For Hartmann's procedure (referred to herein as P-H) the constant $\big((k-1)/(1-P^*)\big)$ of the P-procedure is replaced by $1/\theta$ where $\theta = 2\big(1-(P^*)^{1/(k-1)}\big)$ for $\lambda = \delta^*/2$, and θ solves $1-(P^*)^{1/(k-1)} = \theta-(1/2)\theta^{4/3}$ for $\lambda = \delta^*/4$. For $k > 2$ the constant a_λ is smaller for the P-H procedure than for the P-F procedure.

The P, P-F, and P-H procedures can be *truncated* while still guaranteeing (2.2). The effect of truncation is to decrease the achieved $P\{\text{CS}\}$. At the same time $E\{N\}$, $\text{Var}\{N\}$ and $E\{T\}$, $\text{Var}\{T\}$ are reduced; here T is the total number of observations required to terminate sampling. We do not study truncation of these procedures in the present article.

The P, P-F, and P-H procedures never require more than $W_\lambda+1$ stages to terminate sampling. The effectiveness of these procedures can be measured not only in terms of $E\{N\}$ but also in terms of $E\{T\}$. The appropriate measure(s) to use will depend on the particular practical situation at hand. For example, *stages* can often be equated to *time*; this would be the situation if only one *vector* of observations could be obtained per *day* (say) as with certain production processes. Thus the choice of measure might depend on whether it is more desirable to minimize the expected duration of the experiment than to minimize the expected total number of observations taken in the experiment. The answer will depend on various relative costs. Hartmann's (Fabian's) modification improves the performance of Fabian's (Paulson's) procedure *uniformly* in $(k; \delta^*, P^*)$ and $\underset{\sim}{\mu}$ for both measures.

4 The Performance of the Procedures

Monte Carlo (MC) estimates of the achieved P{CS}, $E\{N\}$, and $E\{T\}$ are given for $k = 4$, $\delta^* = 0.2$ in Tables 1, 2, 3, 4 for $P^* = 0.75$, 0.90, 0.95, 0.99, respectively. Three configurations of the population means, namely, (i) equally-spaced δ^*-apart (ES(δ^*)), (ii) least favorable (LF), and (iii) equal means (EM) were studied for B, T-B, B-K-S, $(\text{B-K-S})_T$, P using $\lambda = \delta^*/4$ and $\delta^*/2$ (P, $\delta^*/4$; P, $\delta^*/2$), P-F using $\lambda = \delta^*/4$ and $\delta^*/2$ (P-F, $\delta^*/4$; P-F, $\delta^*/2$), and P-H using $\lambda = \delta^*/4$ and $\delta^*/2$ (P-H, $\delta^*/4$; P-H, $\delta^*/2$). All of the results for T-B, B-K-S, and $(\text{B-K-S})_T$, (P, $\delta^*/4$ and $\delta^*/2$), (P-F,

TABLE 1. Estimates of achieved probability of a correct selection ($\bar{p}$ or $\bar{w}$), expected number of stages to terminate sampling ($\bar{n}$), and expected total number of observations to terminate sampling ($\bar{t}$) for three configurations of the population means, and selected procedures. ($k = 4$, $\delta^* = 0.2$, $P^* = 0.75$)

Procedure	Equally-spaced δ^* apart: $\mu_{[i+1]} - \mu_{[i]} = \delta^*$ $(1 \leq i \leq 3)$			Least favorable: $\mu_{[1]} = \mu_{[3]} = \mu_{[4]} - \delta^*$			Equal means: $\mu_{[1]} = \mu_{[4]}$	
	$\bar{p}$	$\bar{n}$	$\bar{t}$	$\bar{p}$ or $\bar{w}$	$\bar{n}$	$\bar{t}$	$\bar{n}$	$\bar{t}$
B ($n = 71$)	0.8796	71	284	0.7508	71	284	71	284
T-B ($\hat{n}_1 = 34$, $\hat{n}_2 = 32$, $\hat{h} = 0.6107$)	0.8729 (0.0030)	65.2 (< 0.1)	237.9 (0.3)	0.7361 (0.0028)	65.9 (< 0.1)	258.9 (< 0.1)	66.0 (< 0.1)	261.8 (< 0.1)
B-K-S	0.8493 (0.0033)	34.4 (0.2)	137.8 (0.7)	0.7759 (< 0.0001)	53.3 (0.2)	213.1 (0.6)	72.0 (0.4)	288.0 (1.8)
(B-K-S)$_T$ ($n_0 = 94$)	0.8439 (0.0033)	33.6 (0.2)	134.5 (0.7)	0.7507 (0.0004)	50.2 (0.1)	200.8 (0.4)	60.5 (0.2)	242.2 (1.0)
(P, $\delta^*/4$)	0.9388 (0.0022)	66.4 (0.3)	191.8 (0.7)	0.8621 (0.0023)	84.5 (0.3)	262.6 (0.7)	110.1 (0.4)	335.7 (1.1)
(P, $\delta^*/2$)	0.9626 (0.0017)	84.4 (0.3)	252.2 (0.8)	0.9065 (0.0019)	104.0 (0.2)	341.3 (0.7)	134.3 (0.4)	437.1 (1.1)
(P-F, $\delta^*/4$) $\theta = 0.10954$	0.9228 (0.0024)	57.9 (0.3)	167.7 (0.6)	0.8294 (0.0024)	73.6 (0.2)	227.1 (0.6)	92.6 (0.4)	280.8 (1.0)
(P-F, $\delta^*/2$)	0.9225 (0.0024)	58.9 (0.2)	176.3 (0.6)	0.8293 (0.0024)	72.8 (0.2)	233.3 (0.5)	87.2 (0.3)	277.7 (0.8)
(P-H, $\delta^*/4$) $\theta = 0.12154$	0.9164 (0.0025)	54.6 (0.3)	158.2 (0.6)	0.8200 (0.0025)	69.3 (0.2)	213.0 (0.6)	85.9 (0.3)	259.6 (0.9)
(P-H, $\delta^*/2$) $\theta = 0.18288$	0.9188 (0.0025)	55.6 (0.2)	166.3 (0.5)	0.8151 (0.0025)	68.4 (0.2)	219.0 (0.5)	80.7 (0.2)	256.3 (0.7)

$\delta^*/4$ and $\delta^*/2$), and (P-H, $\delta^*/4$ and $\delta^*/2$) were obtained by MC sampling. The MC results for B-K-S and (B-K-S)$_T$ in the LF- and EM-configurations were abstracted from B-G (1987). All other MC results are new and were prepared especially for this article.

The P{CS} achieved by the B procedure for the particular sample sizes used was calculated by quadrature. The achieved P{CS} for the other procedures was estimated by the observed proportion of correct selections ($\bar{p}$), except for B-K-S and (B-K-S)$_T$ used in the LF-configuration, where a more precise unbiased estimate ($\bar{w}$) was used. (See B-G (1987), Remark 3.3). $E\{N\}$ and $E\{T\}$ were estimated by the observed average number of stages to terminate sampling ($\bar{n}$) and the observed average total number of observations to terminate sampling ($\bar{t}$), respectively. In each of the four tables, the entries under $\bar{n}$ for the B and T-B procedures are the numbers of "vectors" of observations taken in the single- and two-stage procedures, respectively; thus these entries are comparable to the other entries in the $\bar{n}$ columns.

TABLE 2. Estimates of achieved probability of a correct selection ($\bar{p}$ or $\bar{w}$), expected number of stages to terminate sampling ($\bar{n}$), and expected total number of observations to terminate sampling ($\bar{t}$) for three configurations of the population means, and selected procedures. ($k = 4$, $\delta^* = 0.2$, $P^* = 0.90$)

Procedure	Equally-spaced δ^* apart: $\mu_{[i+1]} - \mu_{[i]} = \delta^*$ $(1 \leq i \leq 3)$			Least favorable: $\mu_{[1]} = \mu_{[3]} = \mu_{[4]} - \delta^*$			Equal means: $\mu_{[1]} = \mu_{[4]}$	
	$\bar{p}$	$\bar{n}$	$\bar{t}$	$\bar{p}$ or $\bar{w}$	$\bar{n}$	$\bar{t}$	$\bar{n}$	$\bar{t}$
B ($n = 151$)	0.9588	151	604	0.9008	151	604	151	604
T-B ($\hat{n}_1 = 78$, $\hat{n}_2 = 79$, $\hat{h} = 0.1932$)	0.9583 (0.0018)	116.6 (0.4)	394.7 (0.8)	0.9035 (0.0019)	135.4 (0.2)	474.7 (0.7)	150.9 (0.2)	545.0 (0.8)
B-K-S	0.9354 (0.0022)	61.6 (0.3)	246.3 (1.3)	0.9124 (<0.0001)	95.8 (0.4)	383.2 (1.4)	167.8 (1.1)	671.1 (4.2)
(B-K-S)$_T$ ($n_0 = 205$)	0.9350 (0.0023)	61.8 (0.3)	247.0 (1.3)	0.9006 (0.0001)	93.1 (0.1)	372.6 (0.4)	136.7 (0.5)	546.6 (2.2)
(P, $\delta^*/4$)	0.9745 (0.0014)	93.2 (0.4)	269.5 (1.0)	0.9333 (0.0016)	119.7 (0.3)	381.5 (0.9)	175.5 (0.6)	542.4 (1.7)
(P, $\delta^*/2$)	0.9840 (0.0011)	116.0 (0.4)	347.3 (1.0)	0.9620 (0.0012)	142.4 (0.3)	477.9 (0.9)	202.2 (0.5)	669.8 (1.5)
(P-F, $\delta^*/4$) $\theta = 0.04022$	0.9693 (0.0016)	87.1 (0.4)	252.3 (0.9)	0.9265 (0.0017)	113.1 (0.3)	358.7 (0.9)	160.8 (0.6)	495.8 (1.6)
(P-F, $\delta^*/2$)	0.9702 (0.0016)	92.3 (0.3)	276.0 (0.8)	0.9205 (0.0017)	114.0 (0.3)	376.3 (0.8)	150.7 (0.4)	492.1 (1.2)
(P-H, $\delta^*/4$) $\theta = 0.04175$	0.9720 (0.0015)	86.1 (0.4)	249.6 (0.9)	0.9247 (0.0017)	110.8 (0.3)	351.0 (0.9)	158.5 (0.6)	489.7 (1.5)
(P-H, $\delta^*/2$) $\theta = 0.06902$	0.9684 (0.0016)	90.2 (0.3)	269.8 (0.8)	0.9208 (0.0017)	112.0 (0.3)	369.8 (0.8)	147.9 (0.4)	482.8 (1.1)

The number of independent MC replications on which each estimate is based for each procedure is 12,000 for the ES(δ^*)-configuration, 24,000 for the LF-configuration and 12,000 for the EM-configuration, except that some of the estimates for (B-K-S)$_T$ in the LF-configuration were based on more than 24,000 independent replications. The number in parentheses below each estimate is the estimated standard error of that estimate.

5 Discussion of Results

In Table 5 we have indicated the procedure that, based on our results, dominates in terms of minimum $\bar{n}$ or minimum $\bar{t}$ for each of the three configurations of the population means; these are given for four P^* values.

First we note that in terms of minimum $\bar{n}$, (B-K-S)$_T$ dominates in the LF-configuration for all P^* values. This is not surprising since in that configuration, B-K-S is equivalent to a sequential probability ratio test for

TABLE 3. Estimates of achieved probability of a correct selection ($\bar{p}$ or $\bar{w}$), expected number of stages to terminate sampling ($\bar{n}$), and expected total number of observations to terminate sampling ($\bar{t}$) for three configurations of the population means, and selected procedures. ($k = 4$, $\delta^* = 0.2$, $P^* = 0.95$)

Procedure	Equally-spaced δ^* apart: $\mu_{[i+1]} - \mu_{[i]} = \delta^*$ $(1 \le i \le 3)$			Least favorable: $\mu_{[1]} = \mu_{[3]} = \mu_{[4]} - \delta^*$			Equal means: $\mu_{[1]} = \mu_{[4]}$	
	$\bar{p}$	$\bar{n}$	$\bar{t}$	$\bar{p}$ or $\bar{w}$	$\bar{n}$	$\bar{t}$	$\bar{n}$	$\bar{t}$
B ($n = 213$)	0.9805	213	852	0.9502	213	852	213	852
T-B ($\hat{n}_1 = 115$, $\hat{n}_2 = 110$, $\hat{h} = 0.1358$)	0.9794 (0.0013)	149.4 (0.5)	530.3 (1.0)	0.9495 (0.0014)	174.5 (0.4)	614.2 (1.0)	210.9 (0.3)	748.8 (1.2)
B-K-S	0.9656 (0.0017)	80.8 (0.5)	323.1 (1.8)	0.9566 (<0.0001)	122.4 (0.4)	489.8 (1.8)	261.2 (1.7)	1044.7 (6.6)
(B-K-S)$_T$ ($n_0 = 290$)	0.9628 (0.0017)	81.0 (0.5)	324.2 (1.8)	0.9504 (0.0002)	120.6 (0.3)	482.6 (1.2)	202.9 (0.8)	811.7 (3.0)
(P, $\delta^*/4$)	0.9882 (0.0010)	113.1 (0.5)	327.5 (1.1)	0.9644 (0.0012)	144.2 (0.4)	467.2 (1.1)	229.3 (0.8)	718.2 (2.1)
(P, $\delta^*/2$)	0.9916 (0.0008)	140.6 (0.5)	421.5 (1.1)	0.9796 (0.0009)	169.7 (0.4)	577.9 (1.1)	255.7 (0.6)	855.8 (1.8)
(P-F, $\delta^*/4$) $\theta = 0.01925$	0.9869 (0.0010)	108.6 (0.5)	314.5 (1.1)	0.9621 (0.0012)	138.9 (0.4)	448.9 (1.1)	220.0 (0.7)	685.8 (2.0)
(P-F, $\delta^*/2$)	0.9856 (0.0011)	116.7 (0.4)	349.3 (1.0)	0.9603 (0.0013)	142.8 (0.3)	479.1 (0.9)	201.5 (0.5)	668.2 (1.5)
(P-H, $\delta^*/4$) $\theta = 0.01959$	0.9818 (0.0012)	107.2 (0.5)	310.9 (1.1)	0.9599 (0.0013)	138.8 (0.4)	447.8 (1.1)	216.3 (0.7)	675.9 (2.0)
(P-H, $\delta^*/2$) $\theta = 0.03390$	0.9854 (0.0011)	115.2 (0.4)	345.9 (1.0)	0.9604 (0.0013)	141.8 (0.3)	476.3 (0.9)	200.0 (0.5)	662.2 (1.5)

the corresponding *identification* problem (see B-K-S (1968), Section 4.3.1), and the effect of *truncation* is effectively to eliminate the "excess," i.e., achieved P{CS} $-P^*$. The domination of (B-K-S)$_T$ in terms of minimum $\bar{n}$ extends here to the ES(δ^*)-configuration (and we conjecture that it would continue to hold for even more favorable configurations). (B-K-S)$_T$ also dominates in all three configurations for both $\bar{n}$ and $\bar{t}$ when $P^* = 0.75$; it is at this (the lowest) P^*-value that the excess is greatest in the LF-configuration, and as noted above, truncation essentially eliminates that excess.

Note

In Table 4 the recorded value of $\bar{t}$ in the ES(δ^*)-configuration for $P^* = 0.99$ is smaller for (P-F, $\delta^*/4$) than for (P-H, $\delta^*/4$) although the difference is not statistically significant. However, $E\{T\}$, of which $\bar{t}$ is an estimate, must be smaller for (P-H, $\delta^*/4$) than for (P-F, $\delta^*/4$) since the continuation region for the former is contained in the continuation region for the latter. The

TABLE 4. Estimates of achieved probability of a correct selection ($\bar{p}$ or $\bar{w}$), expected number of stages to terminate sampling ($\bar{n}$), and expected total number of observations to terminate sampling ($\bar{t}$) for three configurations of the population means, and selected procedures. ($k = 4$, $\delta^* = 0.2$, $P^* = 0.99$)

Procedure	Equally-spaced δ^* apart: $\mu_{[i+1]} - \mu_{[i]} = \delta^*$ $(1 \leq i \leq 3)$			Least favorable: $\mu_{[1]} = \mu_{[3]} = \mu_{[4]} - \delta^*$			Equal means: $\mu_{[1]} = \mu_{[4]}$	
	$\bar{p}$	$\bar{n}$	$\bar{t}$	$\bar{p}$ or $\bar{w}$	$\bar{n}$	$\bar{t}$	$\bar{n}$	$\bar{t}$
B ($n = 361$)	0.9964	361	1444	0.9901	361	1444	361	1444
T-B ($\hat{n}_1 = 220$, $\hat{n}_2 = 158$, $\hat{h} = 0.08243$)	0.9963 (0.0006)	236.2 (0.4)	912.5 (0.9)	0.9894 (0.0007)	256.5 (0.4)	964.8 (1.0)	348.2 (0.6)	1242.5 (1.9)
B-K-S	0.9938 (0.0007)	123.6 (0.7)	494.3 (2.6)	0.9914 (<0.0001)	176.5 (0.4)	705.9 (1.6)	542.5 (3.5)	2169.8 (13.9)
(B-K-S)$_T$ ($n_0 = 480$)	0.9921 (0.0008)	122.4 (0.6)	489.6 (2.5)	0.9902 (<0.0001)	175.8 (0.4)	703.0 (1.6)	370.8 (1.2)	1483.3 (4.7)
(P, $\delta^*/4$)	0.9969 (0.0005)	156.2 (0.6)	455.1 (1.4)	0.9916 (0.0006)	196.9 (0.5)	656.6 (1.4)	364.7 (1.1)	1163.7 (3.1)
(P, $\delta^*/2$)	0.9981 (0.0004)	194.2 (0.6)	585.3 (1.4)	0.9956 (0.0004)	231.7 (0.4)	806.7 (1.4)	383.2 (0.8)	1310.0 (2.4)
(P-F, $\delta^*/4$) $\theta = 0.003610$	0.9958 (0.0006)	153.8 (0.6)	447.9 (1.4)	0.9920 (0.0006)	194.7 (0.5)	648.0 (1.4)	356.1 (1.1)	1137.6 (3.0)
(P-F, $\delta^*/2$)	0.9977 (0.0004)	170.9 (0.5)	514.3 (1.3)	0.9919 (0.0006)	206.0 (0.4)	710.9 (1.3)	328.5 (0.7)	1115.0 (2.1)
(P-H, $\delta^*/4$) $\theta = 0.003623$	0.9974 (0.0005)	154.6 (0.6)	449.7 (1.4)	0.9925 (0.0006)	192.9 (0.5)	643.4 (1.4)	357.3 (1.1)	1139.3 (3.1)
(P-H, $\delta^*/2$) $\theta = 0.006689$	0.9970 (0.0005)	171.2 (0.5)	514.8 (1.2)	0.9912 (0.0006)	205.9 (0.4)	709.9 (1.3)	328.7 (0.7)	1112.9 (2.1)

observed difference in $\bar{t}$ means is due to sampling errors. Thus we have "credited" (P-H, $\delta^*/4$) with the smaller value of $\bar{t}$ for $P^* = 0.99$ in the ES(δ^*)-configuration and have similarly "credited" (P-H, $\delta^*/2$) with the smaller value of $\bar{n}$ for $P^* = 0.99$ in the EM-configuration. A similar reversal for $\bar{n}$ occurs in the ES(δ^*)-configuration between (B-K-S) and (B-K-S)$_T$ for $P^* = 0.95$ and 0.90, and for $\bar{t}$ in the ES(δ^*)-configuration for $P^* = 0.90$. In each of these reversals we have "credited" (B-K-S)$_T$ with the smaller $\bar{n}$ or $\bar{t}$ value.

Also noteworthy is the fact that in terms of minimum $\bar{t}$, (P-H, $\delta^*/2$) dominates in the EM-configuration for the largest P^*-values, namely, $P^* =$ 0.90, 0.95, 0.99. This is a consequence of the fact that for each P^*, (P-H, $\delta^*/2$) has the smallest predetermined upper bound on the maximum number of stages to terminate sampling. In terms of minimum $\bar{t}$, (P-H, $\delta^*/4$) dominates in the LF-configuration for the largest P^*-values, namely, $P^* = 0.90$, 0.95, 0.99, and in the ES(δ^*)-configuration for $P^* = 0.95$, 0.99; elimination of populations is easier in the earlier stages with (P-H, $\delta^*/4$)

TABLE 5. Procedure having minimum $\bar{n}$ or minimum $\bar{t}$ for each of three configurations of the population means ($k = 4$, $\delta^* = 0.2$, selected P^*)

	ES(δ^*)-configuration		LF-configuration		EM-configuration	
P^*	$\bar{n}$	$\bar{t}$	$\bar{n}$	$\bar{t}$	$\bar{n}$	$\bar{t}$
0.99	(B-K-S)$_T$	(P-H, $\delta^*/4$)	(B-K-S)$_T$	(P-H, $\delta^*/4$)	(P-H, $\delta^*/2$)	(P-H, $\delta^*/2$)
0.95	(B-K-S)$_T$	(P-H, $\delta^*/4$)	(B-K-S)$_T$	(P-H, $\delta^*/4$)	(P-H, $\delta^*/2$)	(P-H, $\delta^*/2$)
0.90	(B-K-S)$_T$	(B-K-S)$_T$	(B-K-S)$_T$	(P-H, $\delta^*/4$)	(B-K-S)$_T$	(P-H, $\delta^*/2$)
0.75	(B-K-S)$_T$	(B-K-S)$_T$	(B-K-S)$_T$	(B-K-S)$_T$	(B-K-S)$_T$	(B-K-S)$_T$

than with (P-H, $\delta^*/2$).

Further MC studies carried out for $k = 5$, $\delta^* = 0.4$ established that (B-K-S)$_T$, (P-H, $\delta^*/4$) and (P-H, $\delta^*/2$) dominated in terms of minimum $(\bar{n}, \bar{t})$ in essentially the same regions as those described in Table 5. The results of these studies are available from the authors on request.

In our analysis of the performance of the various procedures we have not focused on the achieved P{CS} provided only that (A2.2) is guaranteed. Thus, for example, (P-H, $\delta^*/2$) achieved a P{CS} of 0.815 (approximately) in the LF-configuration for $P^* = 0.75$. This large overprotection could be eliminated by truncation with consequent substantial reductions in $E\{N\}$ and $E\{T\}$ in all configurations. However, the preparation of a table of such truncation numbers would be excessively expensive (and in particular, much more costly than the corresponding tables for (B-K-S)$_T$ in B-G (1987,1989)).

Remark A5.1 In Table 1, the MC estimate of the achieved P{CS} for the T-B procedure, i.e., $\bar{p} = 0.7361$, is significantly lower (statistically) than the specified $P^* = 0.75$. It has been brought to our attention by Professor Ajit Tamhane that this result is explained by the fact that the constants $(\hat{c}_1, \hat{c}_2, \hat{d})$ for $k = 4$, $P^* = 0.75$ in Table II of T-B (1979) on which $(\hat{n}_1, \hat{n}_2, \hat{h})$ are based are in error because they do not satisfy the condition $(\hat{c}_1)^2 + (\hat{c}_2)^2 > (c_{k,P^*})^2$; this condition is necessary (but not sufficient) in order for the two-stage procedure T-B to guarantee the same $\{\delta^*, P^*\}$ condition that the corresponding single-stage procedure B guarantees. The violation of this condition results in $\hat{n}_1$ and/or $\hat{n}_2$ being too small, i.e., $\hat{n}_1 + \hat{n}_2 < n$. Thus for $k = 4$, $P^* = 0.75$, $\delta^* = 0.2$, we have $\hat{n}_1 + \hat{n}_2 = 34 + 32 < n = 71$ which results in the specified P^* value not being attained in the LF-configuration. Several other entries in Table II of T-B (1979) for $P^* = 0.75$ suffer from the same error (although the discrepancies are quite small).

6 Concluding Remarks

We note from Table 5 that neither the single-stage procedure (B) nor the two-stage procedure (T-B) dominates either in terms of $\bar{n}$ or $\bar{t}$ for any configuration or any P^*-value. However, both of these procedures play important roles in real-life situations in which truly sequential experimentation is not a feasible possibility. This is the case, for example, in agriculture where a stage is usually a growing season. Most such experiments would thus require a single-stage procedure; if a two-stage procedure is feasible, then T-B might well be preferable to B, since the former uniformly dominates the latter in terms of $E\{T\}$ while still guaranteeing (2.2).

We mention that at the time that T-B (1979) was written, the LF-configuration for that two-stage procedure had not been determined for $k > 2$. As a consequence, a conservative lower bound on the P{CS} was used which, although it guaranteed (2.2), resulted in a larger $E\{T\}$ than would have been the case if the LF-configuration had been known, and corresponding values of $(\hat{n}_1, \hat{n}_2, \hat{h})$ based on that bound had been calculated. However, Miescke and Sehr (1980) established the LF-configuration for $k = 3$, and more recently, Sehr (1988), and independently Bhandari and Chaudhuri (1987) have established the LF-configuration for all k. Calculations based on the now-known (hitherto conjectured) LF-configuration could be expected to yield substantial decreases in $E\{T\}$ (relative to the present $E\{T\}$-values), especially for large k and/or low P^*. But calculations based on the exact $P\{\text{CS} \mid \text{LF}\}$ given by equation (5.6) of T-B (1977) with δ replaced by δ^* would be prohibitively expensive, at least at the present time; this calculation involves the numerical evaluation of a four-fold integral.

For minimizing $E\{N\}$ when sequential experimentation is feasible, (B-K-S)$_T$ would appear to be preferable except for configurations which are suspected as being close to the EM-configuration when P^* is high, in which situation (P-H, $\delta^*/2$) should be used; however, tables of n_0-values for (B-K-S)$_T$ have been prepared only for $k = 2(1)5$. For $k > 5$ we recommend the use of (P-H, $\delta^*/4$) for all but the EM-configuration, and (P-H, $\delta^*/2$) for the latter configuration. The beauty of (P-H, $\delta^*/4$) or (P-H, $\delta^*/2$) is that in order to implement the procedure it is only necessary to determine the appropriate θ-value for the particular (k, P^*) of interest; this is a trivial calculation.

Acknowledgments: This research was partially supported by the U.S. Army Research office through the Mathematical Sciences Institute of Cornell University. The authors wish to thank Mr. Mark Hartmann for providing us with a copy of his article on an improvement of the Paulson-Fabian procedure; he has approved our including his improved procedure among our calculations. We also acknowledge the assistance of Professor Tom Santner who calculated for us the exact P{CS}-values achieved by the single-stage procedure, and made many constructive suggestions. We are indebted to Professor Ajit Tamhane who solved the mystery reported on in Remark 5.1, and for helpful comments.

References

Bechhofer, R.E. (1954). A single-sample multiple decision procedure for ranking means of normal populations with known variances. *Ann. Math. Statist.* 25, 16–39.

Bechhofer, R.E. and Goldsman, D.M. (1987). Truncation of the Bechhofer-Kiefer-Sobel sequential procedure for selecting the normal population which has the largest mean. *Comm. Statist. — Simula. Computa.* B16(4), 1067–1091.

Bechhofer, R.E. and Goldsman, D.M. (1989). Truncation of the Bechhofer-Kiefer-Sobel sequential procedure for selecting the normal population which has the largest mean (III): supplementary truncation numbers and resulting performance characteristics. *Comm. Statist. — Simula. Computa.* B 18(1), 63–81.

Bechhofer, R.E., Kiefer, J. and Sobel, M. (1968). *Sequential Identification and Ranking Procedures (with special reference to Koopman-Darmois populations)*, Chicago, The University of Chicago Press.

Bhandari, S.K. and Chaudhuri, A.R. (1987). On two conjectures about two-stage selection procedures. To appear in *Sankhyā*, Ser. B.

Fabian, V. (1974). Note on Anderson's sequential procedures with triangular boundary. *Ann. Statist.* 2, 170–176.

Hartmann, M. (1988). An improvement on Paulson's sequential ranking procedure. *Comm. Statist. — Sequential Analysis* 7 (4), 363–372.

Miescke, K.–J. and Sehr, J. (1980). On a conjecture concerning the least favorable configuration of certain two-stage selection procedures. *Comm. Statist. — Theor. Meth.* A9, 1609–1617.

Paulson, E. (1964). A sequential procedure for selecting the population with the largest mean from k normal populations. *Ann. Math. Statist.* 35, 174–180.

Sehr, J. (1988). On a conjecture concerning the least favorable configuration of a two-stage selection procedure. *Comm. Statist. — Theor. Meth.* A17 (10), 3221–3235.

Tamhane, A.C. and Bechhofer, R.E. (1977). A two-stage minimax procedure with screening for selecting the largest normal mean. *Comm. Statist. — Theor. Meth.* A6, 1003–1033.

Tamhane, A.C. and Bechhofer, R.E. (1979). A two-stage minimax procedure with screening for selecting the largest normal mean (II): a new

PCS lower bound and associated tables. *Comm. Statist. — Theor. Meth.* A8, 337-358.

18

Parametric Empirical Bayes Rules for Selecting the Most Probable Multinomial Event

Shanti S. Gupta[1]
TaChen Liang[2]

ABSTRACT Consider a multinomial population with $k(\geq 2)$ cells and the associated probability vector $\underset{\sim}{p} = (p_1, \ldots, p_k)$. Let $p_{[k]} = \max_{1 \leq i \leq k} p_i$. A cell associated with $p_{[k]}$ is called the most probable event. We are interested in selecting the most probable event. Let i denote the index of the selected cell. Under the loss function $L(\underset{\sim}{p}, i) = p_{[k]} - p_i$, this statistical selection problem is studied via a parametric empirical Bayes approach. Two empirical Bayes selection rules are proposed. They are shown to be asymptotically optimal at least of order $0\big(\exp(-c_i n)\big)$ for some positive constants c_i. $i = 1, 2$, where n is the number of accumulated past experiences (observations) at hand. Finally, for the problem of selecting the least probable event associated with $p_{[1]}$ under the loss $p_i - p_{[1]}$, two empirical Bayes selection rules are also proposed. The corresponding rates of convergence are found to be at least of order $0\big(\exp(-c_i n)\big)$ for some positive constants c_i, $i = 3, 4$.

1 Introduction

Consider a multinomial population with $k \geq 2$ cells and the associated probability vector $\underset{\sim}{p} = (p_1, \ldots, p_k)$ where $\sum_{i=1}^{k} p_i = 1$. Let $p_{[1]} \leq \cdots \leq p_{[k]}$ denote the ordered values of the parameters $p_1, \ldots, p_k$. It is assumed that the exact pairing between the ordered and the unordered parameters is unknown. Any event associated with $p_{[k]}$ is considered as the most probable event. A number of statistical procedures based on single samples or sequential sampling rules have been considered in the literature in the classical framework for selecting the most probable event. Bechhofer, Elmaghraby and Morse (1959) have considered a fixed sample procedure through the indifference zone approach. Gupta and Nagel (1967), Panchapakesan (1971) and Gupta and Huang (1975) have studied this selection problem using a

[1]Department of Statistics, Purdue University
[2]Department of Mathematics, Wayne State University

subset selection approach. Cacoullos and Sobel (1966), Alam (1971), Alam, Seo and Thompson (1971), Ramey and Alam (1979, 1980) and Bechhofer and Kulkarni (1984) have considered sequential selection procedures.

We now consider a situation in which one repeatedly deals with the same selection problem independently. In such instances, it is reasonable to formulate the component problem in the sequence as a Bayes decision problem with respect to an unknown (or partially known) prior distribution on the parameter space, and then, use the accumulated observations to improve the decision rule at each stage. This is the empirical Bayes approach due to Robbins (1956, 1964 and 1983).

Empirical Bayes rules have been derived for subset selection goals by Deely (1965). Recently, Gupta and Hsiao (1983) and Gupta and Leu (1988) have studied empirical Bayes rules for selecting good populations with respect to a standard or a control with the underlying populations being uniformly distributed. Gupta and Liang (1986, 1988) have studied empirical Bayes rules for the problem of selecting the best binomial population or selecting good binomial populations. Many such empirical Bayes procedures have been shown to be asymptotically optimal in the sense that the risk for the nth decision problem converges to the optimal Bayes risk which could have been obtained if the prior distribution was fully known and the Bayes procedure with respect to this prior distribution was used.

Note that the above mentioned empirical Bayes rules use the so-called nonparametric empirical Bayes approach. That is, one assumes that the form of the prior distribution is unknown. However, in many cases, an experimenter may have some prior information about the parameters of interest, and he would like to use this information to make appropriate decisions. Usually, it is suggested (for example, see Robbins (1964)), that the prior information be quantified through a class of subjectively plausible priors. In view of this situation, in this paper, it is assumed that the parameters of interest in a multinomial distribution follow some conjugate prior distribution with unknown hyperparameters. Under this statistical framework, two empirical Bayes selection rules are proposed. They are shown to be asymptotically optimal at least of order $0\big(\exp(-c_i n)\big)$ for some positive constants c_i, $i = 1, 2$, where n is the number of accumulated past experiences (observations) at hand. Finally, for the problem of selecting the least probable event associated with $p_{[1]}$ under the loss $p_i - p_{[1]}$, two empirical Bayes selection rules are also proposed. The corresponding rates of convergence are found to be at least of order $0\big(\exp(-c_i n)\big)$ for some positive constants c_i, $i = 3, 4$.

2 Formulation of the Problem under the Empirical Bayes Approach

Consider a multinomial population with $k(\geq 2)$ cells, where the cell π_i has probability p_i, $i = 1, \ldots, k$. Let X_i denote the observations that arise in the cell π_i based on $N(\geq 2)$ independent trials. Thus, for given $\underset{\sim}{p} = (p_1, \ldots, p_k)$, $\underset{\sim}{X} = (X_1, \ldots, X_k)$ has the probability function

$$f(\underset{\sim}{x}|\underset{\sim}{p}) = \frac{N!}{\prod_{i=1}^{k}(x_i!)} \prod_{i=1}^{k} p_i^{x_i}, \tag{2.1}$$

where, $x_i = 0\ ,1,\ \ldots,\ N$ and $\sum_{i=1}^{k} x_i = N$.

For each $\underset{\sim}{p}$, let $p_{[1]} \leq \cdots \leq p_{[k]}$ denote the ordered parameters $p_1, \ldots, p_k$. It is assumed that there is no a priori knowledge about the exact pairing between the ordered and the unordered parameters. Any cell π_i associated with $p_{[k]}$ is considered as the most probable event. Our goal is to derive empirical Bayes rules to select the most probable event.

Let $\Omega = \{\underset{\sim}{p} \mid \underset{\sim}{p} = (p_1, \ldots, p_k),\ 0 < p_i < 1 \text{ and } \sum_{i=1}^{k} p_i = 1\}$ be the parameter space. It is assumed that $\underset{\sim}{p}$ has a Dirichlet prior distribution G with hyperparameters $\underset{\sim}{\alpha} = (\alpha_1, \ldots, \alpha_k)$, where all α_i are positive but unknown. That is, $\underset{\sim}{p}$ has a probability density function of the form

$$g(\underset{\sim}{p}) = \frac{\Gamma(\alpha_0)}{\prod_{i=1}^{k}\Gamma(\alpha_i)} \prod_{i=1}^{k} p_i^{\alpha_i - 1}, \qquad 0 < p_i < 1,\ \sum_{i=1}^{k} p_i = 1, \tag{2.2}$$

where $\alpha_0 = \sum_{i=1}^{k} \alpha_i$.

Let $\mathcal{A} = \{\, i \mid i = 1, \ldots, k\,\}$ be the action space. When action i is taken, it means that the cell π_i is selected as the most probable event. For the parameter $\underset{\sim}{p}$ and action i, the loss function $L(\underset{\sim}{p}, i)$ is defined as

$$L(\underset{\sim}{p}, i) = p_{[k]} - p_i, \tag{2.3}$$

the difference between the most probable and the selected event.

Let $\mathcal{X}$ be the sample space of $\underset{\sim}{X} = (X_1, \ldots, X_k)$. A selection rule $d = (d_1, \ldots, d_k)$ is a mapping from $\mathcal{X}$ into $[0,1]^k$ such that for each $\underset{\sim}{x}\epsilon\mathcal{X}$, the function $d(\underset{\sim}{x}) = \big(d_1(\underset{\sim}{x}), \ldots, d_k(\underset{\sim}{x})\big)$ is such that $0 \leq d_i(\underset{\sim}{x}) \leq 1,\ i = 1, \ldots, k$, and $\sum_{i=1}^{k} d_i(\underset{\sim}{x}) = 1$. Note that $d_i(\underset{\sim}{x}),\ i = 1, \ldots, k$ is the probability of selecting cell π_i as the most probable event given $\underset{\sim}{X} = \underset{\sim}{x}$.

Let D be the class of all selection rules as defined above. For each $d\epsilon D$, let $r(G,d)$ denote the associated Bayes risk. Then $r(G) = \inf_{d\epsilon D} r(G,d)$ is the minimum Bayes risk.

For each $\underset{\sim}{x} \epsilon \mathcal{X}$, let

$$A(\underset{\sim}{x}) = \{\, i \mid x_i + \alpha_i = \max_{1 \le j \le k} (x_j + \alpha_j) \,\}. \tag{2.4}$$

Consider the selection rule $d_G = (d_{1G}, \ldots, d_{kG})$ defined below: for each $i = 1, \ldots, k$,

$$d_{iG} = d_{iG}(\underset{\sim}{x}) = \begin{cases} |A(\underset{\sim}{x})|^{-1} & \text{if } i \epsilon A(\underset{\sim}{x}), \\ 0 & \text{otherwise,} \end{cases} \tag{2.5}$$

where $|A|$ denotes the cardinality of the set A.

It should be noted that in (2.5) any selection rule $d = (d_1, \ldots, d_k)$ satisfying the condition $\sum_{i \in A(\underset{\sim}{x})} d_i(\underset{\sim}{x}) = 1$ is a Bayes selection rule.

A straightforward computation shows that the selection rule d_G is a randomized Bayes selection rule in the class D. Since the values of the hyperparameters $(\alpha_1, \ldots, \alpha_k)$ are unknown, it is impossible to apply this Bayes selection rule d_G for the selection problem at hand. As we mentioned above, we study this selection problem via an empirical Bayes approach.

For each $j = 1, 2, \ldots$, let $\underset{\sim}{X}_j = (X_{1j}, \ldots, X_{kj})$ denote the random observations arising from N independent trials at stage j. Let $\underset{\sim}{P}_j = (P_{1j}, \ldots, P_{kj})$ denote the (random) parameters at stage j. Conditional on $\underset{\sim}{P}_j$, $\underset{\sim}{X}_j$ has a probability function of the form of (2.1). It is assumed that independent observations $\underset{\sim}{X}_1, \ldots, \underset{\sim}{X}_n$ are available, and $\underset{\sim}{P}_j$, $1 \le j \le n$, have the same prior probability density function of the form (2.2), though not observable. We also let $\underset{\sim}{X}_{n+1} = \underset{\sim}{X} = (X_1, \ldots, X_k)$ denote the present observations.

Two empirical Bayes selection rules are proposed depending on whether the value of the parameter α_0 is known or unknown. Note that α_0 is the sum of all the parameters α_i, $1 \le i \le k$. In the case that α_0 is known, the individual values of α_i, $1 \le i \le k$, are still unknown.

First, for each $i = 1, \ldots, k$, and each $n = 1, 2, \ldots$, we let

$$\bar{X}_i(n) = \frac{1}{n} \sum_{j=1}^{n} X_{ij},\, M_i(n) = \frac{1}{n} \sum_{j=1}^{n} X_{ij}^2,\, Z_i(n) = [N\bar{X}_i(n) - M_i(n)]\bar{X}_i(n),$$
$$Y_i(n) = [M_i(n) - \bar{X}_i(n)]N - (N-1)\big(\bar{X}_i(n)\big)^2. \tag{2.6}$$

When α_0 is known, let

$$\hat{\alpha}_{in} = \alpha_0 \bar{X}_i(n) N^{-1}, \tag{2.7}$$

and let

$$A_n(\underset{\sim}{x}) = \{\, i \mid x_i + \hat{\alpha}_{in} = \max_{1 \le j \le k} (x_j + \hat{\alpha}_{jn}) \,\}. \tag{2.8}$$

We then define an empirical Bayes selection rule $\tilde{d}_n = (\tilde{d}_{1n}, \ldots, \tilde{d}_{kn})$ as follows: for each $i = 1, \ldots, k$, $\underset{\sim}{x} \epsilon \mathcal{X}$,

$$\tilde{d}_{in}(\underset{\sim}{x}) = \begin{cases} |A_n(\underset{\sim}{x})|^{-1} & \text{if } i \epsilon A_n(\underset{\sim}{x}), \\ 0 & \text{otherwise.} \end{cases} \tag{2.9}$$

Let $\mu_{i1} = E[\bar{X}_i(n)]$ and $\mu_{i2} = E[M_i(n)]$. Then, following a direct computation, we have $\mu_{i1} = N\alpha_i\alpha_0^{-1}$, $\mu_{i2} = N\alpha_i\alpha_0^{-1} + (N^2 - N)\alpha_i(\alpha_i + 1)\alpha_0^{-1}(\alpha_0 + 1)^{-1}$. Hence, $\alpha_i = L_{i1}L_{i2}^{-1}$, where $L_{i1} = (N\mu_{i1} - \mu_{i2})\mu_{i1}, L_{i2} = (\mu_{i2} - \mu_{i1})N - (N-1)\mu_{i1}^2$. Thus, $Z_i(n), Y_i(n)$, and $Z_i(n)/Y_i(n)$ are moment estimators of L_{i1}, L_{i2}, and $\alpha_i = L_{i1}L_{i2}^{-1}$, respectively. Note that L_{i1} and L_{i2} are both positive, which can be verified directly by the definition of μ_{i1} and μ_{i2}. Also, $Z_i(n) \geq 0$. However, it is possible that $Y_i(n) \leq 0$. Hence, for the case when α_0 is unknown, we first let

$$\Delta_{in}(x_i) = \begin{cases} x_i + Z_i(n)/Y_i(n) & \text{if } Y_i(n) > 0, \\ x_i & \text{otherwise.} \end{cases} \tag{2.10}$$

Also, let

$$A_n^*(\underset{\sim}{x}) = \{\, i \mid \Delta_{in}(x_i) = \max_{1 \leq j \leq k} \Delta_{jn}(x_j) \,\}. \tag{2.11}$$

We then define an empirical Bayes selection rule $d_n^* = (d_{1n}^*, \ldots, d_{kn}^*)$ as follows: for each $i = 1, \ldots, k$, $\underset{\sim}{x} \epsilon \mathcal{X}$,

$$d_{in}^*(\underset{\sim}{x}) = \begin{cases} |A_n^*(\underset{\sim}{x})|^{-1} & \text{if } i \epsilon A_n^*(\underset{\sim}{x}), \\ 0 & \text{otherwise.} \end{cases} \tag{2.12}$$

In the next section, we will study the optimality of the two sequences of empirical Bayes selection rules $\{\tilde{d}_n\}$ and $\{d_n^*\}$.

3 Asymptotic Optimality of Selection Rules $\{\tilde{d}_n\}$ and $\{d_n^*\}$

Consider an empirical Bayes selection rule $d_n(\underset{\sim}{x})$. Let $r(G, d_n)$ be the Bayes risk associated with the selection rule $d_n(\underset{\sim}{x})$. Then $r(G, d_n) - r(G) \geq 0$, since $r(G)$ is the minimum Bayes risk. The nonnegative difference is always used as a measure of optimality of the selection rule d_n.

Definition 3.1 *A sequence of empirical Bayes rules $\{d_n\}_{n=1}^{\infty}$ is said to be asymptotically optimal at least of order β_n relative to the prior distribution G if $r(G, d_n) - r(G) \leq 0(\beta_n)$ as $n \to \infty$, where $\{\beta_n\}$ is a sequence of positive values satisfying $\lim_{n\to\infty} \beta_n = 0$.*

Asymptotic optimality of $\{\tilde{d}_n\}$

We first consider the case where α_0 is known. Note that $\hat{\alpha}_{in}$ is an unbiased estimator of α_i; also $\sum_{i=1}^k \hat{\alpha}_{in} = \alpha_0$ for each $n = 1, 2, \ldots$.

For each $\underset{\sim}{x} \epsilon \mathcal{X}$, let $A(\underset{\sim}{x})$ be as defined in (2.4) and let $B(\underset{\sim}{x}) = \{1, 2, \ldots, k\} \setminus A(\underset{\sim}{x})$. Thus, for each $\underset{\sim}{x} \epsilon \mathcal{X}$, $i \epsilon A(\underset{\sim}{x})$, $j \epsilon B(\underset{\sim}{x})$, $x_i + \alpha_i > x_j + \alpha_j$. Following straightforward computation, we can show

$$0 \leq r(G, \tilde{d}_n) - r(G)$$

$$\leq \sum_{\underset{\sim}{x} \epsilon \mathcal{X}} \sum_{i \epsilon A(\underset{\sim}{x})} \sum_{j \epsilon B(\underset{\sim}{x})} P\{x_i + \hat{\alpha}_{in} \leq x_j + \hat{\alpha}_{jn}\}. \tag{3.1}$$

Now, for $i \epsilon A(\underset{\sim}{x})$, $j \epsilon B(\underset{\sim}{x})$,

$$\begin{aligned}
P\{x_i + \hat{\alpha}_{in} &\leq x_j + \hat{\alpha}_{jn}\} \\
&= P\left\{\frac{1}{n}\sum_{m=1}^{n}\left(\frac{1}{N}(X_{im} - X_{jm}) - \frac{1}{\alpha_0}(\alpha_i - \alpha_j)\right)\right. \\
&\qquad \left. < -(x_i + \alpha_i - x_j - \alpha_j)\alpha_0^{-1}\right\} \\
&\leq P\left\{\frac{1}{n}\sum_{m=1}^{n}\left(\frac{1}{N}(X_{im} - X_{jm}) - \frac{1}{\alpha_0}(\alpha_i - \alpha_j)\right) < -\varepsilon_{ij}\right\} \\
&\leq \exp\{-n2^{-1}\varepsilon_{ij}^2\} \\
&\leq \exp\{-nc_1\},
\end{aligned} \tag{3.2}$$

where

$$\begin{aligned}
\varepsilon_{ij} &= \min_{x_i, x_j} \left\{ \begin{array}{l} |x_i + \alpha_i - x_j - \alpha_j|\alpha_0^{-1} \mid x_i, x_j = 0, 1, \ldots, N, \\ 0 \leq x_i + x_j \leq N,\ x_i + \alpha_i - x_j - \alpha_j \neq 0 \end{array} \right\} \\
&> 0 \qquad \text{since } N \text{ is a finite number,}
\end{aligned}$$

and

$$c_1 = 2^{-1}\min\{\, \varepsilon_{ij}^2 \mid i, j = 1, \ldots, k,\ i \neq j\,\} > 0.$$

In (3.2), the second inequality is obtained using the fact that

$$E\left[\frac{1}{N}(X_{im} - X_{jm}) - \frac{1}{\alpha_0}(\alpha_i - \alpha_j)\right] = 0,$$

$$-1 - \frac{1}{\alpha_0}(\alpha_i - \alpha_j) \leq \frac{1}{N}(X_{im} - X_{jm}) - \frac{1}{\alpha_0}(\alpha_i - \alpha_j) \leq 1 - \frac{1}{\alpha_0}(\alpha_i - \alpha_j)$$

and then making use of Theorem 2 of Hoeffding (1963).

By noting that $\mathcal{X}$ is a finite space, from (3.1) and (3.2), we have the following theorem.

Theorem 3.2 *Let $\{\tilde{d}_n\}$ be the sequence of empirical Bayes selection rules defined in (2.9). Then $r(G, \tilde{d}_n) - r(G) \leq 0(\exp(-c_1 n))$ for some positive constant c_1.*

Asymptotic optimality of $\{d_n^*\}$

For each $\underset{\sim}{x} \epsilon \mathcal{X}$, let $A(\underset{\sim}{x})$ and $B(\underset{\sim}{x})$ be as defined in the previous sections. For the selection rule d_n^*, one can obtain the following result

$$\begin{aligned}
0 &\leq r(G, d_n^*) - r(G) \\
&\leq \sum_{\underset{\sim}{x} \epsilon \mathcal{X}} \sum_{i \epsilon A(\underset{\sim}{x})} \sum_{j \epsilon B(\underset{\sim}{x})} P\{\Delta_{in}(x_i) \leq \Delta_{jn}(x_j)\}.
\end{aligned} \tag{3.3}$$

Since $\mathcal{X}$ is finite, we only need to consider the behavior of $P\{\Delta_{in}(x_i) \leq \Delta_{jn}(x_j)\}$ for each $\underset{\sim}{x}\epsilon\mathcal{X}$. Now

$$\begin{aligned} P\{\Delta_{in}(x_i) \leq \Delta_{jn}(x_j)\} \\ = P\{\Delta_{in}(x_i) \leq \Delta_{jn}(x_j) \text{ and } (Z_i(n) \leq 0 \\ \text{or } Z_j(n) \leq 0 \text{ or } Y_i(n) \leq 0 \text{ or } Y_j(n) \leq 0)\} \\ + P\{\Delta_{in}(x_i) \leq \Delta_{jn}(x_j) \text{ and } (Z_i(n) > 0, \\ Z_j(n) > 0, Y_i(n) > 0 \text{ and } Y_j(n) > 0)\}. \end{aligned} \tag{3.4}$$

Before we go further to study the associated asymptotic behaviors of the above probabilities appearing on the right hand side of (3.4), we need the following lemma.

Lemma 3.3 *Let $b > 0$ be a constant. Then,*

a) $P\{Z_i(n) - L_{i1} < -b\} \leq 0(\exp(-b_i n))$;

b) $P\{Z_i(n) - L_{i1} > b\} \leq 0(\exp(-b_i n))$;

c) $P\{Y_i(n) - L_{i2} < -b\} \leq 0(\exp(-b_i n))$;

d) $P\{Y_i(n) - L_{i2} > b\} \leq 0(\exp(-b_i n))$;

where $b_i = b^2[2N^4(N + \mu_{i1})^2]^{-1} > 0$.

Proof: The techniques used to prove these four inequalities are similar. Here, we give the proof of part a) only.

Note that $Z_i(n) = [N\bar{X}_i(n) - M_i(n)]\bar{X}_i(n) \geq 0$. Hence, $P\{Z_i(n) - L_{i1} < -b\} = 0$ if $L_{i1} - b \leq 0$. So, we assume that $b > 0$ is small enough so that $L_{i1} - b > 0$. Then,

$$\begin{aligned} P\{Z_i(n) - L_{i1} < -b\} \\ = P\{N[(\bar{X}_i(n))^2 - \mu_{i1}^2] - [M_i(n)\bar{X}_i(n) - \mu_{i2}\mu_{i1}] < -b\} \\ \leq P\{\bar{X}_i(n) - \mu_{i1} < -b(2N(N + \mu_{i1}))^{-1}\} \\ + P\{\bar{X}_i(n) - \mu_{i1} > b(4N^2)^{-1}\} + P\{M_i(n) - \mu_{i2} > b(4\mu_{i1})^{-1}\} \\ \leq \exp\{-nb^2[2N^4(N + \mu_{i1})^2]^{-1}\} \\ + \exp\{-nb^2[8N^4]^{-1}\} + \exp\{-nb^2[8N^4\mu_{i1}]^{-1}\} \\ \leq 0(\exp(-nb_i)). \end{aligned} \tag{3.5}$$

Note that in (3.5), the first inequality is obtained from the fact that $0 \leq \bar{X}_i(n) \leq N, 0 \leq M_i(n) \leq N^2$ and an application of Bonferroni's inequality; the second inequality follows from an application of Theorem 2 of Hoeffding (1963) and the last inequality is obtained from the definition of b_i.

Hence, the proof of part a) is complete.

By the positivity of L_{i1} and L_{i2}, and by Lemma 3.3,

$$P\left\{\begin{array}{l}\Delta_{in}(x_i) \le \Delta_{jn}(x_j) \text{ and} \\ (Z_i(n) \le 0 \text{ or } Z_j(n) \le 0 \text{ or } Y_i(n) \le 0 \text{ or } Y_j(n) \le 0)\end{array}\right\}$$
$$\le 0\Big(\exp\big(-n \min(b_i, b_j)\big)\Big)$$
$$= 0\big(\exp(-nb_{ij})\big), \text{ where } b_{ij} = \min\,(b_i, b_j). \tag{3.6}$$

Therefore, we then only need to consider the asymptotic behavior of $P\{\Delta_{in}(x_i) \le \Delta_{jn}(x_j)$ and $(Z_i(n) > 0, Z_j(n) > 0, Y_i(n) > 0$ and $Y_j(n) > 0)\}$.

Let $Q_{ij} = (x_i - x_j)L_{i2}L_{j2} + L_{i1}L_{j2} - L_{i2}L_{j1}$. Then $Q_{ij} > 0$ if $i\epsilon A(\underset{\sim}{x})$ and $j\epsilon B(\underset{\sim}{x})$. Therefore,

$$P\left\{\begin{array}{l}\Delta_{in}(x_i) \le \Delta_{jn}(x_j) \text{ and} \\ (Z_i(n) > 0,\ Z_j(n) > 0,\ Y_i(n) > 0 \text{ and } Y_j(n) > 0)\end{array}\right\}$$
$$\le P\{(x_i - x_j)[Y_i(n)Y_j(n) - L_{i2}L_{j2}] < -Q_{ij}/3\}$$
$$+ P\{Z_i(n)Y_j(n) - L_{i1}L_{j2} < -Q_{ij}/3\}$$
$$+ P\{Y_i(n)Z_j(n) - L_{i2}L_{j1} > Q_{ij}/3\}. \tag{3.7}$$

With repeated applications of Bonferroni's inequality, we have the following inequalities:

$$P\{(x_i - x_j)[Y_i(n)Y_j(n) - L_{i2}L_{j2}] < -Q_{ij}/3\}$$
$$\le P\{Y_i(n) - L_{i2} < -Q_{ij}(6N^4)^{-1}\}$$
$$+ P\{Y_j(n) - L_{j2} < -Q_{ij}(6NL_{i2})^{-1}\}$$
$$\text{if } x_i > x_j; \tag{3.8a}$$

$$P\{(x_i - x_j)[Y_i(n)Y_j(n) - L_{i2}L_{j2}] < -Q_{ij}/3\}$$
$$\le P\{Y_i(n) - L_{i2} > Q_{ij}(6N^4)^{-1}\}$$
$$+ P\{Y_j(n) - L_{j2} > Q_{ij}(6NL_{i2})^{-1}\}$$
$$\text{if } x_i < x_j; \tag{3.8b}$$

$$P\{(x_i - x_j)[Y_i(n)Y_j(n) - L_{i2}L_{j2}] < -Q_{ij}/3\} = 0$$
$$\text{if } x_i = x_j; \tag{3.8c}$$

$$P\{Z_i(n)Y_j(n) - L_{i1}L_{j2} < -Q_{ij}/3\}$$
$$\le P\{Z_i(n) - L_{i1} < -Q_{ij}(6N^3)^{-1}\}$$
$$+ P\{Y_j(n) - L_{j2} < -Q_{ij}(6L_{i1})^{-1}\}; \tag{3.9}$$

and

$$P\{Y_i(n)Z_j(n) - L_{i2}L_{j1} > Q_{ij}/3\}$$
$$\le P\{Y_i(n) - L_{i2} > Q_{ij}(6N^3)^{-1}\}$$
$$+ P\{Z_j(n) - L_{j1} > Q_{ij}(6L_{i2})^{-1}\}. \tag{3.10}$$

Then, by Lemma 3.3 and from Equations (3.7) through (3.10), we conclude that

$$P\left\{\begin{array}{l}\Delta_{in}(x_i) \le \Delta_{jn}(x_j) \text{ and} \\ (Z_i(n) > 0,\ Z_j(n) > 0,\ Y_i(n) > 0 \text{ and } Y_j(n) > 0)\end{array}\right\}$$
$$\le 0\big(\exp(-na_{ij})\big) \text{ for some } a_{ij} > 0. \tag{3.11}$$

Now, from (3.4), (3.6) and (3.11), for each $\underset{\sim}{x}\epsilon\mathcal{X}$, $i\epsilon A(\underset{\sim}{x})$ and $j\epsilon B(\underset{\sim}{x})$,

$$P\{\Delta_{in}(x_i) \le \Delta_{jn}(x_j)\} \le 0\Big(\exp\big(-n \min(b_{ij}, a_{ij})\big)\Big). \tag{3.12}$$

Now, let $c_2 = \min_{i \ne j}\{\min(b_{ij}, a_{ij})\}$. Then $c_2 > 0$.

Based on the preceding, we have the following result.

Theorem 3.4 *Let $\{d_n^*\}$ be the sequence of empirical Bayes selection rules defined in (2.12). Then $r(G, d_n^*) - r(G) \le 0\big(\exp(-c_2 n)\big)$ for some positive constant c_2.*

Remark: Another selection problem related to the multinomial distribution is to select the least probable event; that is, to select the cell associated with $p_{[1]}$. If we consider the loss function

$$L(\underset{\sim}{p}, i) = p_i - p_{[1]}, \tag{3.13}$$

the difference between the selected and the least probable event, then under the statistical model described in Section 2, a uniformly randomized Bayes selection rule is $d_G = (d_{1G}, \ldots, d_{kG})$, where, for each $i = 1, \ldots, k$,

$$d_{iG} = d_{iG}(\underset{\sim}{x}) = \begin{cases} |\tilde{A}(\underset{\sim}{x})|^{-1} & \text{if } i\epsilon\tilde{A}(\underset{\sim}{x}), \\ 0 & \text{otherwise,} \end{cases} \tag{3.14}$$

and

$$\tilde{A}(\underset{\sim}{x}) = \{\, i \mid x_i + \alpha_i = \min_{1\le j\le k} (\alpha_j + x_j) \,\}. \tag{3.15}$$

Let $\hat{\alpha}_{in}, \Delta_{in}(x_i)$ be defined as in (2.7) and (2.10), respectively. When α_0 is known, we let

$$\tilde{A}_n(\underset{\sim}{x}) = \{\, i \mid x_i + \hat{\alpha}_{in} = \min_{1\le j\le k} (x_j + \hat{\alpha}_{jn}) \,\}, \tag{3.16}$$

and define a randomized selection rule $\tilde{d}_n(\underset{\sim}{x}) = \big(\tilde{d}_{1n}(\underset{\sim}{x}), \ldots, \tilde{d}_{kn}(\underset{\sim}{x})\big)$ as follows:

$$\tilde{d}_{in}(\underset{\sim}{x}) = \begin{cases} |\tilde{A}_n(\underset{\sim}{x})|^{-1} & \text{if } i\epsilon\tilde{A}_n(\underset{\sim}{x}), \\ 0 & \text{otherwise.} \end{cases} \tag{3.17}$$

When α_0 is unknown, we let

$$\tilde{A}_n^*(\underset{\sim}{x}) = \{\, i \mid \Delta_{in}(x_i) = \min_{1\le j\le k} \Delta_{jn}(x_j) \,\}, \tag{3.18}$$

and define a randomized selection rule $d_n^*(\underset{\sim}{x}) = \big(d_{1n}^*(\underset{\sim}{x}), \ldots, d_{kn}^*(\underset{\sim}{x})\big)$ as follows:

$$d_{in}^*(\underset{\sim}{x}) = \begin{cases} |\tilde{A}_n^*(\underset{\sim}{x})|^{-1} & \text{if } i\epsilon\tilde{A}_n^*(\underset{\sim}{x}), \\ 0 & \text{otherwise.} \end{cases} \tag{3.19}$$

Following a discussion analogous to that given earlier for the most probable event, we can see that $\{\tilde{d}_n\}$ and $\{d_n^*\}$ are both asymptotically optimal and have the following convergence rates:

$$0 \le r(G, \tilde{d}_n) - r(G) \le 0\big(\exp(-c_3 n)\big),$$
$$0 \le r(G, d_n^*) - r(G) \le 0\big(\exp(-c_4 n)\big),$$

for some positive constants c_3 and c_4, where $r(G)$ now denotes the minimum Bayes risk with respect to the loss function (3.13).

Acknowledgments: This research was partially supported by the Office of Naval Research Contract N00014-84-C-0167 and NSF Grant DMS-8606964 at Purdue University. Reproduction in whole or in part is permitted for any purpose of the United States Government.

References

Alam, K. (1971). On selecting the most probable category. *Technometrics* **13**, 843–850.

Alam, K., Seo, K. and Thompson, J. R. (1971). A sequential sampling rule for selecting the most probable multinomial event. *Ann. Inst. Statist. Math.* **23**, 365–374.

Bechhofer, R. E., Elmaghraby, S. and Morse, N. (1959). A single- sample multiple-decision procedure for selecting the multinomial event which has the highest probability. *Ann. Math. Statist.* **30**, 102–119.

Bechhofer, R. E. and Kulkarni, R. V. (1984). Closed sequential procedures for selecting the multinomial events which have the largest probabilities. *Commun. Statist.-Theor. Meth.* **A13(24)**, 2997–3031.

Cacoullos, T. and Sobel, M. (1966). An inverse-sampling procedure for selecting the most probable event in a multinomial distribution. *Multivariate Analysis* (Ed. P. R. Krishnaiah), Academic Press, New York, 423–455.

Deely, J. J. (1965). Multiple decision procedures from an empirical Bayes approach. Ph.D. Thesis (Mimeo. Ser. No. 45), Department of Statistics, Purdue University, West Lafayette, Indiana.

Gupta, S. S. and Hsiao, P. (1983). Empirical Bayes rules for selecting good populations. *J. Statist. Plan. Infer.* **8**, 87–101.

Gupta, S. S. and Huang, D. Y. (1975). On subset selection procedures for Poisson populations and some applications to the multinomial selection problems. *Applied Statistics* (Ed. R. P. Gupta), North-Holland, Amsterdam, 97–109.

Gupta, S. S. and Leu, L. Y. (1988). On Bayes and empirical Bayes rules for selecting good populations. To appear in Probability and Information Theory 3, a volume in the *Journal of Statistical Planning and Inference*, (Eds. M. Behara and D. Bierlein).

Gupta, S. S. and Liang, T. (1986). Empirical Bayes rules for selecting good binomial populations. *Adaptive Statistical Procedures and Related Topics* (Ed. J. Van Ryzin), IMS Lecture Notes-Monograph Series, Vol. 8, 110–128.

Gupta, S. S. and Liang, T. (1988). Empirical Bayes rules for selecting the best binomial population. *Statistical Decision Theory and Related Topics-IV* (Eds. S. S. Gupta and J. O. Berger), Vol. 1, 213–224, Springer-Verlag, New York.

Gupta, S. S. and Nagel, K. (1967). On selection and ranking procedures and order statistics from the multinomial distribution. *Sankhyā Ser. B* **29**, 1–17.

Hoeffding, W. (1963). Probability inequalities for sums of bounded random variables. *J. Amer. Statist. Assoc.* **58**, 13–30.

Panchapakesan, S. (1971). On a subset selection procedure for the most probable event in a multinomial distribution. *Statistical Decision Theory and Related Topics* (Eds. S. S. Gupta and J. Yackel), Academic Press, New York, 275–298.

Ramey, Jr. J. T. and Alam, K. (1979). A sequential procedure for selecting the most probable multinomial event. *Biometrika* **66**, 171–173.

Ramey, Jr. J. T. and Alam, K. (1980). A Bayes sequential procedure for selecting the most probable multinomial event. *Commun. Statist.-Theor. Meth.* **A9(3)**, 265–276.

Robbins, H. (1956). An empirical Bayes approach to statistics. *Proc. Third Berkeley Symp. Math. Statist. Probab.* **1**, 157–163, University of California Press.

Robbins, H. (1964). The empirical Bayes approach to statistical decision problems. *Ann. Math. Statist.* **35**, 1–20.

Robbins, H. (1983). Some thoughts on empirical Bayes estimation. *Ann. Statist.* **11**, 713–723.

19

Bayesian Estimation in Two-Way Tables With Heterogeneous Variances

Irwin Guttman[1]
Ulrich Menzefricke[2]

ABSTRACT Consider a two-way table with one observation per cell and heterogeneous variances across the columns. We assume there is available proper prior knowledge about these variances and obtain the joint posterior distribution for the variances, where, however, the normalizing constant has to be evaluated numerically. We use this joint posterior distribution to examine the precision (inverse of the variance) in a given column as a fraction of the sum of all column precisions. An example is discussed.

1 Introduction

We examine the analysis of observations y_{ij} arising from a two-way table according to the model

$$y_{ij} = \gamma + \mu_i + \beta_j + \varepsilon_{ij} \quad (i = 1, \ldots, n; j = 1, \ldots, r), \tag{1.1}$$

where γ is the grand mean, μ_i is due to a row effect, β_j is due to a column effect, and the residuals ε_{ij} are independently and normally distributed with mean 0 and precisions η_j, i.e., variances $1/\eta_j$, which differ across columns. Grubbs (1948) analyzes a similar model from a sampling-theory point of view, and he discusses an example involving

> "individual burning times of power train fuses as measured by each of three observers on 30 rounds of ammunition which were fired from a gun. The fuses were all set for a burning time of ten seconds.
>
> The burning time of a fuse is defined as the interval of time which elapses from the instant the projectile leaves the gun muzzle

[1]Department of Statistics, University of Toronto, Toronto, Ontario, Canada, M5S 1A1

[2]Faculty of Management, University of Toronto, Toronto, Ontario, Canada, M5S 1V4

until the fuse functions the projectile. The times given are measured by means of electric clocks. A switch on the gun muzzle starts three different electric shocks as the gun is fired and each observer stops his clock the instant he sees the flash or burst of an individual round. Each timer, of course, stops his clock independently of the other two timers."

Among other things, Grubbs was interested in whether the three observers' measurement error variances were different or whether the observers were equally precise. To this end he obtained unbiased estimators of the variances. He does not, however, examine appropriate tests of significance.

Grubbs' estimators have received attention in the non-Bayesian literature since then, a recent reference being Brindley and Bradley (1985), who also cited other applications with heterogeneous variances. Brindley and Bradley's contribution was to examine the exact distribution of certain test statistics when $r = 3$, and they give some approximate results when $r > 3$.

In this paper we use a Bayesian approach to obtain the joint posterior distribution of the precisions $\eta_1, \ldots, \eta_r$ and examine, in particular, the joint distribution of $\lambda_1, \ldots, \lambda_{r-1}$, where $\lambda_j = \eta_j / \sum_{i=1}^{r} \eta_i$. The parameters λ_j represent the fraction of the precision due to the jth column effect in our model, (1.1). When $\lambda_j = 1/r$, $j = 1, \ldots, r-1$, the precisions are all equal and the model of heterogeneous precisions simplifies to that of the well known model with homogeneous precisions. In section 2 we derive the joint posterior distributions, and in section 3 we discuss an example.

2 The Joint Posterior Distribution

As a first step, we derive the likelihood function and write it in a form convenient for our purposes. Let

$$Y = \{y_{ij}\} \tag{2.1}$$

$$\underline{\beta} = (\beta_1, \ldots, \beta_r) \qquad \sum_{j=1}^{r} \beta_j = 0$$

$$\underline{\mu} = (\mu_1, \ldots, \mu_n) \qquad \sum_{i=1}^{n} \mu_i = 0$$

$$\underline{\eta} = (\eta_1, \ldots, \eta_r)$$

Because of the normality assumption for the independent ε_{ij}'s, we have that the likelihood function, say ℓ, is such that

$$\ell(\gamma, \underline{\beta}, \underline{\mu}, \underline{\eta} | Y) \propto \left[\prod_{j=1}^{r} \eta_j^{n/2} \right] \exp(-Q/2), \tag{2.2}$$

where

$$Q = \sum_{j=1}^{r}\sum_{i=1}^{n} \eta_j (y_{ij} - \gamma - \mu_i - \beta_j)^2. \tag{2.3}$$

Using standard analysis of variance and Lagrangian techniques, we may write Q as

$$Q = \sum_{j=1}^{r}\sum_{i=1}^{n} S_{ij}\delta_{ij} + \alpha \sum_{i=1}^{n}(\hat{\mu}_i - \mu_i)^2 + n\sum_{j=1}^{r}\eta_j(\hat{\gamma} + \hat{\beta}_j - \gamma - \beta_j)^2 \tag{2.4a}$$

with

$$\bar{y}_j = \sum_i y_{ij}/n, \quad S_{jk} = \sum_i (y_{ij} - \bar{y}_j)(y_{ik} - \bar{y}_k),$$

$$\hat{\mu}_i = \alpha^{-1}\sum_{j=1}^{r}\eta_j(y_{ij} - \bar{y}_j),\ \hat{\beta}_j = \bar{y}_j - \hat{\gamma},\ \hat{\gamma} = \bar{\bar{y}} = \sum\sum y_{ij}/rn \tag{2.4b}$$

and where

$$\alpha = \sum_{j=1}^{r}\eta_j,\ \delta_{ij} = \eta_j - \eta_j^2/\alpha \quad \text{if } i = j \text{ and } \delta_{ij} = -\eta_i\eta_j/\alpha,\ i \neq j. \tag{2.4c}$$

In order to obtain the posterior distribution for $\underline{\beta}$, $\underline{\mu}$, $\underline{\eta}$, and γ, the likelihood of (2.2) must be combined with a prior. We will examine the case where

$$p(\gamma, \underline{\beta}, \underline{\mu}, \underline{\eta}) = p(\gamma, \underline{\beta}, \underline{\mu})p(\underline{\eta}) \tag{2.5}$$

with $(\gamma, \underline{\beta}, \underline{\mu})$ independent a priori of $\underline{\eta}$,

$$p(\gamma, \underline{\beta}, \underline{\mu}) \propto c \tag{2.6}$$

the so-called diffuse prior, and where the η_j's are independent gamma distributions with

$$p(\eta_j) \propto \eta_j^{\nu/2-1}\exp(-\nu S_{0j}\eta_j/2). \tag{2.7}$$

Note the common value ν. (Different values of ν for each η_j are easily incorporated, as is the case of a multivariate normal prior for γ, $\underline{\beta}$, $\underline{\mu}$.)

We can now combine the diffuse prior for γ, $\underline{\beta}$ and $\underline{\mu}$ and that for η_j ($j = 1, \ldots, r$) with the likelihood in (2.2) to get a posterior distribution. After integrating out γ, $\underline{\beta}$ and $\underline{\mu}$, we have the marginal posterior distribution for $\eta_1, \ldots, \eta_r$,

$$p(\underline{\eta}|Y) \propto \alpha^{-(n-1)/2}\left(\prod_{j=1}^{r}\eta_j^{(n+\nu-1)/2-1}\right)\exp\left\{-\frac{1}{2}[\cdot]\right\}, \tag{2.8}$$

where $[\cdot] = \sum_{j=1}^{r}(S_{jj} + \nu S_{0j})\eta_j - \frac{1}{\alpha}\sum_{i,j=1}^{r}\eta_i\eta_j S_{ij}$, and $\alpha = \eta_1 + \cdots + \eta_r$. [Note that the dimension of $(\gamma, \underline{\beta}, \underline{\mu})$ is $r+n-1$, since $\sum_{i=1}^{n}\mu_i = \sum_{j=1}^{r}\beta_j = 0$.]

We remark that a proof of (2.8) utilizing the matrix formulation of (1.1), $\underset{\sim}{y} = X\underset{\sim}{\tau} + \underset{\sim}{\varepsilon}$, for X, $\underset{\sim}{\tau}$ and $\underset{\sim}{\varepsilon}$ specific to (1.1), is given in Guttman and Menzefricke (1988).

Obtaining the normalizing constant for (2.8) in closed form does not seen possible, but some progress can be made by transforming from $\eta_1, \ldots, \eta_r$ to $\lambda_1, \ldots, \lambda_{r-1}$, α, where $\lambda_j = \eta_j/\alpha$. The Jacobian of this transformation is α^{r-1}. Integrating out α, we get

$$p(\lambda_1, \ldots, \lambda_{r-1}|Y) \propto \left(\prod_{j=1}^{r} \lambda_j^{(n+\nu-1)/2-1}\right) / \tilde{Q}^{(r-1)(n-1)/2-r\nu/2}, \tag{2.9}$$

where

$$\tilde{Q} = \sum_{j=1}^{r} (S_{jj} + \nu S_{0j})\lambda_j - \sum_{i,j=1}^{r} S_{ij}\lambda_i\lambda_j, \tag{2.10}$$

and where, of course. $\lambda_r = 1 - \lambda_1 - \cdots - \lambda_{r-1}$.

Note that α is the total precision across all columns and so λ_j is the fraction of the total precision attributable to column j. Of course $\sum_{j=1}^{r} \lambda_j = 1$.

Before turning to an example, let us examine the case when there are only 2 columns, i.e., $r = 2$. Using the prior for η_1 and η_2 in (2.7) and transforming to λ_1 and $\alpha = \eta_1 + \eta_2$, see the discussion before (2.9), we can find the prior for λ_1,

$$p(\lambda_1) \propto \frac{[\lambda_1(1-\lambda_1)]^{\nu/2-1}}{[S_{01}\lambda_1 + S_{02}(1-\lambda_1)]^{\nu}}, \tag{2.11}$$

and we note that, as $\nu \to 0$,

$$p(\lambda_1) \propto [\lambda_1(1-\lambda_1)]^{-1} \tag{2.12}$$

which is improper.

When $r = 2$, the posterior density in (2.9) reduces to

$$p(\lambda_1|Y) \propto \{\lambda_1(1-\lambda_1)\}^{(n+\nu-1)/2-1} / \tilde{Q}^{(n-2\nu-1)/2}, \tag{2.13}$$

where

$$\begin{aligned} \tilde{Q} &= \sum_{j=1}^{2} (S_{jj} + \nu S_{0j})\lambda_j - \sum_{i,j=1}^{2} S_{ij}\lambda_i\lambda_j, \\ &= \nu S_{02} + \nu(S_{01} - S_{02})\lambda_1 + (S_{11} + S_{22} - 2S_{12})\lambda_1(1-\lambda_1). \end{aligned} \tag{2.14}$$

This posterior distribution can be easily be evaluated using numerical methods. When the prior distribution in (2.7) is diffuse, i.e., when $\nu = 0$, then the posterior in (2.14) is

$$p(\lambda_1|Y) \propto [\lambda_1(1-\lambda_1)]^{-1},$$

i.e. it is improper and of the same form as the prior in (2.12). The authors conjecture that the posterior is also improper for $r \geq 3$, but cannot formulate a rigorous proof of this.

3 An Example

For numerical purposes, it is convenient to rewrite the expression $\tilde{Q}$ in (2.10) as

$$\begin{aligned}\tilde{Q} &= \sum_{j=1}^{r}(S_{jj} + \nu S_{0j})\lambda_j - \sum_{i,j=1}^{r} S_{ij}\lambda_i\lambda_j \qquad (3.1)\\ &= \nu S_{0r} + \sum_{j=1}^{r-1} \left(s_{jj} + \nu(S_{0j} - S_{0r})\right)\lambda_j - \sum_{i,j=1}^{r-1} s_{ij}\lambda_i\lambda_j,\end{aligned}$$

where $s_{ij} = S_{ij} + S_{rr} - S_{ir} - S_{jr}$ $(i = 1, \ldots, r-1; j = 1, \ldots, r-1)$.

To illustrate the results of section 2 we use the data in Grubbs (1948). Using the $n = 29$ complete observations in Grubbs' data set, where $r = 3$, we have

$$\{s_{ij}\} = \frac{1}{10000}\begin{pmatrix} 8.8916 & 2.4850 \\ 2.4850 & 3.1084 \end{pmatrix}.$$

Grubbs' estimates for the three variances are $(.0253)^2$, $(.0079)^2$ and $(.0157)^2$. Grubbs' would then estimate the $\hat{\lambda}_j$ $(j = 1, 2, 3)$ by $\hat{\lambda}_1 = 0.072$, $\hat{\lambda}_2 = 0.741$, and $\hat{\lambda}_3 = 0.187$.

In any practical application, proper prior distributions for η_j *must* be specified. For our analysis, we chose prior distributions such that $S_{01} = S_{02} = S_{03} = (0.0173)^2 = 0.0003$ in (2.7), i.e. we assume that, a priori, the three precisions have the same expected values. We note that ν effectively indexes the prior sample size, see (2.7). In order to show the effect on ν on posterior inferences, we let $\nu = 0.01$, 0.1, 1 and 5.

Perspective plots of the joint distributions for the four cases are given in Figure 1 (contours of these surfaces are given in Guttman and Menzefricke (1988)). The shape of the plots is quite sensitive to choice of ν, particularly when ν is very small.

Table 1 gives the posterior mean and standard deviation for λ_1 and λ_2 corresponding to the distinct values for ν. As ν increases, the posterior means get closer to the prior values, .33. Note that Grubbs' estimates of 0.072, 0.741, and 0.187 for λ_1, λ_2, and λ_3 are somewhat comparable to the results for the case when $\nu = 0.01$, a situation for which prior information is very diffuse with respect to the precisions η_j.

Acknowledgments: The authors are grateful to L.J. Gleser, two referees, and an Associate Editor of the Olkin Volume for remarks leading to an improved version of the manuscript. We also acknowledge research support by the Natural Sciences and Engineering Research Council of Canada under grants A8743 (for I. Guttman) and A5578 (for U. Menzefricke).

FIGURE 1. Perspective plots of the joint posterior distribution $p(\lambda_1, \lambda_2|Y)$, based on the Grubbs data for various values of ν.

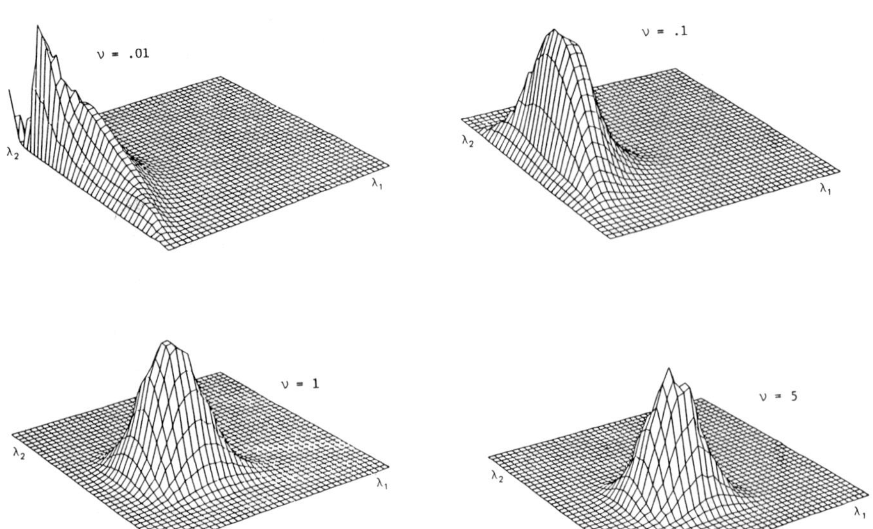

TABLE 1. Descriptors of the marginal posterior distributions of λ_1 and λ_2

ν	mean		standard deviation	
	λ_1	λ_2	λ_1	λ_2
0.01	0.093	0.61	0.048	0.20
0.1	0.15	0.48	0.060	0.12
1	0.28	0.37	0.077	0.081
5	0.32	0.34	0.068	0.068

References

Brindley, D.A. and Bradley, R.A. (1985). "Some new results on Grubbs' estimators." *Journal of the American Statistical Association* **80**, 711–714.

Grubbs, F.E. (1948). "On estimating precision of measuring instruments and product variability." *Journal of the American Statistical Association* **43**, 243–264.

Guttman, I. and Menzefricke, U. (1988). "Bayesian estimation in the unequal variance case for 2–way layouts." Technical Report #13, Department of Statistics, University of Toronto, Toronto, Ontario, Canada, M5S 1A1.

20

Calibrating For Differences

George Knafl[1]
Jerome Sacks[2]
Cliff Spiegelman[3]

ABSTRACT Suppose that an approximate linear model, or nonparametric regression, relates instrument readings y to standards x. A method is derived for constructing interval estimates of displacements $x_1 - x_2$ between standards based on corresponding instrument readings y_1, y_2, and the results of a calibration experiment.

1 Introduction

Calibration (see Rosenblatt and Spiegelman, 1981) is commonly used to relate instrument readings to standards, thereby providing meaningful measurements. Scientists and engineers often want displacements or differences that are measured by using readings at two sites (or times). Examples include change in the volume level of liquid in a nuclear processing tank (Knafl, Sacks, Spiegelman and Ylvisaker, 1984) and, in many chemical measurements, the difference between a sample measurement and a "blank" measurement that corrects for background interferences. When the relationship between the instrument reading and the standard is a straight line, then the calibration is relatively easy to carry out. More complex relationships between standards and readings are more challenging. Our purpose here is to propose methods to perform the calibration of differences in the more realistic and less ideal situations where the relationships between standards and readings depart from linearity.

A typical set-up is that an experiment is performed with observations

$$y_j = f(x_j) + \sigma\epsilon_j, \qquad j = 1, \ldots, n \tag{1.1}$$

---

[1]DePaul University. Research partially supported by NSF Grant DMS 85-30793 and DMS 87-03124.

[2]University of Illinois at Urbana-Champaign. Research partially supported by NSF Grant DMS 85-03793 and ONR Contract N00014-85-K-0357. Computing supported in part by AFOSR Grant 87-0041.

[3]Texas A&M University. Research partially supported by ONR Contract N00014-83-K-0005 and N00014-84-K-0350.

where the x_j's are constant (standards), ϵ_j's are independent $N(0,1)$ random variables, σ is unknown, and f is an unknown function in a family of regression functions $\mathcal{F}$. This calibration experiment or training sample is used to establish the nature of the relationship between x and y (estimate f) as well as to estimate σ.

After the experiment is performed, pairs of new observations (y_{1i}^*, y_{2i}^*), $i = 1, 2 \ldots$ are taken where $y_{pi}^* = f(x_{pi}^*) + \epsilon_{pi}^*$ satisfies the original model but the constants $x_{pi}^* (p = 1, 2)$ are not known. The problem we address is to find interval estimates I_i^* of $x_{2i}^* - x_{1i}^*$ satisfying the uncertainty statement made specific in Section 2 below. This is a special multivariate calibration problem. For an extensive treatment of other multivariate calibration problems see Brown (1982).

The calibration is traditionally discussed in the context of an exact linear model: $\mathcal{F}$ is a set of linear combinations of a finite number of functions, for example, polynomials of fixed degree. Our discussion, however, is in the context of an approximate linear model or nonparametric regression as in Sacks and Ylvisaker (1978), and Knafl, et. al. (1984).

In this context

$$f(x) = \sum_{j=1}^{k} \beta_j f_j(x,t) + r(x,t) \qquad \text{for } t \in \mathcal{X}, x \in \mathcal{X} \tag{1.2}$$

where $|r(x,t)| \leq M(x,t)$, M is known. $f_1, \ldots, f_n$ are also known, and $\mathcal{X}$ is the calibration region. A typical example is $k = 2$, $f_1 = 1$, $f_2 = x - t$, $M(x,t)$ is proportional to $(x-t)^2$ and $r(x,t)$ results from Taylor expansion around given t. The exact linear models are a special case of (1.2) and result from taking $M(x,t) = 0$. See Sacks and Ylvisaker (1978), or Knafl, et. al. (1984) to see how this expansion is used.

Our first step is to discuss the case of straight-line regression (Section 2). In Sections 3 and 4 we describe how to obtain useful intervals when the model has the more general form (1.1), (1.2). The technical aspects, reserved for the Appendix, make it clear that detailed computations are needed to provide adequate confidence intervals. The examples presented in Section 5 indicate the general utility of the proposed methodology.

2 Straight-line Calibration

When the calibration curve f in (1.1) is a straight-line, $f(x) = \beta_1 + \beta_2 x$, there is a natural way to proceed to calibrate for differences. Start with the calibration experiment, let $\hat{\beta}_2$, $\hat{\sigma}$ be the usual estimates of β_2, σ obtained from least-squares theory; set $1/D^2 = \sum_{i=1}^{n} (x_i - \bar{x})^2$ and note that

$$(\hat{\beta}_2 - \beta_2)/\hat{\sigma} D \tag{2.1}$$

has the t-distribution with $n-2$ degrees of freedom. Let (u, v) be a single set of x values (unknown) at which new observations y_u, y_v are made and write

$$\begin{aligned}\tilde{y} &= y_v - y_u = \beta_2(v-u) + \sigma(\epsilon_v - \epsilon_u) \\ &= \beta_2\tilde{x} + 2^{1/2}\sigma\tilde{\varepsilon} \qquad \text{(say)},\end{aligned} \tag{2.2}$$

with $\tilde{x} = v - u$. Set $c_2 = t_{n-2}(1-\delta/2)$, the $1-\delta/2$ quantile of the t_{n-2} distribution, and use (2.1) to get a confidence band for the line $\{\beta_2 x, -\infty < x < \infty\}$:

$$\hat{\beta}_2 x \pm c_2\hat{\sigma}D|x|. \tag{2.3}$$

For simplicity of the exposition to follow assume that Figure 1 is the common case. This is a slight restriction because the only case not covered by Figure 1 is the situation where $\hat{\beta}_2 - c_2\hat{\sigma}D < 0$ while $\hat{\beta}_2 + c_2\hat{\sigma}D > 0$ — a situation of little practical interest for precision instruments and one which leads to confidence "intervals" for $\tilde{x}$ which are infinite and disjoint. A further simplification we make is that $\hat{\beta}_2 > 0$. Again, this is typical in the calibration of precision instruments.

Using these simplifications we let c_1 be a positive constant, to be determined later, and expand the confidence band of (2.3) to

$$\hat{\beta}_2 x \pm (c_2\hat{\sigma}D|x| + c_1\hat{\sigma}) \tag{2.4}$$

(see the dashed line in Figure 1). Now define

$$I = \left[\frac{\tilde{y} - c_1\hat{\sigma}}{\hat{\beta}_2 + \hat{\sigma}c_2 D}, \frac{\tilde{y} + c_1\hat{\sigma}}{\hat{\beta}_2 - \hat{\sigma}c_2 D}\right] \tag{2.5}$$

(see Figure 1). The goal is to choose c_1 to ensure that I contains $\tilde{x}$ with high probability.

This uncertainty requirement has to be made precise. To do so let G be the event

$$\left|\hat{\beta}_2 - \beta_2\right| < c_2\hat{\sigma}D. \tag{2.6}$$

Then (2.3) states that

$$P[G] \geq 1 - \delta \qquad \text{for all } \beta, \sigma,$$

where $\beta = (\beta_1, \beta_2)$. The uncertainty statement we adopt is similar to the one used in Carroll, Sacks and Spiegelman (1988) namely,

$$P_{\beta,\sigma,\tilde{x}}[I \text{ contains } \tilde{x} \mid \hat{\beta}_2, G] \geq 1 - \alpha \tag{2.7}$$

for all β, σ, $\tilde{x}$ (further conditioning on $\hat{\beta}_1$ can also be done). This statement is related to the Scheffé (1973) uncertainty statement for calibration; see Carroll et. al. (1988) for further details. The interpretation of (2.6) and (2.7) is:

FIGURE 1. Calibrating Differences in Straight-line Model

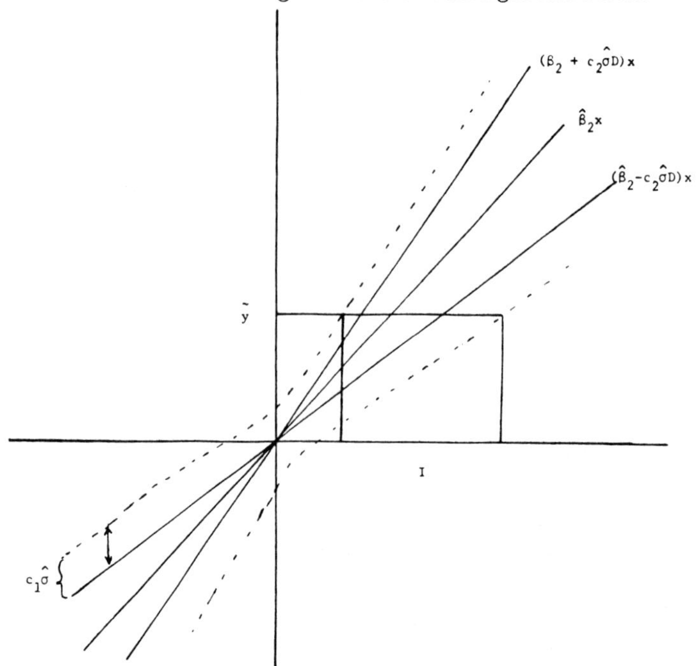

(a) The initial calibration experiment is successful with probability at least $1-\delta$,

(b) Given a successful outcome of the initial calibration experiment, the probability of coverage of the true $\tilde{x}$ is at least $1-\alpha$.

Therefore the expected percent of intervals that cover the true x values, given a good outcome of the initial calibration experiment, is $\geq 1-\alpha$. In order to choose c_1 (see (2.5)) to ensure (2.7) we note that

$$\begin{aligned}
&P[I \text{ contains } \tilde{x} \mid \hat{\beta}_2, G] \\
&\quad = P\left[|\tilde{y}-\hat{\beta}_2\tilde{x}| \leq c_2\hat{\sigma}D|\tilde{x}| + c_1\hat{\sigma} \,\middle|\, \hat{\beta}_2, G\right] \\
&\quad = P\left[|\beta_2\tilde{x} + 2^{1/2}\sigma\tilde{\varepsilon} - \hat{\beta}_2\tilde{x}| \leq c_2\hat{\sigma}D|\tilde{x}| + c_1\hat{\sigma} \,\middle|\, \hat{\beta}_2, G\right] \\
&\quad = P\Big[-c_1\hat{\sigma} + (\hat{\beta}_2-\beta_2)\tilde{x} - c_2\hat{\sigma}D|\tilde{x}| < 2^{1/2}\sigma\tilde{\varepsilon} \\
&\qquad\qquad \leq c_1\hat{\sigma} + (\hat{\beta}_2-\beta_2)\tilde{x} + c_2\hat{\sigma}D|\tilde{x}| \Big| \hat{\beta}_2, G\Big] \\
&\quad \geq P\left[c_1\hat{\sigma} - 2|\tilde{x}|c_2D\hat{\sigma} < 2^{1/2}\sigma\tilde{\varepsilon} < c_1\hat{\sigma} \,\middle|\, \hat{\beta}_2, G\right]. \qquad (2.8)
\end{aligned}$$

From the arguments in Carroll et. al. (1988) we get the right side of (2.8) to be at least as large as

$$P\left[-2^{-1/2}c_1 - 2^{1/2}|\tilde{x}|c_2 D < T_{n-2} < 2^{-1/2}c_1\right] \tag{2.9}$$

where T_{n-2} has the t-distribution with $n-2$ degrees of freedom.

There are two options we consider for handling (2.9). The first comes from noting that (2.9) is minimized when $\tilde{x} = 0$ in which case the choice $c_1 = 2^{1/2}t_{n-2}(1-\alpha/2)$ assures the desired uncertainty statement of (2.7). Clearly, this choice of c_1 is conservative. A second approach is to replace $|\tilde{x}|$ by $|\tilde{y}/\hat{\beta}_2|$, treat it as a constant, and find c_1 by

$$P\left[-2^{-1/2}c_1 - 2^{1/2}|\tilde{y}/\hat{\beta}_2|c_2 D < T_{n-2} < 2^{-1/2}c_1\right] = 1-\alpha. \tag{2.10}$$

Although c_1 is now data-dependent, it is easily calculated and should provide through (2.7) an adequate, if inaccurate, uncertainty statement.

In Section 5 we present a comparison of the numbers obtained here using the conservative choice $c_1 = 2^{1/2}t_{n-2}(1-\alpha/2)$ with those obtainable by applying the methods of Sections 3 and 4. The uncertainty statements we use below in Sections 3 and 4 are natural extensions of (2.6) and (2.7) with somewhat modified definitions of c_1 and c_2.

3 Model Robust Calibration

Recall the model of (1.1) and (1.2) and, as in Section 2, let y_u and y_v denote new observations at unknown sites u and v. Here, however, we do not assume a straight line model but our purpose is still to give a confidence interval for all the $v-u$ that occur. We will follow the formulation in Knafl, Sacks, Spiegelman, and Ylvisaker (1984) but incorporate the simplifications stemming from Carroll, Sacks and Spiegelman (1988).

In the model (1.1), (1.2) take $\mathcal{X}$ to be a compact interval of R^1. We use the calibration experiment to estimate f by choosing $\{c_i(x)\}$ and forming

$$\hat{f}(x) = \sum_{i=1}^{n} c_i(x)y_i. \tag{3.1}$$

The weights $\{c_i(x)\}$ which determine the linear estimates in (3.1) are not specified for now; one set which we use below is obtained as in Sacks and Ylvisaker (1978). Given the $\{c_i(x)\}$ set

$$\begin{aligned} B(f,x) &= \sum c_i(x)f(x_i) - f(x) \\ B(x) &= \sup_{f\in\mathcal{F}} |B(f,x)| \\ D(x) &= \left(\sum c_i^2(x)\right)^{1/2} \end{aligned} \tag{3.2}$$

The estimate $\hat{f}(x)$ is normally distributed with mean $f(x)+B(f,x)$ and variance $\sigma^2 D^2(x)$. From (3.2), the bias $B(f,x)$ of f is bounded by $B(x)$.

Our confidence interval calculations require assumptions about an estimate of the scale σ. In the case of an exact linear model there is no difficulty, we use the standard estimate of scale from least-squares theory and $\hat{\sigma}$ is then independent of the least-squares estimate $\hat{f}$ of f. In the case of the approximately linear models we employ below, the relationship between $\hat{f}$ and $\hat{\sigma}$ is more complicated. In particular, $\hat{\sigma}$ and $\hat{f}$ need not be independent. In important situations, however, there will be an estimate $\hat{\sigma}$ which is independent of $\hat{f}$. This occurs when $\hat{\sigma}$ is based on replicated data and $\mathcal{F}$ is an approximately linear model in error-scaled form (see Sacks and Ylvisaker, 1978) i.e., for each $t \in \chi$ and $f \in \mathcal{F}$

$$|f(x) - \sum \beta_j f_j(x,t)| \leq \sigma M(x,t) \tag{3.3}$$

where the f_j's and M are specified functions. For example

$$|f(x) - f(t) - f'(t)(x-t)| \leq \sigma m(x-t)^2 \tag{3.4}$$

with m a known constant gives rise to such an error-scaled class (see Knafl et. al. 1984 for an example where such an m is known).

In addition, when replicated data are available, we can define $\hat{\sigma}$ so that the distribution of $\sum c_i(x)\epsilon_i$ conditioned on $\hat{\sigma}$ is $N(0, D^2(x))$ and $\nu\hat{\sigma}^2/\sigma^2$ is chi-square with ν degrees of freedom (ν determined by the numbers of replicates). These latter properties were used in Knafl et. al. (1984) where it was noted that care has to be exercised in choosing $\hat{\sigma}$.

The properties we need for $\hat{\sigma}$ then are

(a) $\hat{\sigma}$ independent of $\hat{f}$,

(b) $\nu\hat{\sigma}^2/\sigma^2$ is chi-square with ν degrees of freedom , (3.6)

(c) the conditional distribution of $\sum c_i(x)\epsilon_i$ given $\hat{\sigma}$ is $N(0, D^2(x))$.

The model (3.3) is useful because it permits a choice for $\hat{\sigma}$ satisfying (3.6) which in turn allows the use of the methods in Carroll et. al. (1988). Moreover, it is useful when M cannot be specified in advance and cross-validation is used to estimate M, as described in Knafl et. al. (1984).

With (3.3) and (3.6) assumed, we seek a $1-\delta$ confidence band for f. Following the methods employed in Knafl, et. al. (1984) and Knafl, Sacks, and Ylvisaker (1985), we first find a confidence band for f on a finite grid S of points of $\mathcal{X}$. By linearly interpolating from S to all of $\mathcal{X}$ we then get a confidence band over all of $\mathcal{X}$. Formally, we obtain

$$P_{f,\sigma}\left[|\hat{f}(x) - f(x)| \leq B_0(S) + B(x) + c_2 D(x)\hat{\sigma}, \text{ all } x \in \mathcal{X}\right] \geq 1-\delta \tag{3.8}$$

all $\sigma > 0$, all $f \in \mathcal{F}$. Here, $B_0(S)$ is a constant depending on the mesh of S and the family $\mathcal{F}$. The functions B and D are as in (3.2), and c_2 is a constant chosen to ensure the validity of the coverage probability in (3.8). The case $M = 0$ corresponds to the straight line model and reduces the problem to the context of Section 2 because $B_0(S)$ and $B(x)$ are both 0 and c_2 is $t_{n-2}(1-\delta/2)$. The details of how c_2 and $B_0(S)$ are calculated in general can be found in the references just cited.

For estimating a new v when y_v is observed we follow the procedures in these references, expand the confidence band to

$$\hat{f}(x) \pm (B_0(S) + B(x) + c_2\hat{\sigma}D(x) + c_1\hat{\sigma}) \tag{3.9}$$

and obtain I as in Figure 2. Formally,

FIGURE 2. Calibration for general f

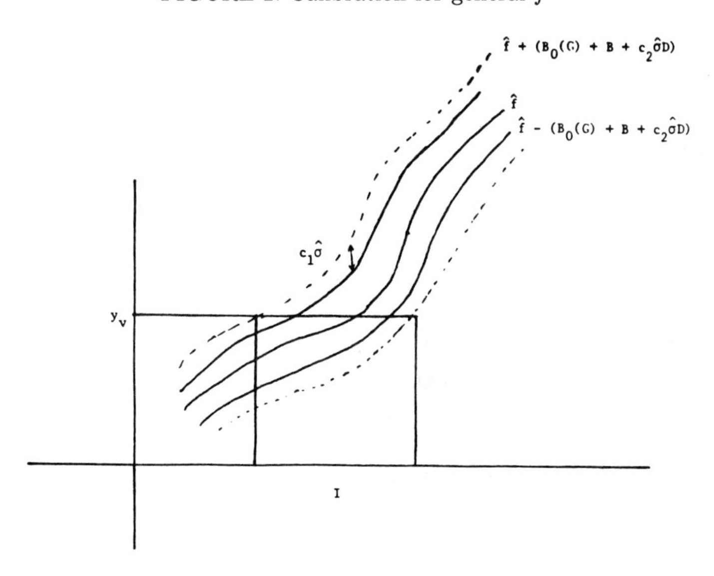

$$I = \{x | y_v \text{ is in the interval defined by (3.9)}\}. \tag{3.10}$$

The constant c_1 is chosen to guarantee

$$P[I \text{ contains } v | Z = z, G] \geq 1 - \alpha \tag{3.11}$$

where G is the event in the probability statement of (3.8) and

$$Z = \sup_{x \in X} |\hat{f}(x) - f(x)|/D(x).$$

We note that Z in (3.11) plays the role of $\hat{\beta}_2$ in (2.7).

The statement (3.11) is discussed in Carroll et. al. (1988); in the context of Section 2 it is equivalent to (2.7). The tactics described there for finding c_1 to satisfy (3.11) lead to calculating

$$P\left[|y_v - \hat{f}(v)| \leq U_0(v) + c_1\hat{\sigma} | Z = z, G\right] \tag{3.12}$$

where $U_0(v) = B_0(S) + B(v) + c_2\hat{\sigma}D(v)$. Under the assumption of (3.6) we can get (3.12) to be at least as large as

$$P[-2U_0(v) - c_1\hat{\sigma} < \sigma\epsilon < c_1\hat{\sigma}]. \tag{3.13}$$

If $U_0(v)$ is known, then c_1 can easily be found to make (3.13) $= 1 - \alpha$. Since v is unknown, we replace $U_0(v)$ by something calculable; namely, $U_0(v)$ where $v = \hat{f}^{-1}(y_v)$. This sets the stage for dealing with differences.

4 Calibrating Differences

With the setting as described in Section 3, let y_u, y_v be two new observations. Our concern is to find a confidence interval for $v-u$ and we approach this problem as follows:

Fix u and let I be a confidence interval for v. Let $\hat{u} = \hat{f}^{-1}(y_u)$ estimate u and take $J = I - \hat{u}$. We will use J as a confidence interval for $v - u$.

The disparate treatment of u, v leads to an asymmetry which can be overcome, but at the expense of added computation and of little apparent gain in the examples we examined.

The problem to be faced is how to construct I (see (3.10)), so that J has adequate coverage probability. In conformity with (3.11), we would ask that

$$P[J \text{ contains } v - u | Z = z, G] \geq 1 - \alpha \tag{4.1}$$

where G, as before, is the event in (3.8), so that $P[G] \geq 1 - \delta$.

For technical reasons we need (4.9) below but we cannot verify that (4.9) holds on G. However, we can replace G by a more restrictive G_1 on which (4.9) does hold. To define G_1 set

$$\tau = \frac{[c(v) - c(u)] \cdot \epsilon}{\frac{\hat{\sigma}}{\sigma}\|c(v) - c(u)\|} \tag{4.2}$$

where $c(u)$ is the vector $(c_1(u), \ldots, c_n(u))$, $\|\ \|$ is ordinary Euclidean norm, and the $\cdot$ is the usual inner product. The random variable τ has, according to (3.6), the t_ν-distribution. We let

$$G_1 = G \bigcap \{|\tau| \leq c_2\} \tag{4.3}$$

and choose c_2 so that

$$1 - P_{f,\sigma}[G] + P[|t_\nu| > c_2] < \delta$$

which assures

$$P_{f,\sigma}[G_1] \geq 1 - \delta, \tag{4.4}$$

by the Bonferroni inequality. Recall that G is the event in (3.8) and the c_2 defined there is not the one that assures (4.4). The latter is larger because G_1 is more restrictive than G.

We replace G by G_1 and now demand that

$$P_{f,\sigma,u,v}[J \text{ contains } v-u|Z=z,G_1] \geq 1-\alpha. \tag{4.5}$$

Having chosen c_2 to assure (4.4), we are left with the need to choose c_1 which enters, as indicated in (3.10), in the definition of I and hence J.

In order to find c_1 we make a series of approximations to estimate the left side of (4.5). We sketch an outline of the argument here; the details are in the Appendix. We use the notation $P_1[\cdot]$ to stand for $P_{f,\sigma,u,v}[\cdot|Z=z,G_1]$. Approximate the left side of (4.5) to be

$$P_1\left[|\sigma(\epsilon_v-\epsilon_u)-\hat{A}_T-d|\leq c_1\hat{\sigma}+\hat{U}_0\right] \tag{4.6}$$

where $T=u-\hat{u}$, $\hat{A}_T$ is an approximation to (see (A1.3))

$$A_T=\left[\hat{f}(v-T)-\hat{f}(v)\right]-\left[\hat{f}(u-T)-\hat{f}(u)\right] \tag{4.7}$$

(A_T is 0 if $u=\hat{u}$ so $\hat{A}_T$ is usually small), d is the difference

$$d=\left[\hat{f}(v)-f(v)\right]-\left[\hat{f}(u)-f(u)\right], \tag{4.8}$$

and $\hat{U}_0$ (see (A1.5)) is an approximation to $U_0(v-T)$ (see after (3.12) for the definition of U_0).

The probability in (4.6) is then bounded below by minimizing over the possible values of d. The range of d when G_1 is satisfied is obtained from (4.8), (3.2) and (4.2) to be

$$\begin{aligned}|d| &= |B(f,v)-B(f,u)+\sigma(c(v)-c(u))\cdot\epsilon| \\ &\leq B(v)+B(u)+\sigma\|c(v)-c(u)\|\ |\tau| \\ &\leq B(v)+B(u)+c_2\hat{\sigma}\|c(v)-c(u)\| \\ &= d_0 \qquad \text{(say)}\ .\end{aligned} \tag{4.9}$$

It is easy to confirm that the values of d which minimize (4.6) are $d=\pm d_0$. This leads to the approximation from below of the left side of (4.5) by

$$\min_{d=\pm\hat{d}_0}\int P_1\left[|\sigma(\epsilon_v-\epsilon_u)-\hat{A}_t-d|<c_1\hat{\sigma}+\hat{U}_0\right]dQ(t) \tag{4.10}$$

where $\hat{d}_0$ is an estimate of d_0 (A1.4) and Q is the conditional distribution of T given $Z=z,G_1$ ((4.10) is based on (A1.6)).

The integrand in (4.10) is approximated in (A1.7). The distribution of Q is also estimated. This finally leads to solving

$$\min_{d=\pm\hat{d}_0}\ \inf_{|h|\leq\hat{U}_0} H(h,c_1,d)=1-\alpha \tag{4.11}$$

for c_1 (see (A1.11)) where $H(h, c, d)$, defined in (A1.10), is an approximation to (4.10). The approximations used at each step appear innocuous and are, for the most part, replacements of unknowns by estimated values based on the data. Extensive Monte Carlo studies would be useful for affirming our belief that the actual probabilities are close to the nominal values.

The calculation of c_1 completes the specification of J, the confidence interval for $v - u$ as defined at the beginning of this section.

5 Examples

The first example arises from atomic absorption spectroscopy and is an example discussed in Carroll et. al. (1988). It was found there that a straight line was an adequate model to use and we therefore adopt this model. Table 1 gives the calibrations for individual y's.

TABLE 1. Individual Calibrations for Spectroscopy Data

$\hat{\sigma} = 0.0035$, $\alpha = 0.10$, $\delta = 0.10$		
y_u	$\hat{u}$	Length of Conf. Interval
0.06	0.016	0.022
0.075	0.038	0.022
0.08	0.045	0.022
0.10	0.074	0.021
0.15	0.146	0.021
0.20	0.218	0.021

In order to correct for unknown contaminants, a neutral sample is often measured so that $\hat{v}$ can be corrected. A common correction is $\hat{v} - \hat{u}$ where $\hat{u}$ is the value from the neutral sample. This is called a correction for blank. As can be seen in Table 2, the method of Section 2 produces confidence intervals of smaller length than does the method of Section 4 when applied to the straight-line model.

This is expected, since the method of Section 4 does not take advantage of the straight-line model to subtract the y's and get rid of the effect of the intercept β_1. Figure 1 indicates that the methods of Section 2 ought to be best when $v - u$ is small and this is borne out in Table 2. The differences between the two methods do not appear great, especially when $y_v - y_u$ is not close to 0, suggesting that the method of Section 4 is adequate and conservative.

The second example we examine is the nuclear processing tank data discussed in Knafl, Sacks, Spiegelman, and Ylvisaker (1984). The details of that data set can be found there. Our concern with calibrating differences arises from the need to estimate the amount of nuclear material put in

TABLE 2. Calibration for Differences — Spectroscopy Data

$\hat{\sigma} = 0.0035,\ \alpha = 0.10,\ \delta = 0.10$				
			Length of Conf. Interval	
y_v	y_u	$\hat{v} - \hat{u}$	Method of Section 2	Method of Section 4
0.06	0.08	0.029	0.026	0.032
0.075	0.15	0.108	0.029	0.032
0.10	0.20	0.144	0.030	0.033

the tank or taken out of the tank. These amounts are then compared with plant records so that any unauthorized transfers of material can be detected promptly.

The methods of Section 3 were used to produce the data in Table 3, while the methods of Section 4 produced the data in Table 4. The model fit is a modification of the type of model specified in (3.2) or (3.3) but one close in spirit. We refer the reader to Knafl et. al. (1984) for the details. The variation in the lengths of the confidence intervals in Table 4 is a consequence of the fluctuation of the underlying regression function.

TABLE 3. Individual Calibrations for Tank Data

$\hat{\sigma} = 2.784,\ \alpha = 0.10,\ \delta = 0.10$		
y_u	$\hat{u}$	Length of Confidence Interval for u
1033	0.568	0.0105
1052	0.576	0.0104
1062	0.581	0.0106
1100	0.597	0.0116
1400	0.726	0.0109
2815	1.325	0.0100

TABLE 4. Calibration of Differences for Tank Data

$\hat{\sigma} = 2.784,\ \alpha = 0.10,\ \delta = 0.10$			
y_u	y_v	$\hat{v} - \hat{u}$	Length of Confidence Interval
1052	1062	0.0043	0.008
1100	1400	0.1292	0.0169
1033	2815	0.7572	0.0136

A1 Appendix

Recall the definition of I from (3.10) and, from Section 4, that

$$J = I - \hat{u}, \quad \hat{u} = \hat{f}^{-1}(y_u).$$

Note the definition of U_0 following (3.12), let $T = u - \hat{u}$ and recall the notation P_1 (see above (4.6)). Then

$$\begin{aligned} P_1[J \supset v - u] &= P_1[I \supset v - T] \\ &= P_1\left[|y_v - \hat{f}(v - T)| \leq U_0(v - T) + c_1\hat{\sigma}\right] \end{aligned} \tag{A1.1}$$

The definition of A_T and d at (4.7) and (4.8) can be used in (A1.1) to get

$$P_1[J \supset v - u] = P_1\left[|\sigma(\epsilon_v - \epsilon_u) - A_T - d| \leq U_0(v - T) + c_1\hat{\sigma}\right]. \tag{A1.2}$$

We approximate A_T by

$$\hat{A}_T = (\hat{f}(\hat{v} - T) - \hat{f}(\hat{v})) - (\hat{f}(\hat{u} - T) - \hat{f}(\hat{u})); \tag{A1.3}$$

that is, we replace u, v by their estimated values. We do the same with d_0, the upper bound on d given in (4.9) and replace it by

$$\hat{d}_0 = B(\hat{v}) + B(\hat{u}) + c_2\hat{\sigma}\|c(\hat{v}) - c(\hat{u})\|. \tag{A1.4}$$

We also replace $U_0(v - T)$ by

$$\hat{U}_0 = U_0(\hat{v} - T). \tag{A1.5}$$

Subject to the approximations due to replacing u, v by $\hat{u}, \hat{v}$, we get an *estimated* lower bound on (A1.2) by taking its minimum over all $|d| \leq d_0$ and replacing d_0, A_T, $U_0(v - T)$ by $\hat{d}_0$, $\hat{A}_T$, $\hat{U}_0$:

$$P_1[J \supset v - u] \geq \min_{|d| \leq \hat{d}_0} P_1\left[|\sigma(\epsilon_v - \epsilon_u) - \hat{A}_T - d| \leq \hat{U}_0 + c_1\hat{\sigma}\right]. \tag{A1.6}$$

If the estimates $\hat{d}_0$, $\hat{A}_T$, $\hat{U}_0$ are treated as exact, the minimum in (A1.6) is achieved at the end points $d = \pm\hat{d}_0$ giving rise to (4.10).

Let $t_\nu = 2^{-1/2}\sigma(\epsilon_v - \epsilon_u)/\hat{\sigma}$. Then t_ν has the t-distribution with ν degrees of freedom. We treat $\hat{A}_t/\hat{\sigma}$, $\hat{d}/\hat{\sigma}$, $\hat{U}_0/\hat{\sigma}$ as exact and bound the integrand in (4.10) by

$$p(t, d) = P\left[|t_\nu - (d + \hat{A}_t)2^{-1/2}/\hat{\sigma}| \leq \frac{\hat{U}_0 2^{-1/2}}{\hat{\sigma}} + 2^{-1/2}c_1\right] \tag{A1.7}$$

by use of the results in Carroll, Sacks, and Spiegelman (1988).

We next estimate Q by discretizing: Let $0 \leq q_{-1} < q_0 < q_1 < \cdots$. Let $s_j = -q_j$ for $j \geq -1$. Use (A1.7) and the discretization to replace (4.10) by

$$\sum_{j=-1}^{\infty} p(q_j, d)[Q(q_j + 1) - Q(q_j)] + \sum_{j=1}^{\infty} p(s_j, d)[Q(s_j) - Q(s_{j+1})]. \quad \text{(A1.8)}$$

Set $G(u,t) = \hat{f}(u-t) - \hat{f}(u)$. Then

$$\begin{aligned} Q(t_{j+1}) - Q(q_j) &= P[q_j < u - \hat{u} < q_{j+1}] \\ &= P[\hat{f}(u - q_{j+1}) < y_u < \hat{f}(u - q_j)] \\ &= P[G(u, q_{j+1}) < \sigma\epsilon_u + f(u) - \hat{f}(u) < G(u, q_j)] \\ &\cong P[G(\hat{u}, q_{j+1}) < \sigma\epsilon_u + f(u) - \hat{f}(u) < G(\hat{u}, q_j)]. \end{aligned} \quad \text{(A1.9)}$$

Put $h = f(u) - \hat{f}(u)$ and note that, on G_1, $|f(u) - \hat{f}(u)| \leq U_0(u) \cong U_0(\hat{u})$. Then (A1.8) becomes (approximately)

$$\begin{aligned} &\sum_{j=1}^{\infty} p(q_{j+1}, d) P[G(\hat{u}, q_{j+1}) < \sigma\epsilon_u + h < G(\hat{u}, q_j)] \\ &+ \sum_{j=-1}^{\infty} p(s_j, d) P[G(\hat{u}, s_j) < \sigma\epsilon_u + h < G(\hat{u}, s_{j+1})] \\ &= H(h, c_1, d) \qquad \text{(say)}. \end{aligned} \quad \text{(A1.10)}$$

We then solve

$$\min_{d=\pm\hat{d}_0} \inf_{|h|\leq U_0(\hat{u})} H(h, c_1, d) = 1 - \alpha \quad \text{(A1.11)}$$

for c_1.

The calculations in (A1.10) can be implemented by dividing through by $\hat{\sigma}$ and treating $G(\hat{u}, s_j)/\hat{\sigma}, U_0(\hat{u})/\hat{\sigma}$ as exact. Our calculations reveal that, typically, the minimizing h in (A1.11) occurs at one of the endpoints, $\pm U_0(\hat{u})$.

References

Brown, P.J. (1982). Multivariate Calibration (with Discussion). *Journal of the Royal Statistics Society Series B* **44**, 287–321.

Carroll, R., Sacks, J., Spiegelman, C. (1988). A quick and easy multiple use calibration curve procedure. *Technometrics* **30**, 137–142.

Knafl, G., Sacks, J., and Ylvisaker, D. (1985). Confidence bands for regression. *J. of Amer. Stat. Assoc.* **80**, 683–691.

Knafl, G., Sacks, J., Spiegelman, C., and Ylvisaker, D. (1984). Nonparametric calibration. *Technometrics* **26**, 233–242.

Lieberman, G.J., Miller, R.G., and Hamilton, M.A. (1967). Unlimited simultaneous discrimination intervals in regression. *Biometrics* **54**, 133–145.

Rosenblatt, J.R., and Spiegelman, C.H. (1981). Discussion of a Bayesian analysis of the linear calibration problem, by William G. Hunter and Warren F. Lamboy. *Technometrics* **23**, 329–333.

Sacks, J., and Ylvisaker, D. (1978). Linear estimation for approximately linear models. *Annals of Statistics* **6**, 1122–1137.

Scheffe, H. (1973). A statistical theory of calibration. *Annals of Statistics* **1**, 1–37.

21

Complete Class Results For Linear Regression Designs Over The Multi-Dimensional Cube

Friedrich Pukelsheim[1]

ABSTRACT Complete classes of designs and of moment matrices for linear regression over the multi-dimensional unit cube are presented. An essentially complete class of designs comprises the uniform distributions on the vertices with a fixed number of entries being equal to unity, and mixtures of neighboring such designs. The corresponding class of moment matrices is minimally complete. The derivation is built on information increasing orderings, that is, a superposition of the majorization ordering generated by the permutation groups, and the Loewner ordering of symmetric matches.

1 Introduction

In a brilliant paper C. -S. Cheng (1987) recently determined optimal designs over the k-dimensional unit cube $[0,1]^k$ for the linear model

$$E[Y] = x'\theta, \quad V[Y] = \sigma^2.$$

In this setting the experimenter chooses the regression vector x in the cube $[0,1]^k$ prior to running the experiment, and then observes the response Y. The response is assumed to have expected value and variance as given above; furthermore, responses at different design points x, and replicated observations at the same point x, are all taken to be uncorrelated. As pointed out by Cheng this model has interesting applications in Hadamard transform optics.

The optimal designs of Cheng (1987) are the j-vertex designs ξ_j and mixtures of $(j+1)-$ and j-vertex designs, defined as follows. A j-vertex of the unit cube $[0,1]^k$ is a vector x with j entries equal to unity and the remaining $k-j$ entries equal to zero, for $j = 0, \dots, k$ (See Figure 1.) There

[1]Cornell University, On leave from the Institut für Mathematik der Universität Augsburg.

are $\binom{k}{j}$ many j-vertices. The j-vertex design ξ_j is the design that has the j-vertices for support, and assigns uniform mass $1/\binom{k}{j}$ to each of them. For mixtures of the form $\alpha\xi_{j+1}+(1-\alpha)\xi_j$ the following notation is convenient, in that it provides a continuous parameterization. Given $j, j = 0, \ldots, k-1$, define the design

$$\xi_s = (s-j)\xi_{j+1} + (1-(s-j))\xi_j \qquad \text{for } s \in (j, j+1).$$

In terms of s we have that j is the integer part of $s, j = \text{ int } s$. In other words, the two integers $j+1$ and j closest to s specify the number of vertices supporting ξ_s, and the fractional part $s-j$ determines the weight for mixing ξ_{j+1} and ξ_j. For example, $\xi_{2.11} = 0.11\xi_3 + 0.89\xi_2$, and $\xi_{7.4} = 0.4\xi_8 + 0.6\xi_7$.

Under the p-mean criteria considered by Cheng (1987) the class of optimal designs is

$$\mathcal{C} = \left\{ \xi_s : s \in \left[\text{int}\frac{k+1}{2}, k \right] \right\},$$

starting from the 'median vertex design' $\xi_{\text{int}(k+1)/2}$ and running through the j-vertex designs ξ_j and mixtures ξ_s up to the design ξ_k that assigns all mass to the vector with each entry equal to unity. It is notationally convenient to define

$$m = \text{int}\frac{k+1}{2};$$

this is the largest median of the set of numbers $0, \ldots, k$. As usual the class of all designs on $[0,1]^k$ is denoted by Ξ.

FIGURE 1. Corners of the unit cube with j entries 1 and the remaining entries 0 are called j-vertices. For the cube in dimension 3 the figure shows the 0-, 1-, 2-, and 3-vertices.

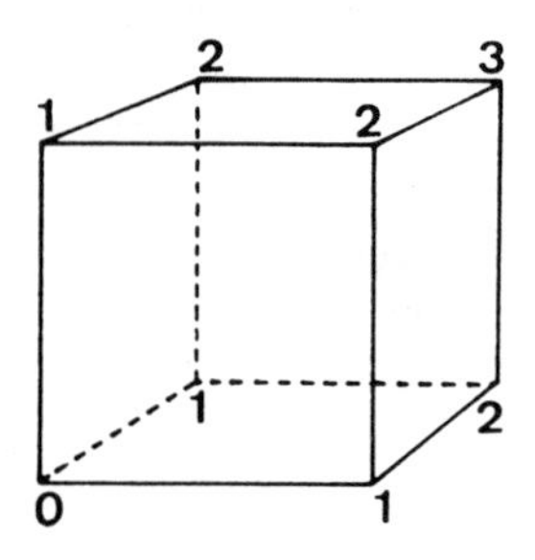

In Section 2 we show that Cheng's class $\mathcal{C}$ is essentially complete, and that the corresponding class of moment matrices $M(\mathcal{C})$ is minimally complete, with respect to the information ordering generated by the permutation group Perm(k). As a consequence, the class $\mathcal{C}$ contains an optimal design whenever the optimality criterion is given by an information function ϕ that is permutationally invariant.

Cheng (1987) studied the subclass of p-means ϕ_p. In Section 3 we present some graphs showing how the optimal support parameter $s(p)$ and the optimal value $v(p)$ change with the order $p \in [-\infty, 1]$ of the mean ϕ_p, and with the dimensionality k.

2 Complete Class Results

The performance of a design ξ hinges on its moment matrix

$$M(\xi) = \int_{[0,1]^k} xx' d\xi.$$

These matrices are of order $k \times k$, and nonnegative definite. Our complete class results refer to the information increasing ordering generated by the group Perm(k) of $k \times k$ permutation matrices. The general theory is surveyed in Pukelsheim (1987). We here only recall such details as are necessary for the present discussion.

A matrix B is said to be *more centered than* a moment matrix A whenever

$$B \in \text{conv}\{QAQ' : Q \in \text{Perm}(k)\},$$

that is, B lies in the convex hull of the orbit of A when the group Perm(k) acts through congruence. A moment matrix M is said to be *at least as informative as* another moment matrix A when in the Loewner ordering one has $M \geq B$ for some matrix B that is more centered than A. A moment matrix is said to be *more informative than* another moment matrix A when M is at least as informative as A, but does not lie in the orbit of A.

Theorem 2.1 *The class of designs $\mathcal{C}$ is essentially complete; that is, for all designs η in Ξ there exists a design ξ_s in $\mathcal{C}$ such that $M(\xi_s)$ is at least as informative as $M(\eta)$. The corresponding class of moment matrices $M(\mathcal{C})$ is minimally complete; that is, for all moment matrices A not in $M(\mathcal{C})$ there exists a moment matrix M in $M(\mathcal{C})$ such that M is more informative than A and there is no proper subclass of $M(\mathcal{C})$ with the same property.*

Proof: Let η be a design not in $\mathcal{C}$. First symmetrization leads to an invariant design $\bar{\eta}$, then a Loewner improvement produces a better design ξ, and another Loewner improvement yields a design ξ_s in the class $\mathcal{C}$.

I. Averaging η leads to a design $\bar{\eta}$ that is permutationally invariant. Its moment matrix $\bar{A}$ is the average of the moment matrix A of η,

$$\bar{A} = \frac{1}{k!} \sum_{Q \in \text{Perm}(k)} QAQ',$$

and therefore more centered than A.

It may happen that η has an invariant moment matrix A without η itself being invariant. In this case $\bar{A} = A$, so that the passage from A to $\bar{A}$ means no improvement whatsoever. The optimal balanced incomplete block designs of Corollary 3.5 in Cheng (1987) provide an instance of this.

II. Being invariant the design $\bar{\eta}$ must be a mixture of j-vertex designs for $j \leq 0$,

$$\bar{\eta} = \sum_{j \geq 0} \beta_j \xi_j$$

with $\min \beta_j \geq 0$ and $\sum \beta_j = 1$. Let $\bar{J}$ be the $k \times k$ matrix with each entry equal to $1/k$, and set $K = I_k - \bar{J}$; this is an orthogonal pair of orthogonal projections. The moment matrix of ξ_j is

$$M(\xi_j) = \Lambda_j \bar{J} + \lambda_j K, \qquad \text{where } \Lambda_j = \frac{j^2}{k}, \qquad \lambda_j = \frac{j(k-j)}{k(k-1)}.$$

Therefore the moment matrix of $\bar{\eta}$ is

$$M(\bar{\eta}) = \sum_{j \geq 0} \beta_j (\Lambda_j \bar{J} + \lambda_j K).$$

The eigenvalues $\Lambda_j$and λ_j increase as j runs over the initial section from 0 up to m. Hence we introduce new weights α_j that sweep the initial mass into the median m

$$\alpha_j = 0 \text{ for all } j < m, \qquad \alpha_m = \sum_{j \leq m} \beta_j, \qquad \alpha_j = \beta_j \text{ for all } j > m.$$

This produces a design which is a mixture of j-vertex designs for $j \geq m$,

$$\xi = \sum_{j \geq m} \alpha_j \xi_j,$$

with a Loewner improved moment matrix $M(\xi) \geq M(\bar{\eta})$. Furthermore the two moment matrices are distinct, unless the weights β_j vanish for $j < m$.

III. The moment matrix of ξ is $M(\xi) = \Lambda \bar{J} + \lambda K$, with

$$\Lambda = \sum_{j \geq m} \alpha_j k \left(\frac{j}{k}\right)^2, \qquad \lambda = \sum_{j \geq m} \alpha_j \frac{k}{k-1} \frac{j}{k} \left(1 - \frac{j}{k}\right).$$

Thus the eigenvalue pair (Λ, λ) varies over the convex set

$$\operatorname{conv}\{(\Lambda_j . \lambda_j) : j = m, \ldots, k\} = \operatorname{conv}\left\{(kz^2, \frac{k}{k-1} z(1-z)) : z = \frac{m}{k}, \ldots, 1\right\}.$$

In other words, on the curve $x(z) = kz^2$ and $y(z) = \frac{k}{k-1} z(1-z)$ we pick the points (Λ_j, λ_j) corresponding to $z = j/k$ for $j \geq m$, and then form their convex hull.

The geometry exhibits that for every eigenvalue pair (Λ, λ) there exist vertices $(\Lambda_{j+1}, \lambda_{j+1})$ and (Λ_j, λ_j) of the upper boundary of the convex hull containing (Λ, λ) such that with some $\alpha \in [0,1]$, $s = j + \alpha$, we obtain

$$\Lambda \leq \alpha\Lambda_{j+1} + (1-\alpha)\Lambda_j = \Lambda_s, \qquad \lambda \leq \alpha\lambda_{j+1} + (1-\alpha)\lambda_j = \lambda_s.$$

Thus the design ξ_s has a Loewner improved moment matrix, $M(\xi_s) \geq M(\xi)$, and lies in Cheng's class $\mathcal{C}$. Furthermore the two moment matrices are distinct, unless ξ itself lies in $\mathcal{C}$.

IV. As s varies over $[m, k]$ the eigenvalues Λ_s and λ_s strictly increase and decrease, respectively. Therefore a proper subclass of $M(\mathcal{C})$ cannot be complete. □

The eigenvalue improvement in part III of the proof appears to be small, indicating that mixtures of j-vertex designs for $j \geq m$ may perform well even when they are not in the class $\mathcal{C}$.

Every optimality criterion ϕ that is isotonic relative to the Loewner ordering, concave, and permutationally invariant is isotonic also relative to the information increasing ordering: M is at least as informative as A if and only if

$$M \geq \sum \alpha_i Q_i A Q_i',$$

with $Q_i \in \text{Perm}(k)$, and $\min \alpha_i \geq 0$ and $\sum \alpha_i = 1$. The functional properties of ϕ then yield

$$\phi(M) \geq \phi\left(\sum \alpha_i Q_i A Q_i'\right) \geq \sum \alpha_i \phi(Q_i A Q_i') = \phi(A).$$

The same reasoning also establishes that if there exists a design $\xi \in \Xi$ that is ϕ-optimal over Ξ then there actually exists a design $\xi_s \in \mathcal{C}$ with the same optimality property. An optimal design always exists provided the criterion ϕ is upper semicontinuous. The following corollary summarizes this behavior.

Corollary 2.2 *Let ϕ be an optimality criterion that is Loewner-isotonic, concave, and permutationally invariant. If a moment matrix M is at least as informative as another moment matrix A then*

$$\phi(M) \geq \phi(A).$$

Moreover, there exists a design $\xi_s \in \mathcal{C}$ which is ϕ-optimal over Ξ,

$$\phi(M(\xi_s)) = \max_{\xi \in \Xi} \phi(M(\xi)),$$

provided ϕ is upper semicontinuous.

A particular class of criteria to which this corollary applies are the p-means ϕ_p, for $p \in [-\infty, 1]$, studied by Cheng (1987).

3 Optimal Designs for the p-mean Criteria

As it happens the complete class of moment matrices $M(\mathcal{C})$ is in fact exhausted by the moment matrices $M(\xi_{s(p)})$ belonging to ϕ_p-optimal designs $\xi_{s(p)}$, as p varies over $[-\infty, 1]$. This follows from Theorem 3.1 in Cheng (1987); we now briefly recall this result. Cheng subdivides the interval $[-\infty, 1]$ using two interlacing sequences of numbers $f(m), g(m), \ldots, f(k), g(k)$, according to

$$-\infty = f(m) < g(m) < f(j) < g(j) < f(j+1) < g(j+1) < f(k) = g(k) = 1$$

for $j = m+1, \ldots, k-2$. His result can then be stated as follows.

Theorem 3.1 *For every order $p \in [-\infty, 1]$, there exists a support parameter $s(p) \in [m, k]$ such that the design $\xi_{s(p)}$ is ϕ-optimal over Ξ. As a function of p the support parameter $s(p)$ is continuous, equal to j on the interval $[f(j), g(j)]$, and strictly increases from j to $j+1$ on the interval $[g(j), f(j+1)]$, for $j = m, \ldots, k-1$.*

FIGURE 2. The graph shows the ϕ_p-optimal support parameter $s_k(p)$ standardized by the dimension k, as a function of the order p of the mean ϕ_p. Most of the variation takes place when p is positive. The limiting value for large dimensions k is 1/2.

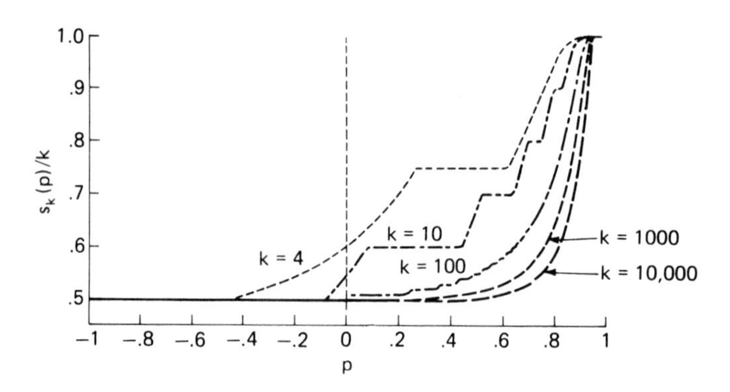

Cheng (1987) actually provides explicit formulae for these quantities, namely

$$f(j) = 1 + \frac{\log\left(1 - \frac{k}{2j-1}\right)}{\log \frac{(k-1)j}{k-j}}, \qquad g(j) = 1 + \frac{\log\left(1 - \frac{k}{2j+1}\right)}{\log \frac{(k-1)j}{k-j}};$$

FIGURE 3. The graph shows the dependence of the optimal value $v(p) = \phi(\xi_{s(p)})$ on the order of p of the mean ϕ_p, for varying dimensions k.

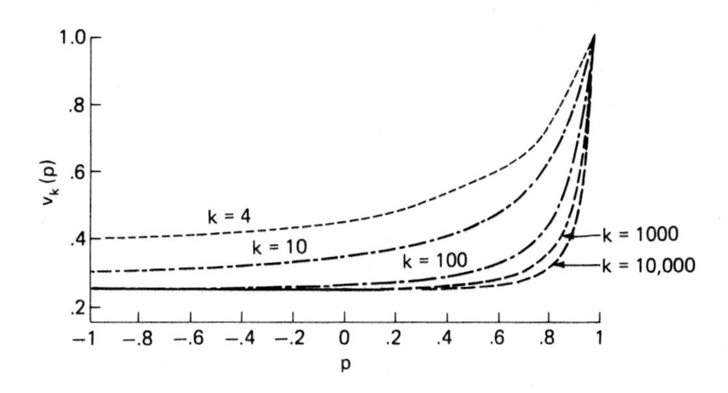

$$s(p) = \frac{j(j+1)\left(k-1+\left\{1-\frac{k}{2j+1}\right\}^{\frac{1}{p-1}}\right)}{(2j+1)(k-1)+(2j+1-k)\left\{1-\frac{k}{2j+1}\right\}^{\frac{1}{p-1}}}, \quad p \in (g(j), f(j+1)).$$

Figures 2 and 3 illustrate how the standardized support parameter $s(p)/k$ and the optimal value $\phi(\xi_{s(p)})$ vary with p. Variation is small for $p < -1$ and is not shown; variation is relatively large for $p > 0$.

Writing $s_k(p)$ in place of $s(p)$ we now show that for large dimensions k the support of the optimal designs tends to the vertices with half of their entries unity,

$$\lim_{k\to\infty} \frac{s_k(p)}{k} = \frac{1}{2}.$$

Let j_k be the integer part of $s_k(p)$, so that $s_k(p) \in [j_k, j_k+1)$, and $p \in [f(j_k), f(j_k+1))$. It suffices to show that j_k/k tends to $1/2$.

From $j_k > m$ we clearly get $\liminf j_k/k \geq 1/2$. We argue that $\limsup j_k/k = 1/2$. Otherwise there is no loss in generality in assuming $\lim j_k/k = \alpha > 1/2$. Then we obtain

$$f(j_k) = 1 + \frac{\log\left(1-\frac{1}{2j_k/k-1/k}\right)}{\log(k-1)\frac{j_k/k}{1-j_k/k}} \to 1 + \frac{\log\left(1-\frac{1}{2\alpha}\right)}{\lim_{k\to\infty}\log(k-1)\frac{\alpha}{1-\alpha}} = 1.$$

Hence eventually p must fall below $f(j_k)$, and this is impossible.

Remark

The criteria used to compare designs in this paper assume that all observations incur a constant cost regardless of the design point at which they are taken. Thus, the total cost is fixed by fixing the total sample size n, and the goal is to allocate proportions of this total sample size to design points in an optimal fashion. For example, a j-vertex design ξ_j assigns approximately $n/\binom{k}{j}$ observations to each of the $\binom{k}{j}$ j-vertices. However, there are many situations where investigators are also concerned with minimizing the number of design points. In this case, the first observation at a new design point is more expensive than all succeeding replications at that point. Criteria reflecting this concern, and designs optimal with respect to such criteria, are a worthy subject for further study.

Acknowledgments: The work of this author is partially supported by the Stiftung Volkswagenwerk, Hannover, and the Mathematical Sciences Institute, Cornell University. This is paper BU-534-M in the Biometrics Unit, Cornell University.

References

Cheng, C. -S. (1987). An application of the Kiefer-Wolfowitz equivalence theorem to a problem in Hadamard Transform Optics. *Ann. Statist.* **15**, 1593–1603.

Pukelsheim, F. (1987). Information increasing orderings in experimental design theory. *Internat. Statist. Rev.* **55**, 203–219.

22

A Unified Method of Estimation in Linear Models with Mixed Effects

C. Radhakrishna Rao[1]

ABSTRACT A unified approach is developed for the estimation of unknown fixed parameters and prediction of random effects in a mixed Gauss-Markov linear model. It is shown that both the estimators and their mean square errors can be expressed in terms of the elements of a g-inverse of a partitioned matrix which can be set up in terms of the matrices used in expressing the model. No assumptions are made on the ranks of the matrices involved. The method is parallel to the one developed by the author in the case of the fixed effects Gauss-Markov model using a g-inverse of a partitioned matrix (Rao, 1971, 1972, 1973, 1985).
A new concept of generalized normal equations is introduced for the simultaneous estimation of fixed parameters, random effects, and random error. All the results are deduced from a general lemma on an optimization problem. This paper is self-contained as all the algebraic results used are stated and proved. The unified theory developed in an earlier paper (Rao, 1988) is somewhat simplified.

1 Introduction

The Gauss-Markov model with fixed and random effects, called the mixed linear model, is written in the form

$$\boldsymbol{Y} = \boldsymbol{X\beta} + \boldsymbol{U\xi} + \boldsymbol{\varepsilon} \tag{1.1}$$

where $\boldsymbol{Y}$ is an n-vector of observations, $\boldsymbol{X}$ is a given $n \times m$ matrix, $\boldsymbol{\beta}$ is an m-vector of unknown fixed parameters, $\boldsymbol{U}$ is a given $n \times p$ matrix, $\boldsymbol{\xi}$ is a p-vector of hypothetical random variables. We make the following assumptions on the first and second order moments of $\boldsymbol{\xi}$ and $\boldsymbol{\varepsilon}$.

$$E(\boldsymbol{\xi}) = \boldsymbol{A}\gamma, \quad E(\boldsymbol{\varepsilon}) = \mathbf{0}, \quad D(\boldsymbol{\xi}) = \boldsymbol{\Gamma}, \quad D(\boldsymbol{\varepsilon}) = \boldsymbol{G}, \quad \text{Cov}(\boldsymbol{\xi}, \boldsymbol{\varepsilon}) = \mathbf{0}, \tag{1.2}$$

[1]Center for Multivariate Analysis, Pennsylvania State University, University Park, PA 16802.

where $D(\boldsymbol{z})$ is the dispersion matrix (covariance matrix) of the random vector $\boldsymbol{z}$.

We develop a simple and a unified approach in the general case, when nothing is assumed about the ranks of the matrices involved, for the estimation of the fixed parameters $\boldsymbol{\beta}$ and the prediction (or estimation) of the hypothetical variables $\boldsymbol{\xi}$ and $\boldsymbol{\varepsilon}$, when the other parameters $\boldsymbol{\gamma}$, $\boldsymbol{\Gamma}$, and $\boldsymbol{G}$ are partly known or completely known. No distributional assumptions have been made except the existence of the first and second moments. First we prove a few algebraic lemmas. The following notations are used.

$\Re^n$ n dimensional Euclidean real vector space.

$R(\boldsymbol{Z})$ vector space spanned by the column vectors of the matrix $\boldsymbol{Z}$.

$\rho(\boldsymbol{Z})$ rank of the matrix $\boldsymbol{Z}$.

$\boldsymbol{Z}^{\perp}$ a matrix of maximum rank such that $\boldsymbol{Z}'\boldsymbol{Z}^{\perp} = \boldsymbol{0}$.

$\boldsymbol{Z}^{-}$ a g-inverse of $\boldsymbol{Z}$, i.e., a matrix satisfying the equation $\boldsymbol{Z}\boldsymbol{Z}^{-}\boldsymbol{Z} = \boldsymbol{Z}$.

$\operatorname{tr} \boldsymbol{Z}$ sum of the diagonal elements of $\boldsymbol{Z}$ when it is a square matrix.

$(\boldsymbol{A} : \boldsymbol{B})$ the matrix obtained by adjoining the columns of the matrix $\boldsymbol{B}$ to those of $\boldsymbol{A}$.

We need the following results on g-inverses, which are well known but reproved in a simple way to make the theory of linear estimation presented here self-contained.

Lemma 1.1 *Let $\boldsymbol{Z}^{-}$ be a g-inverse as defined above. Then:*

$$\boldsymbol{a} = \boldsymbol{Z}^{-}\boldsymbol{b} \text{ is a solution of the consistent equation } \boldsymbol{Z}\boldsymbol{a} = \boldsymbol{b}. \tag{1.3}$$

$$\rho(\boldsymbol{Z}\boldsymbol{Z}^{-}) = \rho(\boldsymbol{Z}) = \operatorname{tr}(\boldsymbol{Z}\boldsymbol{Z}^{-}). \tag{1.4}$$

Proof Since $\boldsymbol{Z}\boldsymbol{a} = \boldsymbol{b}$ is consistent, $\boldsymbol{b} \in R(\boldsymbol{Z})$, i.e., $\boldsymbol{b} = \boldsymbol{Z}\boldsymbol{c}$ for some $\boldsymbol{c}$.

Then $\boldsymbol{Z}\boldsymbol{Z}^{-}\boldsymbol{b} = \boldsymbol{Z}\boldsymbol{Z}^{-}\boldsymbol{Z}\boldsymbol{c} = \boldsymbol{Z}\boldsymbol{c} = \boldsymbol{b}$ which shows that $\boldsymbol{Z}^{-}\boldsymbol{b}$ is a solution.

Now $\boldsymbol{Z}\boldsymbol{Z}^{-}\boldsymbol{Z}\boldsymbol{Z}^{-} = \boldsymbol{Z}\boldsymbol{Z}^{-}$, i.e., $\boldsymbol{Z}\boldsymbol{Z}^{-}$ is idempotent, so that (1.4) follows. □

Lemma 1.2 *Let $\boldsymbol{G}$ be an n.n.d. (non-negative definite) matrix of order $n \times n$, $\boldsymbol{X}$ be an $n \times m$ matrix and*

$$\begin{pmatrix} \boldsymbol{G} & \boldsymbol{X} \\ \boldsymbol{X} & \boldsymbol{0} \end{pmatrix}^{-} = \begin{pmatrix} \boldsymbol{C}_1 & \boldsymbol{C}_2 \\ \boldsymbol{C}_3 & -\boldsymbol{C}_4 \end{pmatrix} \tag{1.5}$$

be any choice of g-inverse. Then:

$$\boldsymbol{X}'\boldsymbol{C}_1\boldsymbol{G} = \boldsymbol{X}'\boldsymbol{C}_1'\boldsymbol{G} = \boldsymbol{0}, \ \boldsymbol{X}\boldsymbol{C}_2'\boldsymbol{X} = \boldsymbol{X} = \boldsymbol{X}\boldsymbol{C}_3\boldsymbol{X}. \tag{1.6}$$

$$\boldsymbol{G}\boldsymbol{C}_1'\boldsymbol{G}\boldsymbol{C}_1\boldsymbol{G} = \boldsymbol{G}\boldsymbol{C}_1\boldsymbol{G}\boldsymbol{C}_1'\boldsymbol{G} = \boldsymbol{G}\boldsymbol{C}_1\boldsymbol{G}\boldsymbol{C}_1\boldsymbol{G} = \boldsymbol{G}\boldsymbol{C}_1'\boldsymbol{G} = \boldsymbol{G}\boldsymbol{C}_1\boldsymbol{G}. \tag{1.7}$$

$$\boldsymbol{X}'\boldsymbol{C}_1\boldsymbol{X} = \boldsymbol{X}'\boldsymbol{C}_1'\boldsymbol{X} = \boldsymbol{0}. \tag{1.8}$$

$$\operatorname{tr} \boldsymbol{G}\boldsymbol{C}_1 = \rho(\boldsymbol{G} : \boldsymbol{X}) - \rho(\boldsymbol{X}). \tag{1.9}$$

Proof First we show that the equation

$$\begin{aligned} \boldsymbol{Ga} + \boldsymbol{Xb} &= \boldsymbol{G\lambda} \\ \boldsymbol{X'a} &= \boldsymbol{X'\mu} \end{aligned} \tag{1.10}$$

is consistent for any vectors $\boldsymbol{\lambda}$ and $\boldsymbol{\mu}$. To do so, we show that any vector orthogonal to the columns of the coefficient matrix on the left-hand side of (1.10) is orthogonal to the vector on the right-hand side of (1.10). Let $(\boldsymbol{\alpha}' : \boldsymbol{\beta}')$ be a row vector such that

$$(\boldsymbol{\alpha}' : \boldsymbol{\beta}') \begin{pmatrix} \boldsymbol{G} & \boldsymbol{X} \\ \boldsymbol{X}' & \boldsymbol{0} \end{pmatrix} = \boldsymbol{0} \Rightarrow \boldsymbol{\alpha}'\boldsymbol{G} + \boldsymbol{\beta}'\boldsymbol{X}' = \boldsymbol{0}, \qquad \boldsymbol{\alpha}'\boldsymbol{X} = \boldsymbol{0}$$
$$\Rightarrow \boldsymbol{\alpha}'\boldsymbol{G\alpha} + \boldsymbol{\beta}'\boldsymbol{X}'\boldsymbol{\alpha} = 0 = \boldsymbol{\alpha}'\boldsymbol{G\alpha} = 0 \Rightarrow \boldsymbol{\alpha}'\boldsymbol{G} = \boldsymbol{0}.$$

The last step follows since $\boldsymbol{G}$ is an n.n.d. matrix. Then,

$$(\boldsymbol{\alpha}' : \boldsymbol{\beta}') \begin{pmatrix} \boldsymbol{G\lambda} \\ \boldsymbol{X}'\boldsymbol{\mu} \end{pmatrix} = 0 \Rightarrow \begin{pmatrix} \boldsymbol{G\lambda} \\ \boldsymbol{X}'\boldsymbol{\mu} \end{pmatrix} \in R \begin{pmatrix} \boldsymbol{G} & \boldsymbol{X} \\ \boldsymbol{X}' & \boldsymbol{0} \end{pmatrix},$$

which establishes the consistency of (1.10). In such a case, using the g-inverse (1.5), we find a solution of (1.10):

$$\hat{\boldsymbol{a}} = \boldsymbol{C}_1 G\boldsymbol{\lambda} + \boldsymbol{C}_2\boldsymbol{X}'\boldsymbol{\mu}, \qquad \hat{\boldsymbol{b}} = \boldsymbol{C}_3 G\boldsymbol{\lambda} - \boldsymbol{C}_4\boldsymbol{X}'\boldsymbol{\mu}.$$

Substituting $\hat{\boldsymbol{a}}$ for $\boldsymbol{a}$ in the second equation of (1.10) and equating the terms involving $\boldsymbol{\lambda}$ and $\boldsymbol{\mu}$ on both sides, we obtain

$$\boldsymbol{X}'\boldsymbol{C}_1\boldsymbol{G} = \boldsymbol{0}, \qquad \boldsymbol{X}'\boldsymbol{C}_2\boldsymbol{X}' = \boldsymbol{X}'. \tag{1.11}$$

Further, the transpose of the g-inverse in (1.5) is also a g-inverse in view of the symmetry of the left-hand side matrix of (1.5). Thus, results analogous to (1.11) hold, giving

$$\boldsymbol{X}'\boldsymbol{C}_1'\boldsymbol{G} = \boldsymbol{0}, \qquad \boldsymbol{X}'\boldsymbol{C}_3'\boldsymbol{X}' = \boldsymbol{X}', \tag{1.12}$$

which prove (1.6). Again, substituting $\hat{\boldsymbol{a}}$ and $\hat{\boldsymbol{b}}$ for $\boldsymbol{a}$ and $\boldsymbol{b}$ in the first equation of (1.5) and equating the terms in $\boldsymbol{\lambda}$ on both sides, we have

$$\boldsymbol{GC}_1\boldsymbol{G} + \boldsymbol{XC}_3\boldsymbol{G} = \boldsymbol{G}. \tag{1.13}$$

Multiplying (1.13) by $\boldsymbol{GC}_1$ and $\boldsymbol{GC}_1'$ and using (1.11) and (1.12) and the fact that $(\boldsymbol{GC}_1\boldsymbol{G})' = \boldsymbol{GC}_1'\boldsymbol{G}$, we get the equalities in (1.7).

It is easy to see that

$$\begin{aligned} \boldsymbol{Ga} + \boldsymbol{Xb} &= \boldsymbol{X\mu} \\ \boldsymbol{X'a} &= \boldsymbol{0} \end{aligned} \tag{1.14}$$

is a consistent equation for any $\boldsymbol{\mu}$, so that $\hat{\boldsymbol{a}} = \boldsymbol{C}_1\boldsymbol{X\mu}$ is a solution. Substituting $\hat{\boldsymbol{a}}$ for $\boldsymbol{a}$ in the second equation of (1.14), we find that $\boldsymbol{X}'\boldsymbol{C}_1\boldsymbol{X\mu} = \boldsymbol{0}$ $\forall\boldsymbol{\mu} \Rightarrow \boldsymbol{X}'\boldsymbol{C}_1\boldsymbol{X} = \boldsymbol{0} = \boldsymbol{X}'\boldsymbol{C}_1'\boldsymbol{X}$, which proves (1.8).

Now

$$\begin{aligned}\rho\begin{pmatrix} \boldsymbol{G} & \boldsymbol{X} \\ \boldsymbol{X}' & \boldsymbol{0}\end{pmatrix}\begin{pmatrix} \boldsymbol{C}_1 & \boldsymbol{C}_2 \\ \boldsymbol{C}_3 & -\boldsymbol{C}_4\end{pmatrix} &= \operatorname{tr}\begin{pmatrix} \boldsymbol{G}\boldsymbol{C}_1 + \boldsymbol{X}\boldsymbol{C}_3 & \boldsymbol{G}\boldsymbol{C}_2 - \boldsymbol{X}\boldsymbol{C}_4 \\ \boldsymbol{X}'\boldsymbol{C}_1 & \boldsymbol{X}'\boldsymbol{C}_2\end{pmatrix} \\ &= \operatorname{tr}(\boldsymbol{G}\boldsymbol{C}_1 + \boldsymbol{X}\boldsymbol{C}_3) + \operatorname{tr}\boldsymbol{X}'\boldsymbol{C}_2 \\ &= \operatorname{tr}\boldsymbol{G}\boldsymbol{C}_1 + \rho(\boldsymbol{X}\boldsymbol{C}_3) + \rho(\boldsymbol{X}'\boldsymbol{C}_2) \\ &= \operatorname{tr}\boldsymbol{G}\boldsymbol{C}_1 + \rho(\boldsymbol{X}) + \rho(\boldsymbol{X}') \qquad (1.15)\end{aligned}$$

since $\boldsymbol{X}\boldsymbol{C}_3$ and $\boldsymbol{X}'\boldsymbol{C}_2$ are idempotent. Note that since $\boldsymbol{G}$ is n.n.d.,

$$\begin{aligned}\boldsymbol{a}'(\boldsymbol{G},\boldsymbol{X}) &= \boldsymbol{c}'(\boldsymbol{X}',\boldsymbol{0}) \\ &\Rightarrow \boldsymbol{a}'\boldsymbol{G} = \boldsymbol{c}'\boldsymbol{X}', \quad \boldsymbol{a}'\boldsymbol{X} = \boldsymbol{0} \\ &\Rightarrow \boldsymbol{a}'\boldsymbol{G}\boldsymbol{a} = \boldsymbol{c}'\boldsymbol{X}'\boldsymbol{a} = \boldsymbol{c}'\boldsymbol{0} = \boldsymbol{0}, \quad \boldsymbol{a}'\boldsymbol{X} = \boldsymbol{0} \\ &\Rightarrow \boldsymbol{a}'(\boldsymbol{G},\boldsymbol{X}) = \boldsymbol{0}.\end{aligned}$$

This shows that the rows of $(\boldsymbol{G},\boldsymbol{X})$ are linearly independent of the rows of $(\boldsymbol{X}',\boldsymbol{0})$. Consequently,

$$\rho\begin{pmatrix} \boldsymbol{G} & \boldsymbol{X} \\ \boldsymbol{X}' & \boldsymbol{0}\end{pmatrix}\begin{pmatrix} \boldsymbol{C}_1 & \boldsymbol{C}_2 \\ \boldsymbol{C}_3 & -\boldsymbol{C}_4\end{pmatrix} = \rho\begin{pmatrix} \boldsymbol{G} & \boldsymbol{X} \\ \boldsymbol{X}' & \boldsymbol{0}\end{pmatrix} = \rho(\boldsymbol{G}:\boldsymbol{X}) + \rho(\boldsymbol{X}). \quad (1.16)$$

Equating (1.15) and (1.16), we have (1.9) and Lemma 1.2 is proved. □

Now we prove the main lemma

Lemma 1.3 *Let* $\boldsymbol{G}$ *and* $\boldsymbol{X}$ *be as in Lemma 1.2,* $\boldsymbol{g} \in R(\boldsymbol{G}:\boldsymbol{X})$ *and* $\boldsymbol{p} \in R(\boldsymbol{X}')$. *Then*

$$\min_{\boldsymbol{X}'\boldsymbol{a}=\boldsymbol{p}} (\boldsymbol{a}'\boldsymbol{G}\boldsymbol{a} + 2\boldsymbol{a}'\boldsymbol{g}) = \boldsymbol{a}_*'\boldsymbol{G}\boldsymbol{a}_* + 2\boldsymbol{a}_*'\boldsymbol{g}, \qquad (1.17)$$

where $\boldsymbol{a}_*$ *is any solution to*

$$\begin{aligned}\boldsymbol{G}\boldsymbol{a} + \boldsymbol{X}\boldsymbol{b} &= -\boldsymbol{g} \\ \boldsymbol{X}'\boldsymbol{a} &= \boldsymbol{p}\end{aligned} \qquad (1.18)$$

With $\boldsymbol{C}_1$, $\boldsymbol{C}_2$, $\boldsymbol{C}_3$, $\boldsymbol{C}_4$ *as defined in Lemma 1.2, one choice of the solution for* $(\boldsymbol{a},\boldsymbol{b})$ *of (1.18) is*

$$\boldsymbol{a}_* = -\boldsymbol{C}_1\boldsymbol{g} + \boldsymbol{C}_2\boldsymbol{p}, \qquad \boldsymbol{b}_* = -\boldsymbol{C}_3\boldsymbol{g} - \boldsymbol{C}_4\boldsymbol{p}, \qquad (1.19)$$

giving the expressions for the minimum in (1.18) as

$$\boldsymbol{g}'\boldsymbol{a}_* - \boldsymbol{p}'\boldsymbol{b}_* = -\boldsymbol{g}'\boldsymbol{C}_1\boldsymbol{g} + \boldsymbol{g}'(\boldsymbol{C}_2 + \boldsymbol{C}_3')\boldsymbol{p} + \boldsymbol{p}'\boldsymbol{C}_4\boldsymbol{p}. \qquad (1.20)$$

Proof Let $\boldsymbol{a}_*$, $\boldsymbol{b}_*$ be any solution of (1.18), and $\boldsymbol{Z} = \boldsymbol{X}^{\perp}$. Then multiplying the first equation of (1.18), with $(\boldsymbol{a},\boldsymbol{b})$ replaced by $(\boldsymbol{a}_*,\boldsymbol{b}_*)$, by $\boldsymbol{Z}'$ and $\boldsymbol{a}_*'$, we obtain

$$\boldsymbol{Z}'\boldsymbol{G}\boldsymbol{A}_* + \boldsymbol{Z}'\boldsymbol{g} = \boldsymbol{0} \qquad (1.21)$$

$$\boldsymbol{a}_*'\boldsymbol{G}\boldsymbol{a}_* + \boldsymbol{a}_*'\boldsymbol{g} = -\boldsymbol{b}_*'\boldsymbol{p}. \qquad (1.22)$$

A general solution of $\boldsymbol{X}'\boldsymbol{a} = \boldsymbol{p}$ is $\boldsymbol{a}_* + \boldsymbol{Z}\boldsymbol{d}$ where $\boldsymbol{d}$ is arbitrary. Then writing $\boldsymbol{a} = \boldsymbol{a}_* + \boldsymbol{Z}\boldsymbol{d}$,

$$\begin{aligned} a'\boldsymbol{G}\boldsymbol{a} + 2a'\boldsymbol{g} &= \boldsymbol{a}_*'\boldsymbol{G}\boldsymbol{a}_* + 2\boldsymbol{a}_*'\boldsymbol{g} + \boldsymbol{d}'\boldsymbol{Z}'\boldsymbol{G}\boldsymbol{Z}\boldsymbol{d} + 2\boldsymbol{d}'(\boldsymbol{Z}'\boldsymbol{G}\boldsymbol{a}_* + \boldsymbol{Z}'\boldsymbol{g}) \\ &= \boldsymbol{a}_*'\boldsymbol{G}\boldsymbol{a}_* + 2\boldsymbol{a}_*'\boldsymbol{g} + \boldsymbol{d}'\boldsymbol{Z}'\boldsymbol{G}\boldsymbol{Z}\boldsymbol{d}, \qquad \text{using (1.21)} \\ &\geq \boldsymbol{a}_*'\boldsymbol{G}\boldsymbol{a}_* + 2\boldsymbol{a}_*'\boldsymbol{g} = \boldsymbol{a}_*'\boldsymbol{g} - \boldsymbol{b}_*'\boldsymbol{p}, \qquad \text{using (1.22)} \end{aligned}$$

with equality when $\boldsymbol{d} = 0$, which proves (1.17). The result (1.20) is obtained by substituting the expressions (1.19) in (1.22). □

Lemma 1.3 plays a crucial role in estimation and prediction problems in linear models. The results are given in Section 2.

Lemma 1.4 *If $\boldsymbol{Y}$ is an n-vector random variable with $E(\boldsymbol{Y}) = \boldsymbol{0}$ and $D(\boldsymbol{Y}) = \sigma^2\boldsymbol{G}$, then an unbiased estimator of σ^2 is*

$$\sigma^2 = \frac{\boldsymbol{Y}'\boldsymbol{G}^-\boldsymbol{Y}}{\rho(\boldsymbol{G})} \tag{1.23}$$

where $\boldsymbol{G}^-$ is any inverse of $\boldsymbol{G}$.

Proof Using (1.4),

$$E(\boldsymbol{Y}'\boldsymbol{G}^-\boldsymbol{Y}) = E\,\mathrm{tr}(\boldsymbol{Y}\boldsymbol{Y}'\boldsymbol{G}^-) = \sigma^2\,\mathrm{tr}\,\boldsymbol{G}\boldsymbol{G}^- = \sigma^2\rho(\boldsymbol{G}), \qquad \text{using (1.4).}$$

□

The basic results on estimation in the Gauss-Markov model with fixed and mixed effects are presented in Sections 2 and 3. The paper provides a synthesis of several well-known results obtained over the last forty years and elaborates on new ideas introduced in a recent paper of the author (Rao, 1988). The object of the paper is to show how all the results can be derived with the use of the above algebraic lemmas, and to emphasize the key role played by the inverse partitioned matrix method introduced in Rao (1971). A new concept of generalized normal equations is introduced.

2 Fixed Effects Linear Model

The Gauss-Markov linear model with fixed effects is

$$\boldsymbol{Y} = \boldsymbol{X}\boldsymbol{\beta} + \boldsymbol{\varepsilon}, \quad E(\boldsymbol{\varepsilon}) = \boldsymbol{0}, \quad D(\boldsymbol{\varepsilon}) = \sigma^2\boldsymbol{G}, \tag{2.1}$$

and the associated problems are those of estimating the unknown parameters $\boldsymbol{\beta}$ and σ^2 and the random error $\boldsymbol{\varepsilon}$. The matrix $\boldsymbol{G}$ is assumed to be known. We use the results of Lemma 1.3 in solving these problems. We denote

$$\begin{pmatrix} \boldsymbol{G} & \boldsymbol{X} \\ \boldsymbol{X}' & \boldsymbol{0} \end{pmatrix}^- = \begin{pmatrix} \boldsymbol{C}_1 & \boldsymbol{C}_2 \\ \boldsymbol{C}_3 & -\boldsymbol{C}_4 \end{pmatrix} \tag{2.2}$$

for any choice of the g-inverse.

BLUE OF $p'\beta$

Consider a linear function $\boldsymbol{a}'\boldsymbol{Y}$ as an unbiased estimator of $\boldsymbol{p}'\boldsymbol{\beta}$. Then

$$E(\boldsymbol{a}'\boldsymbol{Y}) = \boldsymbol{a}'\boldsymbol{X}\boldsymbol{\beta} = \boldsymbol{p}'\boldsymbol{\beta} \quad \forall\boldsymbol{\beta} \Rightarrow \boldsymbol{X}'\boldsymbol{a} = \boldsymbol{p}. \tag{2.3}$$

We find $\boldsymbol{a}$ by minimizing

$$V(\boldsymbol{a}'\boldsymbol{Y}) = \sigma^2\boldsymbol{a}'\boldsymbol{G}\boldsymbol{a} \quad \text{subject to} \quad \boldsymbol{X}'\boldsymbol{a} = \boldsymbol{p}. \tag{2.4}$$

Applying Lemma 1.3 with $\boldsymbol{g} = \boldsymbol{0}$, the BLUE of $\boldsymbol{p}'\boldsymbol{\beta}$ is

$$\boldsymbol{a}_*'\boldsymbol{Y} = \boldsymbol{p}'\boldsymbol{C}_2'\boldsymbol{Y} \tag{2.5}$$

with the minimum variance

$$\sigma^2\boldsymbol{a}_*'\boldsymbol{G}\boldsymbol{a}_* = -\sigma^2\boldsymbol{b}_*'\boldsymbol{p} = \sigma^2\boldsymbol{p}'\boldsymbol{C}_4\boldsymbol{p}, \tag{2.6}$$

using (1.20) and the expressions for $\boldsymbol{a}_*$ and $\boldsymbol{b}_*$ in (1.19).

ESTIMATION (PREDICTION) OF ε

Consider a linear function $\boldsymbol{q}'\boldsymbol{\varepsilon}$ of $\boldsymbol{\varepsilon}$ and let $\boldsymbol{a}'\boldsymbol{Y}$, with $E(\boldsymbol{a}'\boldsymbol{Y}) = \boldsymbol{0} \Rightarrow \boldsymbol{a}'\boldsymbol{X} = \boldsymbol{0}$, be its predictor. Then the mean square error of prediction is

$$\begin{aligned} E(\boldsymbol{q}'\boldsymbol{\varepsilon} - \boldsymbol{a}'\boldsymbol{Y})^2 &= E(\boldsymbol{q}'\boldsymbol{\varepsilon} - \boldsymbol{a}'\boldsymbol{\varepsilon})^2 \\ &= \sigma^2(\boldsymbol{a}'\boldsymbol{G}\boldsymbol{a} - 2\boldsymbol{a}'\boldsymbol{G}\boldsymbol{q} + \boldsymbol{q}'\boldsymbol{G}\boldsymbol{q}). \end{aligned} \tag{2.7}$$

Applying (1.19) with $\boldsymbol{g} = -\boldsymbol{G}\boldsymbol{q}$ and $\boldsymbol{p} = \boldsymbol{0}$, the minimum of (2.7) is attained when the predictor is

$$\boldsymbol{a}_*'\boldsymbol{Y} = -(\boldsymbol{C}_1\boldsymbol{g})'\boldsymbol{Y} = \boldsymbol{q}'\boldsymbol{G}\boldsymbol{C}_1'\boldsymbol{Y} = \boldsymbol{q}'\hat{\boldsymbol{\varepsilon}}. \tag{2.8}$$

The minimum mean square error of prediction is, using (1.19) and (1.20),

$$\begin{aligned} \sigma^2(\boldsymbol{a}_*'\boldsymbol{g} - \boldsymbol{b}_*'\boldsymbol{p} + \boldsymbol{q}'\boldsymbol{G}\boldsymbol{q}) &= \sigma^2(-\boldsymbol{q}'\boldsymbol{G}\boldsymbol{C}_1'\boldsymbol{G}\boldsymbol{q} + \boldsymbol{q}'\boldsymbol{G}\boldsymbol{q}) \\ &= \sigma^2\boldsymbol{q}'(\boldsymbol{G} - \boldsymbol{G}\boldsymbol{C}_1'\boldsymbol{G})\boldsymbol{q}. \end{aligned} \tag{2.9}$$

The results (2.8) and (2.9), which hold for any $\boldsymbol{q}$, imply that the minimum dispersion error predictor of $\boldsymbol{\varepsilon}$ is

$$\hat{\boldsymbol{\varepsilon}} = \boldsymbol{G}\boldsymbol{C}_1'\boldsymbol{Y} = \boldsymbol{G}\boldsymbol{C}_1\boldsymbol{Y} \tag{2.10}$$

with

$$D(\hat{\boldsymbol{\varepsilon}} - \boldsymbol{\varepsilon}) = \sigma^2(\boldsymbol{G} - \boldsymbol{G}\boldsymbol{C}_1\boldsymbol{G}). \tag{2.11}$$

Estimation of σ^2

Now

$$\begin{aligned} E(\hat{\boldsymbol{\varepsilon}}) &= E(\boldsymbol{G}\boldsymbol{C}_1\boldsymbol{Y}) = \boldsymbol{G}\boldsymbol{C}_1\boldsymbol{X}\boldsymbol{\beta} = \mathbf{0}, \qquad \text{by (1.6)}, \\ D(\hat{\boldsymbol{\varepsilon}}) &= D(\boldsymbol{G}\boldsymbol{C}_1\boldsymbol{Y}) = \sigma^2\boldsymbol{G}\boldsymbol{C}_1\boldsymbol{G}\boldsymbol{C}_1'\boldsymbol{G} \\ &= \sigma^2\boldsymbol{G}\boldsymbol{C}_1\boldsymbol{G}, \qquad \text{by (1.7)}. \end{aligned}$$

Then using Lemma 1.4, an unbiased estimator of σ^2 is

$$\hat{\boldsymbol{\varepsilon}}'(\boldsymbol{G}\boldsymbol{C}_1\boldsymbol{G})^-\hat{\boldsymbol{\varepsilon}} = \boldsymbol{Y}'\boldsymbol{C}_1\boldsymbol{G}(\boldsymbol{G}\boldsymbol{C}_1\boldsymbol{G})^-\boldsymbol{G}\boldsymbol{C}_1\boldsymbol{Y} \tag{2.12}$$

with a suitable divisor. Note that $\boldsymbol{Y} \in R(\boldsymbol{G} : \boldsymbol{X})$, so that $\boldsymbol{Y} = \boldsymbol{G}\boldsymbol{\lambda} + \boldsymbol{X}\boldsymbol{\mu}$ for a suitable $\boldsymbol{\lambda}$ and $\boldsymbol{\mu}$. Then

$$\boldsymbol{Y}'\boldsymbol{C}_1\boldsymbol{G}(\boldsymbol{G}\boldsymbol{C}_1\boldsymbol{G})^-\boldsymbol{G}\boldsymbol{C}_1\boldsymbol{Y} = \boldsymbol{\lambda}'(\boldsymbol{G}\boldsymbol{C}_1\boldsymbol{G})(\boldsymbol{G}\boldsymbol{C}_1\boldsymbol{G})^-(\boldsymbol{G}\boldsymbol{C}_1\boldsymbol{G})\boldsymbol{\lambda}, \tag{2.13}$$

since the terms in $\boldsymbol{X}$ vanish, using (1.6). Hence, by the definition of a g-inverse and (1.6), (2.13) reduces to

$$\begin{aligned} \boldsymbol{\lambda}'\boldsymbol{G}\boldsymbol{C}_1\boldsymbol{G}\boldsymbol{\lambda} &= (\boldsymbol{\mu}'\boldsymbol{X}' + \boldsymbol{\lambda}\boldsymbol{G})\boldsymbol{C}_1(\boldsymbol{X}\boldsymbol{\mu} + \boldsymbol{G}\boldsymbol{\lambda}), \quad \text{using (1.6)} \\ &= \boldsymbol{Y}'\boldsymbol{C}_1\boldsymbol{Y}. \end{aligned} \tag{2.14}$$

Now

$$\begin{aligned} E(\boldsymbol{Y}'\boldsymbol{C}_1\boldsymbol{Y}) &= \operatorname{tr} E(\boldsymbol{Y}\boldsymbol{Y}'\boldsymbol{C}_1) = \sigma^2 \operatorname{tr}(\boldsymbol{G} + \boldsymbol{X}\boldsymbol{\beta}\boldsymbol{\beta}'\boldsymbol{X}')\boldsymbol{C}_1 \\ &= \sigma^2 \operatorname{tr} \boldsymbol{G}\boldsymbol{C}_1 + \sigma^2 \operatorname{tr} \boldsymbol{\beta}\boldsymbol{\beta}'\boldsymbol{X}'\boldsymbol{C}'\boldsymbol{X} = \operatorname{tr} \boldsymbol{G}\boldsymbol{C}_1, \qquad \text{using (1.8)} \\ &= \sigma^2 \operatorname{tr} \boldsymbol{G}\boldsymbol{C}_1 = \sigma^2\big(\rho(\boldsymbol{G} : \boldsymbol{X}) - \rho(\boldsymbol{X})\big), \qquad \text{using (1.9)}. \end{aligned}$$

Then an unbiased estimator of σ^2 is

$$\hat{\sigma}^2 = \frac{\boldsymbol{Y}'\boldsymbol{C}_1\boldsymbol{Y}}{\rho(\boldsymbol{G} : \boldsymbol{X}) - \rho(\boldsymbol{X})}. \tag{2.15}$$

If $\boldsymbol{Y}$ is assumed to have a normal distribution, this estimator is the minimum variance unbiased estimator of σ^2.

Normal equations

The expressions for the estimates of $\boldsymbol{p}'\boldsymbol{\beta}$, σ^2, and $\boldsymbol{\varepsilon}$ obtained above suggest a more direct way of obtaining them by first solving the consistent set of equations

$$\begin{aligned} \boldsymbol{G}\boldsymbol{\alpha} + \boldsymbol{X}\boldsymbol{\beta} &= \boldsymbol{Y} \\ \boldsymbol{X}'\boldsymbol{\alpha} &= \mathbf{0} \end{aligned}. \tag{2.16}$$

If $(\hat{\boldsymbol{\alpha}}, \hat{\boldsymbol{\beta}})$ is a solution of (2.6), then we have the following.

(i) The BLUE of an estimable function $\boldsymbol{p}'\boldsymbol{\beta}$ is $\boldsymbol{p}'\hat{\boldsymbol{\beta}}$.

(ii) The minimum dispersion error predictor of $\boldsymbol{\varepsilon}$ is $\hat{\boldsymbol{\varepsilon}} = \boldsymbol{G}\hat{\boldsymbol{\alpha}}$.

(iii) An unbiased estimator of σ^2 is $\hat{\sigma}^2 = \boldsymbol{Y}'\hat{\boldsymbol{\alpha}}/\big(\rho(\boldsymbol{G} : \boldsymbol{X}) - \rho(\boldsymbol{X})\big)$.

We may call (2.16) the generalized normal equations for the simultaneous estimation of $\boldsymbol{\varepsilon}$ and $\boldsymbol{\beta}$.

If $\hat{\boldsymbol{\alpha}}$ and $\hat{\boldsymbol{\beta}}$ are obtained through a g-inverse as defined in (2.2), then we automatically have the following expressions for the precisions of the estimates:

$$V(\boldsymbol{p}'\hat{\boldsymbol{\beta}}) = \sigma^2 \boldsymbol{p}' C_4 \boldsymbol{p}$$
$$D(\hat{\boldsymbol{\varepsilon}} - \boldsymbol{\varepsilon}) = \sigma^2 (\boldsymbol{G} - \boldsymbol{G}C_1\boldsymbol{G}).$$

When $\boldsymbol{G}^{-1}$ exists, we can write the equations (2.16) in terms of the unknowns $\boldsymbol{\varepsilon}$ and $\boldsymbol{\beta}$ to be estimated in the form

$$\begin{aligned} \boldsymbol{\varepsilon} + \boldsymbol{X}\boldsymbol{\beta} &= \boldsymbol{Y} \\ \boldsymbol{X}'\boldsymbol{G}^{-1}\boldsymbol{\varepsilon} &= \mathbf{0} \end{aligned} \qquad (2.17)$$

using the relationship $\boldsymbol{\varepsilon} = \boldsymbol{G}\boldsymbol{\alpha}$. Thus, the equations (2.17) are the appropriate normal equations for $\boldsymbol{\varepsilon}$ and $\boldsymbol{\beta}$ when $\boldsymbol{G}^{-1}$ exists. In such a case, eliminating $\boldsymbol{\varepsilon}$ in the second equation using the first equation in (2.17), we have

$$\boldsymbol{X}'\boldsymbol{G}^{-1}\boldsymbol{X}\boldsymbol{\beta} = \boldsymbol{X}'\boldsymbol{G}^{-1}\boldsymbol{Y} \qquad (2.18)$$

which is the usual normal equation for $\boldsymbol{\beta}$ only. If $\hat{\boldsymbol{\beta}}$ is a solution of (2.18), then from (2.17), $\hat{\boldsymbol{\varepsilon}} = \boldsymbol{Y} - \boldsymbol{X}\hat{\boldsymbol{\beta}}$ is the usual residual.

The equation (2.16), however, is a more natural one which is simple to set up without any initial computations, and which does not involve any assumptions on the ranks of the matrices involved.

Projection operator

The second normal equation of (2.16) implies that $\boldsymbol{\alpha} = \boldsymbol{Z}\boldsymbol{\delta}$ where $\boldsymbol{Z} = \boldsymbol{X}^{\perp}$ and $\boldsymbol{\delta}$ is arbitrary. Substituting for $\boldsymbol{\alpha}$ in the first normal equation of (2.16),

$$\boldsymbol{X}\boldsymbol{\beta} + \boldsymbol{G}\boldsymbol{Z}\boldsymbol{\delta} = \boldsymbol{Y} \qquad (2.19)$$

which provides the decomposition of the observed $\boldsymbol{Y}$ as 'signal + noise' giving estimates of $\boldsymbol{X}\boldsymbol{\beta}$ and ε. Note that $R(\boldsymbol{G}\boldsymbol{Z})$ and $R(\boldsymbol{X})$ are disjoint and $\boldsymbol{Y} \in R(\boldsymbol{G} : \boldsymbol{X}) = R(\boldsymbol{G}\boldsymbol{Z} : \boldsymbol{X})$ w.p.1. Hence the decomposition (2.19) is unique. If $\hat{\boldsymbol{\beta}}$ and $\hat{\boldsymbol{\delta}}$ is a solution of (2.19), then an estimate of $\boldsymbol{p}'\boldsymbol{\beta}$ is $\boldsymbol{p}'\hat{\boldsymbol{\beta}}$ and of $\boldsymbol{\varepsilon}$ is $\hat{\boldsymbol{\varepsilon}} = \boldsymbol{G}\boldsymbol{Z}\hat{\boldsymbol{\delta}}$.

Since $R(\boldsymbol{X})$ and $R(\boldsymbol{G}\boldsymbol{Z})$ are disjoint, although $R(\boldsymbol{G}\boldsymbol{Z} : \boldsymbol{X})$ may not span the whole of $\Re^n$, there exist projection operators $P_{\boldsymbol{X}}$ and $P_{\boldsymbol{G}\boldsymbol{Z}}$ onto $R(\boldsymbol{X})$ and $R(\boldsymbol{G}\boldsymbol{Z})$ in terms of which $\boldsymbol{Y}$ can be decomposed as in (2.19). Then

$$\boldsymbol{X}\hat{\boldsymbol{\beta}} = P_{\boldsymbol{X}}\boldsymbol{Y} \quad \text{and} \quad \hat{\boldsymbol{\varepsilon}} = \boldsymbol{G}\boldsymbol{Z}\hat{\boldsymbol{\delta}} = P_{\boldsymbol{G}\boldsymbol{Z}}\boldsymbol{Y} = (I - P_{\boldsymbol{X}})\boldsymbol{Y}.$$

Rao (1974) and Rao and Yanai (1979) give a detailed discussion of generalized projection operators.

3 Mixed Effects Linear Model

The mixed effects linear model is of the form

$$\boldsymbol{Y} = \boldsymbol{X\beta} + \boldsymbol{U\xi} + \boldsymbol{\varepsilon} \tag{3.1}$$

with

$$E(\boldsymbol{\varepsilon}) = \boldsymbol{0}, \quad E(\boldsymbol{\xi}) = \boldsymbol{A\gamma}, \quad D(\boldsymbol{\varepsilon}) = \boldsymbol{G}, \quad \mathrm{Cov}(\boldsymbol{\varepsilon}, \boldsymbol{\xi}) = \boldsymbol{0}, \quad D(\boldsymbol{\xi}) = \boldsymbol{\Gamma}.$$

We write the model in an alternate form

$$\boldsymbol{Y} = \boldsymbol{X}_*\boldsymbol{\beta}_* + \boldsymbol{U\eta} + \boldsymbol{\varepsilon}, \tag{3.2}$$

where

$$\boldsymbol{X}_* = (\boldsymbol{X} : \boldsymbol{UA}), \quad \boldsymbol{\beta}_*' = (\boldsymbol{\beta}' : \boldsymbol{\gamma}'), \quad \boldsymbol{\eta} = \boldsymbol{\xi} - \boldsymbol{A\gamma}.$$

For purposes of estimation, we first assume that $\boldsymbol{G}$ and $\boldsymbol{\Gamma}$ are known and later comment on their estimation under a given structure.

Estimation of a mixed effect

Let $\boldsymbol{p}'\boldsymbol{\beta}_* + \boldsymbol{q}'\boldsymbol{\eta}$ be a mixed effect to be estimated. If $c + \boldsymbol{a}'\boldsymbol{Y}$ is an unbiased estimator, then

$$E(c + \boldsymbol{a}'\boldsymbol{Y} - \boldsymbol{p}'\boldsymbol{\beta}_* - \boldsymbol{q}'\boldsymbol{\eta}) = \boldsymbol{0} \quad \forall \boldsymbol{\beta} \Rightarrow c = \boldsymbol{0} \quad \text{and } \boldsymbol{X}_*'\boldsymbol{a} = \boldsymbol{p}.$$

The mean square error is

$$E[\boldsymbol{a}'(\boldsymbol{U\eta} + \boldsymbol{\varepsilon}) - \boldsymbol{q}'\boldsymbol{\eta}]^2 = \boldsymbol{a}'\boldsymbol{G}_*\boldsymbol{a} - 2\boldsymbol{a}'\boldsymbol{U\Gamma q} + \boldsymbol{q}'\boldsymbol{\Gamma q},$$

where $\boldsymbol{G}_* = \boldsymbol{U\Gamma U}' + \boldsymbol{G}$. Applying Lemma 1.3, the optimum $\boldsymbol{a}$ is a solution of the equation

$$\begin{aligned} \boldsymbol{G}_*\boldsymbol{a} + \boldsymbol{X}_*\boldsymbol{b} &= \boldsymbol{U\Gamma q} \\ \boldsymbol{X}'\boldsymbol{a} &= \boldsymbol{p}. \end{aligned} \tag{3.3}$$

If

$$\begin{pmatrix} \boldsymbol{G}_* & \boldsymbol{X}_* \\ \boldsymbol{X}_*' & \boldsymbol{0} \end{pmatrix}^- = \begin{pmatrix} \boldsymbol{C}_1 & \boldsymbol{C}_2 \\ \boldsymbol{C}_3 & -\boldsymbol{C}_4 \end{pmatrix} \tag{3.4}$$

for any choice of the g-inverse, then the best linear estimator of $\boldsymbol{p}'\boldsymbol{\beta}_* + \boldsymbol{q}'\boldsymbol{\eta}$ is $\boldsymbol{a}_*'\boldsymbol{Y}$ where

$$\boldsymbol{a}_* = \boldsymbol{C}_1\boldsymbol{U\Gamma q} + \boldsymbol{C}_2\boldsymbol{p} \tag{3.5}$$

and the mean square error is, using (1.19) and (1.20),

$$-\boldsymbol{a}_*'\boldsymbol{U\Gamma q} - \boldsymbol{b}_*'\boldsymbol{p} + \boldsymbol{q}'\boldsymbol{\Gamma q} = \boldsymbol{q}'(\boldsymbol{\Gamma} - \boldsymbol{\Gamma U}'\boldsymbol{C}_1'\boldsymbol{U\Gamma})\boldsymbol{q} + \boldsymbol{p}'\boldsymbol{C}_4\boldsymbol{p} - \boldsymbol{p}'(\boldsymbol{C}_2' + \boldsymbol{C}_3')\boldsymbol{U\Gamma q}. \tag{3.6}$$

Writing $\boldsymbol{p}' = (\boldsymbol{p}_1', \boldsymbol{p}_2')$,

$$\boldsymbol{p}'\boldsymbol{\beta}_* + \boldsymbol{q}'\boldsymbol{\eta} = \boldsymbol{p}_1'\boldsymbol{\beta} + \boldsymbol{p}_2'\boldsymbol{\gamma} + \boldsymbol{q}'\boldsymbol{\eta} = \boldsymbol{p}_1'\boldsymbol{\beta} + (\boldsymbol{p}_2' - \boldsymbol{q}'\boldsymbol{A})\boldsymbol{\gamma} + \boldsymbol{q}'\boldsymbol{\xi}.$$

The formulas (3.5) and (3.6) cover all special cases of linear functions involving one or more of the parameters $\boldsymbol{\beta}$ and $\boldsymbol{\gamma}$ and the random variable $\boldsymbol{\xi}$.

Estimation of the random error

Let $\boldsymbol{r}'\boldsymbol{\varepsilon}$ be a linear function of the random error estimated by $\boldsymbol{a}'\boldsymbol{Y}$. The condition of unbiasedness implies that $\boldsymbol{a}'\boldsymbol{X}_* = \boldsymbol{0}$. The mean square error is

$$E(\boldsymbol{a}'\boldsymbol{Y} - \boldsymbol{r}'\boldsymbol{\varepsilon})^2 = \boldsymbol{a}'\boldsymbol{G}_*\boldsymbol{a} - 2\boldsymbol{a}'\boldsymbol{G}\boldsymbol{r} + \boldsymbol{r}'\boldsymbol{G}\boldsymbol{r}. \tag{3.7}$$

Applying Lemma 1.3, the minimum of (3.7) is attained when the estimator of $\boldsymbol{r}'\boldsymbol{\varepsilon}$ is $\boldsymbol{a}_*'\boldsymbol{Y}$ where $\boldsymbol{a}_* = \boldsymbol{C}_1\boldsymbol{G}\boldsymbol{r}$. The minimum mean square error is, using (1.19),

$$\boldsymbol{r}'(\boldsymbol{G} - \boldsymbol{G}\boldsymbol{C}_1'\boldsymbol{G})\boldsymbol{r}. \tag{3.8}$$

From the above expressions it follows that the minimum dispersion error estimate of $\boldsymbol{\varepsilon}$ is

$$\hat{\boldsymbol{\varepsilon}} = \boldsymbol{G}\boldsymbol{C}_1\boldsymbol{Y} \tag{3.9}$$

with $D(\hat{\boldsymbol{\varepsilon}} - \boldsymbol{\varepsilon}) = \boldsymbol{G} - \boldsymbol{G}\boldsymbol{C}_1\boldsymbol{G}$.

Normal equations

The expressions for the estimators obtained above suggest the following estimation procedure. We set up the generalized normal equations

$$\begin{pmatrix} \boldsymbol{G}_* & \boldsymbol{X} & \boldsymbol{U}\boldsymbol{A} \\ \boldsymbol{X}' & \boldsymbol{0} & \boldsymbol{0} \\ \boldsymbol{A}'\boldsymbol{U}' & \boldsymbol{0} & \boldsymbol{0} \end{pmatrix} \begin{pmatrix} \boldsymbol{\alpha} \\ \boldsymbol{\beta} \\ \boldsymbol{\gamma} \end{pmatrix} = \begin{pmatrix} \boldsymbol{Y} \\ \boldsymbol{0} \\ \boldsymbol{0} \end{pmatrix} \tag{3.10}$$

and obtain a solution $\hat{\boldsymbol{\alpha}}$, $\hat{\boldsymbol{\beta}}$, and $\hat{\boldsymbol{\gamma}}$. Then the estimate of $\boldsymbol{\eta}$ is $\hat{\boldsymbol{\eta}} = \boldsymbol{\Gamma}\boldsymbol{U}'\hat{\boldsymbol{\alpha}}$ and of $\boldsymbol{\varepsilon}$ is $\hat{\boldsymbol{\varepsilon}} = \boldsymbol{G}\hat{\boldsymbol{\alpha}}$. When $\boldsymbol{p}_1'\boldsymbol{\beta} + \boldsymbol{p}_2'\boldsymbol{\gamma}$ is estimable, the estimate is $\boldsymbol{p}_1'\hat{\boldsymbol{\beta}} + \boldsymbol{p}_2'\hat{\boldsymbol{\gamma}}$.

Letting $(\boldsymbol{X} : \boldsymbol{U}\boldsymbol{A}) = \boldsymbol{X}_*$ and $\boldsymbol{\beta}_*' = (\boldsymbol{\beta}', \boldsymbol{\gamma}')$, the equations (3.10) can be written as

$$\begin{aligned} (\boldsymbol{G} + \boldsymbol{U}\boldsymbol{\gamma}\boldsymbol{U}')\boldsymbol{\alpha} + \boldsymbol{X}_*\boldsymbol{\beta}_* &= \boldsymbol{Y} \\ \boldsymbol{X}_*'\boldsymbol{\alpha} &= \boldsymbol{0}. \end{aligned} \tag{3.11}$$

If $\boldsymbol{G}^{-1}$ and $\boldsymbol{\Gamma}^{-1}$ exist, then multiplying the first equation by $\boldsymbol{X}_*'\boldsymbol{G}^{-1}$ and $\boldsymbol{\Gamma}^{-1}\boldsymbol{\Gamma}\boldsymbol{U}'\boldsymbol{\Gamma}^{-1}$ and using the second equation, we obtain the two equations

$$\begin{aligned} \boldsymbol{X}_*'\boldsymbol{G}^{-1}\boldsymbol{U}\boldsymbol{\eta} + \boldsymbol{X}_*'\boldsymbol{G}^{-1}\boldsymbol{X}\boldsymbol{\beta}_* &= \boldsymbol{X}_*'\boldsymbol{G}^{-1}\boldsymbol{Y} \\ (\boldsymbol{\Gamma}^{-1} + \boldsymbol{U}'\boldsymbol{G}^{-1}\boldsymbol{U})\boldsymbol{\eta} + \boldsymbol{U}'\boldsymbol{G}^{-1}\boldsymbol{X}\boldsymbol{\beta}_* &= \boldsymbol{U}'\boldsymbol{G}^{-1}\boldsymbol{Y}. \end{aligned} \tag{3.12}$$

Henderson (1984) derived equations of the type (3.12) when $\boldsymbol{A} = \boldsymbol{0}$. (See also Harville, 1976.) The equations (3.11) provide estimators of $\boldsymbol{\eta}$ and $\boldsymbol{\beta}$ directly.

When $\boldsymbol{G}$ and $\boldsymbol{\Gamma}$ are not both non-singular, or when $\boldsymbol{G}$ and/or $\boldsymbol{\Gamma}$ is a complicated matrix, other methods of solving the equations (3.10) could be explored.

The estimators of $\boldsymbol{\xi}$, $\boldsymbol{\varepsilon}$, $\boldsymbol{\beta}$, and $\boldsymbol{\gamma}$ obtained in Section 3 involve the matrices $\boldsymbol{G}$ and $\boldsymbol{\Gamma}$, which may not be known. In the simplest possible case $\boldsymbol{G}$ and

$\boldsymbol{\Gamma}$ may be of the form $\sigma_1^2 \boldsymbol{V}_1$ and $\sigma_2^2 \boldsymbol{V}_2$ respectively, where $\boldsymbol{V}_1$ and $\boldsymbol{V}_2$ are known and σ_1^2 and σ_2^2 are unknown variance components. In such a case, σ_1^2 and σ_2^2 may have to be estimated using techniques such as the MINQE or maximum likelihood as described in Rao and Kleffe (1988). The estimates of σ_1^2 and σ_2^2 may be substituted for the unknown values in the expressions for the estimators $\boldsymbol{\xi}$, $\boldsymbol{\varepsilon}$, $\boldsymbol{\beta}$, and $\boldsymbol{\gamma}$.

Acknowledgments: This work is supported by Contract N00014-85-K-0292 of the Office of Naval Research and Contract F49620-85-C-0008 of the Air Force Office of Scientific Research. The United States Government is authorized to reproduce and distribute reprints for governmental purposes notwithstanding any copyright notations herein.

References

Harville, D.A. (1976) Extension of Gauss-Markov theorem to include the estimation of random effects. *Ann. of Statistics* 4, 384–395.

Henderson, C.R. (1984). *Applications of Linear Models in Animal Breeding*, University of Guelph.

Rao, C. Radhikrishna (1971). Unified theory of linear estimation. *Sankhyā* A 33, 370–396 and *Sankhyā* A34, 477.

Rao, C. Radhikrishna (1972). A note on the IPM method in the unified theory of linear estimation. *Sankhyā* A34, 285–288.

Rao, C. Radhikrishna (1973). *Linear Statistical Inference and its Applications.* (Second edition), Wiley, New York.

Rao, C. Radhikrishna (1974). Projectors, generalized inverses and BLUE's. *J. Roy. Statist. Soc.* 36, 442–448.

Rao, C. Radhikrishna (1985). A unified approach to inference from linear models. *Proc. First International Tampere Seminar on Linear Models*, Eds. T. Pukkila and S. Puntanen, 361–383.

Rao, C. Radhikrishna (1988). Estimation in linear models with mixed effects: a unified theory. *Proc. Second International Tampere Conference in Statistics*, Eds. T. Pukkila and S. Puntanen, 73–98.

Rao, C. Radhikrishna and Kleffe Jürgen (1988). *Estimation of Variance Components and Applications*, North Holland.

Rao, C. Radhikrishna and Yanai, H. (1979). General definition and decomposition of projectors and some applications. *J. Statist. Planning and Inference* 3, 1–17.

23

Shrinking Techniques for Robust Regression

Richard L. Schmoyer[1]
Steven F. Arnold[2]

ABSTRACT The asymptotic normality of robust estimators suggests that shrinking techniques previously considered for least squares regression are appropriate in robust regression as well. Moreover, the noisy nature of the data frequently encountered in robust regression problems makes the use of shrinking estimators particularly advantageous. Asymptotic and finite sample results and a short simulation demonstrate that shrinking techniques can indeed improve a robust estimator's performance.

1 Introduction

Numerous advances in linear regression have been made in the last several decades. Among these is the formulation of the robust estimation problem, in which an attempt is made to deal with non-normal, heavy-tailed, and contaminated data. Another is the proof, in the classical least squares problem, of the inadmissibility of the usual estimator of the vector of coefficients when the dimension of this vector is greater than two (Stein, 1955), and the subsequent development of shrinking techniques for regression. In this paper, we shall discuss the advantage of applying shrinking techniques to robust estimators.

Suppose that we observe an $n \times 1$ vector $Y = X\beta + e$, where X is an $n \times p$ matrix of rank p, $n > p$, β is an unknown vector in $\Re^p$, and e is a random vector with n independent components each having common distribution function F. In the context of decision theory, we wish to estimate β with the estimator $d(Y)$, subject to a loss $L(d, \beta)$. The risk of d is $R(d, \beta, F) = E_F[L(d, \beta)]$. A good estimator is one for which the risk is small.

Estimators that perform well for varied, and in particular, heavy-tailed F are said to be robust. Typically robust estimators are asymptotically

[1]Oak Ridge National Laboratory, P.O. Box Y, Oak Ridge, TN 37831.

[2]Department of Statistics, The Pennsylvania State University, University Park, PA 16802

normal, in the sense that

$$(X'X)^{1/2}(d-\beta) \xrightarrow{d} Z \sim N(0, VI) \tag{1.1}$$

where $V = V(F)$ is a scalar called the *asymptotic variance*. A reasonable criterion for choosing a robust estimator is that it minimizes the supremum of $V(F)$ as F ranges over a class of distributions (Huber, 1964).

A shrinking estimator is a measurable function δ of an estimator d, which for some specified norm satisfies $\|\delta(d) - \beta^*\| \leq \|d - \beta^*\|$ for some $\beta^* \in \Re^p$. In this case we say that δ shrinks d toward β^*. In general, since we obviously want $\|\delta(d) - \beta\| < \|d - \beta\|$, we try to choose β^* as close as possible to β. For this reason we will refer to β^* as a *guess* of β.

Various forms for shrinking estimators have been proposed for the regression problem with normal errors. These include ridge and Stein estimators (see, for example, Dempster, Schatzoff, and Wermuth, 1977). All are expressible, though perhaps not in closed form, as a function of $\delta(\hat{\beta}, \hat{\sigma}^2)$, where $\hat{\beta}$ is the least squares estimator of β, and $\hat{\sigma}^2$ is the usual unbiased estimator of variance. Because of its asymptotic normality, a robust estimator might reasonably be modified in the same way. More precisely, if d denotes an estimator satisfying (1.1), and $\hat{V}$ is a consistent estimator of its asymptotic variance, then $\delta(d, \hat{V})$ may be a reasonable way to modify d.

Let $\{d_n\}$ be a sequence of estimators of β, and let $\{L_n(d, \beta)\}$ be a sequence of loss functions. Suppose that

$$L_n(d_n, \beta) \xrightarrow{d} W(F),$$

when the errors have distribution function F. We define the *asymptotic risk* of the sequence of estimators $\{d_n\}$, for the sequence $\{L_n\}$ of loss functions and for the error distribution F, by

$$\mathrm{AR}\big(\{d_n\}, (\beta, F)\big) = E_F[W(F)].$$

(Note that some authors call AR the asymptotic distribution risk, reserving the term asymptotic risk for the limit of the finite sample risks.)

In Section 2 of this paper, we derive a very general result, Theorem 2.1, concerning the asymptotic risk of many estimators of the form $\delta_n(d_n, \hat{V}_n)$, for many loss functions. This result shows how the asymptotic risk for $\delta_n(d_n, \hat{V}_n)$ can be computed from $V(F)$ and the risk of a related estimator and loss function in the problem of estimating the mean of a multivariate normal distribution. Theorem 2.1 can be applied to M-estimators, R-estimators, and L-estimators for many loss functions (including non-quadratic ones) and many functions δ_n (including some non-shrinking ones). This result implies that if we apply to a robust estimator a shrinking procedure which is better than the least squares estimator for a particular loss function for the case of normal errors, we get an estimator which has smaller asymptotic risk than the original M-estimator. For quadratic loss functions, the percentage improvement caused by the shrinking procedure has the same basic formula for the M-estimation problem as for the least squares problem. However, the comparison between the models is not completely straightforward, because the scale parameters are different for the two models.

In Section 3, we look at a finite sample result. We show, under fairly general conditions, that if $d = d(Y)$ is a local equivariant estimator of β and $L(d, \beta)$ is a quadratic loss function, then there exists a single shrinking estimator which has smaller risk than $d(Y)$ for all F in a fairly large family C of distributions. This result is similar to a result in Brown (1966), but is more general in that Brown only showed improvement for a single distribution. M-estimators, R-estimators, and L-estimators are all equivariant estimators. One unfortunate aspect of this theorem is that it shows the existence of such an estimator, but does not give an explicit formula for it.

In Section 4, we discuss a simulation of the average risks for a commonly studied M-estimator when it has been shrunk using the positive-part shrinking technique. We look at several commonly studied distributions, all of which have been scaled to have the same inter-quartile range as a standard normal distribution. We let $p = 4$, and $n = 10$, 15, or 20 and do 5000 replications at each situation. We have used the loss function $L(d, \beta) = (d - \beta)' X' X (d - \beta)$ or a closely related, bounded loss function.

Let

$$R^2 = \|X\beta - X\beta^*\|^2$$

be a measure of how accurate our guess is. In our simulation, we let R^2 take on values

$$0, \quad 1, \quad 10, \quad 100, \quad 1000, \quad 100{,}000.$$

(When $R^2 = 0$, we have guessed β perfectly; when $R^2 = 1$, we have come very close; but when $R^2 = 100{,}000$, we have guessed horribly.) In Table 1 below we give the percentage improvement due to shrinking the M-estimator. We also table the percentage improvement under similar conditions for shrinking a least squares estimator for the normal model.

Note first that the difference in percentage improvement between $n = 10$ and $n = 20$ observations is not too large. In addition, note that we can get substantial improvement from shrinking an M-estimator when p is as small as 4 and n is as small as 10, 15, or 20

For the distributions with moderate or light tails (the normal, the contaminated normal, the triangular, and the uniform distributions) shrinking the M-estimator offers roughly the same percentage improvement as shrinking the least squares estimator does. This improvement drops off quickly as our guess gets worse.

For distributions with heavy tails (the t-distributions and the double exponential distributions) the improvement is somewhat broader. When $R^2 = 10$, the improvement for shrinking the M-estimator is nearly twice that for shrinking the least squares estimator.

TABLE 1. Percentage improvement due to shrinking an M-estimator for various R^2

Distribution	R^2 0	1	10	100	1000	100,000
Light						
Triangular $n = 15$	57	44	8	1	0	0
Uniform $n = 15$	57	39	5	1	0	0
Moderate						
Normal $n = 10$	53	41	8	1	0	0
$n = 20$	58	46	9	1	0	0
Contaminated Normal $n = 10$	52	43	12	1	0	0
$n = 20$	58	47	12	1	0	0
Heavy						
T-2 $n = 10$	45	42	28	11	2	0
$n = 20$	57	50	23	3	0	0
T-3 $n = 15$	54	47	17	2	0	0
DE $n = 15$	56	48	18	2	0	0
Very Heavy						
Slash $n = 10$	43	43	42	39	32	2
$n = 20$	47	47	45	41	32	3
Cauchy $n = 15$	48	48	46	42	32	4
Shrinking least squares with normal errors						
$n = 20$	61	48	10	1	0	0

For the distributions with extremely heavy tails (the slash and Cauchy distributions), the improvement from shrinking the M-estimator occurs for a wider range of R^2 than that from shrinking the least squares estimator. When $R^2 = 1000$, there is no improvement from shrinking the least squares estimator, but there still is a 32% improvement from shrinking the M-estimator. Even when $R^2 = 100{,}000$ (so that our guess is terrible), shrinking the M-estimator offers non-negligible improvement for these distributions.

A possible reason for the wide benefit from shrinking M-estimators for very heavy-tailed distributions is given in the asymptotic calculations in Section 2. The asymptotic risk is a function of $\left(n/V(F)\right)^{1/2}(\beta-\beta^*) = \theta(F)$. If $V(F)$ is very large, $\theta(F)$ will be small, even for bad guesses for β^*.

The simulation summarized in Table 1 is only one of several we did, with different shrinking estimators, different loss functions and different assumptions about β and X. The results of all these simulations were qualitatively similar to those given in Table 1. For light and moderate tailed distributions, the gain from shrinking an M-estimator is comparable with the gain from shrinking the least squares estimator for the normal model. However, for heavy tailed distributions, the region of improvement from shrinking an M-estimator is greater than that for shrinking a least-squares estimator, and for distributions with very heavy tails, the region is much greater. See

Schmoyer (1980) for details on these additional simulations.

There is an extensive literature on shrinking estimators for normal problems, and an extensive literature on robust estimators for non-normal data. However, there are not as many papers on applying shrinking procedures to robust estimators. One of the earliest papers applying such shrinking to estimators is by Askin and Montgomery (1980). In that paper, the shrinking was done to increase the stability of the estimators for ill-conditioned X matrixes. This goal is quite different from our purpose which is to decrease the risk (or asymptotic risk) even when the X matrix is not ill-conditioned. Saleh and Sen (1985) consider shrinking M-estimators of location in order to decrease the risk. They approach the problem from an asymptotic perspective similar to that used in Section 2 of this paper. Their results are more limited than ours, in that their results apply to smaller classes of M-estimators, loss functions and shrinking rules than our results do. However, they do find expressions for the limit of the finite sample risk, whereas we find the (easier) risk for the asymptotic distribution. Typically, these two limits should be the same, but it is often quite difficult to establish the necessary uniform integrability conditions to prove their equality. Sen and Saleh (1987) consider the problem of robustly estimating β, when it is suspected to lie in a subspace. They discuss the asymptotic properties of preliminary test and shrunken M-estimators. If the subspace is all of R_p, their model is the same as ours, and they also use asymptotic risk, rather than the limit of finite sample risk, in evaluating procedures. However, their paper considers a much narrower class of robust estimators, loss functions, and shrinking rules than ours does.

2 An Asymptotic Analysis

In this section we state an elementary result which indicates why we would expect that shrinking robust estimators should be sensible, at least for large samples.

To motivate the approach we used to make this calculation, we return briefly to the normal regression model in which we observe $Y \sim N_n(X\beta, I)$. Consider estimating β with the loss function

$$L(d, \beta) = (d - \beta)'X'X(d - \beta).$$

Let $\hat{\beta}$ be the usual unbiased estimator of β. Let $\hat{\beta}_b$ be the James-Stein estimator shrunk to the vector b,

$$\hat{\beta}_b = \left[1 - \frac{(p-2)}{L(b, \hat{\beta})}\right](\hat{\beta} - b) + b.$$

Then

$$R(\hat{\beta}_b, \beta) = E[L(\hat{\beta}_b, \beta)] = p - (p-2)^2 E[(p - 2 + 2K)^{-1}] < p = R(\hat{\beta}, \beta),$$

where $K \sim \text{Poisson}(\gamma)$, $\gamma = L(b, \beta)/2$. When we do asymptotics for regression models, we often assume that $n^{-1}X'X \to A > 0$. In this case, for any fixed β and b,

$$L(b, \beta) = n(b - \beta)(n^{-1}X'X)(b - \beta) \to \infty,$$

and hence

$$\lim_{n \to \infty} R(\hat{\beta}_b, \beta) = p = R(\hat{\beta}, \beta)$$

so that there is no asymptotic improvement from using the James-Stein estimator. A more sensitive development of the asymptotic risk of the James-Stein estimator starts by allowing b to depend on n, in such a way that

$$n^{1/2}(\beta - b_n) \to \theta$$

where θ is an unknown vector which does not depend on n. In other words, in order to get asymptotic improvement with the James-Stein estimator, we are assuming that our "guess" b_n is getting better as the sample size increases. (This approach is similar to that used in defining Pitman asymptotic efficiency in testing problems.) This is the approach we take in this section.

Consider now a non-normal regression model in which we want to estimate the p-dimensional vector β with the sequence of loss functions $L_n(d, \beta)$ of the form

$$L_n(d, \beta) = h\big(n^{1/2}(d - \beta), Q_n\big), \tag{2.1}$$

where $h(a, Q)$ is a continuous function and where Q_n is a sequence of known matrices such that

$$Q_n \to Q \qquad \text{as } n \to \infty.$$

Let $\{d_n\}$ be a sequence of estimators of β. Suppose that for a particular distribution F,

$$L_n(d_n, \beta) \xrightarrow{d} W(F).$$

We define the *asymptotic risk* by

$$\text{AR}(\{d_n\}, \beta, F) = E[W(F)].$$

(Some authors call AR the asymptotic distribution risk.)

We assume that we have a sequence of estimators $\{d_n\}$ such that

$$n^{1/2}(d_n - \beta) \xrightarrow{d} S \sim N_p(0, VA), \tag{2.2}$$

where $A > 0$ is a known matrix and $V = V(F)$ is an unknown scalar which depends on the underlying distribution F and the sequence of estimators

$\{d_n\}$. We also assume that there exists a consistent sequence $\{\hat{V}_n\}$ of estimators of V. Let $\{\beta_n^*\}$ be a sequence of known vectors (guesses) such that

$$\left(\frac{n}{V}\right)^{1/2}(\beta - \beta_n^*) \to \theta$$

where $\theta = \theta(F)$ is a fixed, but unknown vector. (If $n^{1/2}(\beta - \beta_n^*) \to \xi$, then $\theta(F) = V(F)^{1/2}\xi$.) Let

$$U_n = \left(\frac{n}{\hat{V}}\right)^{1/2}(d_n - \beta_n^*).$$

and let $\{R_n\}$ be a sequence of known matrices such that $R_n \to R$. We consider estimators of β of the form

$$\delta_n(d_n, \hat{V}) = \left(\frac{\hat{V}}{n}\right)^{1/2} k(U_n, R_n) + \beta_n^* \tag{2.3}$$

where $k(U, R)$ is a known continuous function of U and R.

Theorem 2.1 *Let* $U \sim N_p(\theta, A)$, $L(d, \theta) = h(d - \theta, Q)$. *Then*

$$L_n(\delta_n, \beta) \xrightarrow{d} L\big(V^{1/2}k(U, R), V^{1/2}\theta\big),$$

$$\text{AR}(\{\delta_n\}, \beta, F) = E[L(V^{1/2}k(U, R), V^{1/2}\theta)].$$

If $L_n(d, \beta) = n(d - \beta)'Q_n(d - \beta)$, *then*

$$\text{AR}(\{\delta_n\}, \beta, F) = VE\big[L\big(k(U, R), \theta\big)\big].$$

Proof Note first that

$$U_n = \left(\frac{n}{\hat{V}}\right)^{1/2}(d_n - \beta_n^*) = \left(\frac{n}{\hat{V}}\right)^{1/2}(d_n - \beta) + \left(\frac{n}{\hat{V}}\right)^{1/2}(\beta - \beta_n^*) \xrightarrow{d} U \sim N(\theta, A).$$

Therefore,

$$n^{1/2}(\delta_n - \beta) = \hat{V}^{1/2}k(U_n, R_n) - n^{1/2}(\beta - \beta_n^*) \xrightarrow{d} V^{1/2}\big(k(U, R) - \theta\big),$$

and

$$\begin{aligned} L_n(\delta_n, \beta) &= h\big(n^{1/2}(\delta_n - \beta), Q_n\big) \xrightarrow{d} h\left(V^{1/2}\big(k(U, R) - \theta\big), Q\right) \\ &= L\big(V^{1/2}k(U, R), V^{1/2}\theta\big). \end{aligned}$$

The formula for the asymptotic risk follows directly. The last equation in the theorem is just a special case of the general formula. □

We illustrate the use of this theorem for shrinking M-estimators. Consider the regression problem in which we observe Y_i independent,

$$Y_i = x_i\beta + e_i,$$

where the x_i are known constant vectors and the e_i are unobserved errors which are i.i.d. with common distribution function F. Let X_n be the $n \times p$ matrix whose ith row is x_i and suppose that

$$R_n = n^{-1}X_n'X_n \to R > 0 \qquad \text{as } n \to \infty.$$

Let d_n be an M-estimator of β based on the first n observations. It is well known that, under fairly general conditions,

$$n^{1/2}(d_n - \beta) \xrightarrow{d} N\big(0, V(F)R^{-1}\big),$$

where $V(F)$ is an unknown constant which depends on F and the M-estimator chosen, and which has a consistent estimator $\hat{V}$. For example, $d_n(Y)$ could be the estimator found by minimizing $r(b) = \sum_{i=1}^{n} \rho(Y_i - x_i b)/\tau$, where τ is a scale parameter and

$$\rho(u) = \begin{cases} u^2/2 & |u| \leq c \\ c|u| - c^2/2 & |u| > c. \end{cases} \tag{2.4}$$

Suppose we want to estimate β with the loss function

$$L_n(d, \beta) = (d-\beta)'X_n'X_n(d-\beta) = h\big(n^{1/2}(d-\beta), R_n\big),$$

where

$$h(u, R) = u'Ru.$$

Let

$$U_n = \left(\frac{n}{\hat{V}}\right)^{1/2}(d_n - \beta_n^*),$$

where β_n^* is a sequence of "guesses" for β. Let

$$\delta_n(d_n) = \left[1 - \frac{(p-2)}{U_n'R_nU_n}\right](d_n - \beta_n^*) + \beta_n^* = \left(\frac{\hat{V}}{n}\right)^{1/2} k(U_n, R_n) + \beta_n^*,$$

where

$$k(U, R) = \left[1 - \frac{(p-2)}{U'RU}\right]U.$$

(This estimator is just the usual James-Stein shrinking formula applied to the M-estimator d_n.) By Theorem 2.1,

$$\mathrm{AR}(\{\delta_n\}, \beta, F) = V(F)E\big(k(U,R) - \theta\big)'R\big(k(U,R) - \theta\big),$$

where

$$U \sim N_p(\theta, R^{-1}).$$

Note that $k(U, R)$ is just the James-Stein estimator for the normal model involving $U \sim N_p(\theta, R^{-1})$. Therefore,

$$\text{AR}(\{\delta_n\}, \beta, F) = V(F)\big(p - (p-2)^2 E[(p-2+2K)^{-1}]\big)$$

where K has a Poison distribution with mean

$$\gamma = \frac{\theta' A \theta}{2} = \lim \frac{(\beta - \beta_n^*)' R (\beta - \beta_n^*)}{2V(F)}.$$

Therefore, we see that

$$\frac{\text{AR}(\{\delta_n\}, \beta, F)}{\text{AR}(\{d_n\}, \beta, F)} = 1 - \frac{(p-2)^2}{p} E[(p-2+2K)^{-1}]$$

so that the percent reduction in asymptotic risk from shrinking an M-estimator in the James-Stein manner has the same formula as the percent reduction from shrinking a least squares estimator in the James-Stein manner. However, the interpretation is not quite so simple, since $\gamma = \lim n(\beta - \beta_n^*)' R(\beta - \beta_n^*)/2\sigma^2$ in the least squares case and $\gamma = \lim n(\beta - \beta_n^*)' R(\beta - \beta_n^*)/2V(F)$ in the case of M-estimators. In fact, for very heavy-tailed distributions, such as the Cauchy distribution, the reduction from shrinking M-estimators may be more than the reduction for shrinking least squares estimators for moderately accurate guesses β_n^* (see the simulation). The reason this occurs is that $V(F)$ is often quite large for heavy-tailed distributions F, so that γ is still quite small, even when β_n^* is only moderately accurate.

Now, consider a family of distribution functions $\boldsymbol{F}$ and a class of sequences of estimators $\boldsymbol{D}$. We say that a sequence of estimators $\{d_n\} \in \boldsymbol{D}$ is minimax, asymptotically, for the family $\boldsymbol{F}$ and class $\boldsymbol{D}$ if

$$\sup_{F \in \boldsymbol{F},\ \beta \in \Re_p} \text{AR}(\{d_n\}, \beta, F) \le \sup_{F \in \boldsymbol{F},\ \beta \in \Re_p} \text{AR}(\{d_n^*\}, \beta, F)$$

for all $\{d_n^*\} \in \boldsymbol{D}$. For a sequence of M-estimators, $\{d_n\}$, and a sequence of quadratic loss functions

$$L_n(d, \beta) = n(d-\beta)' Q_n (d - \beta), \qquad Q_n \to Q,$$

it is easily seen that if $n^{1/2}(d_n - \beta) \xrightarrow{d} N\big(0, V(F)R^{-1}\big)$, then

$$\text{AR}(\{d_n^*\}, \beta, F) = V(F)\,\text{tr}(QR^{-1}).$$

Therefore, a minimax rule for the set of M-estimators is one which minimizes

$$\sup_{F \in \boldsymbol{F}} V(F).$$

Huber (1964) finds some such results for various choices for the family $\boldsymbol{F}$. We now show how to use those results, together with known results for the multivariate normal problem, to find estimators which are minimax in a larger class of estimators. Let $\mathcal{D}$ be the class of M-estimators and let Δ be the class of sequences $\delta_n(d_n)$ satisfying (2.3) for $d_n \in \mathcal{D}$, some matrices $R_n \to R$ and some continuous function $k(U, R)$. Let

$$L(d, \theta) = (d - \theta)' Q (d - \theta).$$

By Theorem 2.1,

$$\mathrm{AR}(\{\delta_n\}, \beta, F) = V(F) E\big[L\big(k(U, R), \theta\big)\big].$$

Let $\{d_n^0\}$ be a minimax sequence of M-estimators for a family $\boldsymbol{F}$, and let $k_0(U, R)$ be a minimax estimator for estimating θ based on $U \sim N(\theta, A)$ with loss function $L(d, \theta)$. Let $\{\delta_n^0\}$ be computed from $k_0(U, R_n)$ and $\{d_n\}$ as defined in Equation (2.3). Then $\{\delta_n^0\}$ is a minimax estimator of β for the family $\boldsymbol{F}$ of distributions and the class Δ.

Similarly, if $\{d_n\}$ has lower asymptotic risk than $\{d_n^*\}$ and if $k(U, R)$ has lower risk than $k^*(U, R)$, then $\{\delta_n\}$ has lower asymptotic risk than $\{\delta_n^*\}$ (where δ_n is computed from d_n and k following (2.3), as is δ_n^* from d_n^* and k^*). In the example discussed above, we know, for the normal problem, that the positive James-Stein estimator, in which we take

$$k(U, R) = \max\left(0, \left[1 - \frac{(p-2)}{U'RU}\right]\right) U \tag{2.5}$$

has smaller risk than the James-Stein estimator does. Therefore, Theorem 2.1 implies that a positive-part James-Stein shrinking of an M-estimator has smaller asymptotic risk than does the James-Stein shrinking of this estimator, which in turn has smaller asymptotic risk than does the original M-estimator.

The reason that we get results so easily for quadratic loss functions is that the asymptotic risk is a product of a function $V(F)$ due to the M-estimator chosen and a function $E\big[L\big(k(U, R), \theta\big)\big]$ based on the shrinking estimator chosen.

We finish this section with some additional comments about Theorem 2.1.

1. It applies to many different robust estimators. The only assumption about the sequence d_n is that is satisfies Equation (2.2). M-estimators, R-estimators, and L-estimators typically satisfy this assumption.

2. The only assumption about the loss function is that it satisfies (2.1). Quadratic loss functions of the form

$$L_n(d, \beta) = (d - \beta)' Q_n (d - \beta)$$

satisfy this assumption as long as $n^{-1}Q_n$ goes to some Q. In studying robust estimators, we often use distributions with very heavy tails. In such situations, it can be helpful to use loss functions of the form $q(L_n(d,\beta))$, where $L_n(d,\beta)$ is a quadratic loss function and q is a bounded function. Such loss functions would also be covered by Theorem 2.1. Examples of such loss functions are

$$L_n^*(d,\beta) = \frac{L_n(d,\beta)}{1+L_n(d,\beta)}, \qquad L_n^{\#}(d,\beta) = B\left(1-\exp\bigl(-B^{-1}L_n(d,\beta)\bigr)\right).$$

One such loss function is used in the simulation to compare estimators when the observations come from heavy-tailed distributions such as the Cauchy distribution.

3. Equation (2.3) is the only assumption about the shrinking procedure δ_n. This assumption allows most Stein and ridge type estimators. For example, the ridge estimator

$$\begin{aligned}\delta_n(d_n) &= (X_n'X_n + ncI)^{-1}X_n'X_n(d_n-\beta_n^*)+\beta_n^* \\ &= \bigl(I + c(n^{-1}X_n'X_n)^{-1}\bigr)(d_n-\beta_n^*)+\beta_n^*\end{aligned}$$

has this form with $R_n = n^{-1}X_n'X_n$, $k(U,R_n) = (I+kR_n^{-1})^{-1}U$. Note that

$$\|\delta_n(d_n)-\beta_n^*\|^2 \le \|d_n-\beta_n^*\|^2 \Leftrightarrow \|k(U,R_n)\|^2 \le \|U\|^2,$$

and hence $\{\delta_n\}$ is a shrinking procedure only when $\|k(U,R_n)\|^2 \le \|U\|^2$. However, Theorem 2.1 is still applicable for non-shrinking estimators.

3 Existence of Improved Estimators

In Section 2 an asymptotic argument was used to motivate shrinking robust estimators. As further motivation we now present a result, which is similar to a theorem of Brown (1966), that shows the existence, under certain conditions, of estimators better than usual equivariant robust estimators in finite-sample problems. The feature that distinguishes our result from Brown's is that we consider the risk, $R(d,\beta,F)$, of an estimator d, as having arguments both $\beta \in \Re^p$ and $F \in C$. For a robust estimator, d, we show the existence of an estimator δ which satisfies $R(\delta,\beta,F) < R(d,\beta,F)$ for all $\beta \in \Re^p$ and all $F \in C$.

Dropping the "n" subscripts used in Section 2, we have $Y = X\beta + e$. We consider estimating β with the quadratic loss function

$$L(d,\beta) = (d-\beta)'Q(d-\beta).$$

Let $d = d(Y)$ denote a translation equivariant estimator of β (e.g., an M-estimator), and let $V = V(F)$ be some measure of the dispersion of $F \in C$. Also let $\Sigma = \Sigma(F) = E_F(d-\beta)(d-\beta)'$, and let ξ denote the largest eigenvalue of Σ. We now give conditions under which there exist constants A and B such that the estimator

$$\delta(D) = \left(I - \frac{B}{A + \|d\|^2} Q^{-1}\right) d \tag{3.1}$$

satisfies

$$R(\delta, \beta, F) < R(d, \beta, F) \qquad \text{for all } \beta \in \Re^p, \ F \in C. \tag{3.2}$$

Theorem 3.1 *Suppose that $R(d, \beta, F) < \infty$ for all $F \in C$, and that there exists V_0 such that $V \leq V_0$ for all $F \in C$. Also suppose that*

(i) $E_F[d] = \beta$;

(ii) there exists $\epsilon > 0$ *such that* $\operatorname{tr}(\Sigma) - 2\xi > \epsilon$ *for all* $F \in C$;

(iii) there exists $N < \infty$ *such that* $E_F \|d - \beta\|^4 / V^2 \leq N$ *for all* $F \in C$.

Then there exists A and B such that the estimator (3.1) satisfies (3.2) for all $\beta \in \Re^p$ and all $F \in C$.

Proof We show the existence of a scalar a for which the estimator

$$\delta(d) = \left(I - \frac{V_0}{a(aV_0 + \|d\|^2)} Q^{-1}\right) d \tag{3.3}$$

satisfies (3.2). Substituting $(A/B)^{1/2}$ for a and $(BA)^{1/2}$ for V_0 in (3.3) establishes (3.2).

Let $\Delta = R(d, \beta, F) - R(\delta, \beta, F)$. Using $\delta(d)$ as in (3.3), $R(\delta, \beta, F)$ is

$$E_F \left[(d-\beta) - \frac{1}{a(a + \|d\|^2/V_0)} Q^{-1} d\right]' Q \left[(d-\beta) - \frac{1}{a(a + \|d\|^2/V_0)} Q^{-1} d\right].$$

By expanding this and subtracting it from $R(d, \beta, F)$, Δ is seen to be

$$E_F \left[\frac{2V_0}{a(aV_0 + \|d\|^2)} d'(d-\beta) - \frac{V_0^2}{a^2(aV_0 + \|d\|^2)^2} d' Q^{-1} d\right]$$

which is greater than or equal to

$$E_F \left[\frac{1}{a(aV_0 + \|d\|^2)} \left(2V_0 d'(d-\beta) - \frac{V_0^2}{qa}\right)\right]. \tag{3.4}$$

where $q > 0$ is the smallest eigenvalue of Q.

Let E_F^0 denote an expectation when $\beta = 0$. Then substituting $d + \beta$ for d in (3.4) gives

$$E_F^0 \left[\frac{1}{a(aV_0 + \|d + \beta\|^2)} \left(2V_0(d + \beta)'d - \frac{V_0^2}{qa} \right) \right]. \tag{3.5}$$

As in Stein (1955), we use the equality $1/(x+y) = 1/x - y/x^2 + y^2/x^2(x+y)$ to expand $1/(aV_0 + \|d + \beta\|^2)$, and with a little more algebra we can write (3.5) as

$$\begin{aligned} E_F^0 &\left[\frac{1}{a(aV_0 + \|\beta\|^2)} \left[2V_0 d'(d + \beta) - \frac{V_0^2}{qa} \right] \right] \\ - E_F^0 &\left[\frac{2d'\beta + \|d\|^2}{a(aV_0 + \|\beta\|^2)^2} \left[2V_0 d'(d + \beta) - \frac{V_0^2}{qa} \right] \right] \\ + E_F^0 &\left[\frac{4(d'\beta)^2 + 4(d'\beta)\|d\|^2 + \|d\|^4}{a(aV_0 + \|\beta\|^2)^2(aV_0 + \|d + \beta\|^2)} \left(2V_0 d'(d + \beta) - \frac{V_0^2}{qa} \right) \right]. \end{aligned} \tag{3.6}$$

By condition (i) the first term in (3.6) is

$$\frac{V_0}{a(aV_0 + \|\beta\|^2)} E_F^0 \left[2\|d\|^2 - \frac{V_0}{qa} \right] = \frac{V_0}{a(aV_0 + \|\beta\|^2)} \left[\mathrm{tr}(\Sigma) - \frac{V_0}{qa} \right]. \tag{3.7}$$

With more algebra, the Cauchy-Schwartz inequality and the fact that $\|\beta\| / (aV_0 + \|\beta\|^2) \leq \left(2(aV_0)^{1/2}\right)^{-1}$, we can show that the second term in (3.6) is at least

$$-\frac{V_0}{a(aV_0 + \|\beta\|^2)} E_F^0 \left[\frac{4(d'\beta)^2}{\|\beta\|^2} + \frac{3\|d\|^3}{(aV_0)^{1/2}} + \frac{2\|d\|^4}{aV_0} + \frac{V_0\|d\|}{qa(aV_0)^{1/2}} \right].$$

Now,

$$E_F^0 \left[\frac{(d'\beta)^2}{\|\beta\|^2} \right] = \frac{\beta'\Sigma\beta}{\|\beta\|^2} \leq \xi, \qquad E_F^0 \left[\frac{\|d\|^k}{V_0^{k/2}} \right] \leq 1 + E_F^0 \left[\frac{\|d\|^4}{V^2} \right] \leq 1 + N,$$

for $k = 1, 2, 3, 4$. Therefore, the second term in (3.6) must be at least

$$\begin{aligned} -\frac{V_0}{a(aV_0 + \|\beta\|^2)} &(4\xi + 3(1 + N)V_0 a^{-1/2} + 2(1 + N)V_0 a^{-1} \\ &+ (1 + N)V_0 q^{-1} a^{-3/2}) \\ &= -\frac{V_0}{a(aV_0 + \|\beta\|^2)} (4\xi + O(a^{-1/2})) \end{aligned}$$

where $O(a^{-1/2})$ is uniform in F and β. (We say that $h(a, F, \beta)$ is $O(a^{-1/2})$ uniformly in F and β if there exists K, not depending on F or β, such that $|h(a, F, \beta)| \leq Ka^{-1/2}$ for a large enough.) In a similar way, we can show that the third term in (3.6) is at least $V_0/\big(a(aV_0 + \|\beta\|^2)\big)O(a^{-1/2})$ where $O(a^{-1/2})$ is uniform in F and β. Therefore, we see that

$$\Delta \geq \frac{V_0}{a(aV_0 + \|\beta\|^2)}[2\,\mathrm{tr}(\Sigma) - 4\xi + O(a^{-1/2})] \geq \frac{V_0}{a(aV_0 + \|\beta\|^2)}[2\epsilon + O(a^{-1/2})].$$

Since $\epsilon > 0$ and $O(a^{-1/2})$ is uniform in F and β, there exists an a such that $\Delta > 0$ for all β and all $F \in C$. □

Condition (i) is satisfied in most problems. In particular, if $d(X\beta) = \beta$ and $E_F(e) = 0$, then this assumption is satisfied because of the location equivariance of d. Condition (ii) is a condition which guarantees that the estimation problem never degenerates into a two-dimensional problem. (See Brown (1973) or Bock (1975) for a discussion of conditions of this type.) Condition (iii) is somewhat harder to interpret, but is similar to assuming that d has bounded kurtosis. (Note that this condition does not imply that the distribution F has bounded kurtosis.) If this condition were not satisfied, the estimator d might be a bad choice for estimating β for that family.

4 Simulation

In this section, we discuss a computer simulation designed to study the performance of the James-Stein positive-part shrinking formula (2.5) as applied to the M-estimator defined by (2.4). We chose $p = \dim(\beta) = 4$, n = number of observations = 10, 15, or 20. We did 5000 replications in each simulation.

The loss functions we considered were all invariant. Since the M-estimator is equivariant, the risk of the positive part estimator depends on the true parameter β and the guess β^* only through $\beta^* - \beta$. Therefore, in simulations, we can take either β^* or β to be 0. We choose to take $\beta = 0$. For each of $R^2 = 0$, 1, 10, 100, 1000, and 100,000, we simulated β^* uniformly on the ellipsoid

$$\|X\beta^*\|^2 = R^2. \tag{4.1}$$

(Small values for R^2 correspond to good guesses β^* for β and large values correspond to bad guesses.)

From the assumption $\beta = 0$ we get $Y = e$. The distribution used in this simulation included 2 light-tailed distributions (the uniform and the triangular ($f(t) = 1 - |t|$, $-1 < t < 1$) distributions), two distributions with moderate tails (the normal distribution and the contaminated normal mixture of 95% $N(0, 1)$ and 5% $N(0, 9)$), three distributions with heavy tails (t distributions with 2 and 3 degrees of freedom and the double exponential distribution) and two distributions with very heavy tails (the Cauchy and slash (ratio of standard normal to independent uniform) distributions). The distributions were all scaled to have the same inter-quartile range so that at least in this one sense they are scale equivalent. The scale was chosen so that if the errors are normal, then their standard deviation is 1. Then, for example, the constraint $\|X\beta^*\|^2 = 100$ corresponds to guessing the observation means with an average accuracy of about $31.6/n^{1/2}$ standard deviations. For $n = 20$ this is about 7 standard deviations.

The matrix X was generated row by row with each row having first entry 1 and remaining entries simulated from a $N(0, \Sigma)$ distribution, where Σ was taken to be a correlation matrix with off-diagonal elements ρ. We used $\rho = 0.99$ for the normal, contaminated normal, t-2, and slash distributions and $\rho = 0.95$ for the remaining distributions.

For the loss functions considered in this study, the risk of the positive-part James-Stein shrinking of the least squares estimator is constant on the sets defined by (4.1). This suggests that the risk of the estimators considered in this study might be nearly constant on these sets. Earlier simulations have indicated this to be true. Hence it may not be too important how we generated X and β^* subject to (4.1).

The M-estimator, d, and its variance estimate are given in Huber (1973), Section 8, and were calculated by the method of iteratively re-weighted least squares. To estimate scale, we used 1.48 times the median absolute deviation of the least squares residuals (and did not iterate the scale estimate).

Performance of the estimators was measured with the loss function

$$L(d, \beta) = (d - \beta)' X' X (d - \beta).$$

However, preliminary simulations indicated that the loss $L(d, \beta)$ was not adequate when the underlying errors are very heavy-tailed. With high-noise errors the variance of this loss function is too great to expect reasonably accurate estimates of the risk in a reasonable number of simulation trials. The problem is with the unboundedness of the quadratic loss functions. For Cauchy and slash errors the loss, L, was modified by calculating

$$L_B = B\big(1 - \exp(-L/B)\big)$$

where L_B denotes the new bounded loss, and B is the bound. B was taken to be 20,000 on the basis of preliminary simulations. For light and moderately heavy-tailed distributions, simulation results for the bounded and unbounded losses were virtually identical. Of course, a bounded loss function may more accurately reflect the true loss characteristics of the problem anyway.

The entire procedure just described was repeated 5,000 times to yield statistically meaningful estimates of the percentage improvement due to shrinking at the different values of R, for the different distributions, etc. Each such estimate was computed by dividing the simulation average of the simulated loss differences, $L(d,0) - L(\delta(d),0)$, by the average of the simulated losses, $L(d,0)$.

Results have already been presented in Table 1. For the distributions with lighter tails (e.g., normal), shrinking is beneficial for small R but is neither detrimental nor beneficial for large R. For heavier tailed distributions, the benefit of shrinking extends to much larger values of R. We believe that the improvement of 2 to 11 percent and more, obtained for values of R around 10 to 31.6, is of substantial practical importance to anyone who is analyzing very noisy data, which is often the setting in which robust procedures are applied.

Table 1 also gives percentage improvements for the positive part James-Stein estimator over the least squares estimator when the errors are normal. These values are similar to those for the shrinking M-estimator in the case of light-tailed error distributions. We have investigated other shrinking versions of the M-estimator considered here (for example, several ridge estimators), and found this relationship to be true for those versions as well. That is, shrinking M-estimators in a situation involving light-tailed errors appear to have about the same effect as does shrinking the least squares estimator in the problem with normal errors. As the error tail weight and variability of the M-estimator increase, so does the potential for improving the M-estimator by shrinking.

Acknowledgments: We wish to thank Leon Gleser for his careful reading of an earlier version of this paper. His many helpful suggestions have greatly improved its readability. Research sponsored by the Office of Health and Environmental Research, U.S. Department of Energy, under contract DE-AC05-840R21400 with Martin Marietta Energy Systems, Inc.

References

Askin, R.G. and Montgomery, D.C. (1980). Augmented robust estimators. *Technometrics*, **22**, 333–341

Bock, M.E. (1975). Minimax estimators of the mean of a multivariate normal distribution. *Ann. Statist.*, **3**, 209–218.

Brown, L.D. (1966). On the admissibility of invariant estimators of one or more location parameters. *Ann. Math. Statis.*, **27**, 1087–1136.

Brown, L.D. (1975). Estimation with incompletely specified loss functions (the case of several location parameters). *J. Amer. Statis. Assoc.*, **70**, 417–427.

Dempster, A.P., Schatzoff, M., and Wermuth, N. (1977). A simulation study of alternatives to ordinary least squares. *J. Amer. Statis. Assoc.*, **72**,

77–91.

Huber, P.J. (1964). Robust estimation of a location parameter. *Ann. Math. Statis.*, **35**, 73–101.

Huber, P.J. (1973). Robust regression: asymptotics, conjectures, and Monte Carlo. *Ann. Statis.*, **1**, 799-821.

James, W. and Stein, C. (1960). Estimation with quadratic loss. *Proc. Fourth Berkeley Symp. on Math. Statis. and Prob.*, **1**, 361–379.

Saleh, A.K.M.E. and Sen, P.K. (1985). On shrinkage M-estimators of location parameters. *Commun. Statist. A — Theory Methods*, **14**, 2313–2329.

Sen, P.K. and Saleh, A.K.M.E. (1987). On preliminary test and shrinkage M-estimation in linear models. *Ann. Statist.*, **15**, 1580–1592.

Stein, C. (1955). Inadmissibility of the usual estimator for the mean of a multivariate normal distribution. *Proc. Third Berkeley Sym. on Math. Statis. and Prob.*, **1**, 197–206.

Stein, C. (1966). An approach to the recovery of inter-block information in balanced incomplete block designs. *Research Paper in Statistics*, F.N. David (ed.), John Wiley & Sons, London.

24

Asymptotic Mean Squared Error of Shrinkage Estimators

T. W. F. Stroud [1]

ABSTRACT When hyperparameters are estimated in a Bayesian model which produces shrinkage estimators of group means, it is well known that the mean square errors of the estimated means are underestimated if hyperparameter estimates are simply substituted for hyperparameter values in mean square formulas. In this article, a method for approximating the mean square error is described for the case where the estimators of the hyperparameters are obtained by maximum likelihood, and hence are asymptotically normal. Under this approach, the Bayesian model is interpreted as a random-effects model. The method is useful in situations such as small area estimation under stratified sampling (with a large number of strata).

1 Introduction

The principle of shrinkage toward a common value in the simultaneous estimation of means of several populations of a similar nature has been a topic of interest for the last three decades, e.g., James and Stein (1961), Box and Tiao (1968), Efron and Morris (1975), Leonard (1976), Dempster, Rubin and Tsutakawa (1981), Reinsel (1985), Peixoto and Harville (1986). An important issue is evaluating the risk of such estimators. When this is done, it is usually done either by using the chi-square (or Wishart) distribution of the sum of squares resulting from the normality of the observations, e.g. Lehmann (1983, p.300–301), Reinsel (1985) and Fuller and Harter (1987), or by invoking a convenient hyperprior for the between-group variance, e.g. Morris (1983a). In this paper an alternative approach is presented, namely the use of large-sample theory for the case where the number of groups becomes large, This technique is applicable to a parametric empirical Bayes situation (e.g., Dempster, Rubin and Tsutakawa, 1981) where global parameters such as the grand mean and the assumed variance of the group means are estimated by maximum likelihood, and can be used in rather complicated situations.

A first approximation to the mean squared error of an empirical Bayes

---

[1] Queen's University at Kingston

estimator of the vector of means, where a consistent estimate of hyperparameters has been used, is the posterior covariance matrix using the estimated hyperparameter values. An empirical Bayes estimator usually shrinks toward a grand mean based on the data, which makes intuitive sense, but the posterior covariance matrix based on estimated hyperparameter values reflects only part of the uncertainty of the estimates of the individual means; it does not reflect the uncertainty due to estimation of the between-group covariance matrix or of the grand mean. It is essentially this point that is made in Dempster, Rubin and Tsutakawa (1981, Section 3, last two paragraphs). The purpose of this article is to try to capture the total uncertainty of estimation of the individual means by using large sample theory.

Because the motivating application (see below) was a problem in finite population sampling, the results are expressed in terms of estimating averages of unobserved data. The standard problem of estimating means of distributions can be obtained simply by letting the finite population sizes go to infinity.

Efron and Morris (1972) evaluate the "relative savings loss" (a variation of mean squared error) of an empirical Bayes estimator for essentially the same model as is dealt with in this article. For their estimator, which is not based on maximum likelihood, the relative savings loss has a particularly simple form for that problem. The method of this article, on the other hand, features maximum likelihood estimation of hyperparameters and can be extended, with perhaps some variations in the derivations, to more complicated problems such as the one we now mention.

The results presented here were motivated by an application (Stroud, 1987) in small area estimation (see Platek et al (1987) for a treatment of this topic). Estimates of numbers employed, unemployed and not in the labor force, based on Canadian Labor Force Survey data, were desired for a large number of Canadian geographical areas which did not correspond to the strata used in the sampling plan. A Bayesian model was formulated for the potential observations in primary sampling units (which were subsets of the strata, and also subsets of the desired areas). Hyperparameters of the Bayesian model were estimated, and the resulting estimated posterior means were aggregated into estimated totals for the desired areas. It was found from simulations using census data that these estimated area totals were superior to expansion estimates (see e.g. Cochran, 1977, p. 21) based on the separate area samples.

In this paper, finite population inference is treated from the superpopulation perspective, under which finite population totals are said to be "predicted" rather than "estimated." The finite population values are random variables, of which a subset (the sample) is randomly observed. We consider the parameters to be the group means, which are subject to a prior involving hyperparameters which are estimated (the parametric empirical Bayes approach). This setup is equivalent to a frequentist superpopulation formulation in which the population values obey a random-effects model. Ghosh and Meeden (1986) and Ghosh and Lahiri (1987) have taken a different, but related, setup and obtained slightly different results.

For simplicity, we avoid the usual complex designs in this paper and consider only stratified random sampling, where the strata are the groups in whose means we are interested. Also for simplicity, the observations in each group are in equal numbers and obey a multivariate normal model with known and equal covariance matrices. See Berger (1985, Section 3.5.4) (and, for related discussion, Section 4.5.3) for the univariate version of the empirical Bayes inference.

In Section 2 the Bayesian model incorporating unknown hyperparameters is described, and the mean squared prediction error of population totals is formulated. Section 3 contains formulas for hyperparameter estimation and the consequent estimation of posterior means and prediction of stratum totals. The approximation of the mean squared prediction error using large sample theory is developed in Section 4 using the Fisher information matrix which is presented in Section 5.

2 The Model and the Prediction Error Vector

Using the superpopulation approach to finite population sampling, we assume that the population for group (stratum) h ($h = 1, \ldots, m$) consists of N_h independent, identically distributed random column vectors with the $\mathcal{N}_p(\underset{\sim}{\mu}_h, \underset{\sim}{\Sigma})$ distribution, of which n from each group are randomly observed. For $h = 1, \ldots, m$, let the average of the n observed values be $\underset{\sim}{z}_h$ and let the average of the $N_h - n$ unobserved values be $\underset{\sim}{z}^*_h$. Then, given the $\underset{\sim}{\mu}_h$, we have

$$\underset{\sim}{z}_h \sim \mathcal{N}_p(\underset{\sim}{\mu}_h, n^{-1}\underset{\sim}{\Sigma}),$$
$$\underset{\sim}{z}^*_h \sim \mathcal{N}_p(\underset{\sim}{\mu}_h, (N_h - n)^{-1}\underset{\sim}{\Sigma}),$$

all independent.

Now Bayesian structure is imposed. Let the $\underset{\sim}{\mu}_h$ be independent, identically distributed random column vectors with the $\mathcal{N}_p(\underset{\sim}{\zeta}, \underset{\sim}{\Omega})$ distribution. The mean squared error we consider is unconditional with respect to $\underset{\sim}{\mu}_h$. For this we need the unconditional distribution of $(\underset{\sim}{z}_h, \underset{\sim}{z}^*_h)$, which is jointly normal with

$$\begin{aligned} E(\underset{\sim}{z}_h) &= E(\underset{\sim}{z}^*_h) = \underset{\sim}{\zeta} \\ V(\underset{\sim}{z}_h) &= \underset{\sim}{\Omega} + n^{-1}\underset{\sim}{\Sigma} \\ V(\underset{\sim}{z}^*_h) &= \underset{\sim}{\Omega} + (N_h - n)^{-1}\underset{\sim}{\Sigma} \\ C(\underset{\sim}{z}_h, \underset{\sim}{z}^*_h) &= \underset{\sim}{\Omega}, \end{aligned} \tag{2.1}$$

where $V(\cdot)$ denotes variance-covariance matrix and $C(\cdot,\cdot)$ denotes cross covariance matrix. The vectors $(\underset{\sim}{z}_h, \underset{\sim}{z}_h^*)'$ are independent across h.

We wish to predict the population total of stratum h. (A minor extension would be to predict the total of an area defined as a collection of strata.) Denote the total of stratum h by $\underset{\sim}{Y}_h$: then

$$\underset{\sim}{Y}_h = n\underset{\sim}{z}_h + (N_h - n)\underset{\sim}{z}_h^*.$$

Let the predicted total $\hat{\underset{\sim}{Y}}_h$ be of the form

$$\hat{\underset{\sim}{Y}}_h = n\underset{\sim}{z}_h + (N_h - n)\underset{\sim}{\psi}_h, \tag{2.2}$$

where $\underset{\sim}{\psi}_h$ is the empirical Bayes estimate of $\underset{\sim}{\mu}_h$ defined early in Section 3. Then the prediction error vector is $\hat{\underset{\sim}{Y}}_h - \underset{\sim}{Y}_h = (N_h - n)(\underset{\sim}{\psi}_h - \underset{\sim}{z}_h^*)$.

Although it might at first glance seem more normal to use the notation $\hat{\underset{\sim}{\mu}}_h$ for the empirical Bayes estimate of $\underset{\sim}{\mu}_h$ we are reserving the caret for a different category of estimate, namely of the global parameters $\underset{\sim}{\zeta}$ and $\underset{\sim}{\Omega}$.

The unconditional (with respect to $\underset{\sim}{\mu}_h$) mean squared error can be defined with respect to an arbitrary positive definite matrix $\underset{\sim}{A}$ as $(N_h - n)^2 E(\underset{\sim}{\psi}_h - \underset{\sim}{z}_h^*)'\underset{\sim}{A}(\underset{\sim}{\psi}_h - \underset{\sim}{z}_h^*)$, which is equal to $(N_h - n)^2 \operatorname{tr}\{\underset{\sim}{A}E(\underset{\sim}{\psi}_h - \underset{\sim}{z}_h^*)(\underset{\sim}{\psi}_h - \underset{\sim}{z}_h^*)'\}$. We therefore define the mean squared error matrix (MSE) as the matrix given by

$$\text{MSE} = (N_h - n)^2 E(\underset{\sim}{\psi}_h - \underset{\sim}{z}_h^*)(\underset{\sim}{\psi}_h - \underset{\sim}{z}_h^*)'. \tag{2.3}$$

The trace of this quantity, without the factor of $(N_h - n)^2$ and with $\underset{\sim}{z}_h^*$ replaced by $\underset{\sim}{\mu}_h$, is sometimes referred to as the empirical Bayes risk (Morris, 1983b) and sometimes simply as the mean squared error (Peixoto and Harville, 1986).

3 Estimation of Hyperparameters and of Group Means

The predicted total given by (2.2) depends on the estimate $\underset{\sim}{\psi}_h$ of the group mean $\underset{\sim}{\mu}_h$. If the hyperparameters $(\underset{\sim}{\zeta}, \underset{\sim}{\Omega})$ were known, we would use the posterior mean vector

$$E(\underset{\sim}{\mu}_h | \underset{\sim}{z}_h) = (\underset{\sim}{\Omega}^{-1} + n\underset{\sim}{\Sigma}^{-1})^{-1}(\underset{\sim}{\Omega}^{-1}\underset{\sim}{\zeta} + n\underset{\sim}{\Sigma}^{-1}\underset{\sim}{z}_h).$$

We propose to use the empirical Bayes estimator,

$$\underset{\sim}{\psi}_h = (\hat{\underset{\sim}{\Omega}}^{-1} + n\underset{\sim}{\Sigma}^{-1})^{-1}(\hat{\underset{\sim}{\Omega}}^{-1}\hat{\underset{\sim}{\zeta}} + n\underset{\sim}{\Sigma}^{-1}\underset{\sim}{z}_h), \tag{3.1}$$

where $\hat{\underset{\sim}{\zeta}}$ and $\hat{\underset{\sim}{\Omega}}$ are maximum likelihood estimators of $\underset{\sim}{\zeta}$ and $\underset{\sim}{\Omega}$, respectively, based on the marginal distribution (2.1) of the $\underset{\sim}{z}_h$. Since marginally the $\underset{\sim}{z}_h$ can be regarded as a multivariate normal sample with unknown mean vector $\underset{\sim}{\zeta}$ and covariance matrix $\underset{\sim}{\Omega} + n^{-1}\underset{\sim}{\Sigma}$, it follows that

$$\hat{\underset{\sim}{\zeta}} = \left(\sum_{h=1}^{m} \underset{\sim}{z}_h\right) \Big/ m, \qquad \hat{\underset{\sim}{\Omega}} = \frac{\sum_{h=1}^{m} \underset{\sim}{z}_h \underset{\sim}{z}_h' - m\hat{\underset{\sim}{\zeta}}\hat{\underset{\sim}{\zeta}}'}{m} - n^{-1}\underset{\sim}{\Sigma}, \tag{3.2}$$

provided that $\hat{\underset{\sim}{\Omega}}$ is positive definite. Thus $\underset{\sim}{\psi}_h$ is given by (3.1), where $\hat{\underset{\sim}{\zeta}}$ and $\hat{\underset{\sim}{\Omega}}$ are given by (3.2). Note that $\hat{\underset{\sim}{\Omega}}$ will be positive definite with high probability when m is large.

The remainder of this section is devoted to obtaining formulas for $E(\underset{\sim}{\psi}_h)$ and $E(\underset{\sim}{\psi}_h \underset{\sim}{\psi}_h')$ in terms of expectations of quantities involving $\hat{\underset{\sim}{\zeta}}$ and $\hat{\underset{\sim}{\Omega}}$, as these expressions are needed in Section 4. These formulas can be derived from formulas for the means of $\sum_{h=1}^{m} \underset{\sim}{\psi}_h$ and $\sum_{h=1}^{m} \underset{\sim}{\psi}_h \underset{\sim}{\psi}_h'$, since by symmetry $E(\underset{\sim}{\psi}_h) = (1/m)E\left(\sum_{h=1}^{m} \underset{\sim}{\psi}_h\right)$, and similarly for $E(\underset{\sim}{\psi}_h \underset{\sim}{\psi}_h')$. It follows from (3.1) and (3.2) that

$$\sum \underset{\sim}{\psi}_h = m\hat{\underset{\sim}{\zeta}}. \tag{3.3}$$

Here, and below, the sign $\sum$ refers to the summation over h from 1 to m. Also, since from (3.2) we have

$$\sum \underset{\sim}{z}_h \underset{\sim}{z}_h' = m(\hat{\underset{\sim}{\zeta}}\hat{\underset{\sim}{\zeta}}' + \hat{\underset{\sim}{\Omega}} + n^{-1}\underset{\sim}{\Sigma}),$$

it follows from (3.1) and (3.2) that

$$\begin{aligned}
\sum \underset{\sim}{\psi}_h \underset{\sim}{\psi}_h' &= (\hat{\underset{\sim}{\Omega}}^{-1} + n\underset{\sim}{\Sigma}^{-1})^{-1} \left\{\sum (\hat{\underset{\sim}{\Omega}}^{-1}\hat{\underset{\sim}{\zeta}} + n\underset{\sim}{\Sigma}^{-1}\underset{\sim}{z}_h)(\hat{\underset{\sim}{\Omega}}^{-1}\hat{\underset{\sim}{\zeta}} + n\underset{\sim}{\Sigma}^{-1}\underset{\sim}{z}_h)'\right\} \\
&\quad (\hat{\underset{\sim}{\Omega}}^{-1} + n\underset{\sim}{\Sigma}^{-1})^{-1} \\
&= (\hat{\underset{\sim}{\Omega}}^{-1} + n\underset{\sim}{\Sigma}^{-1})^{-1} \left\{m\hat{\underset{\sim}{\Omega}}^{-1}\hat{\underset{\sim}{\zeta}}\hat{\underset{\sim}{\zeta}}'\hat{\underset{\sim}{\Omega}}^{-1} + m\hat{\underset{\sim}{\Omega}}^{-1}\hat{\underset{\sim}{\zeta}}\hat{\underset{\sim}{\zeta}}'(n\underset{\sim}{\Sigma}^{-1})\right. \\
&\quad \left. + mn\underset{\sim}{\Sigma}^{-1}\hat{\underset{\sim}{\zeta}}\hat{\underset{\sim}{\zeta}}'\hat{\underset{\sim}{\Omega}}^{-1} + mn^2\underset{\sim}{\Sigma}^{-1}(\hat{\underset{\sim}{\zeta}}\hat{\underset{\sim}{\zeta}}' + \hat{\underset{\sim}{\Omega}} + n^{-1}\underset{\sim}{\Sigma})\underset{\sim}{\Sigma}^{-1}\right\} \\
&\quad (\hat{\underset{\sim}{\Omega}}^{-1} + n\underset{\sim}{\Sigma}^{-1})^{-1} \\
&= m\hat{\underset{\sim}{\zeta}}\hat{\underset{\sim}{\zeta}} + m(\hat{\underset{\sim}{\Omega}}^{-1} + n\underset{\sim}{\Sigma}^{-1})^{-1}(n\underset{\sim}{\Sigma}^{-1})(\hat{\underset{\sim}{\Omega}} + n^{-1}\underset{\sim}{\Sigma})(n\underset{\sim}{\Sigma}^{-1}) \\
&\quad (\hat{\underset{\sim}{\Omega}}^{-1} + n\underset{\sim}{\Sigma}^{-1})^{-1}.
\end{aligned} \tag{3.4}$$

Using the fact that $(A^{-1} + B^{-1})^{-1} = A(A+B)^{-1}B = B(A+B)^{-1}A$, we obtain

$$\Sigma \underset{\sim}{\psi}_h \underset{\sim}{\psi}_h' = m\left\{\hat{\underset{\sim}{\zeta}}\hat{\underset{\sim}{\zeta}}' + \hat{\underset{\sim}{\Omega}}(\hat{\underset{\sim}{\Omega}} + n^{-1}\underset{\sim}{\Sigma})^{-1}\hat{\underset{\sim}{\Omega}}\right\}. \tag{3.5}$$

It follows from (3.3) and (3.5) that

$$E(\underset{\sim}{\psi}_h) = E(\hat{\zeta}), \tag{3.6}$$

$$E(\underset{\sim}{\psi}_h \underset{\sim}{\psi}_h') = E(\underset{\sim}{\zeta}\underset{\sim}{\zeta}') + E\left\{\underset{\sim}{\hat{\Omega}}(\underset{\sim}{\hat{\Omega}} + n^{-1}\underset{\sim}{\Sigma})^{-1}\underset{\sim}{\hat{\Omega}}\right\}. \tag{3.7}$$

Finally, we will need $E(\underset{\sim}{z}_h \underset{\sim}{\psi}_h')$. We have

$$\begin{aligned}
E(\underset{\sim}{z}_h \underset{\sim}{\psi}_h') &= E\left\{(\underset{\sim}{z}_h \underset{\sim}{\hat{\zeta}}' \underset{\sim}{\hat{\Omega}}^{-1} + n \underset{\sim}{z}_h \underset{\sim}{z}_h' \underset{\sim}{\Sigma}^{-1})(\underset{\sim}{\hat{\Omega}}^{-1} + n\underset{\sim}{\Sigma}^{-1})^{-1}\right\} \\
&= \frac{1}{m} E\left\{\left[(\Sigma \underset{\sim}{z}_h)\underset{\sim}{\hat{\zeta}}' \underset{\sim}{\hat{\Omega}}^{-1} + n(\Sigma \underset{\sim}{z}_h \underset{\sim}{z}_h')\underset{\sim}{\Sigma}^{-1}\right](\underset{\sim}{\hat{\Omega}}^{-1} + n\underset{\sim}{\Sigma}^{-1})^{-1}\right\} \\
&= E(\underset{\sim}{\hat{\zeta}\hat{\zeta}'}) + E(\underset{\sim}{\hat{\Omega}}).
\end{aligned} \tag{3.8}$$

4 Approximating the Mean Square Prediction Error

From (2.3), we see that to compute the MSE we need formulas for $E(\underset{\sim}{\psi}_h \underset{\sim}{\psi}_h')$, $E(\underset{\sim}{z}_h^* \underset{\sim}{\psi}_h')$ and $E(\underset{\sim}{z}_h^* \underset{\sim}{z}_h^{*\prime})$. In this section, we compute approximations to the first two of these quantities. These expressions are functions of $\underset{\sim}{\zeta}$ and $\underset{\sim}{\Omega}$. An estimated MSE could thus be obtained by substituting $(\underset{\sim}{\hat{\zeta}}, \underset{\sim}{\hat{\Omega}})$ for $(\underset{\sim}{\zeta}, \underset{\sim}{\Omega})$. $E(\underset{\sim}{z}_h^* \underset{\sim}{z}_h^{*\prime})$ is obtained directly from (2.1).

The approximations are based on the asymptotic distribution of $(\hat{\zeta}, \hat{\Omega})$, and thus approach the correct values as m, the number of groups, becomes large. We assume that n is fixed.

The main result, stated below as a theorem, gives an expression for the asymptotic value of MSE$/(N_h - n)^2$. By the statement "the asymptotic value of $\underset{\sim}{A}_m$ equals $\underset{\sim}{B} + m^{-1}\underset{\sim}{C}$," where $\underset{\sim}{A}_m$, $\underset{\sim}{B}$, and $\underset{\sim}{C}$ are matrices, we mean that

$$m(\underset{\sim}{A}_m - \underset{\sim}{B}) \to \underset{\sim}{C} \quad \text{as } m \to \infty.$$

The result is subject to the usual uniform integrability conditions required to infer convergence of moments from convergence in distribution. The theorem is stated following the derivation of the quantities referred to above as $\underset{\sim}{B}$ and $\underset{\sim}{C}$.

From (3.7) and (3.8), we see that we need approximations to $E(\underset{\sim}{\hat{\zeta}\hat{\zeta}'})$, $E\left\{\underset{\sim}{\hat{\Omega}}(\underset{\sim}{\hat{\Omega}} + n^{-1}\underset{\sim}{\Sigma})^{-1}\underset{\sim}{\hat{\Omega}}\right\}$ and $E(\underset{\sim}{\hat{\Omega}})$. Notice from (3.2) that $E(\underset{\sim}{\hat{\Omega}})$ can be obtained directly from $E(\underset{\sim}{\hat{\zeta}\hat{\zeta}'})$ and $E(\underset{\sim}{z}_h \underset{\sim}{z}_h')$; the latter is given by (2.1). We first derive an approximation to $E(\hat{\zeta}\hat{\zeta}')$, then to $E\left\{\underset{\sim}{\hat{\Omega}}(\underset{\sim}{\hat{\Omega}} + n^{-1}\underset{\sim}{\Sigma})^{-1}\underset{\sim}{\hat{\Omega}}\right\}$.

The basic idea is as follows. Let $\underset{\sim}{x} = g(\hat{\underset{\sim}{\zeta}})$ and $\underset{\sim}{y} = h(\hat{\underset{\sim}{\zeta}})$ be functions of dimension q and s, respectively, of the maximum likelihood estimator $\hat{\underset{\sim}{\zeta}}$. Denote $\underset{\sim}{\xi} = g(\underset{\sim}{\zeta})$ and $\underset{\sim}{\eta} = h(\underset{\sim}{\zeta})$. As $m \to \infty$, the asymptotic approximation to $E(\underset{\sim}{x}\underset{\sim}{y}')$ is

$$E(\underset{\sim}{x}\underset{\sim}{y}') \cong \underset{\sim}{\xi}\underset{\sim}{\eta}' + \frac{1}{m}\left(\frac{\partial \underset{\sim}{\xi}}{\partial \underset{\sim}{\zeta}}\right) \mathcal{I}_1^{-1} \left(\frac{\partial \underset{\sim}{\eta}}{\partial \underset{\sim}{\zeta}}\right)',$$

where, by the asymptotic properties of maximum likelihood estimation, the distribution of $m^{\frac{1}{2}}(\hat{\underset{\sim}{\zeta}} - \underset{\sim}{\zeta})$ converges to the distribution $\mathcal{N}_p(0, \mathcal{I}_1^{-1})$. Here $\mathcal{I}_1$ is the part of the Fisher information matrix pertaining to $\underset{\sim}{\zeta}$, and, as shown in the next section, is equal to $(\Omega + n^{-1}\underset{\sim}{\Sigma})^{-1}$.

Applying this method to $E(\hat{\underset{\sim}{\zeta}}\hat{\underset{\sim}{\zeta}}')$, we find that

$$E(\hat{\underset{\sim}{\zeta}}\hat{\underset{\sim}{\zeta}}') = \hat{\underset{\sim}{\zeta}}\hat{\underset{\sim}{\zeta}}' + \frac{1}{m}(\underset{\sim}{\Omega} + n^{-1}\underset{\sim}{\Sigma}). \tag{4.1}$$

Here, in the simple case of common n and $\underset{\sim}{\Sigma}$, the equality is exact due to the simple form for $\hat{\underset{\sim}{\zeta}}$ given by (3.2). However, the method may be extended to the case of unequal $n_h, \underset{\sim}{\Sigma}_h$.

We now extend this method from vectors to matrices, for the purpose of applying it to functions of $\hat{\underset{\sim}{\Omega}}$. Let $\underset{\sim}{\Gamma}$ and $\underset{\sim}{\Delta}$ be matrix-valued functions of $\underset{\sim}{\Omega}$ of dimension $q \times r$ and $s \times r$, respectively, and let $\hat{\underset{\sim}{\Gamma}}$ and $\hat{\underset{\sim}{\Delta}}$ be their values taken at $\hat{\underset{\sim}{\Omega}}$. As $m \to \infty$ the asymptotic approximation to $E(\hat{\underset{\sim}{\Gamma}}\,\hat{\underset{\sim}{\Delta}}')$ is

$$E(\hat{\underset{\sim}{\Gamma}}\hat{\underset{\sim}{\Delta}}') \cong \underset{\sim}{\Gamma}\underset{\sim}{\Delta}' + \frac{1}{m}\sum_{\alpha=1}^{r}\left(\frac{\partial \underset{\sim}{\Gamma}_\alpha}{\partial \underset{\sim}{\Omega}_c}\right) \underset{\sim}{\Phi} \left(\frac{\partial \Delta_\alpha}{\partial \underset{\sim}{\Omega}_c}\right)', \tag{4.2}$$

where, for $\alpha = 1$ to r, $\underset{\sim}{\Gamma}_\alpha$ and $\underset{\sim}{\Delta}_\alpha$ denote the αth column of $\underset{\sim}{\Gamma}, \underset{\sim}{\Delta}$ respectively, and Φ is the limiting covariance matrix of $m^{\frac{1}{2}}(\hat{\underset{\sim}{\Omega}}_c - \underset{\sim}{\Omega}_c)$. Here $\underset{\sim}{\Omega}_c$ refers to the column vector of distinct components of $\underset{\sim}{\Omega}$, of which there are $p(p+1)/2$. Note that $\underset{\sim}{\Phi}$ is the inverse of the Fisher information matrix $\mathcal{I}_2$, which is given in Section 5. We note that $\underset{\sim}{\Phi} = \underset{\sim}{H}(\underset{\sim}{\Omega} + n^{-1}\underset{\sim}{\Sigma})$, where the matrix-valued function $\underset{\sim}{H}(\cdot)$ is given by

$$[\underset{\sim}{H}(\underset{\sim}{A})]_{ij,kl} = a_{ik}a_{jl} + a_{il}a_{jk}, \; 1 \le i \le j \le p; \; 1 \le k \le l \le p,$$

arranged in lexicographic order. The right side of (4.2) is evaluated by computing each component of each of the matrices $(\partial \underset{\sim}{\Gamma}_\alpha / \partial \underset{\sim}{\Omega}_c)\underset{\sim}{\Phi}(\partial \underset{\sim}{\Delta}_\alpha / \partial \underset{\sim}{\Omega}_c)'$ appearing in (4.2). Call the matrix $\underset{\sim}{B}_\alpha$ and the (ι, κ)-th component $b_{\iota\kappa}$. It is shown in Stroud (1971, p. 1423) that

$$b_{\iota\kappa} = 2\,\mathrm{tr}\left(\frac{d\gamma_{\iota\kappa}}{d\underset{\sim}{\Omega}}\right)\underset{\sim}{A}\left(\frac{d\delta_{\kappa\alpha}}{d\underset{\sim}{\Omega}}\right)\underset{\sim}{A}, \tag{4.3}$$

where $\underset{\sim}{A} = \underset{\sim}{\Omega}+n^{-1}\underset{\sim}{\Sigma}$, $\gamma_{\iota\alpha}$ and $\delta_{\kappa\alpha}$ are the (ι,α)-th and (κ,α)-th components of $\underset{\sim}{\Gamma}, \underset{\sim}{\Delta}$ respectively, and for any scalar function of a symmetric matrix, $y = g(\underset{\sim}{\Omega})$, we define $dy/d\underset{\sim}{\Omega}$ to be a symmetric matrix of the same dimension as $\underset{\sim}{\Omega}$, whose components are defined by

$$\left(\frac{dy}{d\underset{\sim}{\Omega}}\right)_{ii} = \frac{\partial y}{\partial \omega_{ii}}, \qquad (4.4)$$

$$\left(\frac{dy}{d\underset{\sim}{\Omega}}\right)_{ij} = \frac{1}{2}\frac{\partial y}{\partial \omega_{ij}} \quad \text{for } i \neq j.$$

Thus to evaluate (4.2) it remains to obtain the matrices $d\gamma_{\iota\alpha}/d\underset{\sim}{\Omega}$ and $d\delta_{\kappa\alpha}/d\underset{\sim}{\Omega}$. Comparing (4.2) and the last term of (3.7), define $\underset{\sim}{\Gamma} = \underset{\sim}{\Omega}(\underset{\sim}{\Omega} + n^{-1}\underset{\sim}{\Sigma})^{-1}$ and $\underset{\sim}{\Delta} = \underset{\sim}{\Omega}$. We see from (4.4) that $d\delta_{\kappa\alpha}/d\underset{\sim}{\Omega}$ is a matrix with 1 in the (κ,κ) position if $\kappa = \alpha$, or 1/2 in the(κ,α) and (α,κ) positions if $\kappa \neq \alpha$, with zeros elsewhere. We may write this as

$$\frac{d\delta_{\kappa\alpha}}{d\underset{\sim}{\Omega}} = \frac{1}{2}(\underset{\sim}{E}_{\kappa\alpha} + \underset{\sim}{E}'_{\kappa\alpha}), \qquad (4.5)$$

where by $\underset{\sim}{E}_{\kappa\alpha}$ we mean a matrix with 1 in the (κ,α) position and zeros elsewhere.

To evaluate $d\gamma_{\iota\alpha}/d\underset{\sim}{\Omega}$, write

$$\begin{aligned} d\underset{\sim}{\Gamma} &= (d\underset{\sim}{\Omega})(\underset{\sim}{\Omega} + n^{-1}\underset{\sim}{\Sigma})^{-1} + \underset{\sim}{\Omega}\, d(\underset{\sim}{\Omega} + n^{-1}\underset{\sim}{\Sigma})^{-1} \\ &= (d\underset{\sim}{\Omega})(\underset{\sim}{\Omega} + n^{-1}\underset{\sim}{\Sigma})^{-1} - \underset{\sim}{\Omega}(\underset{\sim}{\Omega} + n^{-1}\underset{\sim}{\Sigma})^{-1}(d\underset{\sim}{\Omega})(\underset{\sim}{\Omega} + n^{-1}\underset{\sim}{\Sigma})^{-1}; \end{aligned}$$

$d\gamma_{\iota\alpha}/d\underset{\sim}{\Omega}$ is then just the derivative of the (ι,α)-th component of $d\underset{\sim}{\Gamma}$ with respect to $d\underset{\sim}{\Omega}$. This kind of use of matrix differentials is described in Deemer and Olkin (1951).

The extension of (4.5) to arbitrary linear functions of $d\underset{\sim}{\Omega} \equiv \underset{\sim}{X}$ is a slight generalization of a formula in Dwyer and MacPhail (1948). Denoting the function by $Y = \underset{\sim}{A}\underset{\sim}{X}\underset{\sim}{B}$, where $\underset{\sim}{X}$ is symmetric, the required formula is

$$\frac{dy_{\iota\alpha}}{dX} = \frac{1}{2}(\underset{\sim}{A}'\underset{\sim}{E}_{\iota\alpha}\underset{\sim}{B}' + \underset{\sim}{B}\underset{\sim}{E}'_{\iota\alpha}\underset{\sim}{A}).$$

It follows after some simplification that

$$\frac{d\gamma_{\iota\alpha}}{d\underset{\sim}{\Omega}} = \frac{1}{2n}(\underset{\sim}{\Omega} + n^{-1}\underset{\sim}{\Sigma})^{-1}(\underset{\sim}{\Sigma}\underset{\sim}{E}_{\iota\alpha} + \underset{\sim}{E}'_{\iota\alpha}\underset{\sim}{\Sigma})(\underset{\sim}{\Omega} + n^{-1}\underset{\sim}{\Sigma})^{-1} \qquad (4.6)$$

To summarize, referring to (4.2) we have

$$E\left\{\hat{\underset{\sim}{\Omega}}(\hat{\underset{\sim}{\Omega}} + n^{-1}\underset{\sim}{\Sigma})^{-1}\hat{\underset{\sim}{\Omega}}\right\} \cong \underset{\sim}{\Omega}(\underset{\sim}{\Omega} + n^{-1}\underset{\sim}{\Sigma})^{-1}\underset{\sim}{\Omega} + \frac{1}{m}\sum_{\alpha=1}^{r}\underset{\sim}{B}_{\alpha}, \qquad (4.7)$$

where the (ι, κ) component of $\underset{\sim}{B}_{\alpha}$, denoted by $b_{\iota\kappa}$, is given by (4.3), in which the partial derivative matrices are given in (4.5) and (4.6).

Referring to (2.3), the mean squared prediction error MSE may be calculated from

$$MSE/(N_h - n)^2 = E(\underset{\sim}{\psi}_h \underset{\sim}{\psi}_h') - E(\underset{\sim}{\psi}_h \underset{\sim}{z}_h^{*\prime}) - E(\underset{\sim}{z}_h^* \underset{\sim}{\psi}_h') + E(\underset{\sim}{z}_h^* \underset{\sim}{z}_h^{*\prime}). \quad (4.8)$$

$E(\underset{\sim}{\psi}_h \underset{\sim}{\psi}')$ is given by (3.7), and

$$E(\underset{\sim}{z}_h^* \underset{\sim}{z}_h^{*\prime}) = \underset{\sim\sim}{\zeta\zeta'} + \underset{\sim}{\Omega} + (N_h - n)^{-1}\underset{\sim}{\Sigma}, \quad (4.9)$$

using (2.1). Finally, since $\underset{\sim}{z}_h^*$ may be regarded as $\underset{\sim}{\mu}_h$ plus uncorrelated error, we have

$$\begin{aligned} E(\underset{\sim}{z}_h^* \underset{\sim}{\psi}_h') &= E(\underset{\sim}{\mu}_h \underset{\sim}{\psi}_h') \\ &= E\{E(\underset{\sim}{\mu}_h \underset{\sim}{\psi}_h' | \underset{\sim}{z}_1, \ldots, \underset{\sim}{z}_m)\} \\ &= E\{E(\underset{\sim}{\mu}_h | \underset{\sim}{z}_h) \underset{\sim}{\psi}_h'\} \end{aligned}$$

where

$$E(\underset{\sim}{\mu}_h | \underset{\sim}{z}_h) = (\underset{\sim}{\Omega}^{-1} + n\underset{\sim}{\Sigma}^{-1})^{-1} \underset{\sim}{\Omega}^{-1} \underset{\sim}{\zeta} + (\underset{\sim}{\Omega}^{-1} + n\underset{\sim}{\Sigma}^{-1})^{-1} n\underset{\sim}{\Sigma}^{-1} \underset{\sim}{z}_h.$$

Hence, since $E(\underset{\sim}{\psi}_h) = E(\hat{\underset{\sim}{\zeta}}) = \underset{\sim}{\zeta}$,

$$E(z_h^* \psi_h') = (\underset{\sim}{\Omega}^{-1} + n\underset{\sim}{\Sigma}^{-1})^{-1} \underset{\sim}{\Omega}^{-1} \underset{\sim\sim}{\zeta\zeta'} + (\underset{\sim}{\Omega}^{-1} + n\underset{\sim}{\Sigma}^{-1})^{-1} \underset{\sim}{\Sigma}^{-1} E(\underset{\sim}{z}_h \underset{\sim}{\psi}_h') \quad (4.10)$$

where

$$E(\underset{\sim}{z}_h \underset{\sim}{\psi}_h') = E(\underset{\sim\sim}{\hat{\zeta}\hat{\zeta}'}) + E(\hat{\underset{\sim}{\Omega}})$$

and

$$E(\hat{\underset{\sim}{\Omega}}) = E(\underset{\sim}{z}_h \underset{\sim}{z}_h') - E(\underset{\sim\sim}{\hat{\zeta}\hat{\zeta}'}) - n^{-1}\underset{\sim}{\Sigma}$$

from (3.2), so that

$$\begin{aligned} E(\underset{\sim}{z}_h \underset{\sim}{\psi}_h') &= E(\underset{\sim}{z}_h \underset{\sim}{z}_h') - n^{-1}\underset{\sim}{\Sigma} \\ &= \underset{\sim\sim}{\zeta\zeta'} + \underset{\sim}{\Omega}. \end{aligned} \quad (4.11)$$

where $E(\underset{\sim}{z}_h \underset{\sim}{z}_h') = \underset{\sim\sim}{\zeta\zeta'} + \underset{\sim}{\Omega} + n^{-1}\underset{\sim}{\Sigma}$ from (2.1). Substituting into (4.10) we have

$$E(\underset{\sim}{z}_h^* \underset{\sim}{\psi}_h') = \underset{\sim\sim}{\zeta\zeta'} + (\underset{\sim}{\Omega}^{-1} + n\underset{\sim}{\Sigma}^{-1})^{-1} n\underset{\sim}{\Sigma}^{-1} \underset{\sim}{\Omega}. \quad (4.12)$$

Substituting from (3.7), (4.1) and (4.9) into (4.8) and using (4.1) and (4.7), we have the following result.

Theorem 4.1 *Assuming the model stated at the beginning of Section 2, where n is fixed and $\{N_1, N_2, \ldots\}$ is a given sequence, and subject to uniform integrability conditions, the asymptotic value of the matrix $MSE/(N_h - n)^2$ as defined by (2.3) is equal to*

$$\underset{\sim}{\Omega}(\underset{\sim}{\Omega} + n^{-1}\underset{\sim}{\Sigma})^{-1}\underset{\sim}{\Omega} + \frac{1}{m}\sum_{\alpha=1}^{r} \underset{\sim}{B}_{\alpha} + \left(1 - \frac{1}{m}\right)\underset{\sim}{\Omega}$$
$$+\left(\frac{1}{mn} + \frac{1}{N_h - n}\right)\underset{\sim}{\Sigma} - 2\underset{\sim}{\Omega}(\underset{\sim}{\Omega}^{-1} + n\underset{\sim}{\Sigma}^{-1})^{-1}n\underset{\sim}{\Sigma}^{-1},$$

where the definition of $\underset{\sim}{B}_{\alpha}$ may be found following relation (4.7).

This is the main result. We conclude with a section setting out the form of the Fisher information matrix, which was used in the above development.

5 Fisher Information Matrix of the Hyperparameters

In Sections 1–4 we are dealing with observations

$$\underset{\sim}{z}_h \sim \mathcal{N}(\underset{\sim}{\zeta}, \underset{\sim}{\Omega} + n^{-1}\underset{\sim}{\Sigma}).$$

In this section we deal with the Fisher information matrix, or, equivalently, the asymptotic covariance matrix of

$$m^{\frac{1}{2}}\begin{bmatrix} \hat{\underset{\sim}{\zeta}} - \underset{\sim}{\zeta} \\ \hat{\underset{\sim}{\Omega}} - \underset{\sim}{\Omega} \end{bmatrix}$$

It turns out to be almost as easy to write down the answer for the more general case:

$$\underset{\sim}{z}_h \sim \mathcal{N}(\underset{\sim}{\zeta}, \underset{\sim}{\Omega} + \underset{\sim}{\Sigma}_h),$$

which we do, for possible future applications. The log likelihood function for $(\underset{\sim}{\zeta}, \underset{\sim}{\Omega})$ is

$$\ell(\underset{\sim}{\zeta}, \underset{\sim}{\Omega}) = -\frac{1}{2}\sum_{h=1}^{m} \log|\underset{\sim}{\Omega} + \underset{\sim}{\Sigma}_h| - \frac{1}{2}\sum_{h=1}^{m}(\underset{\sim}{z}_h - \underset{\sim}{\zeta})'(\underset{\sim}{\Omega} + \underset{\sim}{\Sigma}_h)^{-1}(\underset{\sim}{z}_h - \underset{\sim}{\zeta}). \quad (5.1)$$

The Fisher information matrix is given by

$$\mathcal{I} = \sum_{h=1}^{m} E\left(\frac{d\ell_h}{d\Theta}\right)\left(\frac{d\ell_h}{d\Theta}\right)'$$

where $\left(\frac{d\ell_h}{d\Theta}\right)$ is the column vector of components $\frac{\partial \ell_h}{\partial \zeta_i}$ followed by the components $\frac{\partial \ell_h}{\partial \omega_{ij}}$; $i \leq j = 1, \ldots, p$ in lexicographic order. By ℓ_h we mean the

h-th term in the summation in (5.1). Carry out the process of differentiation, and define λ_{ij}^h to be the (i,j) component of $(\underset{\sim}{\Omega} + \underset{\sim}{\Sigma}_h)^{-1}$. It may be verified that for $h = 1, \ldots, m$ and $i, j, k, l = 1, \ldots, p$ with $i \leq j$ and $k \leq \ell$:

$$E\left(\frac{\partial \ell_h}{\partial \zeta_i}\right)\left(\frac{\partial \ell_h}{\partial \zeta_j}\right) = \lambda_{ij}^h,$$

$$E\left(\frac{\partial \ell_h}{\partial \omega_{ii}}\right)\left(\frac{\partial \ell_h}{\partial \omega_{jj}}\right) = \frac{1}{2}\left(\lambda_{ij}^h\right)^2$$

$$E\left(\frac{\partial \ell_h}{\partial \omega_{ij}}\right)^2 = \left(\lambda_{ij}^h\right)^2 + \lambda_{ii}^h \lambda_{jj}^h,$$

$$E\left(\frac{\partial \ell_h}{\partial \omega_{ij}}\right)\left(\frac{\partial \ell_h}{\partial \omega_{kl}}\right) = \lambda_{ij}^h \lambda_{kl}^h,$$

for cases excluding the preceding two, and

$$E\left(\frac{\partial \ell_h}{\partial \zeta_i}\right)\left(\frac{\partial \ell_h}{\partial \omega_{jk}}\right) = 0.$$

If $\mathcal{I}$ is partitioned as $\begin{bmatrix} \mathcal{I}_1 & 0 \\ 0 & \mathcal{I}_2 \end{bmatrix}$, where $\mathcal{I}_1$ is $p \times p$ and $\mathcal{I}_2$ is $p(p+1)/2$ by $p(p+1)/2$, we see that, for the case of equal $\underset{\sim}{\Sigma}_h$, the inverse of $\mathcal{I}$ is $(1/m)(\underset{\sim}{\Omega} + \underset{\sim}{\Sigma}_h)$ and it may be shown that the component of the inverse of $\mathcal{I}_2$ corresponding to i, j, k, l is $(1/m)(a_{ik}a_{jl} + a_{il}a_{jk})$, where a_{ij} is the (i,j)-th component of $\underset{\sim}{\Omega} + \underset{\sim}{\Sigma}_h$.

Acknowledgments: The practical investigation which motivated this work was undertaken as part of a contract with Statistics Canada. Research was supported by a grant from the Natural Sciences and Engineering Research Council of Canada.

References

Berger, J.O. (1985). *Statistical Decision Theory and Bayesian Analysis, 2nd ed.* Springer-Verlag, New York.

Box, G.E.P. and Tiao, G.C. (1968). Bayesian estimation of means for the random-effect model. *J. Amer. Statist. Assoc.* **63**, 174–181.

Cochran, W.G. (1977). *Sampling Techniques, 3rd. ed.* Wiley, New York.

Deemer, W.L. and Olkin, I. (1951). The Jacobians of certain matrix transformations useful in multivariate analysis, based on lectures by P.L. Hsu. *Biometrika* **38**, 345–367.

Dempster, A.P., Rubin, D.B. and Tsutakawa, R.K. (1981). Estimation in covariance components models. *J. Amer. Statist. Assoc.* **76**, 341–353.

Dwyer, A.S. and MacPhail, M.S. (1948). Symbolic matrix derivatives. *Ann. Math. Statist.* **19**, 517–534.

Efron, B. and Morris, C. (1972). Empirical Bayes on vector observations—An extension of Stein's method. *Biometrika* **59**, 335–347.

Efron, B. and Morris, C. (1975). Data analysis using Stein's estimator and its generalizations. *J. Amer. Statist. Assoc.* **70**, 311–319.

Fuller, W.A. and Harter, R.M. (1987). The multivariate components of variance model for small area estimation. In *Small Area Statistics: An International Symposium*(R. Platek *et al*, eds.) Wiley, New York, 103–123.

Ghosh, M. and Lahiri, P. (1987). Robust empirical Bayes estimation of means from stratified samples. *J. Amer. Statist. Assoc.* **82**, 1153–1162.

Ghosh, M. and Meeden, G. (1986). Empirical Bayes estimation in finite population sampling. *J. Amer. Statist. Assoc.* **81**, 1 1058–1062.

James, W. and Stein, C. (1961). Estimation with quadratic loss. *Proc. Fourth Berkeley Symp. Math. Statist. Probab.*, Univ. California Press **1**, 361–379.

Leonard, T. (1976). Some alternative approaches to multiparameter estimation. *Biometrika* **63**, 69–75.

Lehmann, E.L. (1983). *Theory of Point Estimation.* Wiley, New York.

Morris, C.N. (1983a). Parametric empirical Bayes confidence intervals. In *Scientific Inference, Data Analysis and Robustness* (G.E.P. Box, T. Leonard, C.-F. Wu, eds.), Academic Press, New York, 25–50.

Morris, C.N. (1983b). Parametric empirical Bayes inference: theory and applications. *J. Amer. Statist. Assoc.* **78**, 47–55.

Peixoto, J.L. and Harville, D.A. (1986). Comparisons of alternative predictors under the balanced one-way random model. *J. Amer. Statist. Assoc.* **81**, 431–436.

Platek, R. Rao, J.N.K., Särndal, C.E. and Singh, M.P., eds. (1987). *Small Area Statistics: An International Symposium.* Wiley, New York.

Reinsel, G.C. (1985). Mean squared error properties of empirical Bayes estimators in a multivariate random effects general linear model. *J. Amer. Statist. Assoc.* **80**, 642–650.

Stroud, T.W.F. (1971). On obtaining large-sample tests from asymptotically normal estimators. *Ann. Math. Statist.* **42** 1412–1424.

Stroud, T.W.F. (1987). Small area estimation from Canadian Labor Force Survey data. Contributed paper, annual meeting, Statistical Society of Canada, Québec.

Part D

Approaches to Inference

25

Likelihood Analysis of a Binomial Sample Size Problem

Murray Aitkin[1]
Mikis Stasinopoulos[2]

ABSTRACT The problem of estimating the binomial sample size N from k observed numbers of successes is examined from a likelihood point of view. The direct use of the likelihood function for inference about N is illustrated when p is known, and the problem of inference is considered when p is unknown, and has to be eliminated in some way from the likelihood. Different methods (Bayesian, integrated likelihood, conditional likelihood, profile likelihood) for eliminating the nuisance parameter are found to lead to very different likelihoods in N in an example. This occurs because of a strong ridge in the two-parameter likelihood in N and p. Integrating out the parameter p is found to be unsatisfactory, but reparameterization of the model shows that the inference about N is almost unaffected by the new nuisance parameter. The resulting likelihood in N corresponds closely to the profile likelihood in the original parameterization.

1 Introduction

Carroll and Lombard (1985) considered the problem of estimating the parameter N based on k independent success counts $s_1, \ldots, s_k$ from a binomial distribution with unknown parameters N and p. They extended earlier work by Olkin, Petkau and Zidek (OPZ, 1981) on the moment and maximum likelihood estimators by introducing new estimators of N based on integrating out p from the likelihood with respect to a beta distribution, yielding a beta-binomial distribution for the number of successes. The new estimators maximizing this likelihood compared favorably in mean square error terms with the OPZ estimators. Casella (1986) considered perturbations of the likelihood to decide on the "stability" or instability of the ML estimator.

The emphasis throughout these discussions is on point estimation of N, and the comparison of different estimators through their mean square or

[1]Tel Aviv University
[2]Welcome Research Laboratories

relative mean square errors. In practical data analysis, the reporting of a point estimate without some measure of precision would not be sufficient. Carroll and Lombard noted this problem and commented: "... little is known about the shape of $L(N)$ [the likelihood function of N]..., so the question of finding a confidence interval for N remains to be addressed." Casella considered the likelihood and perturbations of it but used only the maximum likelihood estimate.

A Bayesian analysis of the problem was given by Draper and Guttman (1971). They used independent priors on N and p, taking N to be uniform on a finite grid and p to be beta. The marginal posterior distribution for N had the peculiar feature of a long right-hand tail, with finite mass for $N_{\max}$, the largest value of N allowed a priori. Kahn (1987) pointed out that this tail behavior is entirely determined by the prior distributions for p and N, and not at all by the data, and therefore the choice of $N_{\max}$ is important, affecting both the mean and the median of the posterior distribution of N. Raftery (1988) extended the Bayes analysis by assuming a Poisson (μ) prior for N; this avoids the range restriction of Draper and Guttman. Raftery assumed a uniform prior for p in his examples, independent of N, and a hyperprior for μ proportional to μ^{-1}, so that the joint prior for p and N is proportional to N^{-1}.

The Bayesian analysis has the advantage over the "classical" approach in providing the full posterior distribution for N, not just a posterior mode (maximum integrated likelihood estimate). With a uniform prior for N, the posterior for N is simply proportional over the interval $N \leq N_{\max}$ to the integrated likelihood $L(N)$. With an informative prior $p(N)$, the posterior is proportional to $p(N)L(N)$. For a robust Bayes analysis, $L(N)$ or the posterior should not depend strongly on the prior distribution for p or N. A strong dependence should make us very cautious about drawing inferential conclusions about N.

In this paper we present a likelihood analysis of the problem (see Edwards 1972 for an authoritative exposition of this approach). Other methods for eliminating nuisance parameters in the likelihood approach are available, and have been surveyed by Hinde and Aitkin (1987). We compute and present in Section 2 the likelihood functions for N obtained by different methods of eliminating p. Inferences about N can be dramatically different using different methods, as we show with one of the "unstable" examples considered by Carroll and Lombard.

Examination in Section 3 of the two-parameter likelihood in N and p shows why these differences occur, and we find that the likelihoods obtained by integrating out p are unsatisfactory. In Section 4 we consider the canonical decomposition of the likelihood (Hinde and Aitkin, 1987) and show that elimination of the nuisance parameter by integration is effectively impossible in the example. In Section 5 we reparameterize the problem and show that the nuisance parameter can now be eliminated; the resulting "canonical" likelihood in N is almost identical to the profile likelihood in

Section 2. We show that such a result can be expected frequently.

2 Likelihood Functions for N

Given k independent success counts s_i from the binomial distribution $b(N,p)$, the likelihood function in N and p is

$$L(N,p) = \left[\prod_{i=1}^{k} \binom{N}{s_i}\right] p^T (1-p)^{Nk-T}, 0 \leq p \leq 1, N \geq \max(s_i), T = \sum_{i=1}^{k} s_i.$$

Our object is to make an inferential statement about N. Hence, p here is a nuisance parameter and needs to be eliminated from the likelihood. We will use as an example the "unstable" example of OPZ, with $k = 5$ and $s_i = (16, 18, 22, 25, 27)$.

LIKELIHOOD SECTION

Suppose first that p is *known* to be 0.21. Then the likelihood in N is simply the section of $L(N,p)$ at $p = 0.21$. This is easily computed provided $\log n!$ can be accurately calculated. Figure 1 shows the likelihood normalized to a maximum of 1, that is the *relative* likelihood

$$R(N) = L(N, p_0)/L(\hat{N}, p_0) \qquad \text{at } p_0 = 0.21.$$

The maximum likelihood estimate is $\hat{N} = 103$.

Since N is a discrete parameter, the usual "standard error" of the maximum likelihood estimator based on the observed or expected information is not appropriate, and the asymptotic $\mathcal{X}^2$ distribution of $-2\log$ (likelihood ratio) is not valid with a sample of 5.

The likelihood approach uses the relative likelihood, or likelihood ratio, directly to construct likelihood intervals for the parameter. Thus all values of N with $R(N) \geq \gamma$ constitute a $100\gamma\%$ likelihood interval for N. For example, the 15% likelihood interval for N is $(87, 121)$. All such intervals can be obtained from a tabulation of $R(N)$, but as with Bayes posterior densities, it is usually more informative to plot the likelihood than to summarize it in one or two interval statements.

The "calibration" of the likelihood can be understood in conventional repeated sampling or in Bayesian terms. In a sequential likelihood ratio test of $H_0 : \theta = \theta_0$ against $H_1 : \theta = \theta_1$ with equal Type I and Type II error probabilities α, the value $(1-\alpha)/\alpha$ for the likelihood ratio $L(\theta_0)/L(\theta_1)$ would lead to rejection of H_1 and acceptance of H_0 and the value $\alpha/(1-\alpha)$ to rejection of H_0 and acceptance of H_1. In Bayesian terms, if θ_0 and θ_1 have equal prior probabilities, the posterior probability of θ_0 is α if the likelihood ratio $L(\theta_0)/L(\theta_1)$ is a $\alpha/(1-\alpha)$. Thus, in either case, a "conventionally small" α of 0.05 or 0.01 corresponds to a likelihood ratio $\alpha/(1-\alpha)$ of 1/19 or 1/99. These values give a calibration of the likelihood ratio or relative likelihood which is the same for all models, and does not depend on exact or asymptotic sampling distributions. See Aitkin (1986) for further discussion.

We now consider possible methods for eliminating p from the two-parameter likelihood. Several methods were discussed by Carroll and Lombard. Hinde and Aitkin (1987) gave a full discussion of the elimination of nuisance parameters in general.

Conditional likelihood

If N were known, T would be sufficient for p. By conditioning on T, p is eliminated from the likelihood:

$$C(N) = \Pr(s_1 < \cdots < s_k | T, N) = \left[\prod_i \binom{N}{s_i}\right] \Big/ \binom{Nk}{T}.$$

This generalized hypergeometric likelihood depends only on N, but since the distribution $b(Nk, p)$ of the conditioning variable T itself depends strongly on N, some loss of information about N must result from conditioning on T. Formally,

$$L(N,p) = C(N) \cdot \binom{Nk}{T} P^T (1-p)^{Nk-T}$$

and the second factor is ignored in constructing the conditional likelihood.

Profile likelihood

We substitute for p its MLE given N:

$$\hat{p}(N) = T/Nk$$

$$P(N) = \left[\prod_i \binom{N}{s_i}\right] \hat{p}(N)^T \left[1 - \hat{p}(N)\right]^{Nk-T}.$$

(Note that $\hat{p}(N) \leq 1$ always since $N \geq s_{\max} \geq \bar{s} = T/k$.)

The profile likelihood is maximized at the ordinary MLE $\hat{N}$, and is the section of the likelihood $L(N,p)$ along the hyperbola in the parameter space given by

$$Np = T/k.$$

"Integrated" likelihood

Assume a beta "prior" distribution for p

$$f(p) = \frac{1}{B(a+1, b+1)} p^a (1-p)^b,$$

and integrate p out of the likelihood to give

$$B(N) = \left[\prod_i \binom{N}{s_i}\right] \binom{NK + a + b + 1}{T + a + 1}^{-1}$$

omitting a proportionality constant of $T + a + 1$. Here, a and b can be non-integral; in this case the binomial coefficients are interpreted as gamma function ratios.

Carroll and Lombard calculated the MLEs from this likelihood for the particular cases $a = b = 0$, $a = b = 1$, and examined their efficiencies (in terms of mean square error) compared with those of the adjusted moment and ML estimators proposed by OPZ. They noted also that the conditional likelihood is equivalent to the integrated likelihood with $a = -1, b = 0$; in Bayesian terms this would imply strong prior information that p is small, whereas $a = b = 0$ would be the uniform prior and $a = b = 1$ a quadratic with mode at $p = 0.5$.

The efficiency comparisons of the estimators established that the maximum integrated likelihood estimators with $a = b = 0$ and $a = b = 1$ are generally superior to the OPZ estimators. The difficulty with these comparisons is that they are based on one-point inferential summaries: a single value found by one estimation method is compared with a single value found by another estimation method. But no scientist would be satisfied with a point estimate without some information about precision: what could be learned from the report "$\hat{N} = 50$"? This information is available in the likelihood function resulting from the elimination of p, interpreted in the same way as the likelihood in N and p is known.

The likelihood functions described above are plotted in Figure 1 for the "unstable" example of OPZ, together with the section of the likelihood at $p = 0.21$. There is a lower limit to N which must exceed 26; all the likelihoods increase rapidly from this value.

Figure 1 shows remarkable differences among the likelihoods. The section of the likelihood is very condensed; the profile likelihood has a poorly-defined maximum around 100 and falls off very slowly for large values of N; the conditional likelihood becomes flat for very large values of N; the two integrated likelihoods have a well-defined maximum around 50 but very long tails.

Kahn (1987) showed that the integrated likelihoods behave for large N like $CN^{-(a+1)}$, where C is a function of k and T but not N. The intervals of plausible values of N from these likelihoods (or the intervals of highest posterior density if N is given a uniform prior) thus depend explicitly on the prior parameter a. For the conditional likelihood with $a = -1$, the tail behaves like a constant C, as is clearly visible. The posterior distribution for N given by Raftery (1988) for this example (his Figure 1.1a) behaves like CN^{-2} since N is given a prior proportional to N^{-1}, and p is uniform.

FIGURE 1. Likelihoods for N from unstable example: a) Section at $p = 0.21$, b) Conditional, c) Profile, d) Integrated $a = b = 0$, e) Integrated $a = b = 1$.

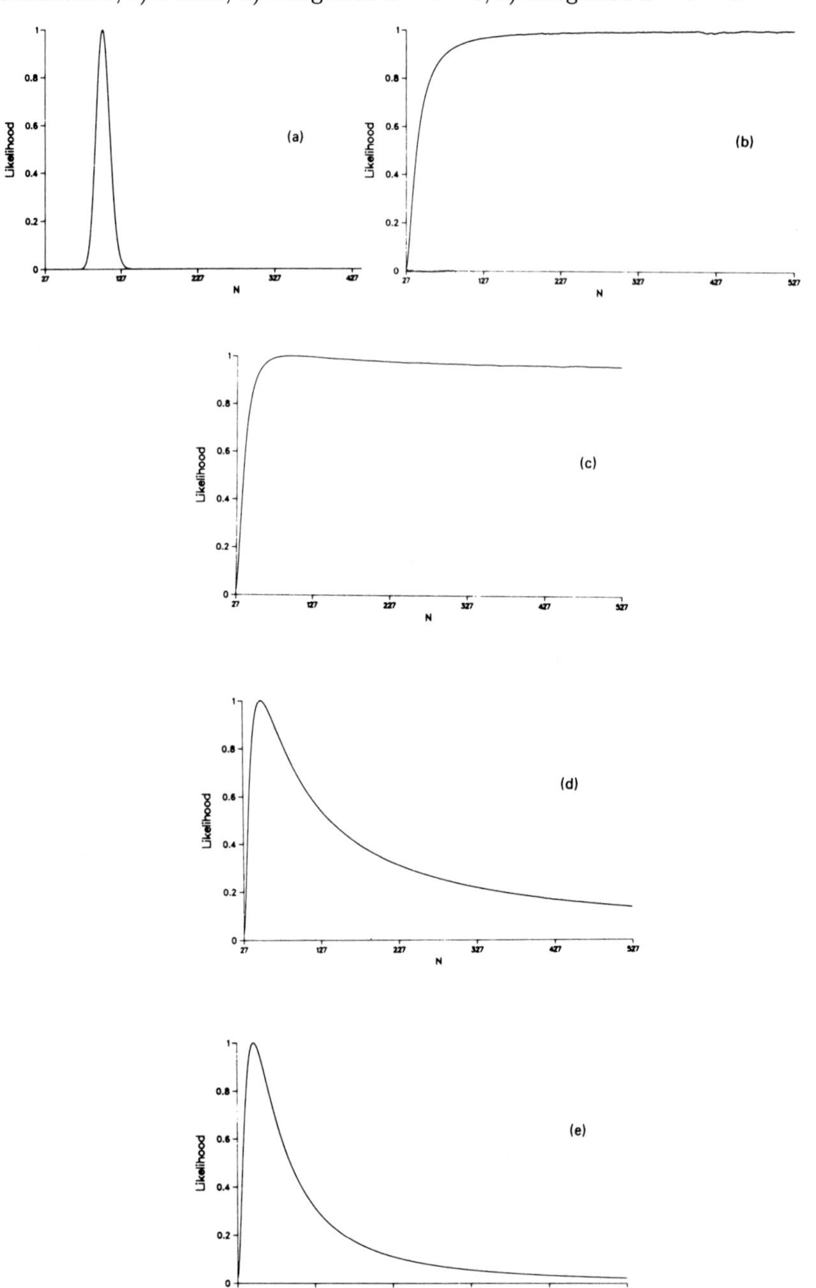

Why do these likelihoods differ so much? Examination of the two parameter likelihood provides the answer.

3 Likelihood Function for N and p

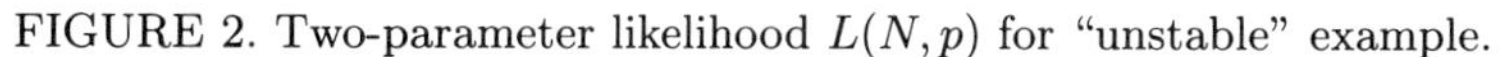

FIGURE 2. Two-parameter likelihood $L(N, p)$ for "unstable" example.

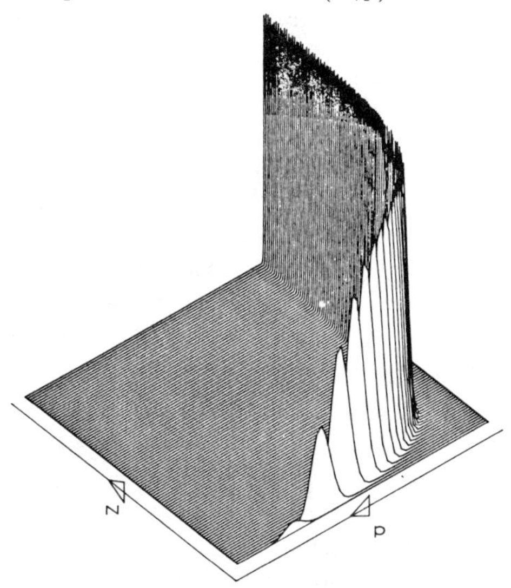

Figure 2 shows a perspective plot of the likelihood in N and p. This function causes particular difficulty to most plotting and contouring routines because the contours are so elongated in the parameter space and the likelihood decreases so rapidly from the ridge where $Np = T/k$. The profile likelihood is the section of the likelihood along this ridge and is essentially flat over a large range of N. The property of "instability" in a given sample of the MLE described by OPZ is that the addition of 1 to the largest s_i results in a very large change to the MLE in the sample. There is no need to perturb the sample to demonstrate this property: it is evident in the likelihood function for the given sample. The MLE is almost arbitrary, and may indeed by affected by the numerical precision with which the factorials in the likelihood are evaluated. OPZ noted that a flat likelihood could be expected when the sample variance of the s_i was close to the sample mean, since in this "near-Poisson" region the parameters N and p are nearly unidentifiable.

As $N \to \infty$, the profile likelihood in N tends to the maximum over λ of the Poisson likelihood

$$L^*(\lambda) = \prod_{i-1}^{k} (1/s_i!) e^{-k\lambda} \lambda^T$$

which is

$$L^*(T/k) = \prod_{i=1}^{k} (1/s_i!) e^{-T} (T/k)^T.$$

For the example, the relative likelihood of "best Poisson" to "best binomial" is

$$\frac{P(\infty)}{P(\hat{N})} = \frac{L^*(T/k)}{P(\hat{N})} = 0.935,$$

showing that the Poisson model is a perfectly reasonable alternative to the binomial. This value of the relative likelihood is approached as an asymptote in Figure 1(c). (A simple confirmation comes from the Poisson homogeneity test, comparing $\sum(s_i - \bar{s})^2/\bar{s} = 85.7/21.6 = 3.94$ to the usual $\mathcal{X}_4^2$ null distribution.) This result completely contradicts the conclusions from the integrated likelihoods (except for $a = -1, b = 0$), since the relative likelihoods go to zero as $N \to \infty$ in these cases.

The great differences between the integrated likelihoods and the conditional likelihood can be easily interpreted in terms of the relative weights assigned to small and large values of p by the prior distribution. The conditional likelihood assigns heavy weight to small values of p; for such values the likelihood is large only for large values of N. The uniform and quadratic priors assign more weight to intermediate values of p and therefore of N. The integrated likelihood in N can be made to have any tail behavior by a suitable choice of prior distribution. In the absence of real evidence about p external to the experiment justifying an informative prior distribution, the use of integrated likelihoods is potentially misleading and cannot be recommended. Even the location of the maximum depends on the prior: with $a = -1$ and $b = 0$ there is no finite maximizing value at all. To quote "$\hat{N} \approx 50$", the MLE from $a = b = 0$ and $a = b = 1$, would certainly be misleading.

This dependence of inference about N on the choice of the prior for p is a consequence of the shape of the likelihood and can be characterized by the eigenvalues in its canonical decomposition, which we now discuss.

4 Canonical Decomposition of the Likelihood

Hinde and Aitkin (1987) described a new approach to the elimination of nuisance parameters in two-parameter likelihoods. They approximated the likelihood $L(\theta, \phi)$ by a separable function $A(\theta)B(\phi)$ of the parameters. A least-squares approximation of $L(\theta, \phi)$ by $A(\theta)B(\phi)$ gives $A(\theta)$ and $B(\phi)$

as the principal eigenfunctions in the eigenfunction expansion of $L(\theta, \phi)$. Computationally, the likelihood function is evaluated as a matrix L over finite grids in each parameter, and the likelihood matrix is approximated by its rank 1 representation in the singular value decomposition. If the squared principal eigenvalue λ_1^2 contains most of the trace of LL', the likelihood is well represented by $A(\theta)B(\phi)$, and inference about θ can be based on the "canonical likelihood" $A(\theta)$. ignoring ϕ. The goodness of the approximation is measured by $\alpha = \lambda_1^2/\text{tr}LL'$; this should be at least 0.95 for a good approximation.

For the unstable example of Section 2, the likelihood function $L(N,p)$ has squared principal eigenvalue λ_1^2 which is only 11% of the trace of LL'. Inference about N thus depends *strongly* on p, and the nuisance parameter p cannot be eliminated by integrating out, since the resulting likelihood in N will depend strongly on the weight function in p that is used in the integration. Thus, the routine use of integrated likelihoods in this problem cannot be recommended.

Both the integrated likelihoods and the accuracy of the canonical approximation depend on the form of the nuisance parameter. Kahn (1987) speculated that reparameterizing the model might help. We now examine the choice of parameterization.

5 The Choice of Nuisance Parameter

The choice of the nuisance parameter in two-parameter likelihoods was discussed by Hinde and Aitkin (1987). They recommended choosing the nuisance parameter ϕ to make the likelihood $L(\theta, \phi)$ as nearly separable as possible. If there exists a parameterization (θ, ψ) such that $L(\theta, \psi) = A(\theta)B(\psi)$ exactly, then the expected information matrix of $\hat{\theta}$ and $\hat{\psi}$ will be diagonal. Cox and Reid (1987) gave a partial differential equation to be satisfied by $\psi = g(\theta, \phi)$ in order to give a diagonal expected information matrix, but this applies only to continuous parameters θ and ϕ. In the discussion of Kalbfleisch and Sprott (1970), Edwards pointed out that if separability holds exactly, then the maximum likelihood estimate $\hat{\psi}(\theta)$ of ψ given θ will not depend on θ. If therefore we can express the partial derivative equation $\partial \log L/\partial \phi = 0$ in the form $g(\theta, \phi) = h(\text{data})$, then the choice $\psi = g(\theta, \phi)$ will have the required property. This choice does not ensure separability; it is a necessary but not sufficient condition.

In the model considered here, we have

$$\frac{\partial \log L}{\partial p} = \frac{T}{p} - \frac{Nk - T}{1-p}$$

so that $\partial \log L/\partial p = 0$ leads to $Np = T/k$ as before. The choice of nuisance parameter $\psi = Np$ will therefore give $\hat{\psi} = T/k$. Here, ψ is just the binomial mean which is estimated by the sample mean $\bar{s} = T/k$. The likelihood function can then be written

$$L(N,\psi) = \left[\prod \binom{N}{s_i}\right] \left(\frac{\psi}{N}\right)^{T} \left(1 - \frac{\psi}{N}\right)^{Nk-T}, \quad N \geq \max(s_i), \quad 0 < \psi \leq N.$$

Note that the parameter space is now not a product space. Figure 3 shows the likelihood function in N and ψ. The strong dependence of the shape of the likelihood section in N on the value of p has now disappeared. The canonical decomposition of $L(N,\psi)$ now gives $\alpha = 0.998$: 99.8% of the trace of LL' is contained in the first eigenvalue λ_1^2. The likelihood function is almost exactly separable; whether we integrate out ψ or maximize over it, almost the same likelihood in N will be obtained. The canonical likelihood for N is shown in Figure 4, together with the profile likelihood. These are very close, and the profile likelihood in N from $L(N,\psi)$ is identical to that from $L(N,p)$ as is easily seen.

FIGURE 3. Two-parameter likelihood $L(N,\psi)$.

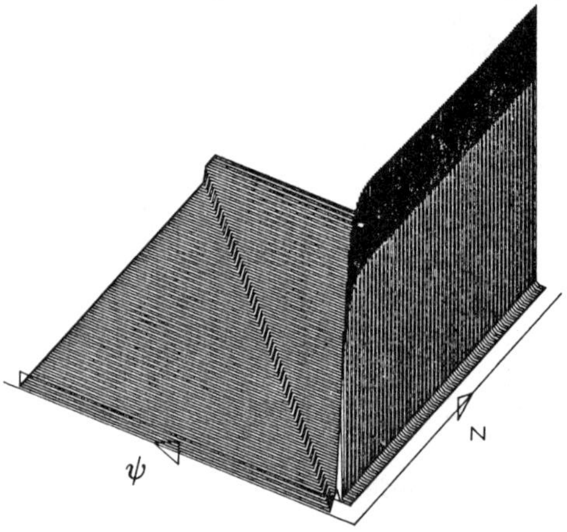

Thus the reparameterization of the likelihood leads to an almost exactly separable likelihood for N and ψ; the resulting likelihood in N is almost identical to the profile likelihood for N in the original parameterization, but is very different from the integrated likelihoods, except for the conditional likelihood.

This discussion has been restricted to one "unstable" example from OPZ. Is examination of the likelihood necessary only in these unstable cases (where the mean and variance of the counts s_i are nearly equal)? Figure 5 shows the likelihood for one of the "stable" examples from OPZ with $k = 20$ and $s_i = (17, 23, 24, 25, 25, 26, 26, 26, 27, 27, 28, 28, 28, 29, 30, 30, 30, 31, 33, 38)$. Here the mean and variance are 27.55 and 17.73, showing considerable under-dispersion. There is a clear maximum of the likelihood at $N = 71$,

FIGURE 4. Canonical and profile likelihoods for N from $L(N,\psi)$.

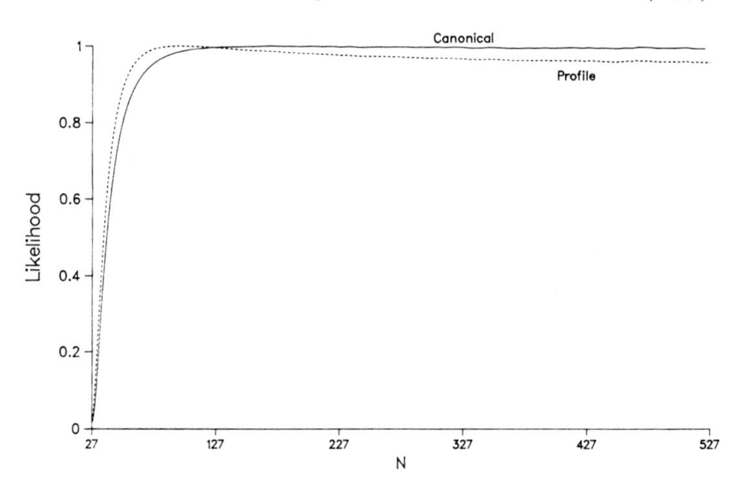

and the maximum falls off along the ridge as N increases. The limit of the profile relative likelihood as $N \to \infty$ is 0.356, so the Poisson is still a plausible alternative to the binomial. The same pronounced curvature in the likelihood is visible, and integration over p still gives integrated likelihoods depending on the prior. The "stability" of the MLE of N is thus no insurance against misleading point inferences, whether from the integrated or the profile likelihoods.

FIGURE 5. Two-parameter likelihood $L(N,p)$ for "stable" example.

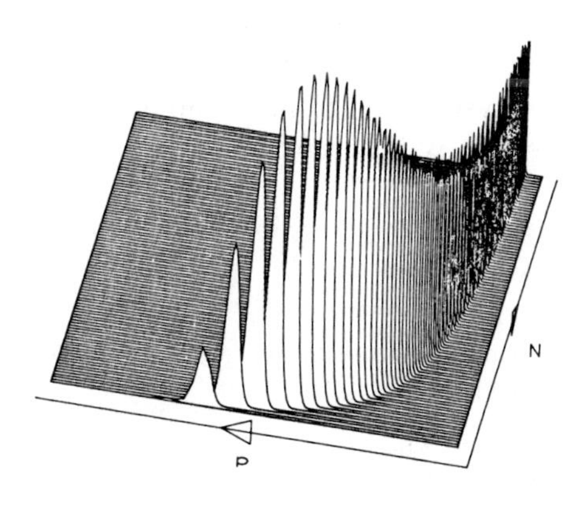

6 Conclusion

The "instability" of the MLE of N is a consequence of the flatness of the likelihood in N in the near-Poisson region when the mean and variance of the counts s_i are nearly equal. The lack of information about N in the unstable cases cannot be corrected by integrating p out of the likelihood; the tail behavior of the integrated likelihoods depends completely on the prior for p.

This example provides further evidence of the importance of careful examination of the likelihood function. The apparently straightforward Bayesian (or integrated likelihood) analysis provides a posterior distribution (integrated likelihood) for N which depends strongly on the specification of the prior. The use of the maximum of the integrated likelihood as a point estimate would not be satisfactory in any scientific analysis without a variability qualification. The integrated likelihood provides a full description of the variability about the MLE, but this description is so sensitive to the prior on p that no confidence can be placed in the conclusions.

Reparameterization solves these difficulties. It no longer matters much whether we integrate or maximize over ψ, since the likelihood $L(N, \psi)$ factors almost exactly. The resulting likelihood in N corresponds closely to the profile likelihood, an unexpected result for the Bayesian view that it is better to integrate out nuisance parameters than to maximize over them.

This result can be expected quite often. If the likelihood function $L(\theta, \phi)$ is not separable, but there is a transformation of the nuisance parameter ϕ to $\psi = g(\theta, \phi)$ such that $L(\theta, \psi)$ *is* separable $\big(L(\theta, \psi) = A(\theta)B(\psi)\big)$, then $A(\theta)$ is exactly the profile likelihood $L\big(\theta, \hat{\phi}(\theta)\big)$. If under the new parameterization the likelihood function is nearly separable, there will be close agreement between the canonical and profile likelihoods, as in this example. This result provides support for the general use of profile likelihoods.

The practical conclusion for the "unstable" example is that nothing prac tical can be said about N except that it is not very small. It could indeed be infinite! This differs very much from the apparent conclusion from the integrated likelihoods that N is around 50.

References

Aitkin, M. (1986). Statistical modelling: the likelihood approach. *The Statistician* **35**, 103–113.

Carroll, R.J. and Lombard, F. (1985). Note on N estimators for the binomial distribution. *J. Amer. Statist. Assoc.* **80**, 423–426.

Casella, G. (1986). Stabilizing binomial n estimators. *J. Amer. Statist. Assoc.* **81**, 172–175.

Cox, D.R. and Reid, N. (1987). Parameter orthogonality and approximate conditional inference (with Discussion). *J. Roy. Statist. Soc. B* **49**, 1–39.

Draper, N. and Guttman, I. (1971). Bayesian estimation of the binomial parameter. *Technometrics* **13**, 667–673.

Edwards, A.W.F. (1972). *Likelihood.* Cambridge University Press.

Hinde, J.P. and Aitkin, M. (1987). Canonical likelihoods: a new likelihood treatment of nuisance parameters. *Biometrika* **74**, 45–58.

Kahn, W.D. (1987). A cautionary note for Bayesian estimation of the binomial parameter n. *Amer. Statist.* **41**, 38–39.

Kalbfleisch, J.D. and Sprott, D.A. (1970). Application of likelihood methods to models involving large numbers of parameters (with Discussion). *J. Roy. Statist. Soc B* **32**, 175–208.

Olkin, I., Petkau, A.J. and Zidek, J.V. (1981). A comparison of n estimators for the binomial distribution. *J. Amer. Statist. Assoc.*, **76**, 637–642.

Raftery, A.E. (1988). Inference for the binomial N parameter: a hierarchical Bayes approach. *Biometrika* **75**, 223–228.

26

Truncation, Information, and the Coefficient of Variation

M.J. Bayarri[1]
M.H. DeGroot[2]
P.K. Goel[3]

ABSTRACT The Fisher information in a random sample from the truncated version of a distribution that belongs to an exponential family is compared with the Fisher information in a random sample from the untruncated distribution. Conditions under which there is more information in the selection sample are given. Examples involving the normal and gamma distributions with various selection sets, and the zero-truncated binomial, Poisson, and negative binomial distributions are discussed. A property pertaining to the coefficient of variation of certain discrete distributions on the non-negative integers is introduced and shown to be satisfied by all binomial, Poisson, and negative binomial distributions.

1 Introduction

Consider the basic statistical problem in which a random variable X is distributed over a certain population according to the (generalized) density $g(x \mid \theta)$ and it is desired to make inferences about the unknown value of the parameter θ which lies in the parameter space Ω. In the usual statistical analysis, it is assumed that the observed data form a random sample from the density $g(x \mid \theta)$. In many situations, however, observations are obtained only from certain selected portions of the underlying population, either because experimental conditions make it impossible to obtain data from the entire population or because experimenters choose to restrict the observations in this way in their experimental design.

In this paper we will consider problems in which the observations are restricted to lie in a specified subset S of the sample space of X. Let

$$s(\theta) = \Pr(X \epsilon S \mid \theta) \quad \text{for} \quad \theta \epsilon \Omega \,, \tag{1.1}$$

---

[1]University of Valencia
[2]Carnegie Mellon University
[3]Ohio State University

where it is assumed that $s(\theta) > 0$ throughout Ω. The statistical analysis will then be based on a random sample $Y_1, \ldots, Y_n$ from the following truncated version of the density g:

$$f(y \mid \theta) = \begin{cases} g(y \mid \theta)/s(\theta) & \text{for} \quad y \in S, \\ 0 & \text{otherwise}. \end{cases} \tag{1.2}$$

The model (1.2) is called a *truncation model* or a *selection model*, and a random sample from (1.2) is called a *selection sample.* The set S is called the *selection set.*

Selection samples occur frequently in practice. Several examples are given in Bayarri and DeGroot (1987a,c), together with a discussion of the Bayesian approach to their analysis.

Our central focus here is to compare the experiment in which a random sample $X_1, \ldots, X_n$ is drawn from the unrestricted density $g(x \mid \theta)$ with the experiment in which a selection sample $Y_1, \ldots, Y_n$ is drawn from the truncation model $f(y \mid \theta)$ in order to determine which of these two experiments is more informative about θ. Many different concepts of the information in an experiment have been discussed in the statistical literature, including sufficiency and the comparison of experiments, as developed by Blackwell (1951, 1953); Kullback-Leibler information (Kullback, 1968); and the widely-used Fisher information. Some relationships among these concepts and further references can be found in Goel and DeGroot (1979). Here we will restrict our study to problems in which θ is a real-valued parameter and the comparison is based on the Fisher information in each experiment. The investigation of the information in selection samples was introduced in Bayarri and DeGroot (1987b), where comparisons based on both Fisher information and other concepts of information were carried out. Some of the results to be discussed here were mentioned there with the details omitted. In closely related subsequent work, Patil and Taillie (1987) calculate the Fisher information for a wide variety of weighted distributions, including some truncation models. Their paper, like this one, concentrates on exponential families of distributions.

This paper has two major purposes. The first is to illuminate the effects of truncation on the information in the most widely used distributions in statistical practice, including the normal, gamma, binomial, Poisson, and negative binomial. We shall accomplish this by studying the Fisher information in truncated exponential families. The second purpose is really a happy bonus. We will present a simple and fascinating property that is satisfied by all binomial, Poisson, and negative binomial distributions, but (relatively speaking) few other discrete distributions. This property, which we discovered in our study of the Fisher information in truncated versions of these standard discrete distributions with the zero class missing, is defined as follows:

Suppose that a random variable X has a discrete distribution on the non-negative integers with finite variance, and let

$$\begin{aligned} p_i &= \Pr(X = i) \qquad \text{for } i = 0,\ 1,\ 2,\ \ldots, \\ \mu &= E(X), \qquad \text{and } \sigma^2 = \operatorname{Var}(X). \end{aligned} \tag{1.3}$$

Then the distribution of X is said to have the *CV property* if

$$\frac{p_0}{p_1} = \frac{\sigma^2}{\mu^2}\,. \tag{1.4}$$

The name *CV property* that we have given to (1.4) is inspired by the fact that the right-hand side of (1.4) is the square of the *coefficient of variation* (the CV) of the distribution of X.

It is easy to verify that the distribution of X has the CV property if it is a member of any one of the following "Big 3" families of discrete distributions:

$$\textit{Binomial:} \quad p_i = \binom{n}{i}\theta^i(1-\theta)^{n-i} \quad \text{for} \quad i = 0, 1, \ldots, n\,, \tag{1.5}$$

for some positive integer n and some number θ $(0 < \theta < 1)$. In this family $\mu = n\theta$ and $\sigma^2 = n\theta(1-\theta)$.

$$\textit{Poisson:} \quad p_i = e^{-\theta}\theta^i/i! \quad \text{for} \quad i = 0, 1, 2, \ldots\,, \tag{1.6}$$

for some number $\theta > 0$. In this family $\mu = \sigma^2 = \theta$.

$$\textit{Negative binomial:} \quad p_i = \binom{-r}{i}\theta^r[-(1-\theta)]^i \quad \text{for} \quad i = 0, 1, 2, \ldots\,, \tag{1.7}$$

where r and θ are numbers such that $r > 0$ and $0 < \theta < 1$. In this family $\mu = r(1-\theta)/\theta$ and $\sigma^2 = r(1-\theta)/\theta^2$.

Although isolated examples of other distributions satisfying the CV property (1.4) can be constructed, we do not know of any other widely-used family of discrete distributions all of whose members satisfy it.

In Section 2 we derive a general expression for the Fisher information in a selection sample from a truncation model with an arbitrary selection set when the underlying distribution belongs to an exponential family. Necessary and sufficient conditions are presented for the information in the selection sample to be greater than that in an unrestricted random sample.

In Section 3, we study the information in a truncated normal distribution for various selection sets, first when the mean is unknown and then when the variance is unknown. In some of the examples that are presented, the unrestricted random sample contains more information than the selection sample, and in other examples the reverse is true.

In Section 4, truncated gamma distributions are studied when the selection set is the upper tail of the distribution. It is shown that whether a selection sample or an unrestricted random sample contains more information about the scale parameter depends on the value of the shape parameter.

In Section 5, we study truncated binomial, Poisson, and negative binomial distributions with the zero class missing. It is shown that for each of these three zero-truncated families, the ratio of the Fisher information in a selection sample to the Fisher information in an unrestricted random sample has the special form $\Pr(X \geq 2 \mid \theta)/[\Pr(X \geq 1 \mid \theta)]^2$.

Finally, in Section 6, we discuss the CV property (1.4) and motivate its derivation by using the results in Section 5. Selected theorems are presented describing the properties of some distributions that satisfy the CV property.

2 Fisher information in a truncated exponential family

Under the usual regularity conditions regarding differentiation with respect to θ inside the expectation operator, the Fisher information $I_X(\theta)$ in an observation X with density $g(x \mid \theta)$ is given by

$$I_X(\theta) = E_g\left[-\frac{\partial^2 \log g(X \mid \theta)}{\partial \theta^2}\right]. \tag{2.1}$$

The Fisher information in a random sample $X_1, \ldots, X_n$ from $g(x \mid \theta)$ is simply $nI_X(\theta)$.

It follows from (2.1) that the Fisher information $I_Y(\theta)$ in an observation Y from the truncation model (1.2) is given by

$$I_Y(\theta) = E_f\left[-\frac{\partial^2 \log g(Y \mid \theta)}{\partial \theta^2}\right] + \frac{d^2 \log s(\theta)}{d\theta^2}. \tag{2.2}$$

If $I_X(\theta) \geq I_Y(\theta)$ for all values of $\theta \epsilon \Omega$, then the experiment $\mathcal{E}_X$ in which a random sample is drawn from the unrestricted density $g(x \mid \theta)$ will always yield at least as much Fisher information about θ as the experiment $\mathcal{E}_Y$ in which a selection sample of the same size is drawn from the truncated model (1.2). In this case, we write $\mathcal{E}_X \succeq_F \mathcal{E}_Y$. Similarly, if $I_Y(\theta) \geq I_X(\theta)$ for all values of $\theta \epsilon \Omega$, we write $\mathcal{E}_Y \succeq_F \mathcal{E}_X$. Of course, in an arbitrary problem, it is not necessarily true that either $\mathcal{E}_X \succeq_F \mathcal{E}_Y$ or $\mathcal{E}_Y \succeq_F \mathcal{E}_X$.

In general, it is important to note that the expectation in (2.1) is calculated with respect to the density $g(x \mid \theta)$ whereas the expectation in (2.2) is calculated with respect to the density $f(y \mid \theta)$. However, this distinction becomes irrelevant if the expression inside the square brackets in (2.1) or (2.2) does not actually depend on X or Y.

This situation will arise if θ is the natural parameter of an exponential family of distributions, in which case $g(x \mid \theta) = a(x)e^{\theta u(x)}/c(\theta)$. It follows from (2.1) and (2.2) that

$$I_X(\theta) = \frac{d^2 \log c(\theta)}{d\theta^2} = \mathrm{Var}[u(X) \mid \theta] \tag{2.3}$$

and

$$I_Y(\theta) = I_X(\theta) + \frac{d^2 \log s(\theta)}{d\theta^2} = \mathrm{Var}[u(Y) \mid \theta] \,. \tag{2.4}$$

A similar result is given by Patil and Taillie (1987) for more general types of weighted distributions.

Now consider an arbitrary exponential family of the form

$$g(x \mid \theta) = \frac{a(x)e^{\omega(\theta)u(x)}}{c(\theta)} \,, \tag{2.5}$$

where $\omega(\theta)$ is a one-to-one differentiable function of θ for $\theta\epsilon\Omega$. If the density (2.5) and the corresponding truncation model (1.2) are reparametrized in terms of $\zeta = \omega(\theta)$, and if $I_X^*(\zeta)$ and $I_Y^*(\zeta)$ denote the Fisher information about ζ in X and Y, respectively, then as is well known,

$$I_Z(\theta) = I_Z^*[\omega(\theta)] \left(\frac{d\zeta}{d\theta}\right)^2 \quad \text{for} \quad Z = X, Y \,. \tag{2.6}$$

The following result is now immediate:

Theorem 2.1 *Suppose that $g(x \mid \theta)$ is given by (2.5). Then for any value of $\theta_0\epsilon\Omega$,*

$$I_Y(\theta_0) \lesseqqgtr I_X(\theta_0) \tag{2.7}$$

if and only if

$$\left\{\frac{d^2 \log s[\omega^{-1}(\zeta)]}{d\zeta^2}\right\}_{\zeta=\omega(\theta_0)} \lesseqqgtr 0 \,. \tag{2.8}$$

The interpretation of this theorem is that any one of the three possible relations in (2.7) holds if and only if the corresponding relation in (2.8) holds. Since $\zeta = \omega(\theta)$ is the natural parameter of the exponential family defined by (2.5), we have, in particular, the following corollary:

Corollary 2.2 *Suppose that $g(x \mid \theta)$ is given by (2.5), and let*

$$t(\zeta) = \Pr[X\epsilon S \mid \theta = \omega^{-1}(\zeta)] \,. \tag{2.9}$$

Then $\mathcal{E}_X \succeq_F \mathcal{E}_Y$ if and only if $\log t(\zeta)$ is a concave function of ζ, and $\mathcal{E}_Y \succeq_F \mathcal{E}_X$ if and only if $\log t(\zeta)$ is convex.

We now apply these results to truncated normal distributions.

3 Truncated normal distributions

Suppose that the distribution of X is normal with unknown mean θ $(-\infty < \theta < \infty)$ and known variance which, without loss of generality, we take to be 1. For this exponential family, the natural parameter is θ itself, and $u(x) = x$. Hence, it follows from (2.3) that, as is well known, $I_X(\theta) = 1$.

Example 1.

Suppose that this normal distribution is truncated so that any observation Y must lie in the interval $S = \{y : \tau_1 < y < \tau_2\}$, where τ_1 and τ_2 are given numbers such that $\tau_1 < \tau_2$. We include the possibility of one-sided truncation by allowing either $\tau_1 = -\infty$ or $\tau_2 = \infty$. Thus,

$$s(\theta) = \Pr(X \epsilon S \mid \theta) = \Phi(\tau_2 - \theta) - \Phi(\tau_1 - \theta) , \tag{3.1}$$

where, as usual, Φ denotes the d.f. of the standard normal distribution.

A straightforward calculation shows that $\dfrac{d^2 \log s(\theta)}{d\theta^2} \leq 0$ if and only if

$$\frac{(\tau_1 - \theta)\phi(\tau_1 - \theta) - (\tau_2 - \theta)\phi(\tau_2 - \theta)}{\Phi(\tau_2 - \theta) - \Phi(\tau_1 - \theta)} \leq \left[\frac{\phi(\tau_1 - \theta) - \phi(\tau_2 - \theta)}{\Phi(\tau_2 - \theta) - \Phi(\tau_1 - \theta)}\right]^2 , \tag{3.2}$$

where ϕ is the standard normal p.d.f. If $\tau_1 \leq \theta \leq \tau_2$ then the left-hand side of (3.2) is negative and the inequality is trivially true. Suppose next that $\theta < \tau_1$. Then (3.2) is equivalent to

$$d_1\phi(d_1) - d_2\phi(d_2) \leq \frac{\phi(d_1) - \phi(d_2)}{M(d_1, d_2)} \tag{3.3}$$

where

$$d_2 = \tau_2 - \theta > d_1 = \tau_1 - \theta > 0 \tag{3.4}$$

and $M(\cdot, \cdot)$ is the generalized Mills' ratio defined by

$$M(\lambda_1, \lambda_2) = \frac{\Phi(\lambda_2) - \Phi(\lambda_1)}{\phi(\lambda_1) - \phi(\lambda_2)} \quad \text{for} \quad 0 \leq \lambda_1 < \lambda_2 \leq \infty . \tag{3.5}$$

The inequality (3.3) now follows from the fact that $d_1\phi(d_1) - d_2\phi(d_2) \leq d_1[\phi(d_1) - \phi(d_2)]$ and the inequality $\lambda_1 M(\lambda_1, \lambda_2) < 1$ for $0 \leq \lambda_1 < \lambda_2 \leq \infty$ (Fang and He, 1984). Finally, the same argument can be applied when $\theta > \tau_2$, so $\dfrac{d^2 \log s(\theta)}{d\theta^2} \leq 0$ for all values of θ $(-\infty < \theta < \infty)$. Thus, by Corollary 2.2, $\mathcal{E}_X \succeq_F \mathcal{E}_Y$. □

It is noteworthy that, as we have just shown in Example 1, an unrestricted random sample provides greater Fisher information for all possible values of θ than a selection sample from any given interval, bounded or unbounded, irrespective of its location or its length. This same conclusion could have been reached by using the expression for the variance of the truncated normal distribution in Example 1, as given for example by Schneider (1986, Sec. 2.2), and then applying (2.4). However, the calculations needed to reach the conclusion of Example 1 would be the same as those that we have presented, starting from (3.2). Indeed, the calculations given in Example 1 can be regarded as a proof of the fact that the variance of any truncated normal distribution for which the selection set S is an interval, bounded or unbounded, is smaller than the variance of the untruncated normal distribution.

Example 2.

Suppose now that the observation Y is restricted to the set $S = \{y : y < \tau_1 \text{ or } y > \tau_2\}$, where $\tau_1 < \tau_2$ are specified finite real numbers. Then

$$s(\theta) = 1 - \Phi(\tau_2 - \theta) + \Phi(\tau_1 - \theta)\,, \tag{3.6}$$

and a straightforward calculation shows that

$$\begin{aligned} \frac{d^2 \log s(\theta)}{d\theta^2} &= \frac{(\tau_2 - \theta)\phi(\tau_2 - \theta) - (\tau_1 - \theta)\phi(\tau_1 - \theta)}{s(\theta)} \\ &\quad - \left[\frac{\phi(\tau_2 - \theta) - \phi(\tau_1 - \theta)}{s(\theta)}\right]^2 . \end{aligned} \tag{3.7}$$

We will show that (3.7) is negative for some values of θ and positive for others.

It can be seen that (3.7) will be negative for any value of θ such that

$$(\tau_2 - \theta)\phi(\tau_2 - \theta) < (\tau_1 - \theta)\phi(\tau_1 - \theta)\,. \tag{3.8}$$

Since the function $u\phi(u)$ is decreasing in u for $u > 1$, the inequality (3.8) will hold for any value of θ such that $\theta < \tau_1 - 1$.

On the other hand, consider the point $\theta_0 = (\tau_1 + \tau_2)/2$ and let

$$d = \tau_2 - \theta_0 = \theta_0 - \tau_1 > 0\,. \tag{3.9}$$

Also, let

$$M(d) = \frac{1 - \Phi(d)}{\phi(d)} \tag{3.10}$$

denote Mills' ratio [the limit of (3.5) as $\lambda_2 \to \infty$ with $\lambda_1 = d$]. Then the value of (3.7) at $\theta = \theta_0$ is simply $d/M(d) > 0$. Thus, in this example, neither $\mathcal{E}_X \succeq_F \mathcal{E}_Y$ nor $\mathcal{E}_Y \succeq_F \mathcal{E}_X$.

It follows from (2.4) that

$$I_Y(\theta_0) - I_X(\theta_0) = \frac{d}{M(d)} > 0\,. \tag{3.11}$$

Furthermore, if the truncation points τ_1 and τ_2 are moved further and further apart, so that $\tau_2 - \tau_1 \to \infty$, then $d \to \infty$ and, from (3.11),

$$I_Y(\theta_0) - I_X(\theta_0) \to \infty \, . \tag{3.12}$$

In other words, if $\tau_2 - \tau_1$ is large, a selection drawn from *outside* the interval $\tau_1 \leq y \leq \tau_2$ will yield *much* more Fisher information than an unrestricted random sample if θ is near the middle of the excluded interval. □

To conclude this section, we will consider an example in which the distribution of X is normal with a known mean which, without loss of generality, we take to be 0, and an unknown precision θ ($\theta > 0$). Recall that the precision of a normal distribution is the reciprocal of its variance.

Example 3.

In this example

$$g(x \mid \theta) = \theta^{1/2}\phi(\theta^{1/2}x) \, , \tag{3.13}$$

so θ is again the natural parameter of the exponential family and $u(x) = -x^2/2$. It follows from (2.3) that

$$I_X(\theta) = \frac{1}{2\theta^2} \, . \tag{3.14}$$

Now suppose that this normal distribution is truncated so that any observation Y must lie in the selection set $S = \{y : y > \tau\}$, where τ is a given positive number. Here,

$$s(\theta) = 1 - \Phi(\theta^{1/2}\tau) \, . \tag{3.15}$$

After some calculation, it can be found that $\dfrac{d^2 \log s(\theta)}{d\theta^2} > 0$ if and only if

$$\frac{1}{M(\gamma)} < \gamma + \frac{1}{\gamma}, \qquad \text{where } \gamma = \theta^{1/2}\tau. \tag{3.16}$$

Since the inequality in (3.16) holds for all $\gamma > 0$ (Gordon, 1941), it follows that $\mathcal{E}_Y \succeq_F \mathcal{E}_X$. □

Thus, a selection sample from the right tail of a normal distribution contains less information about the mean than an unrestricted random sample when the variance is known (as follows from Example 1 with $\tau_2 = \infty$), but more information about the variance than an unrestricted random sample when the mean μ is known, provided that $\tau > \mu$ (as follows from Example 3). By symmetry, a similar conclusion may be made for samples from the left tail of a normal distribution.

4 Truncated gamma distributions

In this section we will consider an example in which X has a gamma distribution and the selection set S is again the right-hand tail of the distribution. This example is interesting because the results depend critically on the value of the shape parameter of the gamma distribution.

Example 4.

Suppose that X has a gamma distribution for which the shape parameter α is known and the scale parameter θ is unknown $(\theta > 0)$; that is,

$$g(x \mid \theta) = \frac{\theta^\alpha}{\Gamma(\alpha)} x^{\alpha-1} e^{-\theta x} \quad \text{for} \quad x > 0 \,. \tag{4.1}$$

For this exponential family, θ is the natural parameter, $u(x) = -x$ and $I_X(\theta) = \alpha/\theta^2$.

Now suppose again that the selection set is $S = \{y : y > \tau\}$, where τ is a given positive number. Let $g_1(x)$ and $G_1(x)$ denote the p.d.f. and the corresponding d.f. of the gamma distribution given by (4.1) with $\theta = 1$, and let

$$h_1(x) = \frac{g_1(x)}{1 - G_1(x)} \quad \text{for} \quad x > 0 \tag{4.2}$$

denote the hazard-rate or failure-rate function for the distribution G_1. Since θ is a scale parameter in (4.1), then $g(x \mid \theta) = \theta g_1(\theta x)$ and $G(x \mid \theta) = G_1(\theta x)$. It is easy to verify that

$$\frac{d^2 \log s(\theta)}{d\theta^2} = -\tau^2 h_1'(\tau\theta) \,, \tag{4.3}$$

where $h_1'(x) = dh_1(x)/dx$. Therefore, by (2.4), $I_Y(\theta) \geq I_X(\theta)$ at a given value of θ if and only if $h_1'(\tau\theta) \leq 0$. It is known (Barlow and Proschan, 1975, Chapter 3) that a gamma distribution has an increasing hazard-rate function if $\alpha > 1$ and a decreasing hazard-rate function if $0 < \alpha < 1$. Therefore, if $\alpha > 1$, then $\mathcal{E}_X \succeq_F \mathcal{E}_Y$, whereas if $\alpha < 1$, then $\mathcal{E}_Y \succeq_F \mathcal{E}_X$. If $\alpha = 1$, the gamma distribution reduces to the exponential distribution for which the hazard-rate function is constant and both of the relations $\mathcal{E}_X \succeq_F \mathcal{E}_Y$ and $\mathcal{E}_Y \succeq_F \mathcal{E}_X$ hold. In fact, if $\alpha = 1$ then both X and $Y - \tau$ have an exponential distribution with parameter θ, so the experiments $\mathcal{E}_X$ and $\mathcal{E}_Y$ are equivalent. □

5 Truncated discrete distributions

Two truncation models that have been widely treated in the statistical literature are the truncated binomial and Poisson distributions in which the zero class is missing (see, for example, David and Johnson, 1952; Irwin, 1959; Cohen, 1960; Dahiya and Gross, 1973; Sanathanan, 1977; Blumenthal and Sanathanan, 1980; and Blumenthal, 1981). These distributions are called *zero-truncated* (or, sometimes, *positive* or *decapitated*) binomial and Poisson distributions (see, e.g., Patil, Boswell, Joshi, and Ratnaparkhi, 1984). In this section we will derive the Fisher information in zero-truncated binomial, Poisson, and negative binomial distributions by fundamental calculations (that is, brute force) in order to motivate our discovery of the CV property (1.4) introduced in Section 1.

Example 5.

Suppose that X has a binomial distribution with parameters n and θ, as defined by (1.5), where θ is unknown ($0 < \theta < 1$). It is well known that

$$I_X(\theta) = \frac{n}{\theta\bar{\theta}}, \tag{5.1}$$

where $\bar{\theta} = 1 - \theta$.

The zero-truncated random variable Y has the following discrete p.f.:

$$f(y \mid \theta) = \binom{n}{y} \frac{\theta^y \bar{\theta}^{n-y}}{s(\theta)} \quad \text{for} \quad y = 1, \ldots, n, \tag{5.2}$$

where

$$s(\theta) = 1 - \bar{\theta}^n. \tag{5.3}$$

We will calculate $I_Y(\theta)$ directly from (5.2) and the basic definition (2.1) of Fisher information. It is found that

$$\frac{\partial^2 \log f(y \mid \theta)}{\partial \theta^2} = \frac{y}{\theta^2} + \frac{n-y}{\bar{\theta}^2} - \frac{n\bar{\theta}^{n-2}(n-1+\bar{\theta}^n)}{(1-\bar{\theta}^n)^2}. \tag{5.4}$$

Since

$$E(Y \mid \theta) = \frac{n\theta}{1-\bar{\theta}^n}, \tag{5.5}$$

we obtain after much algebra,

$$I_Y(\theta) = I_X(\theta)\frac{1 - n\bar{\theta}^{n-1} + (n-1)\bar{\theta}^n}{(1-\bar{\theta}^n)^2} \tag{5.6}$$

or, equivalently,

$$\frac{I_Y(\theta)}{I_X(\theta)} = \frac{\Pr(X \geq 2 \mid \theta)}{[\Pr(X \geq 1 \mid \theta)]^2}. \tag{5.7}$$

We know that the right-hand side of (5.7) is less than 1 because it can be rewritten as

$$\frac{\Pr(X \geq 2 \mid X \geq 1, \theta)}{\Pr(X \geq 1 \mid X \geq 0, \theta)} \tag{5.8}$$

and, as is well known, the binomial distribution has the property that in reliability theory is called "new better than used" (see, e.g., Barlow and Proschan, 1975). Hence, $\mathcal{E}_X \succeq_F \mathcal{E}_Y$. □

We will now carry out a similar calculation for the zero-truncated Poisson distribution.

Example 6.

If X has a Poisson distribution with unknown mean θ ($\theta > 0$), as defined by (1.6), then

$$I_X(\theta) = \frac{1}{\theta} . \tag{5.9}$$

The p.f. of the zero-truncated random variable Y is

$$f(y \mid \theta) = \frac{e^{\theta}\theta^{y}}{y!s(\theta)} \quad \text{for} \quad y = 1, 2, \dots , \tag{5.10}$$

where

$$s(\theta) = 1 - e^{-\theta} . \tag{5.11}$$

It is found that

$$\frac{-\partial^2 \log f(y \mid \theta)}{\partial\theta^2} = \frac{y}{\theta^2} - \frac{e^{-\theta}}{(1 - e^{-\theta})^2} . \tag{5.12}$$

Since

$$E(Y) = \frac{\theta}{1 - e^{-\theta}} , \tag{5.13}$$

we obtain

$$I_Y(\theta) = I_X(\theta)\frac{1 - e^{-\theta} - \theta e^{-\theta}}{(1 - e^{-\theta})^2} . \tag{5.14}$$

It is a striking fact that (5.14) can again be expressed in the form (5.7). Since the Poisson distribution is also known to have the property called "new better than used," we can again conclude that $\mathcal{E}_X \succeq_F \mathcal{E}_Y$. □

Finally, we turn to the zero-truncated negative binomial distribution.

Example 7.

If X has a negative binomial distribution with parameters r and θ, as defined by (1.7), where θ is unknown ($0 < \theta < 1$), then

$$I_X(\theta) = \frac{r}{\theta^2\bar{\theta}} \tag{5.15}$$

and it can again be shown by straightforward calculations that (5.7) holds for the zero-truncated random variable Y. In this example, however, as noted by Patil and Taillie (1987),

$$I_X(\theta) \lesseqqgtr I_Y(\theta) \quad \text{for all } \theta \ (0 < \theta < 1) \tag{5.16}$$

if and only if

$$r \lesseqgtr 1\,. \tag{5.17}$$

Thus, if $r < 1$, then $\mathcal{E}_Y \succeq_F \mathcal{E}_X$; if $r > 1$, then $\mathcal{E}_X \succeq_F \mathcal{E}_Y$; and if $r = 1$, then $I_X(\theta) \equiv I_Y(\theta)$. In this case, both X and $Y - 1$ have a geometric distribution with parameter θ. □

We will now look more closely at just why the relation (5.7) is satisfied in all three of the examples presented in this section, and we will develop the connection between (5.7) and the CV property.

6 The CV property

Suppose that a random variable X has a discrete distribution on the non-negative integers with mean μ and variance σ^2, and $p_0 = \Pr(X = 0) < 1$. If the random variable Y is the zero-truncated version of X, then

$$E(Y^k) = \frac{E(X^k)}{1 - p_0} \quad \text{for} \quad k = 1, 2\,. \tag{6.1}$$

It follows from an easy calculation that (Patil and Taillie, 1987)

$$\mathrm{Var}(Y) = \sigma^2 \frac{1 - p_0 - (\mu^2/\sigma^2)p_0}{(1 - p_0)^2}\,. \tag{6.2}$$

Also, if the p.f. of X can be expressed in the form (2.5) with $u(x) = x$, then it follows from (2.3), (2.4), and (2.6) that

$$\frac{I_Y(\theta)}{I_X(\theta)} = \frac{\mathrm{Var}(Y \mid \theta)}{\mathrm{Var}(X \mid \theta)}\,, \tag{6.3}$$

where Y is any truncated version of X restricted to a fixed selection set S. In particular, if the distribution of X is binomial, Poisson, or negative binomial, as in Examples 5, 6, and 7, then (6.3) holds. Furthermore, it was shown in Section 5 that if Y is the zero-truncated version of X, then for each of these three families, the relation (5.7) holds. The CV property (1.4) now follows from (5.7), (6.3), and (6.2) for every distribution belonging to any one of these "Big 3" families of discrete distributions: binomial, Poisson, and negative binomial.

Of course, once the CV property (1.4) has been put forward, it is trivial to verify directly from (1.5)–(1.7) that every distribution in the "Big 3" satisfies the property. Our purpose in giving the preceding discussion has been to motivate both our interest in the property and the method by which we discovered it. It should be emphasized that although we developed the CV property by studying Fisher information in parametric families of distributions, the property itself pertains to an individual discrete distribution on the non-negative integers. As we will show, there are other discrete distributions outside the "Big 3" families that satisfy this property. However, we do not know any other "standard" distributions satisfying it. For example, it is not satisfied by discrete uniform distributions, hypergeometric distributions, or beta-binomial mixtures, except in the trivial situation in which the distribution is restricted to the two values $X = 0$ and $X = 1$. Characterization problems for families of distributions satisfying the CV property are discussed in Goel and DeGroot (1988).

In the remainder of this paper we will present a few selected results pertaining to the question of which distributions satisfy the CV property. Throughout this discussion we will use the notation given in (1.3), and will restrict ourselves to discrete distributions on the non-negative integers for which both μ and σ^2 are finite and $p_0 p_1 > 0$. This last inequality, of course, implies that $\mu > 0$ and $\sigma^2 > 0$. The *support* of a discrete distribution of this type is the set of non-negative integers i such that $p_i > 0$.

We begin by noting that every distribution with support $\{0, 1\}$ satisfies the CV property. Of course, every such distribution is a binomial distribution with $n = 1$, that is, a Bernoulli distribution. Now consider distributions with support $\{0, 1, 2\}$.

Theorem 6.1 *A distribution with support* $\{0, 1, 2\}$ *satisfies the CV property if and only if it is binomial.*

Proof Since the support is $\{0, 1, 2\}$, it follows that $\mu = p_1 + 2p_2$ and

$$\begin{aligned} \sigma^2 &= p_1 + 4p_2 - \mu^2 \\ &= p_1(1 - p_1) + 4p_0 p_2. \end{aligned} \tag{6.4}$$

Thus, the CV property is satisfied if and only if

$$\frac{p_0}{p_1} = \frac{p_1(1 - p_1) + 4p_0 p_2}{(p_1 + 2p_2)^2} \tag{6.5}$$

or equivalently

$$p_0(p_1^2 + 4p_1 p_2 + 4p_1^2) = p_1^2(1 - p_1) + 4p_0 p_1 p_2 \,. \tag{6.6}$$

If we write the term $1 - p_1$ on the right-hand side of (6.6) as $p_0 + p_2$, and then cancel $p_0 p_1^2 + 4p_0 p_1 p_2$ from each side of (6.6), we find that the CV property is satisfied if and only if

$$p_1^2 = 4p_0 p_2 \,. \tag{6.7}$$

However, (6.7) is precisely the condition that a distribution on the set $\{0, 1, 2\}$ is a binomial distribution with $n = 2$. □

The next result shows that the distribution of a sum of i.i.d. random variables will satisfy the CV property if and only if the distribution of each random variable in the sum satisfies it.

Theorem 6.2 *Suppose that the random variables $X_1, \ldots, X_m$ are i.i.d., and let $X = X_1 + \cdots + X_m$. Then the distribution of X satisfies the CV property if and only if the distribution of X_1 satisfies it.*

Proof We will reserve the notation p_0, p_1, μ, and σ^2 for the distribution of X_1. Then

$$\begin{aligned} &\Pr(X = 0) = p_0^m , \quad \Pr(X = 1) = m p_0^{m-1} p_1 , \\ &E(X) = m\mu , \qquad\quad \mathrm{Var}(X) = m\sigma^2 . \end{aligned} \tag{6.8}$$

The theorem follows immediately from (6.8) and the definition (1.4) of the CV property. □

We will conclude this paper by exhibiting a class of distributions outside the "Big 3" families that satisfy the CV property.

Theorem 6.3 *A distribution supported on the three points $\{0, 1, n\}$ satisfies the CV property if and only if there exists a number θ $(0 < \theta < 1)$ such that*

$$p_0 = \theta^2 , \tag{6.9}$$

$$p_1 = \frac{n}{n-1}\, \theta\bar{\theta} , \tag{6.10}$$

$$p_n = \bar{\theta}^2 + \frac{n-2}{n-1}\, \theta\bar{\theta} . \tag{6.11}$$

Remark. Before we give the proof of Theorem 6.3, we shall describe one possible stochastic mechanism that leads to a distribution of the form specified in the theorem. Suppose that X_1 is a Bernoulli random variable with

$$\Pr(X_1 = 0) = \theta, \quad \Pr(X_1 = 1) = \bar{\theta} , \tag{6.12}$$

and that X_2 is a discrete random variable whose conditional distribution given X_1 is as follows:

$$\Pr(X_2 = 0 \mid X_1 = 0) = \theta , \qquad \Pr(X_2 = 1 \mid X_1 = 0) = \bar{\theta},$$

$$\Pr(X_2 = 0 \mid X_1 = 1) = \frac{\theta}{n-1},$$

$$\Pr(X_2 = n-1 \mid X_1 = 1) = 1 - \frac{\theta}{n-1}. \tag{6.13}$$

Then it is easily verified that the distribution of $X_1 + X_2$ is specified by (6.9)–(6.11). It should be noted that X_1 and X_2 are independent only for $n = 2$. In this case they are i.i.d. Bernoulli random variables, and the distribution in Theorem 6.3 is simply a binomial distribution.

Proof of Theorem 6.3 For a distribution with support $\{0, 1, n\}$,

$$E(X) = p_1 + np_n \quad \text{and} \quad E(X^2) = p_1 + n^2 p_n \,. \tag{6.14}$$

We now reparametrize the problem in terms of new parameters λ and θ defined by the relations

$$p_1 = \frac{n}{n-1}\lambda \quad \text{and} \quad p_0 = \theta - \lambda \,. \tag{6.15}$$

If we let $p_n = 1 - p_0 - p_1$ in (6.14), and replace p_0 and p_1 by the expressions given in (6.15), then we obtain

$$E(X) = n\bar{\theta} \quad \text{and} \quad E(X^2) = n^2(\bar{\theta} - \frac{\lambda}{n}) \,. \tag{6.16}$$

Hence, the CV property will be satisfied if and only if

$$\frac{\theta - \lambda}{n\lambda/(n-1)} = \frac{\bar{\theta} - (\lambda/n)}{\bar{\theta}^2} - 1 \tag{6.17}$$

or, equivalently, after some algebra, if and only if

$$(\lambda - \theta\bar{\theta})\left(\frac{\lambda}{n-1} - \bar{\theta}\right) = 0 \,. \tag{6.18}$$

The relation (6.18) is satisfied if and only if $\lambda = \theta\bar{\theta}$ or $\lambda = (n-1)\,\bar{\theta}$.

Suppose first that $\lambda = \theta\bar{\theta}$. Then in order to have both $0 < p_0 < 1$ and $0 < p_1 < 1$, it is necessary and sufficient that $0 < \theta < 1$. In this case we obtain the distribution specified by (6.9) –(6.11).

Suppose next that $\lambda = (n-1)\bar{\theta}$. In this case $p_0 + p_1 = 1$ and $p_n = 0$, so the support of this distribution has less than three points. The theorem now follows. □

It follows from Theorem 6.2 that the sum of any fixed number of i.i.d. random variables, each having the three-point distribution specified in Theorem 6.3, will also satisfy the CV property. Thus, we can identify distributions satisfying the CV property with more than three points in their support. If $n = 2$ in Theorem 6.3, then we merely generate binomial distributions through this process; but if $n \geq 3$, then we obtain some previously unidentified distributions. However, none of these new distributions can have a support of the form $\{0, 1, 2, \ldots, r-1, r\}$ for some integer $r \geq 3$ because the point $r - 1$ must have probability 0.

Acknowledgments: This research was supported in part by the National Science Foundation under grant DMS-8701770 and the Air Force Office of Scientific Research under contract AFOSR-84-0164. We are indebted to William F. Eddy and Mark J. Schervish for many helpful discussions.

REFERENCES

Barlow, R.E., and Proschan, F. (1975). *Statistical Theory of Reliability and Life Testing.* New York: Holt, Rinehart and Winston.

Bayarri, M.J., and DeGroot, M.H. (1987a). Bayesian analysis of selection models. *The Statistician* **36**, 137-146.

Bayarri, M.J., and DeGroot, M.H. (1987b). Information in selection models. *Probability and Bayesian Statistics.* (ed. by R. Viertl). New York: Plenum Press, 39–51.

Bayarri, M.J., and DeGroot, M.H. (1987c). Selection models and selection mechanisms. Technical Report No. 410, Department of Statistics, Carnegie-Mellon University.

Blackwell, D. (1951). Comparison of experiments. *Proceedings of the Second Berkeley Symposium on Mathematical Statistics and Probability*, 93-102. Berkeley, California: University of California Press.

Blackwell, D. (1953). Equivalent comparison of experiments. *Annals of Mathematical Statistics* **24**, 265-272.

Blumenthal, S. (1981). A survey of estimating distributional parameters and sample sizes from truncated samples. In *Statistical Distributions in Scientific Work* (C. Taillie, G.P. Patil, and B. Baldessari, eds.), Vol. 5, 75-86. Dordrecht: Reidel.

Blumenthal, S., and Sanathanan, L.P. (1980). Estimation with truncated inverse binomial sampling. *Communications in Statistics* A9, 997-1017.

Cohen, A.G. (1960). Estimation in truncated Poisson distributions when zeros and some ones are missing. *Journal of the American Statistical Association* **55**, 342-348.

Dahiya, R.G., and Gross, A.J. (1973). Estimating the zero class from a truncated Poisson sample *Journal of the American Statistical Association* **68**, 731-733.

David, F.N. and Johnson, N.L. (1952). The truncated Poisson. *Biometrics* **8**, 275-285.

Fang, K.T., and He, S.D. (1984). The problem of selecting a specified number of representative points from a normal population (in Chinese). *Acta Mathematicae Applicatae Sinica* **7**, 293-307.

Goel, P.K., and DeGroot, M.H. (1979). Comparison of experiments and information measures. *Annals of Statistics* **7**, 1066-1077.

Goel, P.K., and DeGroot, M.H. (1988). On some characterizations of distributions with CV property (Abstract). *The Institute of Mathematical Statistics Bulletin* **17**, 231.

Gordon, R.D. (1941). Values of Mills' ratio of area to bounding ordinate and of the normal probability integral for large values of the argument. *Annals of Mathematical Statistics* **12**, 364-366.

Irwin, J.O. (1959). On the estimation of the mean of a Poisson distribution from a sample with the zero class missing. *Biometrics* **15**, 324-326.

Kullback, S. (1968). *Information Theory and Statistics.* New York: Dover.

Patil, G.P., Boswell, M.T., Joshi, S.W., and Ratnaparkhi, M.V. (1984). *Dictionary and Classified Bibliography of Statistical Distributions in Scientific Work, Vol. 1: Discrete Models.* Fairland, Maryland: International Co-operative Publishing House.

Patil, G.P., and Taillie, C. (1987). Weighted distributions and the effects of weight functions on Fisher information. Technical Report, Center for Statistical Ecology and Environmental Statistics, Department of Statistics, Pennsylvania State University.

Sanathanan, L.P. (1977). Estimating the size of a truncated sample. *Journal of the American Statistical Association* **72**, 669-672.

Schneider, H. (1986). *Truncated and Censored Samples from Normal Populations.* New York: Marcel Dekker.

27

Asymptotic Error Bounds for Power Approximations to Multinomial Tests of Fit

F.C. Drost[1]
W.C.M. Kallenberg[2]
D.S. Moore[3]
J. Oosterhoff[4]

ABSTRACT The Cressie-Read (1984) class of goodness-of-fit tests is considered. Asymptotic error bounds are derived for two new non-local approximations, the classical noncentral χ^2 approximation, a moment-corrected version of it and normal approximations to the power of these tests.

1 Introduction

Consider the following goodness-of-fit problem. Let $X_1, \ldots, X_n$ be i.i.d. observations with distribution function F^X and suppose the hypothesis H_0 is to be tested that F^X equals a given F. If the range of the X_j is partitioned into k cells $C_1, \ldots, C_k$ and $N_i = \#\{ j : X_j \in C_i, j \leq n \}$, $i = 1, \ldots, k$, multinomial tests of fit can be based on the cell counts N_i and the null probabilities $p_i = P_F(X_j \in C_i)$. In the case of grouped observations the cells $C_1, \ldots, C_k$ may already be given at the start.

Cressie and Read (1984) have considered the class of statistics

$$R^\lambda(N/n : p) = 2nI^\lambda(N/n : p)$$

where $N = (N_1, \ldots, N_k)'$, $p = (p_1, \ldots, p_k)'$ and I^λ is a directed divergence

[1]Econometric Institute, University of Tilburg, The Netherlands.
[2]Dept. of Applied Mathematics, University of Twente, Enschede, The Netherlands.
[3]Dept. of Statistics, Purdue University, West Lafayette, IN, U.S.A.
[4]Dept. of Applied Mathematics and Computer Science, Free University, Amsterdam, The Netherlands.

between probability vectors p and q defined by

$$I^{\lambda}(q:p) = \{\lambda(\lambda+1)\}^{-1} \sum_{i=1}^{k} q_i\{(q_i/p_i)^{\lambda} - 1\}$$

for real $\lambda \neq 0, -1$. One may define I^{λ} by continuity when $\lambda = 0$ or -1.

The class contains the well known statistics

$$X^2 = \sum_{i=1}^{k} (N_i - np_i)^2/(np_i)$$

$$G^2 = 2\sum_{i=1}^{k} N_i \log\left(N_i/(np_i)\right)$$

$$FT^2 = 4\sum_{i=1}^{k} \left\{N_i^{1/2} - (np_i)^{1/2}\right\}^2$$

$$X_m^2 = \sum_{i=1}^{k} (N_i - np_i)^2/N_i$$

for $\lambda = 1$ (Pearson statistic), $\lambda = 0$ (likelihood ratio statistic), $\lambda = -\frac{1}{2}$ (Freeman-Tukey statistic) and $\lambda = -2$ (Neyman's modified chi-square statistic), respectively. Cressie and Read (1984) recommend the use of R^{λ} with $\lambda = 2/3$. See Moore (1984) for an interpretation of I^{λ} as a measure of lack of fit.

Under H_0 the asymptotic distribution of the statistics R^{λ} is χ^2_{k-1} (chi-square on $k-1$ degrees of freedom), although the approximation for moderate sample sizes is not satisfactory outside the interval $(1/3, 3/2)$, see Larntz (1978), Cressie and Read (1984) and Read (1984 a,b).

Here we are concerned with the power of the tests in the Cressie-Read class under alternative distributions G_n of the X_j's. Let $\pi_{in} = P_{G_n}(X_j \in C_i)$, $i = 1, \ldots, k$ and $\pi_n = (\pi_{1n}, \ldots, \pi_{kn})'$. Under contiguous alternatives where $n^{1/2}(\pi_{in} - p_i)$ is bounded for each i as $n \to \infty$, R^{λ} is asymptotically distributed as noncentral $\chi^2_{k-1}(\delta_n)$ with noncentrality parameter

$$\delta_n = n\sum_{i=1}^{k} (\pi_{in} - p_i)^2/p_i,$$

see Patnaik (1949) and Cressie and Read (1984). More precise expansions have been given by many authors, see e.g. Peers (1971), Hayakawa (1975) and Chibisov and van Zwet (1984). However, these are strictly local results based on local expansions of π_{in} around p_i. To obtain good approximations to the power for small or moderate sample sizes, an approach which avoids such local expansions seems to be more promising, since less approximations are involved.

We therefore consider global expansions of R^λ. Exploiting the limiting normality of the multinomial distribution, the limit distributions of the leading terms of the expansion of R^λ can be derived. Such distributions may serve as approximations to the distribution of R^λ under quite arbitrary sequences of alternatives G_n. Two approximations of this nature are studied in this paper.

To measure the precision of power approximations, asymptotic error bounds are a natural yard-stick. Our main theorem furnishes asymptotic error bounds both for the new global approximations and for the more classical noncentral χ^2 and normal approximations (moment-corrected or not), covering the whole range from contiguous to fixed alternatives. An asymptotic comparison of the accuracy of the approximations is thus made possible.

In Section 2 the new power approximations are derived and our main theorem is stated. Proofs are given in Section 3.

In a companion paper Drost et al. (1989) the practical implications of the results are further explored and a numerical study is undertaken to compare the various approximations for small sample sizes. The conclusions are summarized at the end of Section 2.

2 Approximations and Error Bounds

Let $p_1, \ldots, p_k$ be fixed, $\min_i p_i > 0$, and assume that the π_{in} are bounded away from zero. The distribution of N under G_n will often be denoted by P_{π_n}. Put

$$\begin{aligned} Y_{in} &= (N_i - n\pi_{in})/(n\pi_{in})^{1/2}, \qquad i = 1, \ldots, k \\ r_{in} &= \pi_{in}/p_i \end{aligned}$$

and consider the Taylor expansion of R^λ under G_n, for $\lambda \neq 0, -1$,

$$\begin{aligned} R^\lambda(N/n : p) &= 2\{\lambda(\lambda+1)\}^{-1}\left[\sum_{i=1}^{k} r_{in}^\lambda n\pi_{in}\left\{1 + (n\pi_{in})^{-1/2}Y_{in}\right\}^{\lambda+1} - n\right] \\ &= 2\{\lambda(1+\lambda)\}^{-1}\left[\sum_{i=1}^{k} r_{in}^\lambda \{n\pi_{in} + (\lambda+1)(n\pi_{in})^{1/2}Y_{in} \right. \\ &\qquad + \frac{1}{2}(\lambda+1)\lambda Y_{in}^2 \\ &\qquad \left. + \frac{1}{6}(\lambda+1)\lambda(\lambda-1)(n\pi_{in})^{-1/2}Y_{in}^3 + \cdots\} - n\right] \\ &= A^\lambda(Y_n) + O_P(n^{-1/2}) \end{aligned} \tag{2.1}$$

where

$$A^\lambda(Y_n) = \sum_{i=1}^{k} r_{in}^\lambda \left\{Y_{in} + \lambda^{-1}(n\pi_{in})^{1/2}(1 - r_{in}^{-\lambda})\right\}^2 + 2nI^\lambda(\pi_n : p) - n\lambda^{-2}\sum_{i=1}^{k} \pi_{in}(1 - r_{in}^{-\lambda})^2 r_{in}^\lambda$$

$$= B^\lambda(Y_n) + \sum_{i=1}^{k}(r_{in}^\lambda - \bar{r}_{n\lambda})Y_{in}^2 \tag{2.2}$$

with $\bar{r}_{n\lambda} = \sum(1 - \pi_{in})r_{in}^\lambda/(k-1)$ and

$$B^\lambda(Y_n) = \bar{r}_{n\lambda}\sum_{i=1}^{k}\left\{Y_{in} + (\lambda\bar{r}_{n\lambda})^{-1}(n\pi_{in})^{1/2}(r_{in}^\lambda - 1)\right\}^2 + 2nI^\lambda(\pi_n : p) - n\lambda^{-2}\bar{r}_{n\lambda}^{-1}\sum_{i=1}^{k}\pi_{in}(r_{in}^\lambda - 1)^2. \tag{2.3}$$

The leading part $A^\lambda(Y_n)$ of the expansion is a first candidate for approximating R^λ. The statistic $B^\lambda(Y_n)$ is a useful modification of $A^\lambda(Y_n)$ in that its quadratic terms have equal coefficients. Even simpler is the linear part of R^λ (and A^λ) given by

$$L^\lambda(Y_n) = 2n^{1/2}\lambda^{-1}\sum_{i=1}^{k}\pi_{in}^{1/2}(r_{in}^\lambda - 1)Y_{in} + 2nI^\lambda(\pi_n : p). \tag{2.4}$$

By taking appropriate limits in (2.1)–(2.4), the expansion extends to $\lambda = 0$ or -1. Note that $A^\lambda = R^\lambda$ if $\lambda = 1$ and that $A^\lambda = B^\lambda$ if $\lambda = 0$. To turn A^λ, B^λ and L^λ into approximations to the distribution of R^λ, we rely on the asymptotic normality of the multinomial distribution. Let U_n have a multivariate normal distribution

$$U_n = (U_{1n}, \ldots, U_{kn})' \sim N_k(0, I - \pi_n^{1/2}\pi_n^{1/2'})$$

where $\pi_n^{1/2} = (\pi_{1n}^{1/2}, \ldots, \pi_{kn}^{1/2})'$. Since Y_n is asymptotically distributed as U_n under G_n, replace Y_n by U_n in A^λ, B^λ and L^λ and consider the approximations $A^\lambda(U_n)$, $B^\lambda(U_n)$ and $L^\lambda(U_n)$.

To derive their distribution, we employ the following notation. Let Q_n be the diagonal matrix $Q_n = \operatorname{diag}(r_{1n}^{\lambda/2}, \ldots, r_{kn}^{\lambda/2})$, let μ_n be the vector $\mu_n = n^{1/2}\lambda^{-1}\left(\pi_{1n}^{1/2}(1 - r_{1n}^{-\lambda}), \ldots, \pi_{kn}^{1/2}(1 - r_{kn}^{-\lambda})\right)'$ and let $\theta_{1n}, \ldots, \theta_{kn}$ be the eigenvalues and S_n be the $k \times k$ orthonormal matrix of eigenvectors of the matrix $Q_n(I - \pi_n^{1/2}\pi_n^{1/2'})Q_n$. Put $T_n = \operatorname{diag}(\theta_{1n}, \ldots, \theta_{kn})$ and $\omega_n = S_n'Q_n\mu_n$. Here and in the sequel $Z_1, Z_2, \ldots$ are i.i.d. standard normal random variables.

Apply the orthogonal transformation S'_n to $A^\lambda(U_n)$. Then

$$\begin{aligned}A^\lambda(U_n) &= \|Q_n(U_n+\mu_n)\|^2 + 2nI^\lambda(\pi_n:p) - \|Q_n\mu_n\|^2\\ &= \|S'_nQ_n(U_n+\mu_n)\|^2 + 2nI^\lambda(\pi_n:p) - \|S'_nQ_n\mu_n\|^2\\ &\sim \sum_{\theta_{in}\neq 0}\theta_{in}(Z_i+\omega_{in}/\theta_{in}^{1/2})^2 + \sum_{\theta_{in}=0}\omega_{in}^2 + 2nI^\lambda(\pi_n:p) - \sum_{i=1}^k \omega_{in}^2\end{aligned}$$

since $S'_nQ_n(U_n+\mu_n) \sim N_k(\omega_n, T_n)$. As one of the θ_{in} vanishes, assume $\theta_{kn} = 0$. It follows that $A^\lambda(U_n)$ is distributed as a linear combination of noncentral χ^2's:

$$\sum_{i=1}^{k-1}\theta_{in}(Z_i+\omega_{in}/\theta_{in}^{1/2})^2 + 2nI^\lambda(\pi_n:p) - \sum_{i=1}^{k-1}\omega_{in}^2. \tag{2.5}$$

A similar transformation shows that $B^\lambda(U_n)$ is distributed as

$$\bar{r}_{n\lambda}\chi^2_{k-1}(\delta_n^{(\lambda)}) + \xi_n^{(\lambda)} \tag{2.6}$$

where

$$\begin{aligned}\delta_n^{(\lambda)} &= n(\lambda\bar{r}_{n\lambda})^{-2}\left[\sum_{i=1}^k \pi_{in}(r_{in}^\lambda-1)^2 - \left\{\sum_{i=1}^k \pi_{in}(r_{in}^\lambda - 1)\right\}^2\right]\\ \xi_n^{(\lambda)} &= 2nI^\lambda(\pi_n:p) - \bar{r}_{n\lambda}\delta_n^{(\lambda)}.\end{aligned} \tag{2.7}$$

In a sense (2.6) is the "best noncentral χ^2 approximation" to the distribution of R^λ, since $\bar{r}_{n\lambda}$ is chosen to satisfy $E_{\pi_n}\left(B^\lambda(U_n) - A^\lambda(U_n)\right) = 0$. For $\lambda = 0$ it coincides with $A^\lambda(U_n)$; note that in this case $\bar{r}_{n0} = 1$ and

$$\begin{aligned}\delta_n^{(0)} &= n\sum_{i=1}^k \pi_{in}\log^2 r_{in} - n\left(\sum_{i=1}^k \pi_{in}\log r_{in}\right)^2\\ \xi_n^{(0)} &= 2n\sum_{i=1}^k \pi_{in}\log r_{in} - \delta_n^{(0)}.\end{aligned}$$

A local expansion of π_{in} around p_i yields that locally, as $\max_i|\pi_{in}-p_i| \to 0$,

$$\begin{aligned}\bar{r}_{n\lambda} &= 1 + O(\max_i|\pi_{in}-p_i|)\\ \delta_n^{(\lambda)} &= 2nI^1(\pi_n:p) + O\left(n\max_i|\pi_{in}-p_i|^3\right)\\ \xi_n^{(\lambda)} &= O\left(n\max_i|\pi_{in}-p_i|^3\right).\end{aligned} \tag{2.8}$$

Neglecting the remainder terms, $B^\lambda(U_n)$ thus reduces locally to the classical $\chi^2_{k-1}(\delta_n)$ approximation with

$$\delta_n = 2nI^1(\pi_n:p) = n\sum(\pi_{in}-p_i)^2/p_i.$$

A competing approximation for $\lambda \neq 1$ is $\chi^2_{k-1}(\delta_{n\lambda})$ where $\delta_{n\lambda} = 2nI^\lambda(\pi_n : p)$.

Starting from the classical approximation $\chi^2(2nI^1)$, improved approximations can be obtained by the introduction of moment corrections to get the "right" first two moments of R^λ. Define

$$E^*_{\pi_n} R^\lambda = 2nI^\lambda(\pi_n : p) + \bar{r}_{n\lambda}(k-1) + n^{-1}(\lambda - 1)\left\{\frac{1}{3}\sum r^\lambda_{in}(1 - 3\pi_{in} + 2\pi^2_{in})/\pi_{in} + \frac{1}{4}(\lambda - 2)\sum r^\lambda_{in}(1-\pi_{in})^2/\pi_{in}\right\}$$

and

$$\mathrm{Var}^*_{\pi_n} R^\lambda = (4n\lambda^{-2} - 6 - 4\lambda^{-1})\left\{\sum \pi_{in} r^{2\lambda}_{in} - \left(\sum \pi_{in} r^\lambda_{in}\right)^2\right\} + 4\left\{\sum r^{2\lambda}_{in} - \sum r^\lambda_{in} \sum \pi_{in} r^\lambda_{in}\right\} + 2\left\{\sum (1-\pi_{in}) r^{2\lambda}_{in}\right\}.$$

The quantities $E^*_{\pi_n} R^\lambda$ and $\mathrm{Var}^*_{\pi_n} R^\lambda$ are approximations to $E_{\pi_n} R^\lambda$ and $\mathrm{Var}_{\pi_n} R^\lambda$, cf. Drost et al. (1989). The moment-corrected $\chi^2_{k-1}\left(2nI^1(\pi : p)\right)$ is of the form

$$M^\lambda = a^\lambda_n \chi^2_{k-1}\left(2nI^1(\pi : p)\right) + b^\lambda_n,$$

where

$$a^\lambda_n = \left\{\mathrm{Var}^*_{\pi_n} R^\lambda / \left(2k - 2 + 8nI^1(\pi_n : p)\right)\right\}^{1/2}$$

and

$$b^\lambda_n = E^*_{\pi_n} R^\lambda - a^\lambda_n \left(k - 1 + 2nI^1(\pi_n : p)\right).$$

The linear approximation $L^\lambda(U_n)$ is of course normally distributed with expectation $2nI^\lambda(\pi_n : p)$ and variance

$$4n\lambda^{-2}\left\{\sum \pi_{in}(r^\lambda_{in} - 1)^2 - \left(\sum \pi_{in}(r^\lambda_{in} - 1)\right)^2\right\}.$$

Since the non-linear terms of R^λ are of lower order of magnitude than the linear ones if $n^{1/2} \max_i |\pi_{in} - p_i| \to \infty$, this result expresses the asymptotic normality of R^λ under non-contiguous alternatives. The normal approximation is improved if the moments of R^λ (or of $A^\lambda(U_n)$) are employed, cf. Drost et al. (1989). This approximation is discussed in Broffitt and Randles (1977) for the case $\lambda = 1$.

We now state our main result on the approximation errors. For simplicity we write π for π_n and U for U_n.

Theorem 2.1 *Let $k \geq 3$ and $\lambda \in \Re$. Let $0 < \epsilon < 1/k$ and $\Pi_\epsilon = \{\pi \in \Re^k : \min_i \pi_i \geq \epsilon, \sum \pi_i = 1\}$. Let $\{s_n\}$, $s_n > 0$, be a nondecreasing sequence, let $\Pi(s_n) = \{\pi \in \Pi_\epsilon : \max_i |\pi_i - p_i| \leq s_n/n^{1/2}\}$ and let $\Pi^*(s_n) = \Pi_\epsilon \setminus \Pi(s_n)$.*

(i) As $n \to \infty$

$$\sup_{\pi \in \Pi_\epsilon} \sup_{c>0} \left| P_\pi \left(R^\lambda(N/n : p) > c \right) - P\left(A^\lambda(U) > c \right) \right| = O(n^{-1/2}).$$

(ii) If $s_n/n^{1/2} \to 0$, *then as* $n \to \infty$

$$\sup_{\pi \in \Pi(s_n)} \sup_{c>0} \left| P_\pi \left(R^\lambda(N/n : p) > c \right) - P\left(B^\lambda(U) > c \right) \right| = O(s_n n^{-1/2}).$$

(iii) If $s_n/n^{1/2} \to 0$ *and* $\delta_{n\lambda} = 2nI^\lambda(\pi : p)$, *then as* $n \to \infty$

$$\sup_{\pi \in \Pi(s_n)} \sup_{c>0} \left| P_\pi \left(R^\lambda(N/n : p) > c \right) - P\left(\chi^2_{k-1}(\delta_{n\lambda}) > c \right) \right| = O(s_n n^{-1/2}).$$

(iv) If $s_n/n^{1/4} \to 0$ *and* $\delta_n = 2nI^1(\pi : p)$, $\lambda \neq 1$, *then as* $n \to \infty$

$$\sup_{\pi \in \Pi(s_n)} \sup_{c>0} \left| P_\pi \left(R^\lambda(N/n : p) > c \right) - P\left(\chi^2_{k-1}(\delta_n) > c \right) \right| = O(s_n^2 n^{-1/2}).$$

(v) If $s_n/n^{1/2} \to 0$, *then as* $n \to \infty$

$$\sup_{\pi \in \Pi(s_n)} \sup_{c>0} \left| P_\pi \left(R^\lambda(N/n : p) > c \right) - P(M^\lambda > c) \right| = O(s_n n^{-1/2}).$$

(vi) If $s_n \to \infty$ *and* $s_n/n^{1/2} < 1$, *then as* $n \to \infty$

$$\sup_{\pi \in \Pi^*(s_n)} \sup_{c>0} \left| P_\pi \left(R^\lambda(N/n : p) > c \right) - P\left(L^\lambda(U) > c \right) \right| = O(s_n^{-1}).$$

The error bounds in (iv) and (vi), when larger than $O(n^{-1/2})$, *are sharp.*

Remark 2.2 *The theorem also holds for* $k = 2$ *if* c *is restricted to* $\gamma < c < \infty$ *for any fixed* $\gamma > 0$, *i.e. if the significance level is bounded away from one.*

Remark 2.3 *The bound in (ii) remains valid if we further simplify* $B^\lambda(U)$ *by taking* $\bar{r}_\lambda = 1$ *everywhere.*

Remark 2.4 *The error bound in (vi) also holds for normal approximations based on the moments of* $A^\lambda(U)$ *and remains sharp.*

Remark 2.5 *We have no proof that the error bounds in (i)–(iii) are sharp.*

According to the theorem the error of the A^λ approximation (2.5) is at most $Cn^{-1/2}$ for *all* alternatives $\pi_n \in \Pi_\epsilon$. Hence the A^λ approximation is satisfactory both from a local and a non-local point of view.

The B^λ, $\chi^2_{k-1}(\delta_n)$ and χ^2_{k-1} $(\delta_{n\lambda})$ approximations have the same error bound $Cn^{-1/2}$ as A^λ for contiguous alternatives. This bound is familiar from the work of Hayakawa (1975, 1977) and others on power expansions of likelihood ratio tests under contiguous alternatives. But for more distant alternatives the latter three approximations are less accurate than A^λ. The maximum error of $\chi^2_{k-1}(\delta_n)$ increases much faster as alternatives move away than the maximum error of B^λ and $\chi^2_{k-1}(\delta_{n\lambda})$. Theory thus suggests not to employ the traditional $\chi^2_{k-1}(\delta_n)$ approximation at all for $\lambda \neq 1$. Although B^λ follows the structure of R^λ more closely than $\chi^2_{k-1}(\delta_{n\lambda})$, this is not reflected in the asymptotic error bounds. However, $B^\lambda = A^\lambda$ when $\lambda = 0$, implying an error bound $O(n^{-1/2})$ for B^λ in that case, while $\chi^2_{k-1}(\delta_{n\lambda})$ still has an error bound $O(s_n n^{-1/2})$. The moment-corrected $\chi^2_{k-1}(\delta_n)$ with $\delta_n = 2nI^1(\pi_n : p)$ has the same error bound as B^λ (and $\chi^2_{k-1}(2nI^\lambda)$). Hence this moment-correction seems to recover some of the precision that has got lost by local approximation. Since B^λ coincides with the best approximation A^λ for $\lambda = 0$, B^λ still has an edge on M^λ.

The normal approximation shows different behavior. It is accurate for fixed alternatives (cf. Theorem 3 in Hayakawa (1977)), but breaks gradually down for more local alternatives. Therefore Broffitt and Randles (1977) only recommended normal approximations to estimate large powers. Of course the situation changes if $k = k(n) \to \infty$ as $n \to \infty$ (see Morris (1975)), but here we only consider fixed k.

Based on the asymptotic error bounds of Theorem 2.1 and the numerical study in Drost et al. (1989) our recommendations on the use of power approximations are as follows:

(i) Do not use the $\chi^2_{k-1}(2nI^1)$ approximation for other R^λ tests than X^2.

(ii) For practical purposes B^λ is a good and simple approximation for $-1 < \lambda < 3$ and $n > 20$. The approximation M^λ is a possible alternative.

(iii) For accurate work the A^λ approximation (or exact computation) is best.

(iv) For quick and dirty work the $\chi^2_{k-1}(2nI^\lambda)$ approximation is adequate for $-1/2 < \lambda < 2$ and $n > 50$.

3 Proofs

Before proving Theorem 2.1 we derive some preliminary results.

Let $Z_1, Z_2, \ldots$ be i.i.d. standard normal variables with cdf Φ and density ϕ. Repeatedly we use the order relation

$$P(|Z_1| > \log n) = o(n^{-1/2}) \qquad \text{as } n \to \infty, \tag{3.1}$$

which continues to hold if Z_1 is replaced by a standardized binomial $Bin(n,p)$ variable, uniformly in p bounded away from 0 and 1. We begin with a crucial lemma, which is in the same spirit as Theorem 1 in Cox and Reid (1987). Although many constants and sets in the sequel depend on n, subscripts n are often suppressed to simplify notation. The symbols O and o refer to $n \to \infty$.

Lemma 3.1 *Let a_i, b_i $(i = 1, \ldots, m)$ and c be real numbers, $m \geq 2$, let $\mathcal{P}_i(\cdot)$, $i = 1, \ldots, m$ be polynomials of fixed degree $q \geq 0$ and let $a_0 > 0$ and $d_0 > 0$ be fixed. Uniformly for $a_i > a_0$, $b_i \in \Re$, $c > 0$ and the coefficients of the $\mathcal{P}_i$ bounded by d_0,*

$$P\left(\sum_{i=1}^{m} a_i(Z_i - b_i)^2 + n^{-1/2}\sum_{i=1}^{m} \mathcal{P}_i(|Z_i|) \leq c\right)$$
$$= P\left(\sum_{i=1}^{m} a_i(Z_i - b_i)^2 \leq c\right) + O(n^{-1/2}). \tag{3.2}$$

The relation continues to hold for $m = 1$ if $c > \gamma$ for some fixed $\gamma > 0$.

Proof The present short proof is due to A.W. van der Vaart.

Assume $q \geq 1$, since for $q = 0$ the desired result follows by the mean value theorem from the fact that the noncentral χ^2_m density is bounded (uniformly in the noncentrality), provided neighborhoods of 0 are excluded in the case $m = 1$.

It is first shown that for $m \geq 2$, $u_1 < u_2$ and $v \geq 0$

$$P\left(u_1 \leq \sum_{i=1}^{m} a_i(Z_i - b_i)^2 \leq u_2, |Z_m| \geq v\right)$$
$$\leq (2a_0)^{-1}(u_2 - u_1)\exp\left(-\frac{1}{2}v^2\right). \tag{3.3}$$

For $m = 2$ this holds because the area between the two ellipses $\sum a_i(z_i - b_i)^2 = \max\{0, u_j\}$, $j = 1, 2$, is at most $\pi(a_1 a_2)^{-1/2}(u_2 - u_1)$, while the joint density of (Z_1, Z_2) is smaller than $(2\pi)^{-1}\exp(-\frac{1}{2}v^2)$ on the set $\{(z_1, z_2) : |z_2| \geq v\}$.

For $m = 3$ write the LHS of (3.3) as

$$\int_{|z|\geq v} P\left(u_1 - a_3(z - b_3)^2 \leq \sum_{i=1}^{2} a_i(Z_i - b_i)^2 \leq u_2 - a_3(z - b_3)^2\right)$$
$$\times (2\pi)^{-1/2}\exp\left(-\frac{1}{2}z^2\right) dz. \tag{3.4}$$

Application of (3.3) with $m = 2$ and $v = 0$ yields that (3.4) is smaller than

$$(2a_0)^{-1}(u_2 - u_1)2\,(1 - \Phi(v)) \leq (2a_0)^{-1}(u_2 - u_1)\exp\left(-\frac{1}{2}v^2\right).$$

Repeating the argument for $m > 3$, (3.3) follows.

Let $M \geq (q+1)d_0$, $M \in \mathcal{N}$. Write $S = \sum a_i(Z_i - b_i)^2$ and $T = \sum \mathcal{P}_i(|Z_i|)$. Suppose $0 \leq v < mM$. Then by (3.3) with $v = 0$

$$P(u_1 \leq S \leq u_2, |T| \geq v) \leq (2a_0)^{-1}(u_2 - u_1). \tag{3.5}$$

For $v \geq mM$ proceed as follows. Since $|\mathcal{P}_i(|z_i|)| \leq M(|z_i|^q + 1)$ for all $z_i \in \Re$, (3.3) also implies

$$\begin{aligned} &P(u_1 \leq S \leq u_2, |T| \geq v) \\ &\quad \leq \sum_{j=1}^{m} P\left(u_1 \leq S \leq u_2, |Z_j|^q \geq (mM)^{-1}v - 1\right) \\ &\quad \leq m(2a_0)^{-1}(u_2 - u_1)\exp\left(-\frac{1}{2}((mM)^{-1}v - 1)^{2/q}\right). \end{aligned} \tag{3.6}$$

Hence by (3.5) and (3.6)

$$\begin{aligned} &|P(S + n^{-1/2}T \leq c) - P(S \leq c)| \\ &\quad \leq P(S \leq c, S + n^{-1/2}T > c) + P(S > c, S + n^{-1/2}T \leq c) \\ &\quad \leq \sum_{v=0}^{\infty} P\left(c - n^{-1/2}(v+1) < S \leq c - n^{-1/2}v, T > v\right) \\ &\qquad + \sum_{v=0}^{\infty} P\left(c + n^{-1/2}v < S \leq c + n^{-1/2}(v+1), T < -v\right) \\ &\quad \leq 2(2a_0)^{-1}n^{-1/2}\left\{mM + m\sum_{v=mM}^{\infty}\exp\left(-\frac{1}{2}((mM)^{-1}v - 1)^{2/q}\right)\right\} \\ &\quad = O(n^{-1/2}) \end{aligned}$$

since the last sum converges. This establishes (3.2) for $m \geq 2$.

It remains to consider the case $m = 1$. Fix $\gamma > 0$. The preceding argument continues to hold for $c > \gamma$ if we replace (3.3) by

$$P\left(u_1 \leq a_1(Z_1 - b_1)^2 \leq u_2, |Z_1| \geq v\right) \leq (2\pi a_0\gamma)^{-1/2}(u_2 - u_1)\exp\left(-\frac{1}{2}v^2\right)$$

valid for $u_1 > \frac{1}{4}\gamma$, and use the inequality

$$\begin{aligned} &\sum_{v \geq (1/2)\gamma n^{1/2}} P\left(c - n^{-1/2}(v+1) < a_1(Z_1 - b_1)^2 \leq c - n^{-1/2}v, \mathcal{P}_1(|Z_1|) > v\right) \\ &\qquad \leq \sum_{v \geq (1/2)\gamma n^{1/2}} P\left(\mathcal{P}_1(|Z_1|) > v\right) \\ &\qquad \leq \sum_{v \geq (1/2)\gamma n^{1/2}} P(|Z_1|^q > M^{-1}v - 1) \end{aligned}$$

$$\leq (2/\pi)^{1/2} \sum_{v \geq (1/2)\gamma n^{1/2}} \exp\left(-\frac{1}{4}(M^{-1}v)^{2/q}\right)$$
$$\leq (2/\pi)^{1/2} M \int_{((1/2)\gamma n^{1/2}-1)/M}^{\infty} \exp\left(-\frac{1}{4}v^{2/q}\right) dv$$
$$= o(n^{-1/2}).$$

□

Remark 3.2 *There is nothing sacred about the integers $n \in \mathcal{N}$ in (3.2). They can be replaced by any $s_n > 0$ such that $s_n \to \infty$ as $n \to \infty$.*

Corollary 3.3 *Let $U_1, \ldots, U_k$ be jointly $N_k(0, I - \pi^{1/2}\pi^{1/2'})$ distributed with $\pi \in \Pi_\epsilon$. Replacing $Z_1, \ldots, Z_m$ in Lemma 3.1 by $U_1, \ldots, U_k$, the lemma remains valid in the sense that (3.2) holds for $k \geq 3$ (and also for $k = 2$ if $c > \gamma > 0$).*

Proof By an orthogonal transformation similar to that in Section 2,

$$\sum_{i=1}^{k} a_i (U_i - b_i)^2 \sim \sum_{i=1}^{k-1} \alpha_i (Z_i - \beta_i)^2 + \beta_0^2$$

and

$$\sum_{i=1}^{k} \mathcal{P}_i(|U_i|) = \sum_{i=1}^{k}\sum_{j=0}^{q} d_{ij}|U_i|^j \sim \sum_{i=1}^{k}\sum_{j=0}^{q} d_{ij} \left|\sum_{s=1}^{k-1} g_{is} Z_s\right|^j$$

where the α_i are positive and bounded away from zero, the g_{is} (and d_{ij}) are bounded. Since for $1 \leq j \leq q$

$$\left|\sum_{s=1}^{k-1} g_{is} Z_s\right|^j \leq \sum_{s=1}^{k-1} h_{is}^{(j)} |Z_s|^j,$$

where the $h_{is}^{(j)}$ are again bounded, the desired result follows from Lemma 3.1. □

Lemma 3.4 *Let a_i, d_i $(i = 1, \ldots, m)$ and c be real numbers, $m \geq 1$, and let $a_0 > 0$ and $d_0 > 0$ be fixed. Then, uniformly for $\max_i |a_i| > a_0$, $|d_i| < d_0$ and $c \in \Re$*

$$P\left(\sum_{i=1}^{m} a_i Z_i + n^{-1/2} \sum_{i=1}^{m} d_i Z_i^2 \leq c\right) = P\left(\sum_{i=1}^{m} a_i Z_i \leq c\right) + O(n^{-1/2}).$$

The error bound is sharp; Remark 3.2 again applies.

Proof It is sufficient to prove that

$$P\Big(\sum a_i Z_i \pm n^{-1/2} d_0 \sum Z_i^2 \le c\Big) = P\Big(\sum a_i Z_i \le c\Big) + O(n^{-1/2}). \quad (3.7)$$

Consider an orthogonal transformation $\tilde{Z} = \Psi Z$ where Ψ is an $m \times m$ orthonormal matrix with first row $\|a\|^{-1}(a_1, \ldots, a_m)$. Then

$$P\left(\sum a_i Z_i \pm n^{-1/2} d_0 \sum Z_i^2 \le c\right) = P\left(\|a\|\, \tilde{Z}_1 \pm n^{-1/2} d_0 \sum \tilde{Z}_i^2 \le c\right).$$

By direct calculation, using $\|a\| \ge a_0$, uniformly

$$P\left(\|a\|\, \tilde{Z}_1 \pm n^{-1/2} d_0 \tilde{Z}_1^2 \le c\right) = P\left(\|a\|\, \tilde{Z}_1 \le c\right) + O(n^{-1/2}).$$

Hence, by a convolution argument,

$$\begin{aligned} &P\left(\|a\|\, \tilde{Z}_1 \pm n^{-1/2} d_0 \sum \tilde{Z}_i^2 \le c\right) \\ &\qquad = P\left(\|a\|\, \tilde{Z}_1 \pm n^{-1/2} d_0\, \chi^2_{m-1} \le c\right) + O(n^{-1/2}) \end{aligned}$$

where $\tilde{Z}_1$ and χ^2_{m-1} are independent. A conditioning argument immediately shows that the RHS equals $P(\|a\|\, \tilde{Z}_1 \le c) + O(n^{-1/2})$ and (3.7) is proved. That the error bound is sharp follows by direct calculation for $m = 1$ and hence in general. □

Lemma 3.5 *Let $f_m(x;\delta)$ denote the density of the noncentral $\chi^2_m(\delta)$ distribution, $m \ge 1$. Then*

$$f_m(x;\delta) \le \begin{cases} C_m x^{(m-2)/2} \exp(-(1/2)(x^{1/2} - \delta^{1/2})^2) & \text{for } x > 0,\ \delta > \\ C_m x^{-1/2} \exp(-(1/2)(x^{1/2} - \delta^{1/2})^2) & \text{for } 0 < x < 4\delta \\ C_m x^{-1/2} (x/\delta)^{(m-1)/4} \exp(-(1/2)(x^{1/2} - \delta^{1/2})^2) & \text{for } 4m^2 < \delta < \end{cases}$$

where the positive constants C_m do not depend on x or δ . Conversely,

$$f_m(x;\delta) > C_m \delta^{-1/2} \qquad \text{for } |x^{1/2} - \delta^{1/2}| < 1, \quad \delta > 4.$$

Proof All statements are trivial for $m = 1$; so assume $m \ge 2$.

Let $v = v(x, \delta)$ be a real-valued function satisfying $0 \le v \le \frac{1}{2}x$, and let $b_m = (2^m \pi)^{-1/2} \Gamma((m-1)/2)^{-1}$. Then

$$\begin{aligned} f_m(x;\delta) &= \int_0^x f_{m-1}(x-y;0) f_1(y;\delta)\, dy \\ &\le b_m \int_0^x y^{-1/2} (x-y)^{(m-3)/2} \exp\left(-\frac{1}{2}\delta - \frac{1}{2}x + \delta^{1/2} y^{1/2}\right) dy \\ &\le b_m \exp\left\{-\frac{1}{2}(x^{1/2} - \delta^{1/2})^2\right\} \end{aligned}$$

$$
\begin{aligned}
&\times \left[\exp\left\{ -\delta^{1/2} \left(x^{1/2} - (x-v)^{1/2} \right) \right\} \int_0^{x-v} y^{-1/2}(x-y)^{(m-3)/2}\, dy \right. \\
&\left. + \int_{x-v}^{x} y^{-1/2}(x-y)^{(m-3)/2} \exp\left\{ -\delta^{1/2}(x^{1/2}-y^{1/2}) \right\} dy \right. \\
&\le b_m \exp\left\{ -\frac{1}{2}(x^{1/2}-\delta^{1/2})^2 \right\} \\
&\times \left[\exp\left\{ -\frac{1}{2}(\delta/x)^{1/2} v \right\} \int_0^{x} y^{-1/2}(x-y)^{(m-3)/2}\, dy \right. \\
&\left. + (x-v)^{-1/2} \int_0^{v} z^{(m-3)/2} \exp\left\{ -\frac{1}{2}\left(\frac{\delta}{x}\right)^{1/2} z \right\} dz \right] \\
&\le b_m \exp\left\{ -\frac{1}{2}(x^{1/2}-\delta^{1/2})^2 \right\} \\
&\times \left[\exp\left\{ -\frac{1}{2}(\delta/x)^{1/2} v \right\} x^{(m-2)/2} B\left(\frac{1}{2}, \frac{1}{2}(m-1) \right) \right. \\
&\left. + 2^{1/2} x^{-1/2} (x/\delta)^{(m-1)/4} \int_0^{(\delta/x)^{1/2} v} w^{(m-3)/2} \exp\left(-\frac{1}{2} w \right) dw. \right]
\end{aligned}
$$

The first inequality follows by taking $v \equiv 0$. The second inequality follows from the first one if $x < 4m^2$; otherwise take $v(x,\delta) \equiv 2(m-1)\log x$ (bound the last integral by $2^{(m-1)/2}\Gamma\left(\frac{1}{2}(m-1)\right)$). The third inequality is obtained by taking $v(x,\delta) \equiv (m-1)(x/\delta)^{1/2}\log x$ (bound the last integral as before).

To prove the reverse inequality, assume $\delta > 4$, let $|x^{1/2} - \delta^{1/2}| < 1$ and observe that

$$
\begin{aligned}
f_m(x;\delta) &> \frac{1}{2} b_m \int_0^x y^{-1/2}(x-y)^{(m-3)/2} \exp\left(-\frac{1}{2}\delta - \frac{1}{2}x + \delta^{1/2} y^{1/2} \right) dy \\
&\ge \frac{1}{2} b_m \exp\left\{ -\frac{1}{2}(x^{1/2}-\delta^{1/2})^2 \right\} x^{-1/2} \\
&\quad \times \int_{x-1}^{x} (x-y)^{(m-3)/2} \exp\left\{ -\delta^{1/2}(x^{1/2}-y^{1/2}) \right\} dy \\
&\ge \frac{1}{2} b_m \exp\left\{ -\frac{1}{2}(x^{1/2}-\delta^{1/2})^2 - \delta^{1/2} x^{-1/2} \right\} (\delta^{1/2}+1)^{-1} 2(m-1)^{-1} \\
&\ge \frac{1}{2}(m-1)^{-1} b_m e^{-5/2} \delta^{-1/2}.
\end{aligned}
$$

□

We are now prepared to prove our main theorem.

Proof of Theorem 2.1 (i) Let E_n denote the set $\{y \in \Re^k : \max_i |y_i| < \log n\}$. In view of (3.1) $P_\pi(Y_n \in E_n) = 1 - o(n^{-1/2})$ uniformly for $\pi \in \Pi_\epsilon$.

By Corollary 17.2 in Bhattacharya and Ranga Rao (1976)

$$\sup_B |P_\pi(Y_n \in B) - P(U \in B)| = O(n^{-1/2})$$

where the supremum is taken over all Borel measurable convex sets $B \subset \Re^k$ and where $U = (U_1, \dots, U_k)'$ is distributed as in Corollary 3.3. Note that after a linear transformation the asymptotic covariance matrix of the first $k-1$ components of Y_n is nonsingular. The error bound is uniform in $\pi \in \Pi_\epsilon$. Consider R^λ as a function of Y_n, see (2.1). Since R^λ is a convex function of Y_n on E_n, and E_n itself is a convex set, it follows that

$$\sup \left| P_\pi \left(R^\lambda(Y_n) \le c, Y_n \in E_n\right) - P\left(R^\lambda(U) \le c, U \in E_n\right)\right| = O(n^{-1/2})$$

or

$$\sup |P_\pi \left(R^\lambda(Y_n) \le c\right) - P\left(R^\lambda(U) \le c\right)| = O(n^{-1/2}) \qquad (3.8)$$

where the supremum is taken over $\pi \in \Pi_\epsilon$ and $c > 0$.

Conditionally on $Y_n \in E_n$, the terms in the expansion (2.1) beyond the third power of Y_{in} are uniformly bounded by $\epsilon_n = d_\lambda n^{-1} \log^4 n$, where d_λ is a suitable positive constant. This remains true after replacing Y_n by U. By Corollary 3.3 (with $q = 3$)

$$P\left(A^\lambda(U) + \frac{1}{3} n^{-1/2}(\lambda - 1) \sum r_i^\lambda \pi_i^{-1/2} U_i^3 \pm \epsilon_n \le c\right)$$
$$= P\left(A^\lambda(U) \le c\right) + O(n^{-1/2}).$$

Combining this result with (3.8), (i) is established.

(ii) By (2.2)

$$A^\lambda(U) = B^\lambda(U) + \sum (r_i^\lambda - \bar{r}_\lambda) U_i^2$$

where $\bar{r}_\lambda \to 1$ and $r_i^\lambda - \bar{r}_\lambda = O(\max_i |\pi_i - p_i|)$ as $\max_i |\pi_i - p_i| \to 0$. By Corollary 3.3 (with $q = 2$)

$$P\left(A^\lambda(U) \le c\right) = P\left(B^\lambda(U) \le c\right) + O(\max_i |\pi_i - p_i|).$$

The desired result now follows from part (i). Note that this argument remains valid if one takes $\bar{r}_\lambda = 1$.

(iii) Let

$$\tilde{r}_\lambda = \left[\sum \pi_i (r_i^\lambda - 1)^2 - \left\{\sum \pi_i (r_i^\lambda - 1)\right\}^2\right]^{1/2} \left\{2\lambda^2 I^\lambda(\pi : p)\right\}^{-1/2}. \qquad (3.9)$$

Define $\tilde{B}^\lambda$ by (2.3) with $\tilde{r}_\lambda$ replacing $\bar{r}_\lambda$. Obviously

$$A^\lambda(U) = \tilde{B}^\lambda(U) + \sum (r_i^\lambda - \tilde{r}_\lambda) U_i^2$$

where $\tilde{r}_\lambda = 1 + O(\max_i |\pi_i - p_i|)$ and $r_i^\lambda - \tilde{r}_\lambda = O(\max_i |\pi_i - p_i|)$, cf. (2.8). Similarly to (2.6) and (2.7), $\tilde{B}^\lambda(U)$ is distributed as

$$\tilde{r}_\lambda \, \chi^2_{k-1}(\delta_{n\lambda}) + (1 - \tilde{r}_\lambda)\delta_{n\lambda}.$$

By Corollary 3.3 (with $q = 2$)

$$P_\pi(A^\lambda(U) \le c) = P(\tilde{B}^\lambda(U) \le c) + O(\max_i |\pi_i - p_i|).$$

Moreover, in the notation of Lemma 3.5,

$$\begin{aligned} P_\pi\left(\tilde{B}^\lambda(U) \le c\right) &= P\left(\tilde{r}_\lambda \, \chi^2_{k-1}(\delta_{n\lambda}) + (1 - \tilde{r}_\lambda)\delta_{n\lambda} \le c\right) \\ &= P\left(\chi^2_{k-1}(\delta_{n\lambda}) \le c + (c - \delta_{n\lambda})(1 - \tilde{r}_\lambda)/\tilde{r}_\lambda\right) \\ &= P\left(\chi^2_{k-1}(\delta_{n\lambda}) \le c\right) + (c - \delta_{n\lambda})(1 - \tilde{r}_\lambda)\tilde{r}_\lambda^{-1} f_{k-1}(\theta; \delta_{n\lambda}) \end{aligned} \tag{3.10}$$

where $\theta_n = c + t_n(c - \delta_{n\lambda})(1 - \tilde{r}_\lambda)/\tilde{r}_\lambda, 0 \le t_n \le 1$. The first part of Lemma 3.5 implies

$$f_{k-1}(\theta_n; \delta_{n\lambda}) \le \begin{cases} C\theta_n^{-1/2} \exp\left(-\frac{1}{2}(\theta_n^{1/2} - \delta_{n\lambda}^{1/2})^2\right) & \text{if } |\theta_n^{1/2} - \delta_{n\lambda}^{1/2}| < \frac{1}{2}\delta_{n\lambda}^{1/2} \\ C \exp\left\{-\frac{1}{4}(\theta_n^{1/2} - \delta_{n\lambda}^{1/2})^2\right\} & \text{otherwise.} \end{cases}$$

Since $c - \delta_{n\lambda} = (\theta_n - \delta_{n\lambda})\,(1 + o(1))$, it follows that

$$(c - \delta_{n\lambda}) f_{k-1}(\theta_n; \delta_{n\lambda}) = O(1).$$

Hence the last term in the RHS of (3.10) is of order $O(s_n n^{-1/2})$ and the desired result follows from part (i).

(iv) Define $\tilde{r}_{\lambda 1}$ as $\tilde{r}_\lambda$ in (3.9) with $I^\lambda(\pi : p)$ replaced by $I^1(\pi : p)$ and define $\tilde{B}_1^\lambda(U)$ as $\tilde{B}^\lambda(U)$ with $\tilde{r}_\lambda$ replaced by $\tilde{r}_{\lambda 1}$.

Again

$$A^\lambda(U) = \tilde{B}_1^\lambda(U) + \sum (r_i^\lambda - \tilde{r}_{\lambda 1}) U_i^2$$

and

$$\tilde{B}_1^\lambda(U) \sim \tilde{r}_{\lambda 1} \, \chi^2_{k-1}(\delta_n) + (1 - \tilde{r}_{\lambda 1})\delta_n + \delta_{n\lambda} - \delta_n$$

with $\tilde{r}_{\lambda 1} = 1 + O(\max_i |\pi_i - p_i|)$ and $r_i^\lambda - \tilde{r}_{\lambda 1} = O(\max_i |\pi_i - p_i|)$.

Proceeding as in (iii)

$$\begin{aligned} &P\left(\tilde{r}_{\lambda 1} \, \chi^2_{k-1}(\delta_n) + (1 - \tilde{r}_{\lambda 1})\delta_n + \delta_{n\lambda} - \delta_n \le c\right) \\ &\qquad = P\left(\chi^2_{k-1}(\delta_n) + \delta_{n\lambda} - \delta_n \le c\right) + O(s_n n^{-1/2}). \end{aligned}$$

By a local expansion

$$\delta_{n\lambda} - \delta_n = \frac{1}{3}(\lambda - 1) n \sum (\pi_i - p_i)^3 / p_i^2 + O(n \max_i |\pi_i - p_i|^4), \tag{3.11}$$

implying $(\delta_{n\lambda} - \delta_n)/\delta_n^{1/2} = O(s_n^2 n^{-1/2})$. Since by Lemma 3.5 $f_{k-1}(x;\delta) \leq C\delta^{-1/2}$ (all $x > 0$), another application of the mean value theorem yields

$$P_\pi\left(\tilde{B}_1^\lambda(U) \leq c\right) - P\left(\chi_{k-1}^2(\delta_n) \leq c\right) = O(s_n^2 n^{-1/2}).$$

The desired result follows again from part (i).

To prove that the bound in (iv) is sharp, it suffices to show that for given $\{s_n\}$

$$P\left(\chi_{k-1}^2(\delta_n) + \delta_{n\lambda} - \delta_n \leq \delta_n\right) - P\left(\chi_{k-1}^2(\delta_n) \leq \delta_n\right) > \epsilon s_n^2 n^{-1/2}$$

for some $\epsilon > 0$ and appropriate $\pi_1, \ldots, \pi_k$. First note that $\pi_1, \ldots, \pi_k$ exist such that both δ_n/s_n^2 and $|\delta_{n\lambda} - \delta_n|/(s_n^3 n^{-1/2})$ are bounded away from 0 and ∞, cf. (3.11). Since

$$|\delta_n - (\delta_{n\lambda} - \delta_n)|^{1/2} - \delta_n^{1/2} = o(1) \qquad \text{as } s_n^2 n^{-1/2} \to 0,$$

The second part of Lemma 3.5 and the mean value theorem imply the above inequality.

(v) By definition of M^λ and direct calculation we have

$$\begin{aligned} a_n^\lambda &= 1 + O(s_n n^{-1/2}) \\ b_n^\lambda &= \delta_{n\lambda} - \delta_n + \delta_n(1 - a_n^\lambda) + O(s_n n^{-1/2}). \end{aligned}$$

It is seen in the proof of (iv) that

$$\begin{aligned} P\left(\chi_{k-1}^2(\delta_n) + \delta_{n\lambda} - \delta_n \leq c\right) &= P\left(\tilde{B}_1^\lambda(U) \leq c\right) + O(s_n n^{-1/2}) \\ &= P\left(R^\lambda(N/n : p) \leq c\right) + O(s_n n^{-1/2}). \end{aligned} \tag{3.12}$$

Proceeding as in (iii)

$$\begin{aligned} &P\left(a_n^\lambda \chi_{k-1}^2(\delta_n) + b_n^\lambda \leq c\right) \\ &\quad = P\left(a_n^\lambda \chi_{k-1}^2(\delta_n) + (1 - a_n^\lambda)\delta_n \leq c - b_n^\lambda + (1 - a_n^\lambda)\delta_n\right) \\ &\quad = P\left(\chi_{k-1}^2(\delta_n) \leq c - b_n^\lambda + (1 - a_n^\lambda)\delta_{n\lambda}\right) + O(s_n n^{-1/2}) \\ &\quad = P\left(\chi_{k-1}^2(\delta_n) + \delta_{n\lambda} - \delta_n \leq c + O(s_n n^{-1/2})\right) + O(s_n n^{-1/2}) \\ &\quad = P\left(\chi_{k-1}^2(\delta_n) + \delta_{n\lambda} - \delta_n \leq c\right) + O(s_n n^{-1/2}) \end{aligned} \tag{3.13}$$

by the mean value theorem and the boundedness of the density of the noncentral $\chi_{k-1}^2(\delta_n)$-distribution. Combination of (3.12) and (3.13) yields the desired result.

(vi) By (2.2), (2.4) and (2.5)

$$P\left(A^{\lambda}(U) \leq c\right) = P\left(L^{\lambda}(U) + \sum r_i^{\lambda} U_i^2 \leq c\right)$$
$$= P\left(2\sum_{i=1}^{k-1} \theta_{in}^{1/2} \omega_{in} Z_i + 2nI^{\lambda}(\pi : p) + \sum_{i=1}^{k-1} \theta_{in} Z_i^2 \leq c\right). \tag{3.14}$$

The θ_{in} $(i = 1, \ldots, k-1)$ are bounded away from 0 and ∞ and the first $k-1$ components of $\omega_n = S_n' Q_n \mu_n$ (see Section 2) satisfy $\max_i \omega_{in}^2 > \epsilon_1 n \max_i (r_i^{\lambda/2} - r_i^{-\lambda/2})^2/\lambda^2$ for some $\epsilon_1 > 0$. Hence, dividing both members in the last event of (3.14) by $n^{1/2} \max_i |r_i^{\lambda/2} - r_i^{-\lambda/2}|/|\lambda|$, Lemma 3.4 implies that the RHS of (3.14) equals

$$P\left(2\sum_{i=1}^{k-1} \theta_{in}^{1/2} \omega_{in} Z_i + 2nI^{\lambda}(\pi : p) \leq c\right) + O\left((n^{1/2} \max_i |\pi_i - p_i|)^{-1}\right)$$

where the remainder term is $O(s_n^{-1})$. The desired result now follow from part (i). That the bound is sharp is an easy exercise. □

Acknowledgments: The authors are grateful to A.W. van der Vaart for providing a much shorter and more transparent proof of Lemma 3.1 than the original proof.

REFERENCES

Bhattacharya, R.N. and Ranga Rao, R. (1976), *Normal Approximation and Asymptotic Expansions.* Wiley, New York.

Broffitt, J.D. and Randles, R.H. (1979), A power approximation for the chi-square goodness-of-fit test: simple hypothesis case. *J. Amer. Statist. Assoc.* 72, 604–607.

Chibisov, D.M. and van Zwet, W.R. (1984), On the Edgeworth expansion for the logarithm of the likelihood ratio. I. *Theor. Probability Appl.* 29, 427–451.

Cox, D.R. and Reid, N. (1987), Approximations to noncentral distributions. *Canad. J. Statist.*, 15, 105-114.

Cressie, N. and Read, T.R. (1984), Multinomial goodness-of-fit tests. *J. Roy. Statist. Soc., Ser. B,* 46, 440–464.

Drost, F.C., Kallenberg, W.C.M., Moore, D.S. and Oosterhoff, J. (1989) Power approximations to multinomial tests of fit. *J. Amer. Statist. Assoc.* 84, to appear.

Hayakawa, T. (1975), The likelihood ratio criterion for a composite hypothesis under a local alternative. *Biometrika* 62, 451–460.

Hayakawa, T. (1977), The likelihood ratio criterion and the asymptotic expansion of its distribution. *Ann. Inst. Statist. Math.* 129, Part A, 359–378.

Larntz, K. (1978), Small sample comparisons of exact levels for chi- squared goodness of fit statistics. *J. Amer. Statist. Assoc.* 73, 253–263.

Moore, D.S. (1984), Measures of lack of fit from tests of chi-squared type. *J. Statist. Planning and Inf.* 10, 151–166.

Morris, C. (1975), Central limit theorems for multinomial sums. *Ann. Statist.* 3, 165–188.

Patnaik, P.B. (1949), The non-central χ^2 and F distributions and their applications. *Biometrika* 36, 202–232.

Peers, H.W. (1971), Likelihood ratio and associated test criteria. *Biometrika* 58, 577–587.

Read, T.R. (1984a), Small-sample comparisons for the power divergence goodness-of-fit statistics. *J. Amer. Statist. Assoc.* 79, 929–935.

Read, T.R. (1984b), Closer asymptotic approximations for the distributions of the power divergence goodness-of-fit statistics. *Ann. Inst. Statist. Math.* 36, Part A, 59–69.

28

Estimating the Normal Mean and Variance Under A Publication Selection Model

Larry V. Hedges[1]

ABSTRACT Maximum likelihood estimators of the mean and variance of a normal distribution are obtained under a publication selection model in which data are reported only when the hypothesis that the mean is 0 is rejected. An approximation to the asymptotic variance-covariance matrix for these estimators is given. Also discussed are the marginal distributions of the sample mean and variance under the selection model.

1 Estimating the Normal Mean and Variance under A Publication Selection Model

Statistical analyses involving hypothesis testing have become the dominant mode of quantitative practice in the applied social sciences. Indeed, a rather simplistic conception of statistical practice has led some researchers (and even some journal editors) to believe that the results of empirical research studies are only of interest if some theoretical null hypothesis is rejected at a conventional level of significance, usually 5% (see Greenwald, 1975). The belief that research studies are conclusive only if they lead to rejection of null hypotheses encourages selective (conditional) reporting of research results. One form of selective reporting is a consequence of selective *publication* of research results. Some journals in psychology have at times adopted policies explicitly *requiring* statistical significance at the 5% level for publication (see Melton, 1962), while others have used statistical significance as one of the most important (but not a strictly necessary) criteria for publication (see Greenwald, 1975). These mechanisms appear to be effective since reviews of statistics reported in connection with hypothesis tests in psychology journals suggest that well over 90% of the journal articles surveyed rejected a null hypothesis (see Sterling, 1959; Bozarth and Roberts, 1972). Because unpublished studies are not observed, the extent of selective publication based on statistical significance is difficult to know. There

[1]The University of Chicago

may be many unpublished studies languishing in the file drawer for want of statistically significant results, or there may be only a few (see Rosenthal, 1978).

A second selection mechanism is easier to document. Some researchers describe all hypothesis tests that were undertaken, but report statistical summaries of only the analyses that resulted from rejections of null hypotheses. The increased use of statistical methods for combining the results of replicated research studies (meta-analysis) has documented the magnitude of the problem. For example, in each of two recent meta-analyses of studies on gender differences in cognitive abilities (Hyde, 1981, and Eagly, 1981) over 40% of the studies did not report summary statistics on gender differences because the t-test for gender differences was not statistically significant. This suggest that at least 40% of the studies in those two reviews reported statistical results (e.g., means and variances) conditional on the statistical significance of a t-statistic. If the means and variances are observed conditional on the rejection of a null hypothesis, the observed values of those summary statistics may be seriously biased estimates of the corresponding parameters. A related problem is that of conditional reporting of test statistics after a preliminary test. For a discussion of conditional F-tests following a preliminary F-test in the general linear model with normally distributed errors, see Olshen (1973).

A variety of models of selective publication or reporting of results could be posited. The simplest and most extreme model is that data corresponding to nonsignificant results is reported with probability zero. Estimation of the standardized mean difference conditional on a significant two-sample t-statistic under this model was considered by Hedges (1984) and Hedges and Olkin (1985). Others have considered less extreme models of selective publication in which the probability of reporting is an increasing monotonic function of the value of the test statistic. Estimation of the standardized mean difference conditional on the two sample t-statistic under such a model was considered by Iyengar and Greenhouse (1988). Although the latter model is more realistic, the more extreme model assuming that nonsignificant results are never reported is simpler and provides an upper bound on the biasing effects of selection.

This paper is an examination of the problem of interpreting the sample mean and variance from a normal distribution when they are reported conditional upon rejection of the hypothesis that the mean is zero. First the conditional model is stated. Then the marginal distributions of the sample mean and variance are examined. Next maximum likelihood estimation of the population mean μ and variance σ^2 is considered and an approximation to the asymptotic variance-covariance matrix of the maximum likelihood estimates is given. Finally an example illustrates the procedure.

2 Notation and Model

Suppose that the original data are n independent scores $x_1, \ldots, x_n$ from a normal distribution with unknown mean μ and unknown variance σ^2. That is,

$$x_i \sim N(\mu, \sigma^2), \qquad i = 1, \ldots, n.$$

The experimenter tests the two-sided hypothesis $H_0 : \mu = 0$ versus the alternative $H_1 : \mu \neq 0$ at significance level α using an F-test (or equivalently a two-tailed t-test). Hypothesis H_0 is rejected if

$$\frac{n\bar{x}^2}{s^2} > c^2, \tag{2.1}$$

where $\bar{x} = \sum x_i/n$ and $s^2 = \sum(x_i - \bar{x})^2/(n-1)$ are sample estimates of μ and σ^2, and c is a function of n and α.

Consider the case in which the statistics $\bar{x}$ and s^2 are observed only if (2.1) is satisfied; that is, only if the hypothesis H_0 is rejected at level α. Let $\bar{x}_*$ and s_*^2 denote random variables corresponding to the *observed* mean and variance. Since the statistics are observed only if H_0 is rejected, $n\bar{x}_*^2 > c^2 s_*^2$.

This selection model might be called "restriction to significant results." It differs from selection models such as simple truncation and censoring in that the selection is based on a ratio of the variables and not on either variable alone. For a discussion of selection models involving truncation, see Bayarri and DeGroot (1986a,b); Dawid and Dickie (1977); Hedges and Olkin (1985); and Schneider (1986). Our selection model also differs from censoring models in which a known number of observations are unavailable (see Schneider, 1987).

3 The Distribution of $\bar{x}_*$ and s_*^2

The joint distribution $f(u, v)$ of $\bar{x}$ and s^2 is

$$f(u,v) = \frac{(m/2)^{m/2} v^{m/2-1} \sqrt{n} \exp\{-[n(u-\mu)^2 + mv]/2\sigma^2\}}{\sqrt{2\pi}\sigma^{m+1}\Gamma(m/2)} \tag{3.1}$$

where $m = n - 1$, which implies that the joint distribution $f_*(u, v)$ of $\bar{x}_*$ and s_*^2 is

$$f_*(u,v) = \begin{cases} f(u,v)/A(\mu/\sigma) & \text{if } nu^2 > vc^2 \\ 0 & \text{if } nu^2 \leq vc^2, \end{cases} \tag{3.2}$$

where $A(\mu/\sigma)$ is a normalizing constant equal to the probability that the absolute value of a noncentral t-variate with $n - 1$ degrees of freedom and noncentrality parameter $\sqrt{n}\mu/\sigma$ exceeds c in (2.1).

Integrating (3.2) over the values of s_*^2 for which $f_*(\bar{x}_*, s_*^2)$ is nonzero yields the marginal distribution $f_*(u)$ of $\bar{x}_*$

$$f_*(u) = \frac{\sqrt{n}\Gamma(m/2; nmu^2/2c^2\sigma^2)}{\sqrt{2\pi}A(\mu/\sigma)\sigma} \exp\left(-n(u-\mu)^2/2\sigma^2\right) \tag{3.3}$$

where $\Gamma(a;x)$ is the incomplete gamma ratio given by

$$\Gamma(a;x) = \frac{1}{\Gamma(a)} \int_0^x t^{a-1} e^{-t} dt.$$

Thus the marginal distribution of $\bar{x}_*$ is essentially a normal distribution weighted by an incomplete gamma ratio. Evaluating (3.3) for $\alpha = .05$, and for various values of μ and n, we see that the marginal distribution of $\bar{x}_*$ is generally skewed away from zero. Plots of the density function are given in Figure 1 for $\sigma = 1.0$, $n = 20$ and 80, and $\mu = 0.0$, 0.50, and 1.00. Note that the density function is bimodal for $\mu = 0.0$.

The moments of $\bar{x}_*$ are easily obtained by numerical integration. The expected value of $\bar{x}_*$ is given in Table 1 for $\sigma = 1$, $\alpha = .05$, and various values of μ and n. Comparable values derived from the simulation study of Lane and Dunlap (1978) are also given. As expected, the bias of $\bar{x}_*$ as an estimator of μ is greatest when sample sizes are small and the effect size $\delta = (\mu/\sigma)$ is moderate. When the sample size n is large, the bias in $\bar{x}_*$ is smaller, although it may not be negligible even if the sample size is $n = 100$. It is interesting to note that the absolute magnitude of the bias decreases near $\mu/\sigma = 0$ and that if the mean μ is exactly zero, then $\bar{x}_*$ is unbiased. In contrast, the *relative* bias $E[(\bar{x}_*) - \mu]/\mu$ becomes large near $\mu = 0.0$ and tends to one as μ/σ becomes large. The implication of these results is that the sample mean may substantially overestimate the magnitude of the population mean under the model of restriction to significant results.

4 The Marginal Distribution of the Variance

We obtain the marginal distribution of the variance s_*^2 by integrating over the values of the mean difference for which $f_*(\bar{x}_*, s_*^2)$ is nonzero. This region can be characterized by values of $\bar{x}_*$ such that

$$\bar{x}_* > s_* c/\sqrt{n}, \text{ or } \bar{x}_* < -s_* c/\sqrt{n}.$$

This yields the marginal distribution $f_*(v)$ of s_*^2 given by

$$f_*(v) = \frac{\{\Phi[-z - \sqrt{n}\mu/\sigma] + 1 - \Phi[z - \sqrt{n}\mu/\sigma]\}\,(v/2)^{\frac{m}{2}-1} e^{-mv/2\sigma^2}}{\Gamma(m/2)\sigma^m m^{-m/2} A(\mu/\sigma)} \tag{4.1}$$

where $z = c\sqrt{v}/\sigma, \Phi(x)$ is the standard normal cumulative distribution function and $A(\mu/\sigma)$ is the normalizing constant in (3.2).

TABLE 1. Expected Value of the Sample Mean Conditional on a Significant F-Statistic at the $\alpha = .05$ Level

n	$\mu = .25$		$\mu = .50$		$\mu = 1.00$	
	Simulated	Exact	Simulated	Exact	Simulated	Exact
10	.96	.96	1.39	1.37	1.66	1.65
20	.96	.89	1.14	1.10	1.32	1.30
30	.83	.80	.93	.93	1.15	1.15
40	.72	.73	.83	.83	1.07	1.08
60	—	.62	—	.71	—	1.02
80	—	.56	—	.64	—	1.00
100	—	.51	—	.60	—	1.00

Note: In these data $\sigma = 1.0$. Exact values were obtained by numerical integration. Simulated values were calculated from data given in Lane and Dunlap (1978).

FIGURE 1. Probability density function of the observed mean difference $\bar{x}_*$ for sample size (a) $n = 20$ and (b) $n = 80$ when $\sigma = 1.0$ and only mean differences that are statistically significant at the $\alpha = .05$ level may be observed. Note that the density functions for $\mu = 0.0$ are bimodal.

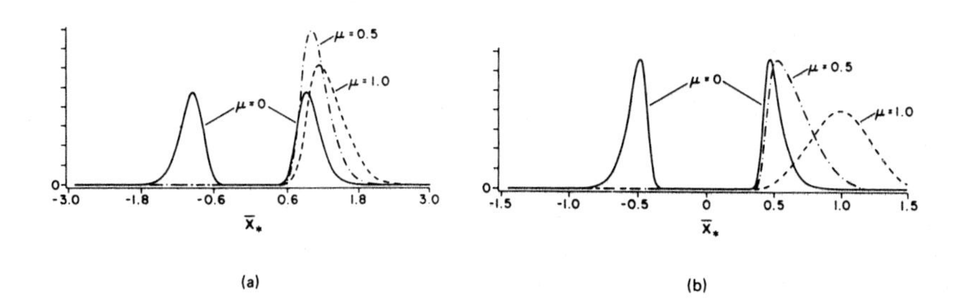

Evaluating the density function (4.1) for $\alpha = .05$ and for various values of n and μ we see that the marginal distribution of s_*^2 is much like a chi-square, but weighted more heavily toward zero. Plots of the density function are given in Figure 2 for $\sigma = 1.00, n = 20$, and 80, and for $\mu = 0.0, 0.50$, and 1.00.

The moments of s_*^2 are easily obtained by numerical integration. The expected value of s_*^2 is given in Table 2 for $\alpha = .05$ and various values of μ and n. The results in Table 2 suggest that s_*^2 underestimates σ^2, especially when μ and n are small. These results also show that the relative bias of s_*^2 as an estimate of σ^2 is smaller than the relative bias of $\bar{x}_*$ as an estimate of μ. For example, the relative bias of s_*^2 given by $[E(s_*^2) - \sigma^2]/\sigma^2$ is less

FIGURE 2. Probability density function of the observed variance s_*^2 for sample size (a) $n = 20$ and (b) $n = 40$ when $\sigma = 1.0$ and only the data corresponding to mean differences that are statistically significant at the $\alpha = .05$ level may be observed.

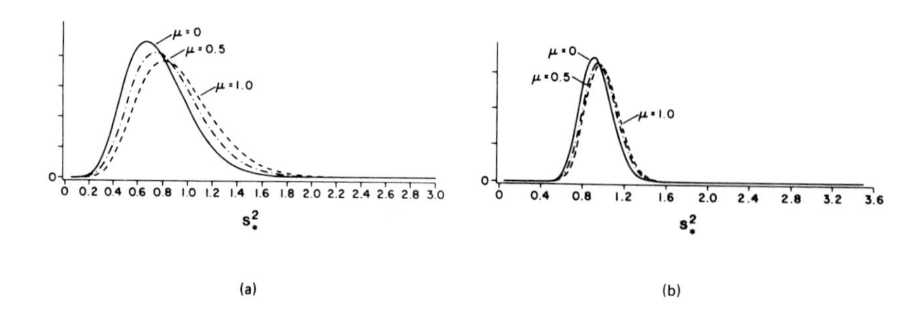

TABLE 2. Expected Value of the Sample Variance Conditional on a Significant F-Statistic at the $\alpha = .05$ Level

n	$\mu = .25$	$\mu = .50$	$\mu = 1.00$
10	.58	.63	.73
20	.80	.85	.92
30	.88	.91	.97
40	.91	.95	.99
50	.95	.97	1.00
80	.96	.98	1.00
100	.97	.99	1.00

Note: In these data $\sigma = 1.0$. These values were obtained by numerical integration.

than 10 percent for $\mu/\sigma = .25$ and $n = 40$ whereas the relative bias of $\bar{x}_*$ is nearly 200 percent in the same situation. Note that, unlike $\bar{x}_*$, s_*^2 is not unbiased when $\mu = 0$.

5 Maximum Likelihood Estimation of μ and σ^2

The log-likelihood of $\bar{x}_*$ and s_*^2 is

$$L = -n\log(\sigma) - [n(\bar{x}_* - \mu)^2 + (n-1)s_*^2]/2\sigma^2 - n\log[A(\mu/\sigma)]. \quad (5.1)$$

Maximization of (5.1) directly is complicated by the last term, which is a function of both μ and σ^2 that is difficult to compute. It is convenient, therefore, to re-parameterize using either μ or σ and $\delta = \mu/\sigma$.

Reparameterizing the likelihood (5.1) in terms of σ and δ yields

$$L = -n\log(\sigma) - [n(\bar{x}_* - \delta\sigma)^2 + (n-1)s_*^2]/2\sigma^2 - n\log[A(\delta)]. \qquad (5.2)$$

Setting the derivative with respect to σ equal to zero and solving for σ yields

$$\hat{\sigma} = \frac{-\hat{\delta}\bar{x}_* + [\hat{\delta}^2\bar{x}_*^2 + 4(\bar{x}_*^2 + (n-1)s_*^2/n)]^{\frac{1}{2}}}{2} \qquad (5.3)$$

Differentiating (5.2) with respect to δ and substituting the expression given in (5.3) for σ yields the likelihood equation

$$\frac{2y}{-\delta y + [\delta^2 y^2 + 4(y^2 + n')]^{\frac{1}{2}}} - \delta - \frac{A'(\delta)}{A(\delta)} = 0, \qquad (5.4)$$

where $A'(\delta) = \partial A(\delta)/\partial\delta$, $y = \bar{x}_*/s_*$, and $n' = (n-1)/n$. A closed form solution of (5.4) for the maximum likelihood estimate of δ is not available, but (5.4) can be solved numerically given any specific values of $\bar{x}_*/s_*$ and n.

Table 3 gives the maximum likelihood estimate $\hat{\theta} = \sqrt{n}\hat{\delta}$ of $\theta = \sqrt{n}\delta$ for positive values of $\bar{x}_*$ and $n = 20(4)40(10)100(50)200$. Enter Table 3 in a column corresponding to a sample size n and move down to find the row corresponding to the sample value of $t_* = \sqrt{n}\bar{x}_*/s_*$. Only positive values are tabulated. For negative values of $\bar{x}_*$, obtain $\hat{\theta}$ by noting that if $\sqrt{n}\bar{x}_*/s_*$ corresponds to $\hat{\theta}_0$ then $-\hat{\theta}_0$ corresponds to $-\sqrt{n}\bar{x}_*/s_*$. Note that the minimum observable positive value of $\sqrt{n}\bar{x}_*/s_*$ is the two-tailed critical value c and that when $\sqrt{n}\bar{x}_*/s_* = c$, the maximum likelihood estimate of θ is not zero. Both c and the value of $\hat{\theta}$ corresponding to c are given in Table 3.

Maximizing the likelihood (5.2) after re-parameterizing in terms of μ and δ, or noting that $\mu = \delta\alpha$, yields the maximum likelihood estimate $\hat{\mu}$ of μ as

$$\hat{\mu} = \hat{\delta}\hat{\sigma}. \qquad (5.5)$$

6 The Asymptotic Distribution of the Estimates

Calculation the expectation of the second derivatives of the likelihood (5.2) and inverting the information matrix yields the asymptotic distribution of $(\hat{\sigma}, \hat{\delta})$ as

$$\sqrt{n}[(\hat{\delta}, \hat{\sigma}) - (\delta, \sigma)] \sim N(0, \Sigma) \qquad (6.1)$$

where

$$\Sigma = \begin{pmatrix} (1+\delta^2/2)\gamma & -\delta_*\sigma/2\beta \\ -\delta_*\sigma/2\beta & (1-\lambda)\sigma^2/2\beta \end{pmatrix}, \qquad (6.2)$$

TABLE 3. Maximum Likelihood Estimator $\hat{\theta}$ as a Function of t_* and n

	Sample Size n													
	20	24	28	32	36	40	50	60	70	80	90	100	150	200
$t_{\min} = c$	2.093	2.069	2.052	2.040	2.030	2.023	2.010	2.001	1.995	1.990	1.987	1.984	1.976	1.972
$\hat{\theta}_{min}$	.53	.52	.51	.51	.51	.50	.50	.50	.49	.49	.49	.49	.48	.48
t_*														
2.10	0.53	0.54	0.54	0.54	0.54	0.54	0.55	0.55	0.55	0.55	0.55	0.55	0.55	0.55
2.20	0.59	0.59	0.59	0.60	0.60	0.60	0.61	0.61	0.61	0.62	0.62	0.62	0.62	0.62
2.30	0.65	0.66	0.66	0.67	0.68	0.68	0.69	0.69	0.70	0.70	0.71	0.71	0.72	0.72
2.40	0.72	0.74	0.75	0.77	0.78	0.78	0.80	0.81	0.82	0.83	0.84	0.84	0.85	0.86
2.50	0.82	0.86	0.88	0.90	0.92	0.94	0.97	0.99	1.01	1.02	1.04	1.04	1.07	1.09
2.60	0.97	1.03	1.08	1.13	1.16	1.19	1.25	1.28	1.31	1.33	1.35	1.36	1.40	1.41
2.70	1.21	1.32	1.40	1.45	1.50	1.53	1.59	1.63	1.65	1.67	1.69	1.70	1.73	1.75
2.80	1.55	1.67	1.74	1.79	1.83	1.86	1.91	1.94	1.96	1.97	1.98	2.00	2.02	2.03
2.90	1.89	1.99	2.05	2.09	2.12	2.14	2.18	2.21	2.22	2.23	2.24	2.25	2.27	2.28
3.00	2.19	2.26	2.31	2.34	2.37	2.39	2.42	2.44	2.45	2.46	2.47	2.47	2.49	2.49
3.10	2.44	2.50	2.54	2.57	2.59	2.60	2.63	2.64	2.65	2.66	2.67	2.67	2.68	2.69
3.20	2.67	2.71	2.75	2.77	2.78	2.80	2.81	2.83	2.83	2.84	2.85	2.85	2.86	2.86
3.30	2.87	2.90	2.93	2.95	2.96	2.97	2.99	3.00	3.00	3.01	3.01	3.01	3.02	3.03
3.40	3.05	3.08	3.10	3.12	3.13	3.13	3.15	3.15	3.16	3.16	3.17	3.17	3.17	3.18
3.50	3.21	3.24	3.26	3.27	3.28	3.28	3.29	3.30	3.31	3.31	3.31	3.31	3.32	3.32
3.60	3.37	3.39	3.40	3.41	3.42	3.43	3.43	3.44	3.44	3.44	3.45	3.45	3.45	3.45
3.70	3.52	3.53	3.54	3.55	3.56	3.56	3.57	3.57	3.57	3.57	3.58	3.58	3.58	3.58
3.80	3.65	3.67	3.68	3.68	3.69	3.69	3.70	3.70	3.70	3.70	3.70	3.70	3.70	3.70
3.90	3.79	3.80	3.80	3.81	3.81	3.81	3.82	3.82	3.82	3.82	3.82	3.82	3.82	3.82
4.00	3.91	3.92	3.93	3.93	3.93	3.93	3.94	3.94	3.94	3.94	3.94	3.94	3.94	3.94

and

$$\begin{aligned}
&\delta_* = \mu_*/\sigma, \ \mu_* = E(\bar{x}_*), \ \sigma_*^2 = E(s_*^2), \\
&\beta = [(1-\lambda)(3\sigma_*^2/\sigma^2 - 2\delta\delta_* - 1) + (2-3\lambda)\delta_*^2]/2, \\
&\gamma = [3(\sigma_*^2/\sigma^2 + \delta_*^2) - 2\delta\delta_* - 1]/[2\beta(1+\delta^2/2)], \\
&\lambda = \{[A'(\delta)]^2 - A''(\delta)A(\delta)\}/[A(\delta)]^2,
\end{aligned} \tag{6.3}$$

and the primes imply derivatives with respect to δ. When there is no selection so that $\lambda = 0, \delta_* = \delta$, and $\sigma_* = \sigma$, then $\beta = \lambda = 1$ and the diagonal elements of (6.2) reduce to the asymptotic variances $\sigma^2/2$ and $(1 + \delta^2/2)$ respectively, of the maximum likelihood estimators of σ and δ under the model with no selection.

The asymptotic joint distribution of $\hat{\mu}$ and $\hat{\sigma}$ is obtained by applying the usual delta method to (6.1) to yield

$$\sqrt{n}[(\hat{\mu}, \hat{\sigma}) - (\mu, \sigma)] \sim N(0, \Sigma) \tag{6.4}$$

where

$$\Sigma = \begin{pmatrix} \sigma_{11} & (\sigma^2/2\beta)(\delta - \delta_* - \delta\lambda) \\ (\sigma^2/2\beta)(\delta - \delta_* - \delta\lambda) & (\sigma^2/2\beta)(1 - \lambda) \end{pmatrix}, \tag{6.5}$$

$$\sigma_{11} = [3(\sigma_*^2 + \mu_*^2) - 4\mu\mu_* + \mu^2(1-\lambda) - \sigma^2]/2\beta,$$

and $\beta, \lambda, \mu_*, \sigma_*$, and δ_* are as in (6.3). Note that if there is no selection and $\lambda = 0$, (6.5) reduces to the usual asymptotic covariance matrix of $\bar{x}$ and s.

Since

$$A(\delta) = 1 - \int_{-c}^{c} f(t|\sqrt{n}\delta)dt,$$

where $f(t|\theta)$ is the probability density function of a non-central t-variate with $n-1$ degrees of freedom and non-centrality parameter θ, λ involves integrals of derivatives of the non-central t-density, which can be difficult to compute. An approximation to the non-central t-distribution studied by Laubscher (1960) can be used to obtain a relatively simple (and reasonably accurate) approximation to λ. Using this approximation

$$A(\delta) \approx 1 - \int_{h(-c)}^{h(c)} \phi[t - h(\sqrt{n}\delta)]dt$$

where

$$h(x) = \sqrt{2}\sinh^{-1}(x/\sqrt{2}),$$

and $\phi(x)$ is the standard normal density function. This approximation yields

$$A'(\delta) \approx A(\delta)b(\eta - \gamma_1)$$

and

$$A''(\delta) \approx A(\delta)b^2[1 + (2\eta + b\delta/2)\gamma_1 - \gamma_2 - \eta^2 - b\delta\eta/2],$$

where

$$b = [n/(1 + n\delta^2/2)]^{\frac{1}{2}}, \qquad \eta = h(\sqrt{n}\delta),$$

and γ_1 and γ_2 are the first and second moments about zero of a truncated normal distribution with mean η, unit variance, and truncation points $\pm h(c)$.

7 Example

A twenty-year longitudinal study of aging twins by Blum, Fosshage and Jarvik (1972) reported changes in measures of intellectual functioning between the first measurement in 1947 and the second measurement in 1967. The sufficient statistics for analysis in this report appear to be reported only when the mean is significantly different from zero at the $\alpha = .05$ level of significance since several analyses yielding nonsignificant results are mentioned, but not reported. One analysis that was reported examined the change over the 20-year period in a memory task involving the recollection of a list of one-digit numbers. The group included $n = 20$ men, whose mean change score was $\bar{x}_* = -4.40$. The standard deviation of the change scores was $s_* = 7.56$. Given that this particular result is selectively reported, what estimates should we make for the population mean and standard deviation to account for the selection? Computing $t_* = -\sqrt{20}\ 4.40/7.56 = -2.60$, and entering Table 3 under $n = 20$, we see that $\hat{\theta} = -.97$, which yields $\hat{\delta} = -.22, \hat{\sigma} = 8.11$, and $\hat{\mu} = -1.78$. Thus the estimated standard deviation is larger (about 7%) than that observed, while the absolute magnitude of the estimated mean is considerably smaller (about 59.5%) that that observed. In this case it appears that the practice of reporting only significant results produced a substantial overestimate of the magnitude of the 20-year change in intellectual functioning.

A1 Details of Computations

Computations for the distributions of $\bar{x}_*$ and s_*^2 were simplified by using the distributions of $\sqrt{n}\bar{x}_*$ and $(n-1)s_*^2$, respectively. The IMSL (1977) subroutine MDGAM was used to compute the incomplete gamma ratio $\Gamma(a; x)$ and the IMSL subroutine MDNOR was used to compute the normal cumulative distribution function $\Phi(x)$. All of the numerical integrations were computed using IMSL subroutines DCADRE and DCSQDU.

Similarly, computations for the distribution of $\sqrt{n}\bar{x}_*/s_*$ were based on the non-central t-distribution with non-centrality parameter $\theta = \sqrt{n}\delta$ and $m = n - 1$ degrees of freedom. The density function for a non-central t-variate with m degrees of freedom and non-centrality parameter θ is given (see e.g., Resnikoff and Lieberman, 1957) by

$$h(x|\theta, m) = \frac{m!}{2^{\frac{m-1}{2}}\Gamma(m/2)\sqrt{\pi m}} e^{-\frac{1}{2}\left(\frac{m\theta^2}{m+x^2}\right)} \left(\frac{m}{m+x^2}\right)^{\frac{m+1}{2}} Hh_m\left(\frac{-\theta x}{\sqrt{m+x^2}}\right)$$

where

$$Hh_m(y) = \int_0^\infty \frac{v^m}{m!} e^{-\frac{1}{2}(v+y)^2}\, dv.$$

For $m < 20$, values of $Hh_m(y)$ can be obtained as

$$Hh_m(y) = P_m(y)Hh_0(y) + Q_m(y)Hh_{-1}(y)$$

where $P_m(y)$ and $Q_m(y)$ are polynomials and

$$Hh_0(y) = \frac{1}{2\pi}\int_0^y e^{-\frac{1}{2}t^2}\, dt,$$

and

$$Hh_{-1}(y) = \frac{e^{-\frac{1}{2}y^2}}{2\pi}$$

A recurrence relationship among the polynomials $P_m(y)$ and $Q_m(y)$ simplifies their computation. For large m it is easier to compute $Hh_m(y)$ by using an asymptotic expansion given by Resnikoff and Leiberman, which is accurate to five decimal places when $m > 20$. This expansion is

$$Hh_m(y) = \frac{1}{m!}t^m e^{-\frac{1}{2}(t+y)^2} \left(\frac{2\pi t^2}{m+t^2} \right)^{\frac{1}{2}} \left[1 - \frac{3m}{4(m+t^2)^2} + \frac{5m^2}{6(m+t^2)^3} \right]$$

where

$$t = \frac{-y + (y^2 + 4m)^{\frac{1}{2}}}{2}.$$

References

Bayarri, M.J. and De Groot, M. (1986a). *Bayesian analysis of selection models.* Technical Report Number 365. Department of Statistics, Carnegie Mellon University.

Bayarri, M.J. and De Groot M. (1986b). *Information in selection models.* Technical Report Number 368. Department of Statistics, Carnegie Mellon University.

Blum, J.E., Fosshage, J.L. and Jarvik, L.F. (1972). Intellectual changes and sex differences in octogenarians: A twenty year longitudinal study of aging. *Developmental Psychology* **7**, 178–187.

Bozarth, J.D. and Roberts, R.R. (1972). Signifying significant significance. *American Psychologist* **27**, 774–775.

Dawid, A.P. and Dickie, J.M. (1977). Likelihood and Bayesian inference from selectively reported data. *Journal of the American Statistical Association* **72**, 845–850.

Eagly, A.H. and Carli L.L. (1981). Sex of researchers and sex-typed communication as determinants of sex differences in influenceability: A meta-analysis of social influence studies. *Psychological Bulletin* **90** 1–20.

Greenwald, A.G. (1975). Consequences of prejudice against the null hypothesis. *Psychological Bulletin* **82**, 1–20.

Hedges, L.V. (1984). Estimation of effect size under non-random sampling: The effects of censoring studies yielding statistically insignificant mean differences. *Journal of Educational Statistics* **9**, 61–85.

Hedges, L.V. and Olkin, I. (1985). *Statistical methods for meta-analysis.* Academic Press, New York.

Hyde, J.S. (1981). How large are cognitive gender differences?: A meta-analysis using ω^2 and d. *American Psychologist* **36**, 892–901.

International Mathematical and Statistical Libraries, Inc. (1977). *IMSL Library 1 (7th ed.).* Houston.

Iyengar, S. and Greenhouse, J.B. (1988). Selection models and the file drawer problem (with discussion). *Statistical Science* **3**, 109–135.

Lane, D.M. and Dunlap, W.P. (1978). Estimating effect sure: Bias resulting from the significance criterion in editorial decisions. *British Journal of Mathematical and Statistical Psychology* **31**, 107–112.

Laubscher, N.F. (1960). Normalizing the non-central t and F distributions. *Annals of Mathematical Statistics* **31**, 1105–1112.

Melton, A.W. Editorial. (1962). *Journal of Experimental Psychology* **64**, 553–557.

Olshen, R.A. (1973). The conditional level of the F-test. *Journal of the American Statistical Association* **68**, 692–698.

Resnikoff, G.J. and Lieberman, G.J. (1957). *Tables of the noncentral t-distribution.* Stanford University Press, Stanford.

Rosenthal, R. (1979). The "file drawer problem" and tolerance for null results. *Psychological Bulletin* **86**, 638–641.

Sterling, T.C. (1959). Publication decisions and their possible effects on inferences drawn from tests of significance — or vice versa. *Journal of the American Statistical Association* **54**, 30–34.

Schneider, H. (1986). *Truncated and censored samples for normal populations.* Marvel-Dekker, New York.

29

Estimating Poisson Error Rates When Debugging Software

Gerald J. Lieberman[1]
Sheldon M. Ross[2]

ABSTRACT Five estimators for the vector of mistake rates of errors discovered in debugging software are proposed and compared for a model in which an unknown number of errors yield numbers of mistakes having independend Poisson distributions.

1 Introduction

Ross (1985) proposed a model in which there is an unknown number, m, of errors contained in a piece of software. He supposed that these errors caused mistakes according to independent Poisson processes with the rate corresponding to the ith such error being (the unknown) λ_i, $i = 1, \ldots, m$. At a fixed time $t = 1$, the output of the software is analyzed and all resulting mistakes determined, and the errors causing these mistakes identified. In this paper we will be interested in estimating the mistake rates of those errors that are discovered. In Section 2 we present five possible estimators for the vector of mistake rates. The motivation for these estimators is then given in Section 3, and in Section 4 we report on the results of extensive simulations done to compare these estimators.

2 The Estimators

Consider m independent Poisson random variables having respective means $\lambda_1, \ldots, \lambda_m$ where both m and the λ_i, $i = 1, \ldots, m$ are assumed unknown. Suppose that k of the Poisson variables are observed to be positive, taking on the set of values $\{N_i, i = 1, \ldots, k\}$. The $m - k$ Poisson random variables

---

[1]Department of Operations Research, Stanford University.

[2]Department of Industrial Engineering and Operations Research, University of California, Berkeley.

taking on value 0 are unobserved (and thus the value $m-k$ is not learned). We are interested in estimating the vector of means, which we will designate by $\lambda_1, \ldots, \lambda_k$ corresponding to the Poisson random variables having respective observed values $N_1, \ldots, N_k$. Let M_i, $i \geq 1$, denote the number of Poisson random variables that are observed to take on the value i. Also, let $N = \sum_i N_i$ and, for $i = 1, \ldots, m$ set $I_i = 1$ if $N_i > 0$ and $I_i = 0$ if $N_i = 0$. We will consider the following five estimators of $(\lambda_1, \ldots, \lambda_k)$:

(i) The estimator $E^{(1)}$ where $E_i^{(1)} = N_i, \ i = 1, \ldots, k$

(ii) The estimator $E^{(2)}$ where $E_i^{(2)} = N_i - M_1/k, \ i = 1, \ldots, k$

(iii) The estimator $E^{(3)}$ where $E_i^{(3)} = cN_i, \ i = 1, \ldots, k$, where $c = (N - M_1)/N$

(iv) The estimator $E^{(4)}$ where $E_i^{(4)} = aN_i + (1-a)\bar{N}, \ i = 1, \ldots, k$, where $\bar{N} = N/k$ and $a = \sum_i I_i(N_i - \bar{N})^2 / (N + \sum_i I_i(N_i - \bar{N})^2)$

(v) The estimator $E^{(5)}$ where $E_i^{(5)} = cE_i^{(4)}, \ i = 1, \ldots, k$, with c as in (iii)

3 Motivation for the Estimators

In Ross (1985) it was supposed that the errors causing the observed mistakes were identified and corrected, and the major problem of interest was to then estimate the sum of the λ's of those errors that remained. Two estimators of this quantity considered in Ross (1985) were

$$M_1 \qquad \text{and} \qquad \sum_{i \geq 1} M_i \frac{ie^{-i}}{1 - e^{-i}},$$

where M_i is equal to the number of errors that were determined to have caused exactly i mistakes. (The estimator M_1 had previously been proposed by Robbins (1968) in a slightly different context.) Whereas simulation indicated that the latter estimator appeared to perform better, a recent paper by Derman and Koh (1968) has analytically shown that whereas the latter estimator will usually dominate when the number of errors m is small to moderate, the former estimator will be preferable for larger values of m. The Derman-Koh paper (1968) concerns itself primarily with some clever mathematics to establish its results, and does not present any intuitive reason for the superiority of M_1 when m is large. However, in knowing their result, it is not too difficult to see why it might be so. We now present such an argument.

Suppose that m is a very large number and all of the λ's are approximately equal to c/m, a very small number. In this case the number of mistakes by time 1 will be approximately Poisson distributed with mean c, and almost all of these mistakes will be singletons — that is, M_1 will also be approximately Poisson with mean c, and the other M_i, $i > 1$, will be zero with high probability. Since the sum of the λ's corresponding to errors having 0 mistakes is

$$c - (\text{number of errors causing mistakes})c/m \simeq c$$

it is clear that the best estimate of this is M_1. (That is, in this situation we are effectively interested in estimating a parameter that is equal to the mean of the observed Poisson random variable M_1.) In addition, when m is very large and there is a mix of values of the λ (with a large number being very small) the data corresponding to the larger values of λ will play a relatively negligible role in the observed values of the estimators, and again the pivotal role will be played by the large set of errors having small mistake rates. For this reason, it intuitively appears that M_1 should again dominate.

We are now in position to present intuitive justifications for the five estimators presented in Section 3.

(a) The estimator $E^{(1)}$ is the maximum likelihood estimator.

(b) The rationale behind the estimators $E^{(2)}$ and $E^{(3)}$ is that since N, the total number of mistakes, is Poisson distributed with mean $\lambda \equiv \sum_{i=1}^{m} \lambda_i$, it follows that it is the best estimator of λ. In addition, since we think that there should be some positive estimator of the sum of those $m - k$ errors that have not yet caused any mistakes — and we will take M_1 as our estimator of this quantity — it follows that the maximum likelihood estimators N_i overestimate the λ_i, $i = 1, \dots, k$. Since the total overestimate of the sum of the λ_i, $i = 1, \dots, k$ is M_1 the estimator $E^{(2)}$ is an attempt to correct this by decreasing each of the estimates by a fixed amount, whereas $E^{(3)}$ attempts to correct by scaling each estimator by the same fraction. (Intuitively, it appears to the authors that scaling would be more appropriate than fixed amount reductions, but we initially consider both possibilities.)

(c) The estimator $E^{(4)}$ is an attempt to estimate a vector of means by using a "weighted towards the average" approach developed in Jun (1988). Specifically, Jun considers a problem in which one has n (a known number) of independent random variables having a common form of distribution with unknown means. If X_i, $i = 1, \dots, n$ are the values of these variables then Jun considers estimators of the means $E[X_i]$ of the form $aX_i + (1 - a)\bar{X}$, where $\bar{X} = \sum X_i/n$. He then determines the best (in the sense of minimizing the sum of squares of the errors) value of a, which will be a function of the unknown parameters of the distributions, and then estimates this quantity by considering the same function evaluated not at the parameters but at their maximum likelihood estimates. In the case of estimating n independent Poisson means, Jun's approach leads to the estimators

$$aX_i + (1-a)\bar{X}$$

with

$$a = \sum (X_i - \bar{X})^2 \Big/ \left(\sum X_i + \sum (X_i - \bar{X})^2 \right).$$

The estimators $E_i^{(4)}$ thus directly make use of this "weighted towards the average" approach without compensating for the "overestimation hypothesis" noted in (b). The estimators $E_i^{(5)}$ modify these estimators (by scaling) to take this overestimation hypothesis into account.

For any of the above estimators E, define the mean square error of E by

$$\text{MSE}(E) = E\left[\sum I_i (E_i - \lambda_i)^2 \right].$$

The mean square error of the maximum likelihood estimator $E^{(1)}$ is computed by the following reasoning:

$$\begin{aligned} \lambda_i &= E[(N_i - \lambda_i)^2] \\ &= E[(N_i - \lambda_i)^2 \mid I_i = 1](1 - e^{-\lambda_i}) + \lambda_i^2 e^{-\lambda_i}. \end{aligned}$$

Hence,

$$\begin{aligned} E[I_i(N_i - \lambda_i)^2] &= E[(N_i - \lambda_i)^2 \mid I_i = 1](1 - e^{-\lambda_i}) \\ &= \lambda_i (1 - \lambda_i e^{-\lambda_i}) \end{aligned}$$

and thus,

$$\text{MSE}(E^{(1)}) = \sum_{i=1}^{m} \lambda_i (1 - \lambda_i e^{-\lambda_i}).$$

Unfortunately, however, we have not been able to obtain a simple closed form expression for any of the other mean square errors and have thus had to resort to simulation to determine which estimator is superior.

Remark 3.1 *Another estimator one might consider would be obtained by estimating λ_i, $i \geq 1$, by utilizing the conditional distribution of a Poisson random variable with mean λ_i given that it is positive. To see that this would not be beneficial, consider any set of estimators E_i of λ_i, $i \geq 1$, where E_i depends on the data only through N_i. Since we are only interested in estimating λ_i when $I_i = 1$ we can suppose that $E_i = 0$ when $I_i = 0$. Let $\sigma_i^2 = E[(E_i - \lambda_i)^2]$ and note that*

$$\begin{aligned} \sigma_i^2 &= \lambda_i^2 e^{-\lambda_i} + E[(E_i - \lambda_i)^2 \mid I_i = 1](1 - e^{-\lambda_i}) \\ &= \lambda_i^2 e^{-\lambda_i} + E[I_i(E_i - \lambda_i)^2]. \end{aligned}$$

Hence, since $\sum \lambda_i^2 e^{-\lambda_i}$ is a constant it follows that the choice of good estimators E_i, $i \geq 1$, in our problem is the same as in the problem where one attempts to estimate all of the Poisson means, and since N_i is an admissible estimate of λ_i in this unconstrained problem there is no necessity to consider other estimators. (In particular, we need not consider an estimator of the form: estimate λ_i as 0 if $N_i = 0$; otherwise use the maximum likelihood estimate based on the conditional distribution of Poisson random variable given that it is positive.)

4 Simulation Analysis of the Estimators

In all the simulations we ran it always turned out that $\text{MSE}(E^{(i)})$ decreased in i, $i = 1, \ldots, 5$. For instance, in two cases we set $m = 100$ and chose the 100 values of λ uniformly between 0 and 3. In both cases we then ran 100 simulations to estimate the mean square errors. The results were quite similar in both cases, giving the following estimates of the mean square errors in one of the cases:

$$M_1 = 121.5, \quad M_2 = 112.8, \quad M_3 = 82.8, \quad M_4 = 51.4, \quad M_5 = 38.8,$$

where $M_i = \text{MSE}(E^{(i)})$. In another simulation where we chose $m = 250$, with 100 of the λ uniformly chosen between 0 and 3, 100 uniformly chosen between 5 and 7, and 50 uniformly chosen between 8 and 10, we obtained the following estimates of the mean square errors:

$$M_1 = 1148.4, \quad M_2 = 1145.4, \quad M_3 = 1099.8, \quad M_4 = 751.4, \quad M_5 = 742.3.$$

In this case it should be noted that most of the errors result from estimating the errors with large values of λ, and thus there is not as much scope for percentage improvement as in the case considered above.

In addition, it is worth noting that in almost all simulation runs (as opposed to just the average of the runs) the sum of squares of errors of the five estimators were decreasing. Table 1 gives the results of the last 19 (of 100) runs when $m = 300$ and 100 of the λ's were uniformly chosen in $(0, 2)$, 100 in $(3, 9)$, and 100 in $(4, 9)$. The tabulated values T_i refer to the sum of squares of the errors for estimator i. $\bar{T}_i$ is the average and V_i is the variance of the T_i's over all 100 runs.

Summarizing, it would seem that compensating for the overestimation by reducing the maximum likelihood estimates by a fixed amount (estimator E^2) results in a slight improvement, whereas compensating by scaling results in a much more substantial improvement. In addition, the greatest improvement is obtained by not using the maximum likelihood estimates directly but rather "weight them towards the average." This (that is, $E^{(4)}$) in itself leads to a great improvement over the other estimates, and an even greater improvement is obtained by combining this with the overestimation compensation (to obtain $E^{(5)}$).

TABLE 1. Simulation Runs

Trial	T_1	T_2	T_3	T_4	T_5
82	647.747	647.473	643.206	229.471	228.074
83	585.470	585.450	581.712	207.167	206.310
84	604.730	605.735	602.105	234.720	234.935
85	834.172	833.265	831.070	341.082	339.573
86	648.429	648.422	646.514	246.160	245.710
87	613.959	614.597	605.235	229.525	228.042
88	604.203	604.636	602.792	231.050	231.075
89	472.874	473.547	470.957	186.796	187.129
90	784.148	783.199	773.775	308.084	304.354
91	628.643	628.761	626.926	240.765	240.475
92	805.946	806.518	796.980	328.493	326.213
93	832.697	830.782	826.141	334.841	331.580
94	603.276	604.875	599.399	233.535	233.934
95	642.532	644.378	640.355	260.696	261.549
96	785.062	784.748	780.390	317.131	315.631
97	719.839	720.195	717.996	284.529	284.282
98	607.406	607.406	607.406	224.901	224.901
99	735.818	735.818	735.818	289.577	289.577
100	774.466	775.525	768.768	317.919	317.073

$\bar{T}_1 = 672.520$, $\bar{T}_2 = 672.606$, $\bar{T}_3 = 669.028$, $\bar{T}_4 = 260.833$, $\bar{T}_5 = 260.002$, $V_1 = 10062.819$, $V_2 = 10036.589$, $V_3 = 9847.313$, $V_4 = 2488.182$, $V_5 = 2447.342$

Acknowledgments: Research of Gerald J. Lieberman supported by the Office of Naval Research under contract N00014-84-K-0244 (NR347-124). Research of Sheldon M. Ross supported by the U.S. Air Force Office of Scientific Research under Grant AFOSR-86-0153.

References

Derman, C., and S. Koh, (1988). "On the Comparison of Two Software Reliability Estimators." *Probability in the Engineering and Informational Sciences*, **2**, 15–21.

Jun, C.H., (1988). "Heuristics Procedures for Simultaneous Estimation of Several Normal Means." *Probability in the Engineering and Informational Sciences*, **2**, 95–113.

Robbins, H. (1968). "Estimating the Probabilities of Unobserved Events." *Annals of Math. Stat.*, **39**, 256–57.

Ross, S.M., (1985). "Statistical Estimation of Software Reliability." *IEEE Trans. Software Engineering*, Vol SE-11, No. 5, 479–83, .

30

A Comparison of the Likelihood Ratio, Wald, and Rao Tests

Albert Madansky[1]

ABSTRACT Three commonly considered methods for forming approximate (large-sample) tests of a simple null hypothesis are: (a) the likelihood ratio test, (b) the Wald "linearization" test, and (c) a quadratic scores test due to Rao. In the context of testing that the variance of a normal distribution is equal to one, it is possible to make detailed finite-sample power comparisons, both "local" and "nonlocal," of these tests. In contrast to at least one assertion made in the literature (Chandra and Joshi, 1983), none of the three tests dominates another, even locally.

1 Introduction

Suppose we have a probability distribution whose natural parameter θ is an m-vector, and we wish to test the simple null hypothesis $\theta = \theta_0$. It may be difficult to obtain an exact test of this hypothesis, and so approximate test procedures have been adduced. The three most commonly considered methods for obtaining asymptotic tests are the likelihood ratio test, the Wald test (called "linearization" in Madansky and Olkin (1965)), and the quadratic score test due to Rao (see, for example, section 6e.2 of Rao (1965)). In Rao (1962) (and also in Rao (1965), but not in later editions of that book) it was conjectured that the Rao test is locally more powerful than the others. Peers (1971) developed the power functions of these three tests to $O(n^{-1/2})$ and showed by example that "no one of these criteria is uniformly superior to the other two." But Chandra and Joshi (1983) considered the case where $m = 1$ and argued that if we view each of these tests as a two-sided test based on the maximum likelihood estimate of $\hat{\theta}$ of θ and we superposed on the tests the condition that the size (i.e., probability content) of each of the tails be equal, then the Rao conjecture is true and that the Wald test is the worst of the three. This note looks at a specific example, testing the simple hypothesis that the variance θ of a normal

[1]Graduate School of Business, University of Chicago

distribution is equal to 1, and shows that the Chandra-Joshi conclusion is incorrect.

The author wishes to thank the referee of an earlier version of this paper, who not only saved me from the embarrassment of a public display of my confused exposition in that version but who also brought to my attention the Peers and Chandra-Joshi papers. The referee also provided me with an unpublished paper by Kallenberg (1983), who considered the same example, calculating the Bahadur slopes for these three tests for a "nonlocal" comparison and also providing a "local" approximation (around $\theta = 1$) to the number of observations required to achieve a fixed power for each of the three tests. He reached the following conclusions:

	comparison	
	"local"	"nonlocal"
$\theta < 1$	$W > LR > R$	$W = LR > R$
$\theta > 1$	$R > LR > W$	$R = LR > W$

where LR denotes likelihood ratio test, W denotes Wald test, R denotes Rao test, and the $>$ $(=)$ between two tests is to be read as "is better than" ("is equivalent to"). In this paper we consider the finite sample properties of these three tests and develop comparable conclusions.

2 The Example

As stated earlier, we consider the problem of testing the hypothesis $H_0 : \theta = \theta_0$ given a random sample $x_1, \ldots, x_n$ from a normal distribution with mean zero and unknown variance θ. Here $\hat{\theta} = s/n$, where $s = \sum_{i=1}^{n} x_i^2$. The three test statistics — the likelihood ratio, Wald, and Rao statistics — are given respectively by

$$L = n \log \theta_0 + n\hat{\theta}/\theta_0 - n \log \hat{\theta} - n,$$
$$W = n(\hat{\theta} - \theta_0)^2/2\hat{\theta}^2,$$

and

$$R = n(\hat{\theta} - \theta_0)^2/2\theta_0^2.$$

We see from the form of each of the tests that they are each two-sided tests based on $\hat{\theta}$, each with its own pair of critical values. Thus, as pointed out by Kallenberg, superposition of the equal-tailed condition required by Chandra-Joshi will reduce the three tests to one common test which is none of the three under consideration.

Table 1 exhibits the critical values for $\hat{\theta}$ for testing $\theta = 1$ against $\theta \neq 1$ based on the large sample distribution (χ_1^2) of L, W, and R. It also exhibits the exact probabilities of rejection of H_0 based on these three approximate tests. We note that under the null hypothesis the probability of rejecting

H_0 using L is smaller than that using W and greater than that using R. Thus L is more conservative than W and less conservative than R when $\theta_0 = 1$.

As another comparison of the tests, we can hold the exact power fixed for a given alternative θ and observe the rate of convergence to zero of the exact level of significance for each of the three tests. Tables 2 and 3 exhibit the convergence to zero of the exact level of significance for each of the three test procedures when the power is held fixed at 0.60 for two different non-null values of θ, $\theta = 0.6$ and $\theta = 1.4$, respectively. Note that for $\theta = 0.6$ the size of the Wald test converges to zero more rapidly than that of the likelihood ratio test, and that for $\theta = 1.4$ the size of the Rao test converges to zero more rapidly than that of the likelihood ratio test.

Bahadur (1967) defined a test to be optimal for a given non-null θ if the rate of convergence to zero of the size is faster than that of any other test procedure with the same power at that alternative θ, and showed that the likelihood ratio test is optimal. It is easy to determine that the function $\rho(\theta)$ measuring the rate of convergence to zero of the size of the likelihood ratio test is given for this example by

$$\rho(\theta) = e^{-(\theta - 1 - \log \theta)/2},$$

that is, the size of the likelihood ratio test tends to zero at the rate $\big(\rho(\theta)\big)^n$. Regression of the logarithm of the levels of Table 2 on n, with no intercept included, produced the following coefficients:

	$\theta = 0.6$	$\theta = 1.4$
likelihood ratio	−0.0622	−0.0328
Wald	−0.0694	−0.0189
Rao	−0.0409	−0.0364

These regression coefficients should be contrasted with $\log \rho(0.6) = -0.0554$ and $\log \rho(1.4) = -0.0318$. Note that the regression coefficients calculated for the likelihood ratio test are reasonably close to the asymptotic rates, but that the magnitude of other sets of regression coefficients do not conform with Bahadur's optimality theorem. That is, the magnitude of the slope for the Wald test when $\theta = 0.6$ exceeds the magnitude of the slope for the likelihood ratio test. Similarly, the magnitude of the slope for the Rao test exceeds that of the likelihood ratio test when $\theta = 1.4$. This is of course an artifact of the values of n used in calculating the regression coefficients, and does not invalidate Bahadur's result.

Table 4 exhibits the power of each of these three tests for various values of θ and n, to yield another comparison of these tests. Note that for small n, of which $n = 3$ is representative, since the critical regions using W and R are one-sided, L has better power for θ corresponding to the tail of the distribution of $\hat{\theta}$ which is never in the critical region, i.e., better power than R for $\theta < 1$ and better power than W for $\theta > 1$. This behavior of the power function of L persists, though not to as marked a degree, for moderate n (represented by $n = 15$) and even large n (like $n = 250$). Thus for a two-sided alternative there is no clear power dominance of any of the three test procedures over each other, and so we conclude that the ordering of L, W, and R based on exact power considerations is as follows:

$$\begin{array}{ll} \theta < 1 & W > LR > R \\ \theta > 1 & R > LR > W. \end{array}$$

Finally, emulating the finite sample calculations performed by Kallenberg, we present in Table 5 the sample sizes required for each of these tests so that, for various values of θ, the power of the test will be 0.5. Here the results are not quite as clearcut. For $\theta < 1$ we have that $W > LR > R$, but for $\theta > 1$ we almost have that $R > LR > W$, except that when $\theta = 1.1$ the ranking is $W > R > LR$.

This simple example had the virtue of producing different critical regions for the Wald, Rao, and likelihood ratio tests, each based on the same test statistic with known small sample distribution, thus enabling a detailed comparison of the behavior of the three tests. From our study we conclude that no one of these three tests dominates the others, and that a casual use of Bahadur slopes for comparison may be misleading.

References

Bahadur, R.R. (1967). An optimal property of the likelihood ratio statistic. *Proceedings of the Fifth Berkeley Symposium on Mathematical Statistics and Probability,* Vol 1. Berkeley and Los Angeles: University of California Press, 13–26.

Chandra, T.K. and Joshi, S.N. (1983). Comparison of the likelihood ratio, Rao's and Wald's tests and a conjecture of C.R. Rao. *Sankhyā*, Series A 45, 226–46.

Kallenberg, W.C.M. On likelihood ratio, Rao's and Wald's tests. Report No. 259. (June) Department of Mathematics and Information Science, Vrije Universiteit, Amsterdam.

Madansky, A. and Olkin, I. (1969). Approximate confidence regions for constraint parameters. *Multivariate Analysis II*, P.R. Krishnaiah, New York, Academic Press, 261–86.

Peers, H.W. (1971). Likelihood ratio and associated test criteria. *Biometrika* 58 (December) 477–587.

Rao, C.R. (1962). Efficient estimates and optimum inference procedure in large samples. *Journal of the Royal Statistical Society*, Series B, 24, 46–72.

Rao, C.R. (1965). *Linear Statistical Inference and its Applications*, New York, Wiley.

TABLE 1. Critical values of $\hat{\theta}$ for nominal $\alpha = 0.05$ and exact level of significance of these critical regions

	likelihood ratio			Wald			Rao		
n	lower	upper	level	lower	upper	level	lower	upper	level
1	0.008	6.751	0.0805	0.265	—	0.3928	0	3.772	0.521
2	0.057	4.403	0.0678	0.338	—	0.2867	0	2.960	0.518
3	0.115	3.546	0.0623	0.385	—	0.2359	0	2.600	0.504
4	0.166	3.088	0.0593	0.419	—	0.2051	0	2.386	0.489
5	0.211	2.797	0.0576	0.447	—	0.1840	0	2.240	0.476
6	0.249	2.593	0.0564	0.469	—	0.1685	0	2.132	0.465
7	0.282	2.441	0.0555	0.488	—	0.1564	0	2.048	0.455
8	0.310	2.323	0.0548	0.505	50.000	0.1467	0.020	1.980	0.447
9	0.336	2.228	0.0542	0.520	13.150	0.1388	0.076	1.924	0.441
0	0.359	2.149	0.0538	0.533	8.100	0.1320	0.123	1.877	0.138
15	0.444	1.896	0.0524	0.583	3.517	0.1096	0.284	1.716	0.442
20	0.501	1.754	0.0517	0.617	2.630	0.0967	0.380	1.620	0.450
25	0.543	1.661	0.0513	0.643	2.244	0.0882	0.446	1.554	0.457
30	0.575	1.595	0.0510	0.664	2.025	0.0823	0.494	1.506	0.462
35	0.602	1.544	0.0508	0.681	1.882	0.0779	0.531	1.496	0.467
40	0.623	1.505	0.0506	0.695	1.780	0.0746	0.562	1.438	0.470
45	0.642	1.472	0.0505	0.708	1.704	0.0719	0.587	1.403	0.472
50	0.657	1.445	0.0504	0.718	1.645	0.0698	0.608	1.392	0.475
100	0.748	1.303	0.0500	0.783	1.383	0.0599	0.723	1.277	0.485
150	0.790	1.244	0.0499	0.815	1.293	0.0565	0.774	1.226	0.488
200	0.817	1.209	0.0498	0.836	1.244	0.0548	0.804	1.196	0.490
250	0.835	1.186	0.0497	0.851	1.213	0.0537	0.825	1.175	0.491
300	0.848	1.169	0.0497	0.862	1.191	0.0530	0.840	1.160	0.492
350	0.859	1.156	0.0497	0.871	1.174	0.0525	0.852	1.148	0.492
400	0.868	1.145	0.0497	0.878	1.161	0.0522	0.861	1.139	0.493
450	0.875	1.136	0.0497	0.884	1.150	0.0519	0.869	1.131	0.493
500	0.881	1.129	0.0497	0.890	1.141	0.0517	0.876	1.124	0.493
1000	0.915	1.090	0.0497	0.919	1.096	0.0506	0.912	1.088	0.495

TABLE 2. Critical values of $\hat{\theta}$ for Power= 0.60 when $\theta = 0.6$ and exact level of significance of these critical regions

	likelihood ratio			Wald			Rao		
n	lower	upper	level	lower	upper	level	lower	upper	level
15	0.629	1.495	0.2434	0.629	2.432	0.1485	0.628	1.372	0.2958
20	0.628	1.495	0.1764	0.629	2.444	0.1056	0.628	1.372	0.2282
25	0.626	1.497	0.1286	0.627	2.461	0.0763	0.627	1.373	0.1775
30	0.626	1.499	0.0943	0.626	2.478	0.0557	0.626	1.374	0.1391
35	0.625	1.501	0.0695	0.625	2.495	0.0409	0.625	1.375	0.1098
40	0.624	1.503	0.0515	0.624	2.510	0.0302	0.624	1.376	0.0873
45	0.624	1.504	0.0382	0.624	2.524	0.0224	0.624	1.376	0.0698
50	0.623	1.506	0.0285	0.623	2.537	0.0167	0.623	1.377	0.0561
100	0.618	1.515	0.0020	0.618	2.624	0.0011	0.618	1.382	0.0080
150	0.615	1.520	0.0002	0.615	2.673	0.0001	0.615	1.385	0.0016
200	0.613	1.523	0.0000	0.613	2.706	0.0000	0.613	1.387	0.0004
250	0.612	1.525	0.0000	0.612	2.730	0.0000	0.612	1.388	0.0001
500	0.609	1.531	0.0000	0.609	2.796	0.0000	0.609	1.391	0.0000
1000	0.606	1.535	0.0000	0.606	2.849	0.0000	0.606	1.394	0.0000

TABLE 3. Critical values of $\hat{\theta}$ for Power= 0.60 when $\theta = 1.4$ and exact level of significance of these critical regions

	likelihood ratio			Wald			Rao		
n	lower	upper	level	lower	upper	level	lower	upper	level
15	0.745	1.308	0.4468	0.789	1.339	0.4877	0.710	1.290	0.4206
20	0.752	1.297	0.3940	0.805	1.318	0.4436	0.715	1.285	0.3615
25	0.754	1.295	0.3431	0.808	1.311	0.4011	0.714	1.286	0.3050
30	0.752	1.297	0.2957	0.809	1.308	0.3609	0.709	1.291	0.2533
35	0.750	1.301	0.2528	0.809	1.309	0.3235	0.704	1.297	0.2081
40	0.747	1.305	0.2148	0.808	1.311	0.2894	0.698	1.302	0.1669
45	0.744	1.310	0.1820	0.807	1.314	0.2586	0.692	1.308	0.1383
50	0.741	1.314	0.1538	0.806	1.317	0.2310	0.687	1.313	0.1125
100	0.722	1.341	0.0291	0.797	1.341	0.0799	0.659	1.341	0.0163
150	0.714	1.353	0.0062	0.793	1.353	0.0316	0.647	1.353	0.0031
200	0.710	1.360	0.0015	0.791	1.360	0.0136	0.640	1.360	0.0008
250	0.707	1.365	0.0004	0.789	1.365	0.0062	0.635	1.365	0.0002
500	0.700	1.376	0.0000	0.785	1.376	0.0002	0.624	1.376	0.0000
1000	0.695	1.383	0.0000	0.783	1.383	0.0000	0.617	1.383	0.0000

TABLE 4. Power of tests

.1	.6712	.9909	0	1.0000	1.0000	.9997			
.2	.3675	.8764	0	.9956	.9998	.8731			
.3	.2342	.7214	0	.8971	.9847	.4910			
.4	.1650	.5902	.0003	.6595	.8883	.2239			
.5	.1241	.4889	.0015	.4221	.7095	.0994	1.0000		1.0000
.6	.0980	.4114	.0047	.2544	.5174	.0457	.9999	.0000	.9998
.7	.0809	.3515	.0109	.1513	.3583	.0234	.9804	.9889	.9724
.8	.0700	.3044	.0207	.0921	.2424	.0173	.6948	.7661	.6449
.9	.0640	.2666	.0340	.0617	.1628	.0240	.2124	.2770	.1766
1.0	.0621	.2359	.0504	.0524	.1096	.0442	.0497	.0537	.0491
1.1	.0637	.2106	.0392	.0607	.0743	.0781	.1917	.1313	.2195
1.2	.0684	.1895	.0898	.0841	.0510	.1245	.5412	.4423	.5800
1.3	.0757	.1716	.1118	.1203	.0357	.1807	.8371	.7702	.8595
1.4	.0851	.1563	.1345	.1664	.0258	.2435	.9625	.9383	.9695
1.5	.0962	.1432	.1578	.2198	.0200	.3098	.9937	.9883	.9951
1.6	.1088	.1318	.1812	.2776	.0175	.3768	.9990	.9981	.9993
1.7	.1225	.1218	.2045	.3375	.0179	.4417	.9998	.9997	.9998
1.8	.1370	.1130	.2275	.3973	.0214	.5031	1.0000	.9999	1.0000
1.9	.1522	.1052	.2502	.4550	.0280	.5604			
2.	.1678	.0982	.2724	.5098	.0380	.6126			
3.	.3247	.0565	.4574	.8511	.2853	.8984			
4.	.4535	.0376	.5828	.9546	.5877	.9715			
5.	.5509	.0274	.6684	.9845	.7836	.9907			
6.	.6243	.0210	.7290	.9940	.8879	.9965			
7.	.6805	.0168	.7735	.9974	.9410	.9985			
8.	.7243	.0138	.8072	.9988	.9679	.9993			
9.	.7592	.0116	.8334	.9994	.9820	.9996			
10.	.7874	.0100	.8542	.9996	.9895	.9998			

TABLE 5. Sample size required for 0.05 level of significance and 0.5 power

θ	LR	W	R
0.5	19	1	30
0.6	33	14	46
0.7	65	38	83
0.8	161	117	187
0.9	704	609	757
1.1	834	943	783
1.2	225	284	200
1.3	108	149	92
1.4	65	98	53
1.5	45	72	35

31

On the Inadmissibility of the Modified Step-Down Test Based on Fisher's Method For Combining Independent p-Values

John I. Marden[1]
Michael D. Perlman[2]

ABSTRACT Marden and Perlman (1988) have shown that the classical step-down procedure for the Hotelling T^2 testing problem is inadmissible in most cases. Mudholkar and Subbaiah (1980) proposed a modified step-down procedure wherein the p-values associated with the sequence of stepwise F tests are combined according to Fisher's combination method. In the present paper it is shown that the modified step-down procedure is inadmissible if at least one step is of dimension one.

1 Introduction

The Hotelling T^2 problem (with covariates) may be expressed in the following canonical form. One observes X and S, independent, where

$$X \sim \mathcal{N}_p(\mu, \Sigma), \quad S \sim \mathcal{W}_p(\Sigma, n), \tag{1.1}$$

respectively, a p-dimensional normal distribution and a p-dimensional Wishart distribution. The covariance matrix Σ is assumed to be positive definite, but otherwise Σ and the mean vector μ are unknown. We also assume that $n \geq p$, so that S is nonsingular with probability one.

[1]Department of Statistics, University of Illinois at Urbana-Champaign, Urbana, Illinois 61801.

[2]Department of Statistics GN-22 , University of Washington, Seattle, Washington 98195.

Partition μ and Σ into $q+1$ blocks:

$$\mu = \begin{pmatrix} \mu_0 \\ \mu_1 \\ \vdots \\ \mu_q \end{pmatrix}, \qquad \Sigma = \begin{pmatrix} \Sigma_{00} & \Sigma_{01} & \cdots & \Sigma_{0q} \\ \Sigma_{10} & \Sigma_{11} & \cdots & \Sigma_{1q} \\ \vdots & \vdots & & \vdots \\ \Sigma_{q0} & \Sigma_{q1} & \cdots & \Sigma_{qq} \end{pmatrix} \tag{1.2}$$

with $\mu_i : p_i \times 1$ and $\Sigma_{ij} : p_i \times p_j$, where $p_0 + p_1 + \cdots + p_q = p$, $p_0 \geq 0, p_1 \geq 1, \ldots, p_q \geq 1$. The Hotelling T^2 testing problem with covariates is that of testing

$$H_0 : \mu = 0 \quad \text{vs.} \quad H_A : \mu_0 = 0. \tag{1.3}$$

Since $\mu_0 = 0$ under both hypotheses, the variables measured by X_0 act as covariates. If $p_0 = 0$, this problem reduces to the ordinary T^2 problem of testing $\mu = 0$ vs. $\mu \neq 0$.

Consider the nested sequence of hypotheses

$$H_0 \equiv H^{(q)} \subset H^{(q-1)} \subset \cdots \subset H^{(1)} \subset H^{(0)} \equiv H_A, \tag{1.4}$$

where

$$H^{(i)} : \mu_j = 0 \qquad \text{for } j = 0, 1, \ldots, i. \tag{1.5}$$

The classical step-down test proceeds by testing $H^{(1)}$ vs. $H^{(0)}$, $H^{(2)}$ vs. $H^{(1)}, \ldots$, $H^{(q)}$ vs. $H^{(q-1)}$, using the likelihood ratio test (LRT) at each step. The overall null hypothesis $H_0 \equiv H^{(q)}$ is accepted if and only if each step-wise LRT accepts its null hypothesis.

The level α_i likelihood ratio test (LRT) for testing $H^{(i)}$ vs. $H^{(i-1)}$, $i = 1, \ldots, q$, accepts $H^{(i)}$ if and only if

$$Y_i \leq a_i \equiv a_i(\alpha_i), \tag{1.6}$$

where, for $i = 0, 1, \ldots, q$,

$$Y_i = \frac{T_i^2 - T_{i-1}^2}{1 + T_{i-1}^2}, \tag{1.7}$$

$$T_i^2 = \begin{pmatrix} X_0 \\ X_1 \\ \vdots \\ X_i \end{pmatrix}' \begin{pmatrix} S_{00} & S_{01} & \cdots & S_{0i} \\ S_{10} & S_{11} & \cdots & S_{1i} \\ \vdots & \vdots & & \vdots \\ S_{i0} & S_{i1} & \cdots & S_{ii} \end{pmatrix}^{-1} \begin{pmatrix} X_0 \\ X_1 \\ \vdots \\ X_i \end{pmatrix}. \tag{1.8}$$

Here, $T_{-1}^2 \equiv 0$, while $T_0^2 \equiv Y_0 \equiv 0$ when $p_0 = 0$; note too that

$$1 + T_i^2 = \prod_{j=0}^{i} (1 + Y_j). \tag{1.9}$$

The joint distribution of $(Y_0, Y_1, \ldots, Y_q)$ is as follows (cf. Marden and Perlman (1988)):

$$\begin{cases} Y_i \mid Y_0, Y_1, \ldots, Y_{i-1} & \sim \quad \chi^2_{p_i}\left(\Delta_i/(1+T^2_{i-1})\right)/\chi^2_{n_i}, \qquad i = 1, \ldots, q; \\ Y_0 & \sim \quad \chi^2_{p_0}/\chi^2_{n_0} \end{cases} \tag{1.10}$$

where

$$n_i = n - \sum_{j=0}^{i} p_j + 1 \tag{1.11}$$

and the two chi-squared random variables appearing in each ratio are independent, with $\chi^2_\nu(\lambda)(\chi^2_\nu)$ denoting a noncentral (central) chi-squared random variable with ν degrees of freedom and noncentrality parameter λ. The parameters $\Delta_1, \ldots, \Delta_q$ in (1.10) are determined by the relations

$$\Delta_i = \tau_i^2 - \tau_{i-1}^2 (\geq 0), \qquad i = 1, \ldots, q, \tag{1.12}$$

where τ_i^2 is defined as T_i^2 in (1.8) but with (X, S) replaced by (μ, Σ) (so $\tau_0^2 \equiv 0$).

Note that the hypothesis $H^{(i)}$ in (1.5) is equivalent to $\Delta_1 = \cdots = \Delta_i = 0$. Thus when $H^{(i)}$ is true, the statistics $Y_0, Y_1, \ldots, Y_q$ are mutually independent F statistics, hence the critical value $a_i(\alpha_i)$ for the step-wise LRT (1.6) is the upper α_i-point of the (non-normalized) F_{p_i, n_i} distribution. The classical step-down test, with acceptance region

$$\{Y_1 \leq a_1, \ldots, Y_q \leq a_q\}, \tag{1.13}$$

therefore has overall significance level α given by

$$1 - \prod_{i=1}^{q} (1 - \alpha_i). \tag{1.14}$$

At this point we restrict consideration to the class of tests for (1.3) that depend on (X, S) only through $(Y_0, Y_1, \ldots, Y_q)$. Under this restriction, the testing problem (1.3) is equivalent to the reduced problem

$$H_0 : \Delta_1 = \cdots = \Delta_q = 0 \quad \text{vs.} \quad H_A : \Delta_1 \geq 0, \ldots, \Delta_q \geq 0 \tag{1.15}$$

with at least one inequality strict. Theorem 3.1 of Marden and Perlman (1988) [hereafter abbreviated as MP (1988)] characterizes the class of all admissible tests for (1.15) based on $(Y_0, Y_1, \ldots, Y_q)$. They applied this minimal complete class theorem to show that the step-down test (1.13) is inadmissible for problem (1.15), hence *a fortiori* inadmissible for (1.3), if $p_0 = 0$ and $q \geq 3$, or if $p_0 > 0$ and $q \geq 2$. The test is also inadmissible if $p_0 = 0$ and $q = 2$ or if $p_0 > 0$ and $q = 1$, unless α is sufficiently small. Thus, except for a few isolated cases, the classical step-down test is inadmissible.

Mudholkar and Subbaiah (1980) proposed a modified step-down test for problem (1.15), using Fisher's method to combine the observed p-values $Q_1, \ldots, Q_q$ associated with the step-wise LRT's based on $Y_1, \ldots, Y_q$, respectively. If $y_1, \ldots y_q$ are the observed values of $Y_1, \ldots, Y_q$, then we define

$$\begin{aligned} Q_i \equiv Q_i(y_i) &= \text{Prob}[Y_i \geq y_i \mid H^{(i)}] \\ &= \int_{y_i/(1+y_i)}^{1} f_i(z)\, dz, \end{aligned} \tag{1.16}$$

where $f_i(z)$ is the central Beta density of $Y_i/(1+Y_i)$ under $H^{(i)}$ given by

$$\begin{cases} f_i(z) = d_i z^{p_i/2-1}(1-z)^{n_i/2-1}, & 0 < z < 1, \\ \quad d_i = \Gamma\left(\frac{1}{2}(p_i+n_i)\right)/\Gamma\left(\frac{p_i}{2}\right)\Gamma\left(\frac{n_i}{2}\right). & \end{cases} \tag{1.17}$$

Since $Q_1, \ldots Q_q$ are independent and uniformly distributed on (0,1) when H_0 is true, Fisher's combination method based on $\prod Q_i$ may be applied to obtain an overall level α test for (1.15). Precisely, Mudholkar and Subbaiah (1980) proposed the test with acceptance region

$$\left\{ -2 \sum_{i=1}^{q} \ln Q_i(Y_i) \leq \chi^2_{2q,\alpha} \right\}, \tag{1.18}$$

where $\chi^2_{2q,\alpha}$ denotes the upper α-point of the χ^2_{2q} distribution. (Note, however, that $Q_1, \ldots, Q_q$ are *not* independent under H_A.)

For the case $p_0 = 0$, Mudholkar and Subbaiah (1980) evaluated the power functions of the modified step-down procedure (1.18) and the overall T^2 test based on $T^2_q \equiv X'S^{-1}X$, and concluded that the two tests are of comparable performance. Since the T^2_q test is known to be admissible (Stein (1956)) and proper Bayes (Kiefer and Schwartz (1965)), this raises the question of the admissibility of the modified step-down test (1.18).

In the present paper, Theorem 3.1 of MP (1988) is restated as Theorem 2.1 below, then is applied to establish a necessary convexity condition for the admissibility of tests based on $(Y_0, Y_1, \ldots, Y_q)$ for problem (1.15). This result, Corollary 2.2, is then used in Section 3 to show that the modified step-down test (1.18) is inadmissible whenever $\min\{p_1, \ldots, p_q\} = 1$. Only partial results are available for the remaining cases; in particular, this test is admissible for (1.15) when $p_0 = 0$ and $p_1 = \cdots = p_q = 2$.

Marden (1982) and Marden and Perlman (1982) studied the problem of combining the p-values associated with F statistics that are independent under *both* the null and the alternative hypotheses. As in the present paper, they first established a minimal complete class theorem, then derived a necessary condition for admissibility. This necessary condition imposes a convexity requirement on the acceptance region of an admissible test, when that region is expressed in terms of a certain set of transformed variables. The method in the present paper is of a related nature.

2 The Complete Class Theorem and a Necessary Condition for Admissibility

In order to state Theorem 3.1 of MP (1988), some notation from that paper must be reviewed. Let $y \equiv (y_0, y_1, \ldots, y_q)$ and $(t_0^2, t_1^2, \ldots, t_q^2)$ denote possible values of $Y \equiv (Y_0, Y_1, \ldots, Y_q)$ and $(T_0^2, T_1^2, \ldots, T_q^2)$, respectively (cf. (1.7) and (1.8)). Now define

$$v_i = y_i \Big/ \prod_{j=0}^{i} (1 + y_j), \qquad i = 0, 1, \ldots, q, \tag{2.1}$$

$$u_i = \sum_{j=0}^{i} v_j, \qquad i = 0, i, \ldots, q, \tag{2.2}$$

$$v = (v_0, v_1, \ldots, v_q), \tag{2.3}$$

$$u = (u_1, \ldots, u_q). \tag{2.4}$$

(Note that u is q-dimensional regardless of whether $p_0 = 0$ or $p_0 > 0$.) The mapping $y \to v$ is $1-1$ and the range of v is

$$\mathcal{V} = \left\{ v \;\middle|\; v_0, v_1, \ldots, v_q > 0,\ 0 < \sum_{i=0}^{q} v_i < 1 \right\}, \tag{2.5}$$

while that of u is

$$\mathcal{U} = \{ u \mid 0 < u_1 < \cdots < u_q < 1 \}. \tag{2.6}$$

For the testing problem (1.15), the class of all tests depending on y is identical to the class of all tests depending on v.

Let $f_\Delta(y)$ denote the density of Y under H_A, where $\Delta = (\Delta_1, \ldots, \Delta_q)$, and denote the likelihood ratio by

$$R_\Delta \equiv f_\Delta / f_0. \tag{2.7}$$

It follows from (1.10) (cf. MP (1988) Section 3) that

$$R_\Delta \equiv R_\Delta(v) = \prod_{i=1}^{q} \exp\{-(1 - u_{i-1})\Delta_i/2\}\, G_i(v_i \Delta_i/2), \tag{2.8}$$

where

$$G_i(z) = \sum_{k=0}^{\infty} \frac{\Gamma(\frac{1}{2}(p_i + n_i) + k)}{\Gamma(\frac{1}{2}(p_i + n_i))} \frac{\Gamma\left(\frac{p_i}{2}\right)}{\Gamma\left(\frac{p_i}{2} + k\right)} \frac{z^k}{k!}. \tag{2.9}$$

Define

$$l_i \equiv l_i(v) = \left.\frac{\partial R_\Delta}{\partial_{\Delta_i}}\right|_{\Delta=0} = \frac{1}{2}\left[\left(\frac{p_i + n_i}{p_i}\right) v_i - (1 - u_{i-1})\right], \quad i = 1, \ldots, q, \tag{2.10}$$

$$d(v; \lambda, \pi_0, \pi_1) = \Sigma\lambda_i l_i + \int_{\{0<\Sigma\Delta_i<1\}} \left[\frac{R_\Delta - 1}{\Sigma\Delta_i}\right] \pi_0(d\Delta) + \int_{\{1\le\Sigma\Delta_i\}} R_\Delta \pi_1(d\Delta), \tag{2.11}$$

where π_0 is a finite measure on $\{0 < \Sigma\Delta_i \le 1\}$, π_i is a locally finite measure on $\{1 \le \Sigma\Delta_i\}$, and $\lambda \equiv (\lambda_1, \ldots, \lambda_q)$ with $\lambda_1, \ldots, \lambda_q \ge 0$.

Finally, let $\mathcal{C}_u$ represent the class of all relatively closed, convex, and nonincreasing subsets of $\mathcal{U}$, and let $\mathcal{C}$ denote the class of all pre-images in $\mathcal{V}$ of all members of $\mathcal{C}_u$ under the mapping $u : \mathcal{V} \to \mathcal{U}$ determined by (2.2). The indicator function of a set A is denoted by I_A.

Theorem 2.1 **(MP (1988))** *A test function $\varphi \equiv \varphi(v)$ is admissible for problem (1.15) if and only if*

$$\varphi(v) = 1 - I_{C\cap A'}(v) \tag{2.12}$$

for a.e. $v \in \mathcal{V}$, where $C \in \mathcal{C}$,

$$A' = \{ v \in \mathcal{V} \mid d(v; \lambda, \pi_0, \pi_1) \le c \} \tag{2.13}$$

for some λ, π_0, π as above, $|c| < \infty$, and $|d(v; \lambda, \pi_0, \pi_1)| < \infty$ for $v \in$ interior (C). □

This theorem states that a test based on $(Y_0, Y_1, \ldots, Y_q)$ is admissible for problem (1.15) if and only if its acceptance region in $\mathcal{V}$ is of the form $C \cap A'$. Corollary 2.2, the main result of this section, presents a necessary convexity condition for the acceptance region of an admissible test for (1.15).

For $i = 0, 1, \ldots, q$, define

$$v_i^* = v_i^{r_i}, \tag{2.14}$$

where $r_0 = 1$ and, for $i \ge 1$,

$$r_i = \inf_{z>0} \left[\mathrm{Var}_z(K_i)/E_z(K_i)\right], \tag{2.15}$$

where K_i denotes an integer-valued random variable with probability mass function proportional to

$$\frac{\Gamma\left(\frac{1}{2}(p_i + n_i) + k\right)}{\Gamma(\frac{1}{2}p_i + k)} \frac{z^k}{k!}, \quad k = 0, 1, 2, \ldots$$

(cf. Marden and Perlman (1980), equations (2.9) and (2.10)). It is shown by Marden and Perlman (1980, 1982) that for $i \ge 1$,

$$\max\left(\frac{1}{2}, \frac{p_i}{p_i + n_i}\right) < r_i < 1. \tag{2.16}$$

Let

$$v^* \equiv v^*(v) = (v_0^*, v_1^*, \ldots, v_q^*) \tag{2.17}$$

and let $\mathcal{V}^*$ denote the image of $\mathcal{V}$ under the 1-1 mapping $v^*(v)$, i.e.,

$$\mathcal{V}^* = \{ v^* \mid v_0^*, \ldots, v_q^* > 0, 0 < \sum_{i=0}^{q} (v_i^*)^{1/r_i} < 1 \}. \tag{2.18}$$

Define $\mathcal{C}^*$ to be the class of all convex and nonincreasing subsets of $\mathcal{V}^*$.

Corollary 2.2 *A necessary condition for the admissibility of a test function $\varphi \equiv \varphi(v)$ for problem (1.15) is that*

$$\varphi(v(v^*)) = 1 - I_{C^*}(v^*) \qquad [a.e.\ v^* \in \mathcal{V}^*] \tag{2.19}$$

for some subset $C^* \in \mathcal{C}^*$, where $v(v^*)$ is the inverse of the mapping $v^*(v)$.

Proof If φ is admissible for (1.15), then by Theorem 2.1, $\varphi(v) = 1 - I_{C \cap A'}(v)$ for some $C \in \mathcal{C}$ and A' as in (2.13). We shall show that $v^*(C) \in \mathcal{C}^*$ and $v^*(A') \in \mathcal{C}^*$, hence since $\mathcal{C}^*$ is closed under intersections, $v^*(C \cap A') \in \mathcal{C}^*$, which verifies (2.19).

Since u_i is linear and increasing in each v_j (see (2.2)), $C \in \mathcal{C}$ is convex and nonincreasing in v. Because $0 < r_i \leq 1$ for each $i \geq 0$, $v^*(C) \in \mathcal{C}^*$.

To see that $v^*(A') \in \mathcal{C}^*$, first express R_Δ as a function of v^*, i.e., define

$$R_\Delta^*(v^*) = R_\Delta(v(v^*)).$$

Then from (2.8),

$$\ln R_\Delta^*(v^*) = \sum_{i=1}^{q} \left[v_0^* + (v_1^*)^{1/r_1} + \cdots + (v_{i-1}^*)^{1/r_{i-1}} - 1 \right] \Delta_i/2$$
$$+ \sum_{i=1}^{q} \ln G_i \left((v_i^*)^{1/r_i} \frac{\Delta_i}{2} \right). \tag{2.20}$$

Since $0 < r_i \leq 1$, $(v_i^*)^{1/r_i}$ is convex and increasing in v_i^*. Marden and Perlman (1980, equations (2.32) and (2.33)) show that $\ln G_i\left((v_i^*)^{1/r_i}\Delta_i/2\right)$ is convex in v_i^*, while it is clearly nondecreasing since $\Delta_i \geq 0$. Thus, $\ln R_\Delta^*(v^*)$ itself is a convex and nondecreasing function of v^*, and the same is true of $l_i \equiv l_i(v(v^*))$ in (2.10), so therefore $d(v(v^*); \lambda, \pi_0, \pi_1)$ in (2.11) is also convex and nondecreasing in v^*. This directly implies that $v^*(A') \in \mathcal{C}^*$. □

3 Inadmissibility of the Modified Step-down Procedure

The following theorem is the main result of this paper.

Theorem 3.1 *Suppose that $q \geq 2$ and $0 < \alpha < 1$. If $\min\{p_1, \ldots, p_q\} = 1$, the level α modified step-down procedure (1.18) is inadmissible for problem (1.15) within the class of all tests based on $(Y_0, Y_1, \ldots, Y_q)$, hence is inadmissible for problem (1.3) among all tests based on (X,S).*

Proof Let A^* denote the acceptance region (1.18) expressed in terms of v^*. We shall show that A^* is *not* equal a.e. to a convex subset of $\mathcal{V}^*$, hence, by Corollary 2.2, the test determined by A^* is inadmissible.

For $i = 1, \ldots, q$ define the functions h on $\mathcal{V}^*$ and g_i on $(0, 1)$ by

$$h(v^*) = \sum_{i=1}^{q} g_i \left(\frac{v_i(v^*)}{1 - u_{i-1}(v^*)} \right) \tag{3.1}$$

$$g_i(x) = -\ln Q_i \left(\frac{x}{1-x} \right). \tag{3.2}$$

Since $v_i/(1 - u_{i-1}) = y_i/(1 + y_i)$, the acceptance region A^* expressed in terms of v^* is given by

$$A^* = \left\{ v^* \in \mathcal{V}^* \;\middle|\; h(v^*) \leq \frac{1}{2}\chi^2_{2q,\alpha} \right\}. \tag{3.3}$$

Note that $h(v^*) < \infty$ for $v^* \in \mathcal{V}^*$, while

$$g_q \left(\frac{v_q(v^*)}{1 - u_{q-1}(v^*)} \right) \to \infty$$

and hence $h(v^*) \to \infty$ as v^* approaches the upper boundary

$$\partial_u \mathcal{V}^* \equiv \left\{ v^* \mid v_0^*, \ldots, v_q^* > 0,\ \sum_{i=0}^{q} (v_i^*)^{1/r_i} = 1 \right\} \tag{3.4}$$

of $\mathcal{V}^*$ from within $\mathcal{V}^*$. Because $\alpha > 0$ implies $\chi^2_{2q,\alpha} < \infty$, the upper boundary

$$\partial_u A^* \equiv \left\{ v^* \in \mathcal{V}^* \;\middle|\; h(v^*) = \frac{1}{2}\chi^2_{2q,\alpha} \right\} \tag{3.5}$$

of A^* must be uniformly bounded below $\partial_u \mathcal{V}^*$, hence there exists $0 < \varepsilon < 1$ such that

$$u_q \equiv u_q(v^*) \equiv \sum_{i=0}^{q} (v_i^*)^{1/r_i} \leq 1 - \varepsilon, \qquad v^* \in \partial_u A^*. \tag{3.6}$$

By assumption, $p_k = 1$ for some $k = 1, \ldots, q$. Suppose first that $1 \leq k \leq q - 1$. Fix values $\widetilde{v}_i^*$ for v_i^*, $i \neq k, q$, and define the section

$$A_s^* \equiv A_s^*(\widetilde{v}_i^*, i \neq k, q) = \left\{ (v_k^*, v_q^*) \mid h(w^*(v_k^*, v_q^*)) \leq \frac{1}{2}\chi^2_{2q,\alpha} \right\}, \tag{3.7}$$

where

$$w^* \equiv w^*(v_k^*, v_q^*) = (\widetilde{v}_0^*, \ldots, \widetilde{v}_{k-1}^*, v_k^*, \widetilde{v}_{k+1}^*, \ldots, \widetilde{v}_{q-1}^*, v_q^*). \tag{3.8}$$

Because A_s^* is the intersection in $\mathcal{V}^*$ of A^* with a hyperplane and since h is continuous, if A_s^* is not convex for some set of values $\{\widetilde{v}_i^*, i \neq k, q\}$ then A^* cannot be equal a.e. to a convex subset of $\mathcal{V}^*$.

Choose $\widetilde{v}_i^* > 0, i \neq k, q$, small enough that A_s^* is nonempty. (Such a section exists since $\alpha < 1$ requires that A^* be nonempty.) Since $h(v^*)$ is strictly increasing in each v_i^*, A_s^* must be of the form

$$A_s^* = \left\{ (v_k^*, v_q^*) \mid 0 < v_q^* \leq a(v_k^*), \quad 0 < v_k^* < \widehat{v}_k^* \right\}, \tag{3.9}$$

where $a(\cdot)$ is a well-defined, continuous, positive, and strictly decreasing function that satisfies

$$h\big(w^*(v_k^*, a(v_k^*))\big) = \frac{1}{2}\chi^2_{2q,\alpha}, \quad 0 < v_k^* < \widehat{v}_k^*, \tag{3.10}$$

and where

$$0 < \widehat{v}_k^* \equiv \sup\left\{v_k^* \mid (v_k^*, v_q^*) \in A_s^*\right\} < 1 \tag{3.11}$$

because A_s^* is nonempty and $\widetilde{v}_i^* > 0$ for $i \neq k, q$. Since h has continuous partial derivatives, $a(\cdot)$ is continuously differentiable. We shall show that

$$\lim_{v_k^* \to 0} a'(v_k^*) = -\infty, \tag{3.12}$$

which implies that A_s^* is not convex, since $a(\cdot)$ is strictly decreasing.

From (3.10),

$$a'(v_k^*) = -\left[\frac{\partial h(v^*)}{\partial v_k^*} \bigg/ \frac{\partial h(v^*)}{\partial v_q^*}\right]_{v^* = w^*\left(v_k^*, a(v_k^*)\right)}, \tag{3.13}$$

hence to establish (3.12) it suffices to show that

$$\lim_{v_k^* \to 0} \frac{\partial h(v^*)}{\partial v_k^*}\bigg|_{v^* = w^*\left(v_k^*, a(v_k^*)\right)} = \infty, \tag{3.14}$$

$$0 \leq \lim_{v_k^* \to 0} \frac{\partial h(v^*)}{\partial v_q^*}\bigg|_{v^* = w^*\left(v_k^*, a(v_k^*)\right)} < \infty. \tag{3.15}$$

By (3.1) and (2.14),

$$\frac{\partial h(v^*)}{\partial v_k^*} = \left[g_k'\left(\frac{v_k(v^*)}{1 - u_{k-1}(v^*)}\right)\left(\frac{1}{1 - u_{k-1}(v^*)}\right) + \sum_{i=k+1}^{q} g_i'\left(\frac{v_i(v^*)}{1 - u_{i-1}(v^*)}\right)\frac{v_i(v^*)}{\left(1 - u_{i-1}(v^*)\right)^2}\right]\frac{1}{r_k}(v_k^*)^{(1/r_k - 1)}. \tag{3.16}$$

From (3.2), (1.16), and (1.17),

$$
\begin{aligned}
g_i'(x) &= \frac{d_i x^{\frac{1}{2}p_i - 1}(1-x)^{\frac{1}{2}n_i - 1}}{Q_i\big(x/(1-x)\big)} \\
&\geq d_i x^{\frac{1}{2}p_i - 1}(1-x)^{\frac{1}{2}n_i - 1} \\
&> 0
\end{aligned}
\tag{3.17}
$$

for $0 < x < 1$, while $v_i(v^*) > 0$, so (recall $p_k = 1$)

$$
\begin{aligned}
\frac{\partial h(v^*)}{\partial v_k^*} &\geq \frac{d_k}{r_k}\left[\frac{v_k(v^*)}{1-u_{k-1}(v^*)}\right]^{-\frac{1}{2}}\left[1 - \frac{v_k(v^*)}{1-u_{k-1}(v^*)}\right]^{\frac{1}{2}n_k - 1}\frac{(v_k^*)^{1/r_k - 1}}{[1-u_{k-1}(v^*)]} \\
&= \frac{d_k}{r_k}[1-u_{k-1}(v^*)]^{-\frac{1}{2}}\left[1 - \frac{(v_k^*)^{1/r_k}}{1-u_{k-1}(v^*)}\right]^{\frac{1}{2}n_k - 1}(v_k^*)^{\frac{1}{2r_k} - 1}.
\end{aligned}
\tag{3.18}
$$

The relation (3.14) now follows from the facts that $r_k > \frac{1}{2}$ and that

$$1 \geq 1 - u_{k-1}(v^*) > 1 - u_q(v^*) \geq \varepsilon > 0 \tag{3.19}$$

when

$$v^* = w^*\big(v_k^*, a(v_k^*)\big) \in \partial_u A^* \tag{3.20}$$

(see (3.6)).

The first inequality in (3.15) is immediate since $h(v^*)$ is increasing in v_q^*. To establish the second inequality, apply (3.16) and (3.17) to obtain

$$
\begin{aligned}
\frac{\partial h(v^*)}{\partial v_q^*} &= \frac{d_q\left[\frac{v_q(v^*)}{1-u_{q-1}(v^*)}\right]^{\frac{1}{2}p_q - 1}\left[1 - \frac{v_q(v^*)}{1-u_{q-1}(v^*)}\right]^{\frac{1}{2}n_q - 1}}{r_q Q_q\left(\frac{v_q(v^*)}{1-u_q(v^*)}\right)[1-u_{q-1}(v^*)]}(v_q^*)^{\frac{1}{r_q} - 1} \\
&= \frac{d_q (v_q^*)^{\frac{p_q}{2r_q} - 1}\left[1 - \frac{v_q(v^*)}{1-u_{q-1}(v^*)}\right]^{\frac{1}{2}n_q - 1}}{r_q Q_q\left(\frac{v_q(v^*)}{1-u_q(v^*)}\right)[1-u_{q-1}(v^*)]^{\frac{1}{2}p_q}}.
\end{aligned}
\tag{3.21}
$$

When (3.20) holds, however,

$$1 > v_q^* = a(v_k^*) \to a(0+) > 0 \qquad \text{as } v_k^* \to 0, \tag{3.22}$$

$$1 > 1 - u_{q-1}(v^*) > 1 - u_q(v^*) \geq \varepsilon > 0 \tag{3.23}$$

(cf. (3.6)), hence

$$1 > 1 - \frac{v_q(v^*)}{1-u_{q-1}(v^*)} > \varepsilon, \tag{3.24}$$

$$\limsup_{v_k^* \to 0} \frac{v_q(v^*)}{1-u_q(v^*)} \leq \frac{\big(a(0+)\big)^{1/r_q}}{\varepsilon} < \infty. \tag{3.25}$$

The relations (3.21)–(3.25) together yield the second inequality in (3.15). (When $n_q \geq 2$, apply the first inequality in (3.24), while when $n_q = 1$, apply the second inequality.)

In order to complete the proof of the theorem, it remains to consider the case where $p_q = 1$. For this purpose we set $k = q - 1$ in the preceding argument and shall show that some section

$$A_s^* \equiv A_s(\widetilde{v}_i^*, i \neq q-1, q) \tag{3.26}$$

of A^* fails to be convex when the $\widetilde{v}_i^*$ are sufficiently small that A_s^* is nonempty. Since $a(\cdot)$ (defined as in (3.9) but with $k = q-1$ now) is strictly decreasing, this will follow from

$$\lim_{v_{q-1}^* \to \widehat{v}_{q-1}^*} a'(v_{q-1}^*) = 0 \tag{3.27}$$

(see (3.11) for the definition of $\widehat{v}_{q-1}^* > 0$). By (3.13) with $k = q - 1$, it suffices to show that

$$\lim_{v_{q-1}^* \to \widehat{v}_{q-1}^*} \left. \frac{\partial h(v^*)}{\partial v_q^*} \right|_{v^* = w^*\left(v_{q-1}^*, a(v_{q-1}^*)\right)} = \infty, \tag{3.28}$$

$$0 \leq \lim_{v_{q-1}^* \to \widehat{v}_{q-1}^*} \left. \frac{\partial h(v^*)}{\partial v_{q-1}^*} \right|_{v^* = w^*\left(v_{q-1}^*, a(v_{q-1}^*)\right)} < \infty. \tag{3.29}$$

For (3.28), first set $p_q = 1$ in (3.21) and verify that (3.18) holds with k replaced by q. Then (3.28) follows (as did (3.14) from (3.18)) from the facts that $r_q > \frac{1}{2}$ and that when $v^* = w^*$, (3.23) holds and

$$v_q^* = a(v_{q-1}^*) \to 0 \qquad \text{as } v_{q-1}^* \to \widehat{v}_{q-1}^*. \tag{3.30}$$

Once again, the first inequality in (3.29) is immediate. For the second inequality, begin by applying (3.16) (with $k = q - 1$) and the first line in (3.17) to evaluate $\partial h(v^*)/\partial v_{q-1}^*$, then use the facts that $p_q = 1$ and $n_{q-1} \geq 2$ to obtain the bound

$$\begin{aligned} \frac{\partial h(v^*)}{\partial v_{q-1}^*} \leq & \frac{d_{q-1}(v_{q-1}^*)^{(p_{q-1}/2r_{q-1})-1}}{r_{q-1}Q_{q-1}\left(\frac{v_{q-1}(v^*)}{1-u_{q-1}(v^*)}\right)[1 - u_{q-2}(v^*)]^{(p_{q-1}/2)}} \\ & + \frac{d_q(v_q^*)^{(1/2r_q)}\left[1 - \frac{v_q(v^*)}{1-u_{q-1}(v^*)}\right]^{\frac{n_q}{2}-1}}{r_qQ_q\left(\frac{v_q(v^*)}{1-u_q(v^*)}\right)\left(1 - u_{q-1}(v^*)\right)^{3/2}}(v_{q-1}^*)^{(1/r_{q-1})-1}. \end{aligned} \tag{3.31}$$

As in (3.21)–(3.25), both of these terms remain bounded below ∞ when $v^* = w^*$ and $v_{q-1}^* \to \widehat{v}_{q-1}^*$ (also apply (3.30)), so the verification of (3.29) (and therefore that of (3.26)) is complete. □

Remark 3.2 *The modified step-down procedure (1.18)* is *admissible for problem (1.15) when $p_0 = 0$ and $p_1 = \cdots = p_q = 2$.* This is seen as follows. Since $p_i = 2$,

$$Q_i = \left(1 - \frac{v_i}{1-u_{i-1}}\right)^{\frac{1}{2}n_i} = \left(\frac{1-u_i}{1-u_{i-1}}\right)^{\frac{1}{2}n_i}, \tag{3.32}$$

so, when $p_0 = 0$, the acceptance region (1.18) is equivalent to

$$\left\{u \in \mathcal{U} \,\middle|\, -2\sum_{i=1}^{q-1} \ln(1-u_i) - n_q \ln(1-u_q) \leq \chi^2_{2q,\alpha}\right\}. \tag{3.33}$$

(We use the fact that $n_{i-1} = n_i + 2$; see (1.11).) Since the region (3.33) belongs to the class $\mathcal{C}_u$ (defined above Theorem 2.1), the corresponding test function $\varphi \equiv \varphi(v)$ has the form $\varphi(v) = 1 - I_C(v)$ for some set $C \in \mathcal{C}$. By Theorem 3.1, therefore, this test is admissible for problem (1.15). (Take $\lambda = 0, \pi_0 = 0, \pi_1 = 0$, and $c = 1$ in (2.13), so that $A' = \mathcal{V}$.) □

Lastly, we discuss the admissibility and/or inadmissibility of the modified step-down procedure (1.18) in the remaining cases not covered by Theorem 3.1 or Remark 3.2.

(i) $q = 1$, $p_0 = 0$: both the classical and modified step-down procedures reduce to the ordinary T^2 test based on $Y_1 \equiv T_1^2$, which is known to be admissible.

(ii) $q = 1$, $p_0 > 0$: both procedures reduce to the LRT based on $Y_1 \equiv (T_1^2 - T_0^2)/(1 + T_0^2)$, which is admissible for $\alpha \leq \alpha^*$ and inadmissible for $\alpha > \alpha^*$, where $0 < \alpha^* \equiv \alpha^*(n, p_1) < 1$. (See Marden and Perlman (1980)).

(iii) $q \geq 2, p_0 > 0, p_1 = \cdots = p_q = 2$; and

(iv) $q \geq 2, p_0 \geq 0, 2 \leq \min\{p_1, \ldots, p_q\} < \max\{p_1, \ldots, p_q\}$: no results for these two cases are available, but it is suspected that the admissibility/inadmissibility of the modified step-down procedure (1.18) may depend on the value of the significance level α. The classical step-down procedure is inadmissible in both these cases, with the exception of the situation where $q = 2$ and $p_0 = 0$, in which case it is admissible only if $\alpha \leq \alpha^{**}$ for some $0 < \alpha^{**} < 1$ (see MP (1988)).

4 Comments

It is possible to combine the p-values $Q_1, \ldots, Q_q$ according to methods other than Fisher's, e.g., by replacing $-\Sigma \ln Q_i$ by $-\Sigma\Phi^{-1}(Q_i)$, where Φ is the standard normal distribution function. Also, one may consider weighted versions of these combination statistics, such as $-\Sigma\delta_i \ln Q_i$ with $\delta_i \geq 0$. (The classical step-down procedure (1.13) is equivalent to a *weighted* version of the Tippett combination statistic $\min\{Q_1, \ldots, Q_q\}$, namely, $\min\{Q_1^{\delta_1}, \ldots, Q_q^{\delta_q}\}$ for appropriate weights δ_i.)

A study of these various procedures has not been completed, but it is suspected that they remain inadmissible in many cases. We believe that the basic difficulty stems from the inappropriateness of testing H_0 vs. H_A by means of the sequence of intermediate problems $H^{(i)}$ vs. $H^{(i-1)}$ (cf. (1.4), (1.5)), for which Y_i is the LRT statistic. MP (1988, Section 6) showed that an alternative step-down procedure based on $\max\{T_1^2, \ldots, T_q^2\}$ or its weighted version *is* admissible for problem (1.15) and, in fact, for the original problem (1.3). Use of this procedure corresponds to testing H_0 vs. H_A by means of the sequence of problems $H^{(i)}$ vs. H_A, $i = 1, \ldots, q$, for which T_i^2 is the LRT statistic when $p_0 = 0$. The operating characteristics of this procedure will be investigated in a subsequent study.

Finally, since the LRT for testing $H^{(i)}$ vs. H_A is based on the statistic $Z_i \equiv (T_i^2 - T_0^2)/(1 + T_0^2)$ rather than on T_i^2 when $p_0 > 0$, a step-down procedure based on $\max\{Z_1, \ldots, Z_q\}$ or its weighted version should also be studied. This too is under investigation.

Acknowledgments: Research of John I. Marden supported in part by National Science Foundation Grant No. MCS 82-01771. Research of Michael D. Perlman supported in part by National Science Foundation Grant No. MCS 86-03489.

REFERENCES

Kiefer, J. and Schwartz, R. (1965). Admissible Bayes character of T^2-, R^2-, and other fully invariant tests for classical multivariate normal problems. *Ann. Math. Statist.* **36** 747–770.

Marden, J.I. (1982). Minimal complete classes of tests of hypotheses with multivariate one-sided alternatives. *Ann. Statist.* **10** 962–970.

Marden, J.I. and Perlman, M.D. (1980). Invariant tests for means with covariates. *Ann. Statist.* **8** 25–63.

Marden, J.I. and Perlman, M.D. (1982). The minimal complete class of procedures for combining independent noncentral F tests. In *Statistical Decision Theory and Related Topics III.* (J. Berger and S.S. Gupta, eds.) **2** 139–181.

Marden, J.I. and Perlman, M.D. (1988). On the inadmissibility of step-down procedures for the Hotelling T^2 problem. Submitted to *Ann. Statist.*

Mudholkar, G.S. and Subbaiah, P. (1980). Testing significance of a mean vector—a possible alternative to Hotelling's T^2. *Ann. Inst. Statist. Math.* **32A** 43–52.

Stein, C. (1956). The admissibility of Hotelling's T^2 Test. *Ann. Math. Statist.* **27** 616–623.

32

On A Statistical Problem Involving the Measurement of Strength and Duration-of-Load for Construction Materials

Sam C. Saunders[1]

ABSTRACT This paper presents a new analysis, applicable to construction materials, for estimating the probabilistic behavior of the duration-of-load, say T_ℓ when the load is of magnitude ℓ, based on the imposed stress ratio of load to characteristic material strength. The stochastic behavior of the logarithm of the duration-of-load, given the strength S exceeds ℓ, is of axiomatic importance. We assume the conditional distribution, presuming strength S were known, to be of the form

$$[\ln T_\ell \mid S = s] - H(s/\ell) \sim \sigma_0 Z,$$

where the r.v. Z is a standard variate appropriately chosen for the material, and the regression H is a monotone increasing function of the form

$$H(x) = a_0 + a_1 x + a_{-1} x^{-1} \quad \text{for } x > 0.$$

Here either a_1 or a_{-1} may be zero.
Estimation procedures are derived to determine the unknown parameters from the type of data that is available from the observable variate $[\ln T_\ell \mid S > \ell]$. The estimated distribution is then used to calculate the safe-life, at an assurance level p, as the largest value ξ such that

$$\Pr[T_\ell > \xi \mid S > \ell] \geq p.$$

A tabulation of ξ_p, as a function of p at a given load ℓ for a material of given characteristic strength, can be used to replace the so-called safety factors.

1 Introduction

An estimate is desired of the distribution of the time to failure, under a given load, for structural components made of a particular building ma-

[1]Department of Pure and Applied Mathematics, Washington State University, Pullman, WA 99164-2930

terial. The relevant data are sparse and come from either of two possible experiments:

In the first experimental procedure, call it experiment I, only the (short-term) strength, call it S, of each specimen is obtained by subjecting it to an increasing load until it fails, or the limit of the testing machine is reached. Thus, after a number of replications, the distribution of the strength can be estimated based on the sample obtained under type I censoring on the right.

In experiment II a specimen is selected at random from the population and subjected to an increasing load until it fails or it sustains a preassigned stress level ℓ. Thus either the strength S is determined, whenever $S < \ell$, or the time to failure, called the duration-of-load and labeled T_ℓ, is measured whenever $S > \ell$; but not both. Of course, samples are obtained at several stress levels.

A widely held assumption based on the Physics of Materials is:

A: The duration-of-load, T_ℓ, of a specimen having strength S under stress ℓ, when $\ell < S$, is a function of the stress ratio ℓ/S and not of the stress itself.

The methods presently utilized for dealing with the dichotomy of experiment II may be unsatisfactory since each observation of the duration-of-load is assigned a short-term strength by an *ad hoc* equal-rank assumption between the two samples, (Murphy, 1983), or by transforming each time to failure into an equivalent value from a constant stress ratio, (Barrett and Foschi, 1982; Madsen and Barrett, 1976). Moreover, if a stochastic model is adopted then the implications of the stress-ratio assumption **A** must be treated by careful argument involving the conditional random variables involved.

The basic problem is that of expressing the unknown distribution of the duration-of-load, given non-failure initially, in terms of those distributions that can be estimated from the data available from experiments I and II, i.e., we must estimate the distribution of the conditional variate $[T_\ell \mid S > \ell]$.

One specialization of assumption **A**, (Murphy, 1983), postulates a linear regression of the form:

$$E\left[\ln T_\ell \mid S = s\right] = b + a\left(\frac{\ell}{s}\right) \tag{1.1}$$

where $a < 0$ and $b > 0$ are constants which must be estimated from data. Of course, no direct observations on $[T_\ell \mid S = s]$ can be obtained because of the dichotomy of experiment II. In Murphy (1983) the material was selected and a large set of observations of the strength of the material was obtained. A load ℓ was fixed from which a series of ordered observations of the log-duration-of-load, $\ln T(\ell)$, were made. The supposed applicable stress ratios were deduced by dividing the jth ordered duration observation by the corresponding equal-rank ordered strength observation. Unfortunately, forming ratios of order-observations from the marginal distributions of jointly distributed random variables will not adequately reproduce the original dependence.

Nevertheless, from such ratios equation (1.1) was fitted by least-squares. This procedure did yield some agreement with the facts when the stress ℓ was assigned near the median strength of the material, as estimated from experiment I, since the expected duration-of-load is very short. In this case, the fitted values of a and b are of approximately the same magnitude but of opposite sign. However equation (1.1) implies duration-of-loads which seem unduly conservative whenever ℓ is near zero, since the expected log-life can be no higher than b. It is shown here that the entailed duration-of-load values, being so unsatisfactorily short when ℓ is small, may be a consequence of this type of analysis of the assumed model of equation (1.1) and not a state of nature.

If one were to assume a linear regression of alternate form, viz.,

$$E\left[\ln T_\ell \mid S=s\right]=b+a\left(\frac{s}{\ell}\right), \tag{1.2}$$

where now $a>0$, $b<0$ are constants to be estimated, the predicted behavior is correct when ℓ is near zero, giving estimates of the duration-of-load which are high and in conformity with the sparse data, but it does not agree well with the mass of data, which can be obtained, because of the time and expense involved, only when ℓ is a sizable fraction of the median strength.

A more inclusive model which generalizes equations (1.1) and (1.2) by assuming the regression is a partial Laurent expansion in the stress ratio ℓ/s, namely,

$$E\left[\ln T_\ell \mid S=s\right]=H(s/\ell)\equiv a_0+a_1\left(\frac{s}{\ell}\right)+a_{-1}\left(\frac{\ell}{s}\right) \tag{1.3}$$

for some constants $a_0, a_1 \leq 0, a_{-1} \leq 0$ which are to be estimated from the data, is examined here in order to unify the treatment.

By recognizing the stochastic variability of all the quantities that can be measured, and incorporating the regression on strength into the joint distribution of T_ℓ, S - from which are derived the marginal distributions of S and the conditional distribution of $[T_\ell \mid S>\ell]$ - a realistic model is obtained. Maximum likelihood estimation procedures are then derived to estimate the unknown parameters of interest using data which reflects the actual testing procedures followed. A quantity of great engineering interest is the safe-life; a quantity which, at a specified stress ratio, the duration-of-load will exceed with a prescribed high level of assurance. An integral formula is given by which the safe-life can be evaluated.

It is claimed that this method of analysis gives results which agree well with the few complete data sets available for structural timbers, (Martin and Saunders, 1988), q.v., and it makes reasonable predictions for the duration-of-load for this material.

2 The Problem of Determining the Proper Distributions

The first problem is that of expressing the unknown distribution of the duration-of-load, given survival at a prescribed stress, in terms of those distributions that can be more easily estimated from the data that are available.

By introducing notation for an indicator function, namely,

$$\langle S \leq \ell \rangle = 1 \quad \text{if } S \leq \ell \quad \text{and 0 otherwise,}$$

we can denote the outcome of Experiment II at stress ℓ as observing both the random variable and the event, respectively,

$$Y_\ell = S\langle S \leq \ell \rangle + T_\ell \langle S > \ell \rangle \quad \text{and} \quad \langle S \leq \ell \rangle.$$

Hence, by the calculus of probability, the distribution of Y_ℓ is

$$\begin{aligned} F_{Y_\ell}(x) &= \Pr[T_\ell \leq x \mid S > \ell] \Pr[S > \ell] + \Pr[S \leq \min(x, \ell)] \\ &= F_{T_\ell|S>\ell}(x)\, \bar{F}_S(\ell) + F_S[\min(x, \ell)] \end{aligned}$$

which after rearrangement gives the conditional distribution of the duration-of-load, viz.,

$$F_{T_\ell|S>\ell}(x) = \frac{F_{Y_\ell}(x) - F_S[\min(x,\ell)]}{\bar{F}_S(\ell)} \quad \text{for} \quad x > 0. \tag{2.1}$$

Thus, correspondingly, we find

$$F_{T_\ell|S\leq\ell}(x) = 1 \quad \text{for} \quad x > 0. \tag{2.2}$$

From Experiment II we can, for various values of ℓ, estimate the distribution F_{Y_ℓ} and also obtain some data which may contribute to our knowledge of F_S. Assuming that F_S is known we can, by Equation (1.1), estimate the conditional distribution $F_{T_\ell|S>\ell}$ for each of the several values of ℓ for which F_{Y_ℓ} is known.

How can we estimate the distribution $F_{T_\ell|S}(x \mid s)$ for each value of s and ℓ ? We know this distribution would be equivalent with the conditional distribution of the duration-of-load given the random stress ratio $\Lambda = \ell/S$, say $F_{T_\ell|\Lambda}(t \mid \lambda)$. This last distribution is, by assumption **A**, the one thought to be critical in engineering practice. The statistical problem then is to determine the conditional distribution $F_{T_\ell|S}$ using the data that are available from replications of Experiments I and II as previously described. But first we must obtain an expression which expresses the unknown distribution in terms of ones that are determinable.

If we let $F_{T_\ell,S}$ be the joint distribution of (T_ℓ, S), then assuming all densities exist

$$\begin{aligned} F_{T_\ell|S>\ell}(x) &= \int_0^x \int_\ell^\infty f_{T_\ell,S}(t,s)\frac{1}{\bar{F}_S(\ell)}\,ds\,dt \\ &= \int_0^x \int_\ell^\infty f_{T_\ell|S}(t \mid s)\frac{f_S(s)}{\bar{F}_S(\ell)}\,ds\,dt \\ &= \int_0^x \int_\ell^\infty f_{T_\ell|S}(t \mid s)\,dF_{S|S>\ell}(s)\,dt \end{aligned} \tag{2.3}$$

and finally we obtain a relationship which does not require densities :

$$F_{T_\ell|S>\ell}(x) = \int_{s=\ell}^\infty F_{T_\ell|S}(x \mid s)\,dF_{S|S>\ell}(s). \tag{2.4}$$

This equation is another version of the conditional distribution of the duration-of-load as given in equation [5] from Martin and Saunders (1982). Thus, to a certain degree of precision F_S can be determined from data obtained from replications of Experiment I and to another degree $F_{T_\ell|S>\ell}$ may be estimated using the results from Experiment II. We must therefore postulate, in accord with information about the mechanics of failure, the unobservable distribution $F_{T_\ell|S}$ as required in equation (2.4) and determine its parameters by correct statistical procedures.

There are many studies which are analogous to the one described here which arise in determining biological sensitivity to certain dosages of toxic material, or radiation. But each requires a different analysis because of the known relationships between "life" and "stress", as well as the type and quantity of data obtainable. An archetypical example is the study of survival of mice under various doses of radiation (Johnson and Johnson, 1980).

3 A General Model Involving Scale and Location Parameters

We make two assumptions, concerning the distribution of strength and the conditional distribution of the time to failure under load ℓ, which are in accord with **A**. These two assumptions are:

B: The survival distribution of strength S is of the form

$$\bar{F}_S(s) = \exp\left\{-Q_1\left(\frac{\ln s - \mu_1}{\sigma_1}\right)\right\} \quad \text{for} \quad s > 0, \tag{3.1}$$

for some known hazard function Q_1. Thus the distribution of $\ln(S)$ is known except for location and scale parameters μ_1 and σ_1 respectively.

C: The conditional distribution of time to failure under load ℓ, call it T_ℓ, given the value of strength $S = s$, will have a distribution with a known hazard Q_0 but with unknown scale parameter σ_0, and with a location parameter which is a function of the stress ratio, call it $H(s/\ell)$, as expressed by

$$\bar{F}_{T_\ell|S}(t \mid s) = \exp\left\{-Q_0\left[\frac{\ln t - H(s/l)}{\sigma_0}\right]\right\} \quad \text{for} \quad t > 0. \tag{3.2}$$

The choice of the hazard functions, Q_0 and Q_1 is left unspecified here so that this analysis may be applicable to as many materials as possible, but we have in mind two canonical cases. When $Q_1(x) = Q_0(x) = e^x$, we call it the *Gumbel* case and when $Q_1(x) = Q_0(x) = -\ln\phi(-x)$, we call it the *Galton* (log-normal) case. Here ϕ is the standard normal distribution.

These distributions can be combined in equation (2.4), which for convenience is here rewritten using survival distributions, as

$$\bar{F}_{T_\ell|S>\ell}(t) = \int_{s=\ell}^{\infty} \bar{F}_{T_\ell|S}(t \mid s)\, dF_{S|S>\ell}(s). \tag{3.3}$$

Thus if we substitute equations (3.1) and (3.2) into (3.3), and for notational simplicity set

$$\varepsilon = Q_1\left(\frac{\ln\ell - \mu_1}{\sigma_1}\right), \qquad \Delta(x) = H\left[\frac{\beta_1}{\ell}\exp(\sigma_1 x)\right] \tag{3.4}$$

where we set

$$\beta_1 = e^{\mu_1}, \tag{3.5}$$

and then by making the change of variable

$$y = Q_1\left(\frac{\ln s - \mu_1}{\sigma_1}\right) - \varepsilon \tag{3.6}$$

we obtain, from assumptions **A** and **B**, the survival distribution for the logarithm of the duration-of-load given survival at load ℓ, namely,

$$\bar{F}_{X(\ell)}(x) = \int_{y=0}^{\infty} \exp\left\{-Q_0\left[\frac{x - c\Delta[Q_1^{-1}(y+\varepsilon)]}{\sigma_0}\right] - y\right\} dy. \tag{3.7}$$

Here, and subsequently, we write for the conditional variate of primary interest

$$X(\ell) = [\ln T_\ell \mid S > \ell].$$

Unfortunately this integral in equation (3.7) cannot be evaluated in closed form, but requires numerical evaluation even in the simplest cases of the Gumbel or Galton formulation.

In terms of random variables, we are assuming in equation (3.2) that in distribution

$$[\ln T_\ell \mid S = s] - H(s/\ell) \sim \sigma_0 Z \tag{3.8}$$

where Z is the standard variate with known hazard function Q_0 given by

$$F_Z(z) = \Pr[Z \le z] = 1 - \exp\{-Q_0(z)\} \quad \text{for} - \infty < z < \infty.$$

We now introduce notation for the mean and variance of Z, viz.,

$$E\,Z = \int_{-\infty}^{\infty} z\, dF_Z(z) \equiv -C; \tag{3.9}$$

$$\mathrm{Var}(Z) = \int_{-\infty}^{\infty} z^2\, dF_Z(z) - C^2 \equiv \delta^2. \tag{3.10}$$

Note that the value of the constants δ^2 and C will change with each choice of Q_0 in the model.

We now have, recalling the definition,

$$\begin{aligned} E\,X(\ell) &= \int_{s=\ell}^{\infty} E[\ln T_\ell \mid S = s]\, dF_{S|S>\ell}(s) \\ &= -\sigma_0 C + E[H(S/\ell) \mid S > \ell]. \end{aligned} \tag{3.11}$$

In order to include the models that have been proposed previously in the literature we assume the function $H(\cdot)$ can be defined as in equation (1.3), namely,

$$H(x) = a_0 + a_1 x + a_{-1} x^{-1} \quad \text{for } x > 0,$$

where three constants (or maybe two, depending on the model) are to be determined.

Thus we have

$$E[H(S/\ell) \mid S > \ell] = \sum_{j=-1}^{1} a_j E\left[\left(\frac{S}{\ell}\right)^j \mid S > \ell\right], \tag{3.12}$$

and we now seek to evaluate these conditional expectations. By definition

$$\begin{aligned} E[S \mid S > \ell] &= \int_{s=\ell}^{\infty} \bar{F}_{S|S>\ell}(s)\, ds \\ &= \int_{s=\ell}^{\infty} \exp\left\{-Q_1\left(\frac{\ln s - \mu_1}{\sigma_1}\right) + \varepsilon\right\} ds, \end{aligned} \tag{3.13}$$

where we have made use of the definition of ε in equation (3.4). By using the transformation of equation (3.6) to change the variable of integration in equation (3.13) and utilizing the definition in equation (3.5), we find

$$\begin{aligned} E[S/\ell \mid S > \ell] &= \frac{\beta_1}{\ell} \int_0^\infty \exp\left\{-y + \sigma_1 Q_1^{-1}(y+\varepsilon)\right\} \, dy \\ &= \int_0^\infty \exp\{-y + \sigma_1 c\psi(y,\varepsilon)\} \, dy, \end{aligned} \tag{3.14}$$

where we have introduced the function, recall equation (3.5),

$$\psi(y,\varepsilon) = Q_1^{-1}(y+\varepsilon) - Q_1^{-1}(\varepsilon) \quad \text{for } \; y, \varepsilon > 0. \tag{3.15}$$

After a similar exercise we find a corresponding result

$$E[\ell/S \mid S > \ell] = \int_0^\infty \exp\{-y - c\sigma_1 \psi(y,\varepsilon)\} \, dy. \tag{3.16}$$

By comparing equation (3.14) and (3.16) we see that by setting

$$K(\varepsilon, x) = \int_0^\infty \exp[x\psi(t,\varepsilon) - t] \, dt \tag{3.17}$$

for $\varepsilon > 0$ and $-\infty < x < \infty$, that both integrals can be defined using this expression.

From equation (3.12) we have the result that

$$E[H(S/\ell) \mid S > \ell] = a_0 + a_{-1}K(\varepsilon, -\sigma_1) + a_1 K(\varepsilon, \sigma_1).$$

Thus we finally obtain from equation (3.11) the desired regression equation

$$E\,X(\ell) = b + a_{-1}K(\varepsilon, -\sigma_1) + a_1 K(\varepsilon, \sigma_1), \tag{3.18}$$

where the coefficients $a_1, a_{-1}, b = a_0 - \sigma_0 C$ are unknown and must be estimated from the data, or determined from the theory of the mechanics of fracture.

Note that this equation depends on the ratio of load ℓ to characteristic strength β_1 only through the modified stress ratio $\varepsilon = Q_1\left[\ln(\ell/\beta_1)^{1/\sigma_1}\right]$ from equation (3.4). It is not in terms of the true stress ratio, ℓ/s, which can be known only to Nature, but cannot be determined by the investigator. Also note how this regression equation (3.18) involves coefficients with multipliers which are conditional expectations rather than simple stress ratios of load to (unknown) strength. It is not surprising that the consequences are quite different for this model than the ones hitherto proposed and studied.

4 Behavior of The Conditional Expectations

Since Q_1 is a hazard function on $\Re$, by the definition given in equation (3.1), the value of ε will vary monotonically as a function of ℓ. Since Q_1^{-1} is a monotone decreasing map from $\Re_+$ onto $\Re$, we have from equation (3.15) that

$$\psi(t,\varepsilon) \leq 0 \quad \text{for all } \; t,\varepsilon > 0;$$

and for all $t > 0$

$$\lim_{\varepsilon \to 0} \psi(t,\varepsilon) = \lim_{\varepsilon \to 0} Q_1^{-1}(\varepsilon) = -\infty. \tag{4.1}$$

If $\ln S$ has an IHR distribution, i.e., $Q_1' = q_1$ is increasing, as it is in the Gumbel case, then $\psi(t,\varepsilon)$ is monotone decreasing as a function of $\varepsilon > 0$ since

$$\operatorname{sgn}\left[\frac{\partial \psi(t,\varepsilon)}{\partial \varepsilon}\right] = q_1[Q_1^{-1}(t+\varepsilon)] - q_1[Q_1^{-1}(\varepsilon)] \leq 0.$$

It follows easily in this case that

$$\lim_{\varepsilon \to \infty} \psi(t,\varepsilon) = 0, \quad \text{for any } t > 0, \tag{4.2}$$

whenever $\lim_{\varepsilon \to \infty} q_1(\varepsilon) = \infty$.

From equation (3.17) there follows by the Dominated Convergence Theorem that

$$\lim_{\varepsilon \to 0} K(\varepsilon, x) = \begin{cases} 0 & \text{if } x < 0, \\ 1 & \text{if } x = 0, \\ \infty & \text{if } x > 0. \end{cases} \tag{4.3}$$

If equation (4.1) is true and $\ln S$ is IHR then by the Monotone Convergence Theorem we have

$$\lim_{\varepsilon \to \infty} K(\varepsilon, x) = 1 \quad \text{for any} \;\; x > 0. \tag{4.4}$$

Note that from equations (3.18) and (4.4) that

$$\lim_{\ell \to \infty} E\,X(\ell) = b + a_1 + a_{-1}. \tag{4.5}$$

Note this limit must be non-negative by physical considerations, and from equations (3.18) and (4.3) we see that

$$\lim_{\ell \to 0} E\,X(\ell) = \operatorname{sgn}(a_1) \cdot \infty. \tag{4.6}$$

Since, as the load approaches zero the expected log-duration-of-load, should from physical reasoning, get large, we must have

$$a_1 > 0. \tag{4.7}$$

We note that in the case of the Galton (log-normal) distribution for strength that $\ln S$ is not IHR. However, it can be shown directly in this case that equation (4.2) is true.

For some materials, it is possible to simplify the model and reduce the number of parameters by taking

$$a_1 = a > 0, \; a_{-1} = -a, \; \text{and} \; b < 0. \tag{4.8}$$

Since, $H(s/\ell)$ is, as we have argued by the mechanics of fracture, a monotone decreasing function of ℓ, we assume that $H(\cdot)$ is an increasing function. Moreover, for a range of values of $x = s/\ell$ smaller than unity, $H(x)$ must be negative.

Recalling equations (1.1), (1.2), and (1.3), respectively we now consider three models:

Model I	$H(s/\ell) = b + a(\ell/s)$	for some $a < 0, \; b > 0$,
Model II	$H(s/\ell) = b + a(s/\ell)$	for some $a > 0, \; b < 0$,
Model III	$H(s/\ell) = b + a(s/\ell) - a(\ell/s)$	for some $a > 0, \; b < 0$.

5 The Variance of $X(\ell) = [\ln T_\ell \mid S > \ell]$ and its behavior

From equation (3.8) we have

$$\mathrm{Var}[\ln T_\ell \mid S] = \sigma_0^2 \delta^2$$

where the constant δ^2 is determined by the the choice of Q_0 in equation (3.10). Here δ^2 equals $\pi^2/6$ or 1 in the Gumbel or Galton cases, respectively.

For notational convenience set, from equation (3.18),

$$m(\ell) \equiv E\,X(\ell) = b + aA(\varepsilon)$$

where ε is the modified stress defined in terms of the ratio of load ℓ to the median strength β_1 in equation (3.4) We now define:

$$A(\varepsilon) = K(\varepsilon, -\sigma_1) \qquad \text{in Model I} \tag{5.1}$$

$$A(\varepsilon) = K(\varepsilon, \sigma_1) \qquad \text{in Model II} \tag{5.2}$$

$$A(\varepsilon) = K(\varepsilon, \sigma_1) - K(\varepsilon, -\sigma_1) \quad \text{in Model III.} \tag{5.3}$$

Then we have

$$\begin{aligned} \mathrm{Var}\,X(\ell) &= E\left\{[\ln T_\ell - m(\ell)]^2 \mid S > \ell\right\} \\ &= E\left\{E\left([\ln T_\ell - m(\ell)]^2 \mid S\right) \mid S > \ell\right\}, \end{aligned}$$

where we have made use of conditional expectation given S. By using the identity in equation (3.8) we obtain

$$\mathrm{Var}\,X(\ell) = E\left\{E[\sigma_0(Z + C) + H(s/\ell) - \sigma_0 C - m(\ell)]^2 \mid S = s \mid S > \ell\right\}.$$

25

Likelihood Analysis of a Binomial Sample Size Problem

Murray Aitkin[1]
Mikis Stasinopoulos[2]

ABSTRACT The problem of estimating the binomial sample size N from k observed numbers of successes is examined from a likelihood point of view. The direct use of the likelihood function for inference about N is illustrated when p is known, and the problem of inference is considered when p is unknown, and has to be eliminated in some way from the likelihood. Different methods (Bayesian, integrated likelihood, conditional likelihood, profile likelihood) for eliminating the nuisance parameter are found to lead to very different likelihoods in N in an example. This occurs because of a strong ridge in the two-parameter likelihood in N and p. Integrating out the parameter p is found to be unsatisfactory, but reparameterization of the model shows that the inference about N is almost unaffected by the new nuisance parameter. The resulting likelihood in N corresponds closely to the profile likelihood in the original parameterization.

1 Introduction

Carroll and Lombard (1985) considered the problem of estimating the parameter N based on k independent success counts $s_1, \ldots, s_k$ from a binomial distribution with unknown parameters N and p. They extended earlier work by Olkin, Petkau and Zidek (OPZ, 1981) on the moment and maximum likelihood estimators by introducing new estimators of N based on integrating out p from the likelihood with respect to a beta distribution, yielding a beta-binomial distribution for the number of successes. The new estimators maximizing this likelihood compared favorably in mean square error terms with the OPZ estimators. Casella (1986) considered perturbations of the likelihood to decide on the "stability" or instability of the ML estimator.

The emphasis throughout these discussions is on point estimation of N, and the comparison of different estimators through their mean square or

[1]Tel Aviv University
[2]Welcome Research Laboratories

We presume the samples to be large enough that the assumption of normality of the sample means can be made. We assume that $\bar{X}_i$ is normally distributed with mean m_i and variance v_i^2 where

$$m_i = b + aA_i \quad \text{and} \quad v_i^2 = \frac{[c + a^2 B_i]}{n_i} \quad \text{for } i = 1, \ldots, k. \tag{6.1}$$

Here the A_i and B_i are calculated values of the $A(\varepsilon_i)$ and $B(\varepsilon_i)$ as determined by the model chosen. The unknown parameters a, b and c can now be estimated using the method of maximum likelihood.

Except for scale or location constants independent of the unknown parameters, the log-likelihood becomes

$$L = \sum_{i=1}^{k} \left\{ \ln(v_i^2) + (\bar{x}_i - m_i)^2 (v_i^2)^{-1} \right\}. \tag{6.2}$$

By setting the first two partial derivatives equal to zero we obtain the two equations:

$$\frac{\partial L}{\partial b} = 0 \quad \text{iff} \quad \sum_{i=1}^{k} \frac{\bar{x}_i - m_i}{v_i^2} = 0, \tag{6.3}$$

$$\frac{\partial L}{\partial c} = 0 \quad \text{iff} \quad \sum_{i=1}^{k} \frac{1}{n_i v_i^2} = \sum_{i=1}^{k} \frac{(\bar{x}_i - m_i)^2}{n_i v_i^4}, \tag{6.4}$$

and then by combining equation (6.4) with $\partial L/\partial a = 0$ we obtain, after some simplification, the third equation:

$$\frac{\partial L}{\partial a} = 0 \quad \text{iff} \quad k = \sum_{i=1}^{k} \frac{(\bar{x}_i - m_i)^2}{v_i^2} + \sum_{i=1}^{k} \frac{(\bar{x}_i - m_i) A_i}{v_i^2}. \tag{6.5}$$

After substituting from equation (6.1), we obtain, by machine computation, the simultaneous solution of these three non-linear equations, which we denote by $\hat{a}$, $\hat{b}$ and $\hat{c}$.

As an initial guess for an iterative procedure to compute the MLE's we obtain the (modified) minimum Chi-square estimates, $\tilde{a}$ and $\tilde{b}$. Let

$$\chi^2 = \sum_{i=1}^{k} \frac{(\bar{x}_i - b - aA_i) n_i}{s_i^2},$$

and then by setting $\partial \chi^2/\partial a = 0$ and $\partial \chi^2/\partial b = 0$ we obtain the two simultaneous linear equations:

$$a \sum \frac{A_i n_i}{s_i^2} + b \sum \frac{n_i}{s_i^2} = \sum \frac{\bar{x}_i n_i}{s_i^2} \tag{6.6}$$

$$a \sum \frac{A_i^2 n_i}{s_i^2} + b \sum \frac{A_i n_i}{s_i^2} = \sum \frac{\bar{x}_i n_i A_i}{s_i^2}, \tag{6.7}$$

the solution of which is $\tilde{a}$, $\tilde{b}$. Now we set $\partial L/\partial c = 0$, then $\tilde{c}$ is the solution of the non-linear equation in the variable c:

$$\sum_{i=1}^{k} \frac{1}{c + (\tilde{a}^2)B_i} = \sum_{i=1}^{k} \frac{n_i(\bar{x}_i - m_i)^2}{[c + (\tilde{a}^2)B_i]^2}. \tag{6.8}$$

7 The Calculation of a Safe Duration-of-Load

Once we have chosen an appropriate distribution, which to fix ideas we assume is the Gumbel distribution, and one of the three forms of the regression function H then the three parameters $\hat{a}$ and $\hat{b}$ and $\hat{c}$ can be calculated. We already know both σ_1, β_1 from the distribution of strength. Thus for any preassigned load ℓ we can solve for a safe life, say ξ_p, at any specified assurance level p near unity. We compute the modified stress ratio which for the Gumbel case is $\varepsilon^{\sigma_1} = \ell/\beta_1$, then using equation (3.6) we seek the largest value ξ such that

$$\int_{y=0}^{\infty} \exp\left\{-Q_0\left[\frac{\ln \xi - c\Delta[Q_1^{-1}(y+\varepsilon)]}{\sigma_0}\right] - y\right\} dy \geq p. \tag{7.1}$$

Of course, with machine computation we can invert the problem and for a specified duration-of-load and a given level of assurance solve for the value of the modified stress ratio ε and then the load ℓ. Hence this calculation can be used to replace and quantify the safety factors that are in use with certain materials and applications.

Acknowledgments: The author thanks Dr. Jonathan W. Martin, of the Center for Building Technology at the National Bureau of Standards, for bringing this problem to his attention and for pointing out its importance in the construction industry.

References

Elandt-Johnson, R.C. and Johnson, N.L. (1980). Chap. 5, Complete Mortality Data. Estimation of Survival Function. *Survival Models and Data Analysis* , John Wiley & Sons.

Foschi, R.O. and Barrett, J.D. (1982). Load-Duration Effects in Western Hemlock Lumber. *Proceedings of the American Society of Civil Engineers,* **108**, No. ST7, 1494-1510.

Madsen, B. and Barrett, J.D. (1976). Time-Strength Relationship for Lumber. *Univ. of British Columbia, Dept. of Civil Engineering Structural Research Series Report #13.*

Martin, J.W. and Saunders, S.C. (1988). The Problem of Strength and Duration-of-load for Structural Timbers. to appear in *The Proceedings of the 1988 International Conference on Timber Engineering*, held in Seattle Washington, Sept.1988.

Martin, J.W. and Saunders, S.C. (1982). Relationship between Life Test Analysis and Tests at a Constant Stress Ratio. from *Structural Uses of Wood in Adverse Environments*, Eds. Meyer and Kellogg, 317-326, Van Nostrand Reinhold Co.

Murphy, J.F. (1983). Load Duration Effects in Western Hemlock Lumber. *Journal of Structural Engineering*, **109**, 2943-2946.

Index